International Conference on Recent Developments in Robotics, Embedded and Internet of Things (ICRDREIOT2020)

AA007097

IOP Conference Series: Materials Science and Engineering
Volume 994

Tamil Nadu, India
16 - 17 October 2020

ISBN: 978-1-7138-3304-8
ISSN: 1757-8981

Printed from e-media with permission by:

Curran Associates, Inc.
57 Morehouse Lane
Red Hook, NY 12571

Some format issues inherent in the e-media version may also appear in this print version.

Institute of Physics
Dirac House, Temple Back
Bristol BS1 6BE UK

Phone: 44 1 17 929 7481
Fax: 44 1 17 920 0979

techtracking@iop.org

Additional copies of this publication are available from:

Curran Associates, Inc.
57 Morehouse Lane
Red Hook, NY 12571 USA
Phone: 845-758-0400
Fax: 845-758-2633
Email: curran@proceedings.com
Web: www.proceedings.com

TABLE OF CONTENTS

ROBOTICS

IOT

EMBEDDED

MEDICAL

NETWORKS

ANTENNA

COMMUNICATION

IMAGE PROCESSING

MATHEMATICS

VLSI

IOP Conf. Series: Materials Science and Engineering **994** (2020) 011001 doi:10.1088/1757-899X/994/1/011001

Preface: Proceeding of the International Conference on Recent Developments in Robotics, Embedded and Internet of Things ICRDREIOT2020

This volume published by IOP Conference Series: Material Science and Engineering proceeding is an outcome of the International Conference on Recent Developments in Robotics, Embedded and Internet of Things ICRDREIOT2020 that assimilates the research work of some prominent researchers in the extensive area of Developments in Robotics, Embedded and Internet of Things.

This volume presents an anthology of works presented at ICRDREIOT2020, covering subjects in the area such as Robotics , Embedded , IoT , Image Processing , Analysis of Medical images, Communication , Networks ,Mathematical concepts and VLSI.

I am also grateful to the Chairman Shri. P. Sriram and Principal Dr.A.Ramesh, Chennai Institute of Technology, Members of ICRDREIOT2020 committee and the researchers who have actively participated in the elaboration of this book.

A special thanks to Dr. Kazuaki Tanaka, Kyushu Institute of Technology, Japan, Dr.R.Mervin, Professor, Institution: Saudi Electronic University, Riyadh. Dr.R.Menaka, Professor, Chennai Institute of Technology, who took the responsibility of being the editors of this book.

Dr.R.Balamurali
Dr.R.Menaka
(Convenor- ICRDREIOT2020)
Chennai Institute of Technology

ICRDREIOT2020-COMMITTEE

COLLEGE COMMITTEE

CHIEF PATRON

- **Shri. P.Sriram**, Chairman, Chennai Institute of Technology, Chennai

PATRONS

- Shri. P.Janakiraman, Vice Chairman, Chennai Institute of Technology, Chennai
- Smt. S.Sridevi, Secretary, Chennai Institute of Technology, Chennai

CO-PATRONS

- Dr. A.Ramesh, Principal, Chennai Institute of Technology, Chennai
- Dr. P.Partheeban, Dean (P&D), Chennai Institute of Technology, Chennai

CONVENER

- Dr. R.Balamurali, Associate Professor ,Department of Electronics and Communication Engineering, CIT
- Dr.R.Menaka, Professor, Department of Electronics and Communication Engineering ,CIT

ORGANIZING SECRETARY

- Dr. R.Dhanagopal, Associate Professor, Department of Electronics and Communication Engineering, CIT

COMMITTEE MEMBERS

- Dr. T.Thandapani, Professor, Department of Electronics and Communication Engineering, CIT
- Dr. Swarnaravidrababu , Professor, Department of Electronics and Communication Engineering, CIT

IOP Conf. Series: Materials Science and Engineering **994** (2020) 011002 doi:10.1088/1757-899X/994/1/011002

- Dr. R.Sureshkumar, Associate Professor, Department of Electronics and Communication Engineering, CIT
- Dr. Annie Grace, Associate Professor, Department of Electronics and Communication Engineering, CIT
- Dr.Kiran George, Assistant Professor, Department of Electronics and Communication Engineering, CIT
- Mr. A.Sivabalan, Assistant Professor, Department of Electronics and Communication Engineering, CIT
- Mrs. N.Archana, Associate Professor, Department of Electronics and Communication Engineering, CIT
- Mr.V.Kishorekumar, Assistant Professor, Department of Electronics and Communication Engineering, CIT
- Mr.A.Senthilkumar, Associate Professor, Department of Electronics and Communication Engineering, CIT

INTERNATIONAL ADVISORY COMMITTEE

- Dr. Mervin, Saudi Electronic University
- Dr. Kazuaki Tanaka, Kyushu Institute of Technology, Japan
- Dr. Xavier Fernando, Ryerson University, Canada
- Mr. Alexander Bours, DiManEx BV, Netherlands
- Mr. Ganesh Raju, Linaro, USA
- Mr. Henk Jonker, DiManEx BV, Netherlands

NATIONAL ADVISORY COMMITTEE

- Dr.Pramod Sharma, Jaipur IEM.
- Dr.A.Sivabalan, NEC
- Dr.P.Dhanajeyam, PEC
- Dr.S.Srinivasan-AU, Madurai
- Dr.Muttan, AU-Chennai.
- Dr.Subharani, PSG Tech., Coimbatore
- Dr.Manikandan, ISRO, Bangalore

IOP Conf. Series: Materials Science and Engineering **994** (2020) 011002 doi:10.1088/1757-899X/994/1/011002

- Dr.Ashokkumar, IITK
- Dr.P.C.Jain, CDAC, Noida
- Dr.N.S.Murthy, NIT, Warangal
- Dr.Kanjana Baskar, VIT Chennai
- Dr.Amitkonar, Jadapur Univ.,Calcutta
- Mr.Deepak Misra, IIST, Tiruvandram
- Dr.C.Gomathy- SRM University
- Dr.Sarathy-IIT-Madras

Peer review declaration

All papers published in this volume of *IOP Conference Series: Materials Science and Engineering* have been peer reviewed through processes administered by the Editors. Reviews were conducted by expert referees to the professional and scientific standards expected of a proceedings journal published by IOP Publishing.

- **Type of peer review: Single-blind**
- **Conference submission management system:** *Email*
- **Number of submissions received:** *84*
- **Number of submissions sent for review:** *72*
- **Number of submissions accepted:** *50*
- **Acceptance Rate (Number of Submissions Accepted / Number of Submissions Received X 100):** *59.52%*
- **Average number of reviews per paper:** *4*
- **Total number of reviewers involved:** *7*
- **Contact person for queries:** *Dr.R.Dhanagopal*

IOP Conf. Series: Materials Science and Engineering **994** (2020) 012001 doi:10.1088/1757-899X/994/1/012001

Imitation model for keystroke dynamics base on state context representation

Dmitry V. Pashchenko[1], Alexey I. Martyshkin[2], Dmitry A. Trokoz[3], Elena A. Balzannikova[4]

[1]Penza State Technological University, Penza, Russia, dmitry.pashchenko@gmail.com
[2]Penza State Technological University, Department of computing machines and systems, Penza, Russia, alexey314@yandex.ru
[3]Penza State Technological University, Department of computing machines and systems, Penza, Russiadmitriy.trokoz@gmail.com
[4]Penza State Technological University, Research Department, Penza, Russia, elenabalzannikova@gmail.com

Abstract. This paper describes different methods of keystroke dynamics. As part of a study on the development of methods for the dynamic analysis of keystroke dynamics, a method is proposed for representing the process of typing on the keyboard in the form of a successive change of states. Based on states it is proposed to define certain sequences called the state context as feature of a biometric image.
Based on this representation, an imitation model of keystroke dynamics was developed, which allows replacing the physical source of biometric data for the purpose of testing and debugging the identification system of keystroke dynamics
Keywords: keystroke dynamics, biometry, imitation model

1. Introduction

The development of modern information and Internet technologies places increasing demands on security systems that perform a number of critical functions:

- preventing unauthorized access to information and systems;
- Prevention of illegal actions on behalf of another user;
- Prevention of the illegal distribution of confidential or personal information

Besides, often to provide security, it is necessary to establish the authorship of a particular electronic document.

Today, to implement these functions, a variety of methods are used based on the possession of secret information (password or PIN), ownership of an attribute (for example, a smart card), using of biometric data for authentication, or using an electronic digital signature to establish authorship.

Of greatest interest in this area are biometric methods, because they provide a number of advantages: the data source is inseparable from the user, biometric data is much more difficult to steal or forge. In addition, a distinctive feature of biometric methods is the ability to dynamically

IOP Conf. Series: Materials Science and Engineering **994** (2020) 012001 doi:10.1088/1757-899X/994/1/012001

identify the user, thereby revealing the substitution of the operator. Other authentication methods (knowledge or attribute) require a single identification only at the time of user login. However, not all biometric methods can provide continuous identification. Methods such as fingerprint scanning, iris scanning or voice recognition are difficult to apply for these purposes. To implement this function, it would be possible to use user recognition by face geometry using a web camera located opposite the monitor, but standard input devices of a personal computer are more suitable for these purposes: a keyboard and a mouse. These identification methods provide a number of additional advantages: the lack of additional equipment and the possibility of a hidden identification procedure.

2. The concept of keystroke dynamics

2.1 Survey

Among the methods of analyzing Keystroke dynamics, there are two main areas: static and dynamic analysis.

Static analysis involves an identification procedure by entering a predetermined sequence of characters: a passphrase. A typical area of application for this type of analysis is the user authorization procedure based on entering the system by using a password, which, in combination with the keystroke dynamics analysis, improves the system security in case of compromise. Building the keystroke dynamics image for each user requires several samples of entering a passphrase, which, basically, does not take much time.

Dynamic analysis allows you to identify the user regardless of the text they type. This approach allows not only to identify the user during login, but also to perform continuous analysis of the Keystroke dynamics during the entire session, in order to recognize the substitution of the active user. In addition, this method allows not to use predefined password while the identification procedure, but to enter a small random text, saving the user and the system from the need to store passwords. This type of analysis requires a longer period of training the system to build the most complete and accurate biometric image.

In addition, it should be noted that such systems can be used both on the corporate network or the Internet, where many users are registered and a large number of biometric images are generated, and for individual use to protect personal information. In the first case, the task of the system consists mainly in classifying the keystroke dynamics image among registered users. To train classifier for a new user can be used existing biometric images as images of “foreign”. Thus, the larger the database of images, the more data is provided to the system for training. In turn, if there is only one registered user, there is no database of “foreign” images. In this case, the task is to assess the similarity measure of the received and stored biometric images and determine the confidence interval, according to which an authentication or identification decision is made.

The main purpose of the research is developing a keystroke dynamics analysis system, based on such methods that support the identification procedure for both a single and a network user.
To do this, it is necessary to consider the concept of keystroke dynamics and determine its main characteristics.

2.2 Keystroke dynamics characteristics

The basic metrics for entering text on the keyboard are timestamps of two types of events: pressing a key and releasing a key. Let’s denote $P_i(t)$ as the key release event at time t and $R_i(t)$ as the key release event at time t.

Based on these events, a number of features can be derived that can be the basis for the analysis of Keystroke dynamics. The primary features are time intervals between a pair of events of various types. Such intervals include: key holding time, delay, and time between keystrokes.

The holding time of the i key can be defined as the interval between releasing and pressing the same key:

$DT_i = R_i(t) - P_i(t)$

The interval between keystrokes is defined as the interval between consecutive events of keystrokes:

$FT_i = P_{i+1}(t) - P_i(t)$

Next, let's define the delay as the interval between releasing the previous key and pressing the next one:

$LT_i = P_{i+1}(t) - R_i(t)$

In the case of intersection of the key holding time, this feature will have a negative value. The described events and primary features can be represented graphically, as shown in Figure 1.

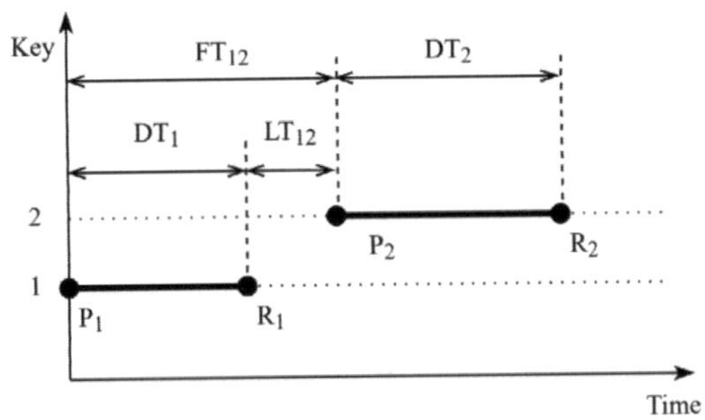

Figure 1 Primary features of keystroke dynamics

Based on a combination of the considered primary features, composite features can be formed: digraphs, trigraphs, etc. The specific form of a composite feature depends on the method used to analyze keystroke dynamics [1].

3. Background research

One of the most common areas of behavioral biometrics is the analysis of keystroke dynamics. The issues of keystroke dynamics analysis were first considered in the works of R. Joyce and G. Gupta in 1990, who used the method of estimating the intervals between keystrokes when entering a fixed passphrase and comparing with a reference sample [2].

Today, there are many studies in the field of analysis of Keystroke dynamics, both among Russian and among foreign scientists[3]. Among Russian developments, the works of R.V. Yeruslanov, Yu.A. Bryukhomitsky, M.N. Kazarin, who in their works use various methods of analysis of keystroke dynamics: poly-Gaussian distribution, estimation of probability density, Haar decomposition[4, 5, 6, 7].

In addition, the research of I.G. Sidorkina and A.N.Savinov , in which the separation of the sample into features with and without overlapping, is applied using the calculation of the Euclidean measure for features and the poly-Gaussian distribution [8].

In the works of D.S. Krutokhvostov and V.E. Khitsenko suggested frequency histograms of using alternative keys and other features based on chi-square statistics. The empirical distribution functions of the legal and analyzed user are compared (by key holding time) based on Kolmogorov-Smirnov statistics[9].

IOP Conf. Series: Materials Science and Engineering **994** (2020) 012001 doi:10.1088/1757-899X/994/1/012001

Among foreign studies, one can note P.Borus, S.Mondal, Syed Idrus, Syed Zulkarnain, H. Barghouthi, who use many methods in their work, among which are [10,11,12, 13]:

- Weighted Euclidean distance estimate for a single key;
- A weighted combination of algorithms (for individual keys): ANN, CPANN and SVM. Weights are optimized by the genetic algorithm
- An algorithm for digraphs based on two distance metrics: pairwise weighted Euclidean and correlation distance.

In the works of A. Bhatia, M. Hanmandl, keystroke dynamics is analyzed by entering a passphrase using methods: SVM, a random forest classifier for transformed characters [14].
It is also worth noting the work of Sampath K. Vuyyuru, Vir V. Phoha, Shrijit S. Joshi, Shashi Phoha, Asok Ray, which use the hidden Markov model apparatus for continuous analysis of keystroke dynamics [15].

4. Statement of the problem

The aim of the research is to develop a system of dynamic user identification by keystroke dynamics that implements a client-server architecture. The system should perform user registration, creating a biometric image of keystroke dynamics, and implementation dynamic user identification according to the temporal characteristics of typing on the keyboard during the entire session [16].

Learning and debugging methods requires a large amount of test data, what basically requires the presence of user. For long-term practical tests it will be extremely difficult to ensure the continual presence of users on whom the recognition system will be debugged. Based on this, it is proposed to use the method of simulation of a biometric image of a Keystroke dynamics based on one or more test samples of user input. The resulting image can then be used to simulate the input of real user data for modeling the system of keystroke dynamics analysis. In addition, in the process of simulation, it is possible to perform the process of debugging the system by adding distortions or noise into the original model to simulate various external factors, which were considered earlier.

Thus, the data approach allows you to generate images of “foreign” to simulate attempts to log in an unauthorized user or substitute an operator.

4.1 System model

A keystroke dynamics analysis system consists of several components. The collection component intercepts keyboard events and transfers them to the preprocessing unit, which converts keyboard events and generates features. This component transfers the generated features to the server component, which builds the image of the keystroke dynamics, and trains the user identification model [17, 18].

Schematically, the proposed user identification system can be represented, as shown in Figure 2.

The data source of the collection component is the keyboard. The simulation model, in turn, generates events that are identical to this component, replacing it. In order for the component of the simulation model to generate events with time characteristics corresponding to a specific user, a pre-received sample of input from this user should be used as the initial data for the model [19].
Figure 3 represents a system model in which a collection component is replaced by a simulation model:

IOP Conf. Series: Materials Science and Engineering **994** (2020) 012001 doi:10.1088/1757-899X/994/1/012001

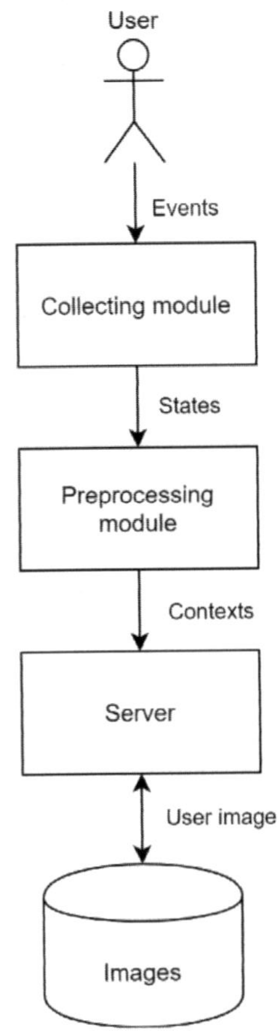

Figure 2 Model of keystroke dynamics analysis system

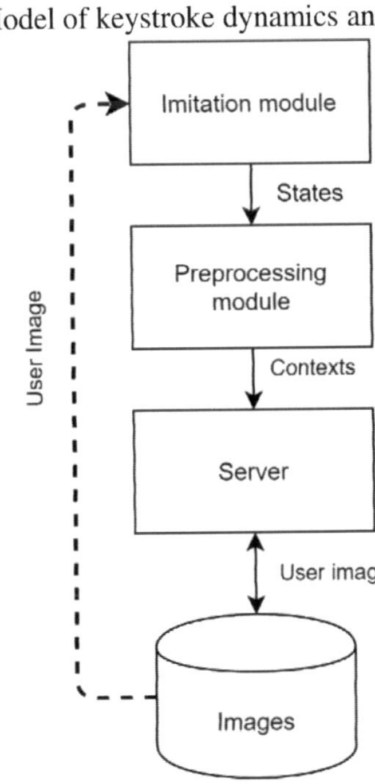

Figure 3 Model of keystroke dynamics analysis system in which a collection component is replaced by a simulation model

Let's consider the methods used to represent events, features, and the biometric image of keystroke dynamics in this system.

5. The concept of state context

In contrast to passphrase based identification in which the sequence of events is predefined, continuous analysis involves the identification based on a sequence of events that occur in an arbitrary way. To solve this problem, it is necessary to define a basic feature consisting of one or more consecutive events, on which the keystroke analysis will be based on. The events of pressing two keys are considered to be a basic sequence. Thus, the basic sequence consists of four events: P_1, R_1, P_2, R_2. The sequence of these events may vary. There are the possible options for following these events, which correspond to different ways of pressing a pair of keys.

The simplest sequence model assumes that there is no intersection of the key holding interval: the keys are pressed one after the other. Graphically, this model can be represented as shown in Figure 4. Thus, the sequence of events for the presented model will be as follows: $P_1 \rightarrow R_1 \rightarrow P_2 \rightarrow R_2$. This model is typical for most cases of pressing symbol keys.

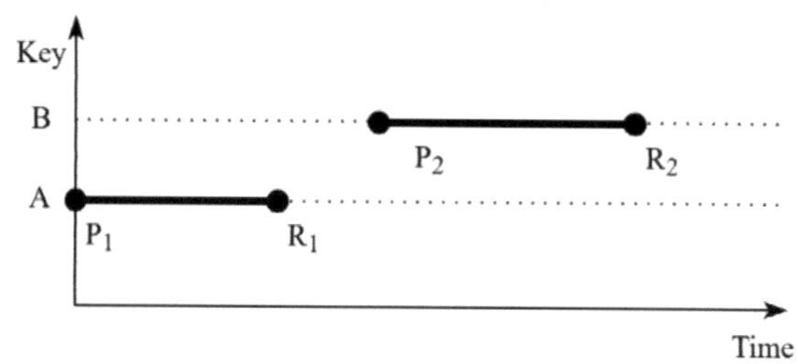

Figure 4 Sequence model without intersection

Other considered models suggest the intersection of the key holding interval. In one case, pressing the second key occurs while holding the first, and releasing after releasing the first. We designate this model as a model with partial intersection. Figure 5 shows a graphical representation of this process. The sequence of events will be as follows: $P_1 \rightarrow P_2 \rightarrow R_1 \rightarrow R_2$. This sequence is often found when using utility keys, such as Alt + Shift.

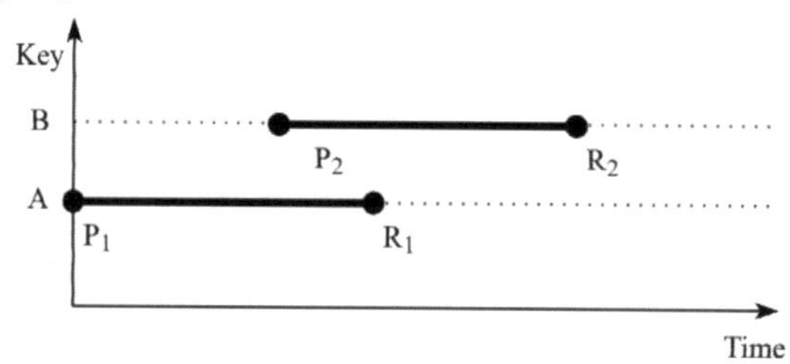

Figure 5 Sequence model with partial intersection

Another option for intersecting intervals is to completely overlap with the holding interval of the first key of the holding interval of the second. Graphically, this process can be represented as shown in Figure 6. The sequence of events for this model will be as follows: $P_1 \rightarrow P_2 \rightarrow R_2 \rightarrow R_1$. A typical example is the typing of an extra key character using the Shift key.

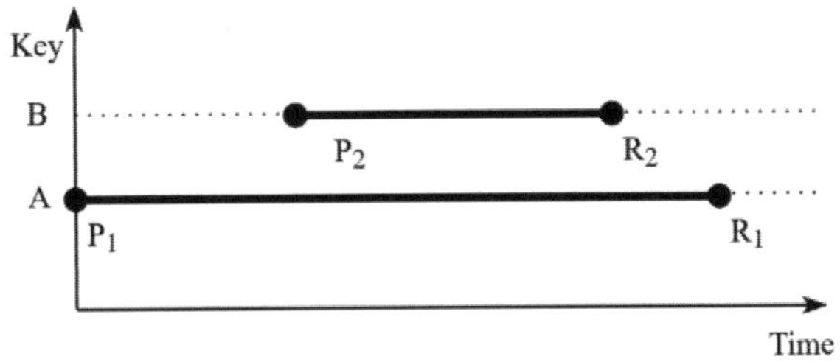

Figure 6 Sequence model with full intersection

Thus, to form the most complete image of the keyboard, it is proposed to define these models and their temporal characteristics. To solve this task, the following model for the presentation of a biometric image of keystroke dynamics is proposed.

5.1 Keyboard state Representation Format

To simplify the keyboard model, represent it as a fixed set of keys, each of which at any given time can be in one of two states: pressed or released. Depending on the specific keyboard, the number of keys may vary. To build a model, we restrict the set of keys to only those that, as a rule, are involved in typing. Such keys include: symbolic and numeric keys (47 keys of the main keyboard), as well as service keys: Space, Enter, Right Ctrl, Left Ctrl, Right Shift, Left Shift, Right Alt, Left Alt, Caps Lock, Tab, Back Space . Thus, the keyboard model will contain 58 keys, based on which the time characteristics will be evaluated.

Represent the state of the keyboard as a vector, each element of which contains the state of a separate key. Set of states, based on the proposed model, consists of two elements:

$$State = \{Pressed, Released\}$$

Therefore, the keyboard state vector will look like this:

$$Keyboard = [State_1, State_2, \dots, State_i, \dots, State_{58}]$$

where i is the key number.

Since the proposed model has 58 elements, each of which can be in one of two states, the total number of possible states of the model is 2^{58}, which is more than a quadrillion states. Obviously, processing such a number of states is practically impossible and the vast majority of states will never occur. Therefore, to simplify the model and reduce the number of possible states, we introduce a limit on the number of keys that can be in pressed state by two. This restriction allows you to almost completely describe the process of typing text on the keyboard. An exception is the use of control combinations, which involve the simultaneous pressing of three or more keys. These combinations can be excluded from the model or considered separately as additional features. Thus, taking into account the described limitation, the number of possible states can be calculated as one initial state when all keys are released, the number of possible states when one key is pressed and the number of states with two simultaneously pressed keys: $1 + C_{58}^1 + C_{58}^2 = 1712$, which is a more acceptable amount for building a model of keystroke dynamics.

Thus, this approach allows us to present the process of typing on the keyboard not as a sequence of events of pressing or releasing an arbitrary key, but as a sequential change in the state of the keyboard, presented in the form of fixed size vectors, each of which is associated with a time stamp. The time intervals between two consecutive states are features on the basis of which a biometric image of keyboard writing can be built.

However, in this paper, a different approach is proposed that would allow one to take into account the described models of keystroke sequences. For example, consider a keyboard model that consists of only two keys: A and B. Based on this, we present the state matrix for this model, which is shown in Table I. Columns A and B indicate the state of the key: 0 - released, 1 - pressed. The set of states of individual keys is designated as state S_i.

Table 1 Matrix of states for the keyboard model consisting of two keys

A	B	State
0	0	S_0
0	1	S_1
1	0	S_2
1	1	S_3

In this case, the sequence of states for two consecutive keystrokes without intersections will be as follows: $S_0 \xrightarrow{\Delta t_1} S_2 \xrightarrow{\Delta t_2} S_0 \xrightarrow{\Delta t_3} S_1 \xrightarrow{\Delta t_4} S_0$. Graphically, this process can be represented as shown in Figure 7.

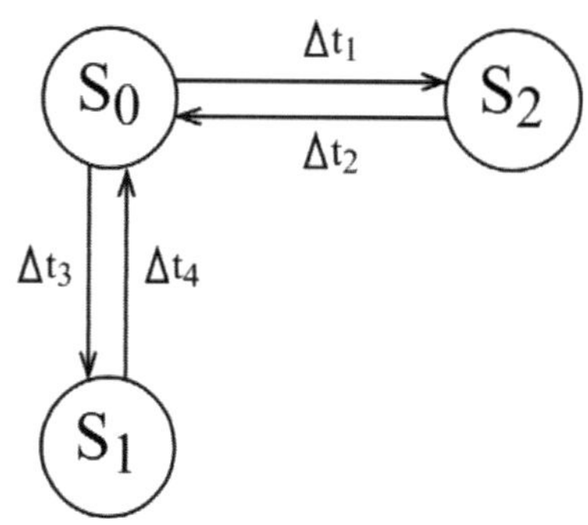

Figure 7 Sequence of states for a model without intersections

For the second model with incomplete overlap, the sequence will have the following form: $S_0 \xrightarrow{\Delta t_1} S_2 \xrightarrow{\Delta t_2} S_3 \xrightarrow{\Delta t_3} S_1 \xrightarrow{\Delta t_4} S_0$. Graphically, this process is presented in Figure 8.

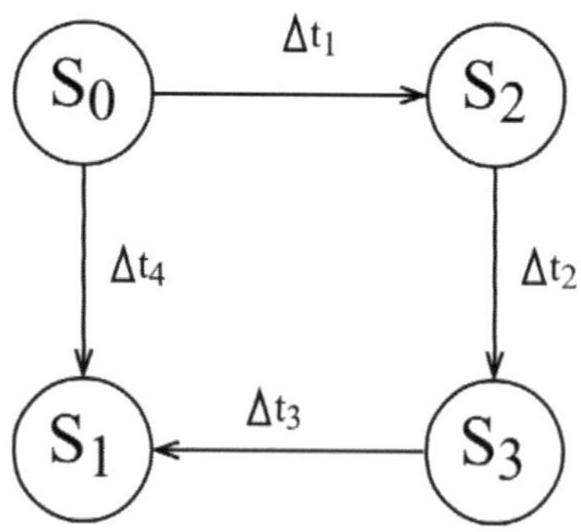

Figure 8 Sequence of states for incomplete intersection model

Sequence of states for full intersection model will looks as: $S_0 \xrightarrow{\Delta t_1} S_2 \xrightarrow{\Delta t_2} S_3 \xrightarrow{\Delta t_3} S_2 \xrightarrow{\Delta t_4} S_0$, what shown in Figure 9.

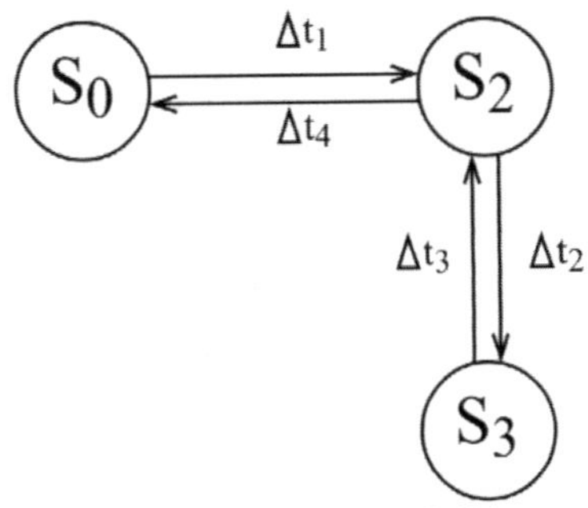

Figure 9 Sequence of states for full intersection model

5.2 The concept of state context
In each of the considered models, the initial and final states S_0 coincide, and the maximum number of states does not exceed 4. Based on this, it is proposed to determine the context of the state chain, which contains 4 consecutive states, between which three time intervals are calculated. Thus, the composite feature of the keyboard handwriting will have the following form:

$$C_{ijkl} = \{\{S_i, S_j, S_k, S_l\}; \{\Delta t_{ji}, \Delta t_{kj}, \Delta t_{lk}\}\},$$

where *i, j, k, l* are the state codes of the keyboard.

This approach will allow not only extracting the primary features of keystroke dynamics, but also to separate them depending on the model of state sequence that were considered earlier. It is assumed that this method allows you to build the most complete and accurate model of the keyboard handwriting of users[13].

5.3 Initation model
As described earlier, according to the developed system architecture, the collection component captures keyboard events and transforms the received information to the set of states, calculating the time interval between them. Based on this, the simulation model should generate a sequence of states with parameters identical to user parameters.
The basic requirements for the simulation model are:

- The sequence of states must be random;
- The vectors of two consecutive states must differ from each other by one bit

The first requirement ensures the independence of the model from the generated text, which allows modeling a set of sequences of arbitrary length.

The second requirement means that only one event has occurred between two consecutive states: the key is pressed or released, which corresponds to the proposed model of keystroke dynamics based on the state contexts presentation.

In order for the simulation model to repeat the temporal characteristics between keyboard events and intersection model of keystrokes, it is proposed to use the keystroke dynamics representation based on the state contexts described earlier.

We denote $S = [S_0, \dots, S_i, \dots]$ as the state vector of unlimited length, where i is the index of the state in the generated sequence. Upon transition to the next state, an output signal is generated that corresponds to the transition time between states. The sequences of states, as well as time intervals, are determined by the parameters of a pre-formed set of contexts of a particular user.

The initial data for the imitation model are formed on a test sample of user text input. From than sample a sequence of states is extracted, and contexts are formed. According to the above definition, the context consists of four successive states, between which the time interval is calculated in the process of user typing text. For contexts with the same sequence of states, the mathematical expectation and standard deviation for each transition are calculated.

Thus, from a set of contexts C for $C_{wxyz} = \{\{s_w, s_x, s_y, s_z\}; \{\Delta t_{xw}, \Delta t_{yx}, \Delta t_{zy}\}\}$, a descriptor $\bar{C}$ is formed as follows:

$$\bar{C}_{wxyz} = \{\{s_0^w; s_1^x; s_2^y; s_3^z\}; \{\mu_0^{\Delta t_{xw}}; \sigma_0^{\Delta t_{xw}}\}; \{\mu_1^{\Delta t_{yx}}; \sigma_1^{\Delta t_{yx}}\}; \{\mu_2^{\Delta t_{zy}}; \sigma_2^{\Delta t_{zy}}\}\}$$

The initial state is the state of the keyboard on which no key is pressed:
$S_i = 0, i = 0$

Next, from the set of descriptors, $\bar{C}_{ijkl}$ is randomly selected, in which the first state s_0^w coincides with S_0. Thus, $S_{i+1} := s_0^w,\ \exists \bar{C}_{wxyz}: s_0^w = 0$. Subsequent generated states correspond to the internal states of the current context. At the initial step, the internal state index k, which is initially assigned the value 1. The subsequent event is formed depending on the current internal state index according to the following rules:

$$S_{i+1} := \begin{cases} s_2^j;\ k := 2, \mathrm{k} = 1 \\ s_3^j;\ k := 3, \mathrm{k} = 2 \\ s_1^{j+1};\ k := 1, \mathrm{k} = 3 \text{ и } \bar{C}^{j+1}: s_0^{j+1} = s_3^j \end{cases}$$

If the index k is 1 or 2, then in this case the internal state of the current context index, the index is incremented. If the index is 3, $\bar{C}$ for which the first internal state is equal to the subsequent state of the current context, is randomly selected and stored. The next generated state is the second internal state of the selected descriptor. Schematically, the process of generating states by forward transitions through context descriptors can be represented, as shown in Figure 10.

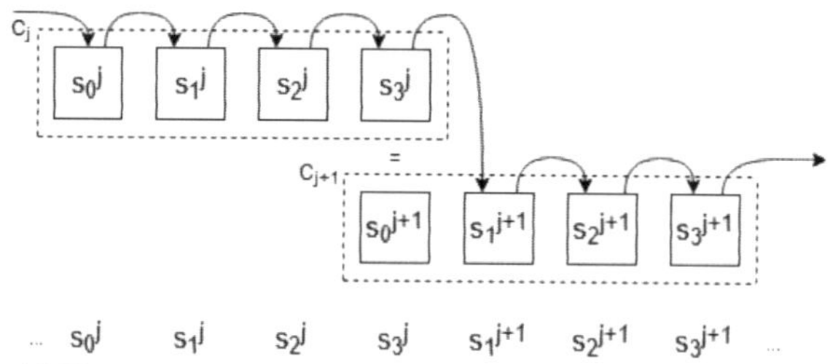

Figure 10 The process of generating states by forward transitions

Thus, the generation of a sequence of states on the basis of contexts ensures compliance with the set conditions for the states model: a random sequence of events of unlimited length is generated, and due to their "overlapping" in the described way, there is only one event between two successive states.

When transitioning between states within the context, a random interval is generated that has a normal distribution with mathematical expectation and variance parameters in the corresponding context descriptor.

$$T_{i+1} = T_i + norm\left(\mu_j^{\Delta t_{k,k-1}}, \sigma_j^{\Delta t_{k,k-1}}\right)$$

During the transition between states, a situation where there is no context descriptors for which s_0^w coincides with S_0 is possible. In this case, a reverse transition occurs within the current context until an initial state is reached that corresponds to the situation when no key is pressed. Based on this, in the process of forward transitions when moving to the next context descriptor, the previous descriptor is pushed onto the stack. In the process of reverse transitions, descriptors are sequentially extracted from it.

The reverse generation process can be represented, as shown in Figure 11.

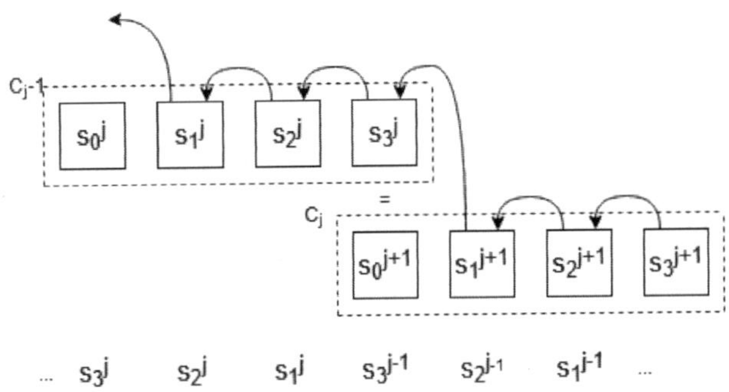

Figure 11 The process of generating states by reverse transitions

Upon reaching the internal state corresponding to the initial one, the forward generation process begins again.
To complicate the simulation model, it is proposed to expand the state space by comparing the transition between two states with several intervals allocated as a result of cluster analysis of the intervals of the original user sample.

4.4 The process of modeling and debugging a model
As a result, a component of imitation modeling was developed according to the proposed mathematical model. According to the test results, this component allowed to generate about 20 events per second, which is about 6-12 times faster than the average continuous typing speed (at a speed of 200-400 characters per second). In addition, tests showed that the resulting model generates states that are approximately 80% present in the model and, as a result, are involved in making an identification decision.

6. Conclusions

Thus, for today, behavioral biometry and analysis of the user's keyboard handwriting is an important and urgent task of ensuring the security of information systems and user information.

As part of a study on the development of methods for the dynamic analysis of keystroke dynamics, a method is proposed for representing the process of typing on the keyboard in the form of a successive change of states. Based on states it is proposed to define certain sequences called the state context as feature of a biometric image.

Based on this representation, an imitation model of keystroke dynamics was developed, which allows replacing the physical source of biometric data for the purpose of testing and debugging the identification system of keystroke dynamics.

The simulation component based on the presentation of state contexts has the following advantages;

- Allows you to repeat the model of key press overlapping as described earlier;
- The generated events form the contexts that are in the user's stored biometric image;
- Allows you to generate a random sequence of events of unlimited length.

However, the main drawback of this approach is that the generated sequence of characters does not correspond to the real typing of meaningful text.

One of the directions for further development is increasing in the performance of the simulation imitation component in order to increase the number of generated events and reduce the time of one debug iteration and optimize the model debugging process [20-21]. In addition, in this area, it is necessary to develop a method for introducing distortions into the generated image in order to debug analysis algorithms for periodic short-term and long-term changes in the user's keyboard handwriting.

References

[1] D. V. Pashchenko, E. A. Balzannikova , I. G. Sergina, "User identification method by means of biometric image of keystroke dynamics with double-chained representation" Вопросы радиоэлектроники, № 12, C 83–89, 2018, DOI 10.21778/2218-5453-2018-12-83-89.

[2] R. Joyce and G. Gupta "Identity Authentication Based on Keystroke Latencies" Communications of the ACM vol. 33 no. 2 pp. 168-176, 1990, DOI: 10.1145/75577.75582

[3] Shen Teh P., Beng Jin Teoh A, Yue S. "A Survey of Keystroke Dynamics Biometrics", The Scientific World Journal –Hindawi Publishing Corporation, 24 p., 2013.

[4] R. V. Eruslanov "Keyboard identification of the user on the basis of the multilinked presentation method", Radio engineering and infocommunication systems, №3 pp 43 – 50, 2009 .

[5] U.A. Bryukhomitsky "Bar graph keystroke dynamics recognition method" News of the Southern Federal University. Technical science, vol.112, №11, 8 p, 2010.

[6] Bryukhomitsky U.A., and Kazarin M.N. "Multi-connected representation of biometric parameters in keyboard monitoring systems", Information counteraction to threats of terrorism. Scientific and practical journal №14, 144-148 pp., 2010

[7] Bryukhomitsky U.A., and Kazarin M.N. "The method of biometric user identification by keyboard handwriting based on the Haar decomposition and Hamming proximity measure" Izvestiya Southern Federal University. Technical science, vol. 33, no. 4, 2003, pp. 141-149.

[8] I. G. Sidorkina, A. N. Savinov. "Three algorithms of control access to the KSSI on the basics of recongnition of keystroke dynamics" Bulletin of the Chuvash University, no. 3, 2013, pp. 293-301.

[9] D.S Krutohvostov, V.E. Khitsenko, "Password authentication and continuous authentication by keystroke dynamics using mathematical statistics", Cybersecurity issues, №5(24), 2017, 91 – 99 pp, DOI: 10.21681/2311-3456-2017-5-91-99.

[10] P. Bours, S. Mondal,. "Continuous Authentication with Keystroke Dynamics". 2015DOI: 10.13140/2.1.2642.5125..

[11] Mondal, Soumik & Bours, Patrick. (2017). Person Identification by Keystroke Dynamics Using Pairwise User Coupling. IEEE Transactions on Information Forensics and Security. PP. 1-1. 10.1109/TIFS.2017.2658539.

[12] Syed Idrus, Syed Zulkarnain & Cherrier, Estelle & Rosenberger, Christophe & Bours, Patrick. (2014). Soft Biometrics for Keystroke Dynamics.

[13] P. Bours, H. Barghouthi "Continuous authentication using keystroke dynamics," Proceedings of the Norsk Informasjonssikkerhetskonferanse (NISK'09). 1-11, 2009.

[14] A Bhatia, Hanmandlu, M. "Keystroke Dynamics Based Authentication Using Information Sets," Journal of Modern Physics, Vol.8 No.9, August 2017, DOI: 10.4236/jmp.2017.89094.

[15] Vuyyuru, Sampath K. et al. "Computer User Authentication using Hidden Markov Model through Keystroke Dynamics." (2006).

[16] Martyshkin, A.I. & Pashchenko, D.V. & Trokoz, D.A.. (2019). Queueing Theory to Describe Adaptive Mathematical Models of Computational Systems with Resource Virtualization and Model Verification by Similarly Configured Virtual Server. 1-6. 10.1109/RUSAUTOCON.2019.8867620.

[17] Alexey, I. Martyshkin and V. Pashchenko Dmitry. "Development and Research of Means for the Collection and Reporting of Open-Source Information for Use in Decision-Making Systems." Journal of Computational and Theoretical Nanoscience 16 (2019): 3103-3114.

[18] Dubravin, Aleksey & Zinkin, Sergey & Paschenko, Dmitry. (2015). Formal and conceptual definitions of the hybrid model of distributed computings in networks. 1-6. 10.1109/SIBCON.2015.7147047.

[19] Pashchenko, Dmitry & Trokoz, Dmitry & Martyshkin, Alexey & Sinev, Mihail & Svistunov, Boris. (2020). Search for a substring of characters using the theory of non-deterministic finite automata and vector-character architecture. Bulletin of Electrical Engineering and Informatics. 9. 10.11591/eei.v9i3.1720.

[20] Pashchenko, Dmitry et al. "Using Algebra of Hyper-Dimensional Vectors for Heuristic Representation of Data While Training Wide Neural Networks." 2019 XXI International Conference Complex Systems: Control and Modeling Problems (CSCMP) (2019): 168-171.

[21] Janarthanan, R., Balamurali, R., Annapoorani, A. and Vimala, V., 2020. Prediction of rainfall using fuzzy logic. Materials Today: Proceedings.

Frequency Analysis of Hydraulic Modeling As A Way to Forecast Flooding on the Allala River

Kastali Abdennour[1], Zeroual Ayoub[2], Kravchenko Lyudmila [3],Khadzhidi Anna*[4], BoukhanefIssam[4]

[1] UniversityHassibBenbouali of Chlef, Laboratory of Chemistry Vegetable-Water-Energy, Hydraulic Dapartment, Algeria
[2] National Higher Hydraulic School, Water Engineering and Environment Laboratory, Blida, Algeria
[3]Don State Technical University, Russia
[4]Kuban State Agrarian University named after I.T. Trubilina, Russia

Email: dtn-khanna@yandex.ru

Abstract.The article presents research on modeling the recurrence of emergency situations caused by flooding on the Allala river. It is proposed to predict the frequency of floods using the Saint-Venant equation for hydrodynamic modeling. The HEC RAS model was used to compare flood intensity with different water level positions in the river. The parameters of the water source were measured on a section of 8 km long. As a result of research, it was found that the existing coastal protection structures on the banks of the Allala river are insufficient to prevent an emergency flood situation at peak river flows.
Keywords: flood, river, simulation, dam, water level.

1. Introduction

The Allala river is a tributary of Algeria's longest river, the Sheliffa, and is located in the Northwest of the country. During the year, the Allala river experiences small water flows that occur only when precipitation occurs, as with most rivers in Algeria. However, in excessively wet years, the amount of precipitation falling on the research territory exceeds the average annual norm for the region and is of a stormwater nature. As a result of this situation, the largest flood in the last twenty years occurred in 2017. As a result of the disaster, coastal territories were affected. Therefore, the aim of the research is to develop a method for predicting flood risks on the Allala river using hydraulic modeling.

Using the Baratin method to quantify confidence intervals, a set of curves was constructed, from which the corresponding curve was selected, which characterized the reconstruction of the annual peak discharge based on the hydrograph of the stage from 1973 to 2017.Based on the results of this stage and the FFA, the peak flood hazard is calculated for the deferred return period. The HEC RAS model simulates the water level for different locations of the Allala river. The simulation results confirmed that the existing embankment system is not sufficient to withstand the risk of flooding. Therefore, it is necessary to build flood models that will help predict flooding both in real time and in floodplain mapping (Prabeer Kumar Parhi) using the results of HEC RAS.

2. Methods

The Sidi Akkacha hydrometric station was established in 1972 and is geographically located at the outlet of the Oued-Allala watershed, which has an area of 296,34 km^2, a perimeter of 94,59 km, an average slope of 2,25 %, and an internal water channel of 31,57 km. The highest flood level recorded at the station was achieved with a discharge of 538 m^3/s, which corresponds to a water level of 6,1 m (March 1979).

2.1 Frequency analysis of data

Frequency analysis is used to determine the number of occurrence of random events and assess the accuracy of flood prediction. In this study, the null statistical hypotheses used for data analysis are the exponential distributions of Weibull and Gumbel. The result shows that the distribution of data from the Sidi-Akacha measuring station follows an exponential distribution with a mathematical expectation of 129; an asymmetry coefficient (CS) = 1.71; an average square deviation (Sd) = 121; an kurtosis coefficient (Ck) = 4.72 and a coefficient of variation (Cv) = 0.936.the adequacy of the models was checked using the Chi–square criterion.

The estimated flood flow for different periods is shown in Figure 1.

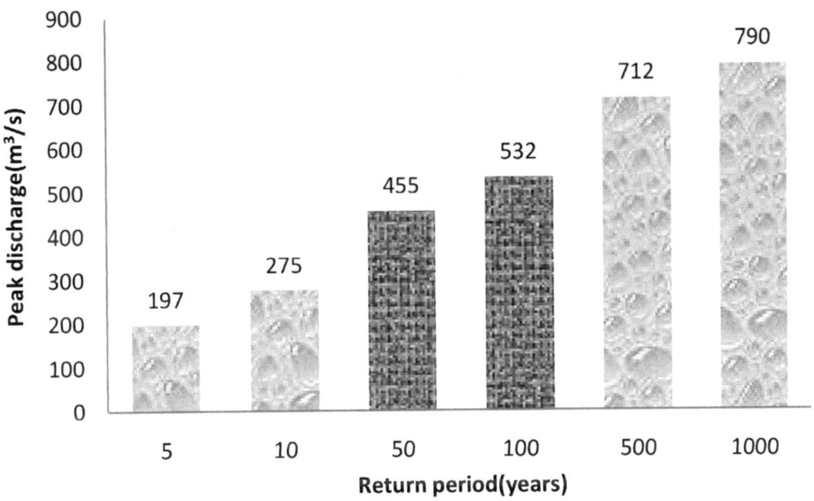

Figure 1. Forecast of peak flood discharges

2.2 The model HEC-RAS

The River Analysis System (HEC-RAS) hydrological engineering centers, developed by the US army corps of Engineers, allow one-dimensional stationary and 1D-and 2D-non-stationary calculations of river hydraulics. It is an integrated software system designed for interactive use of a multi-tasking, multi-user network environment (Brunner, 2016),(Ramakrishnan,2020)

HEC-RAS depends on the finite-difference solutions of the Saint-Venant equations (equations (1) - (2)

$$\frac{\partial A}{\partial t}+\frac{\partial Q}{\partial x}=0\ldots\ldots\ldots \quad (1)$$

$$\frac{\partial Q}{\partial t}+\frac{\partial (Q^2/A)}{\partial x}+gA\frac{\partial H}{\partial x}+gA(S-S_f)=0\ldots \quad (2)$$

where A is the cross-sectional area perpendicular to the flow; m^2

Q-flow rate, m^3/s;
g-acceleration of free fall;
H– water height, m;
S-bed tilt;

IOP Conf. Series: Materials Science and Engineering **994** (2020) 012002 doi:10.1088/1757-899X/994/1/012002

S_f-energy slope;
t– the temporal coordinate;
x-control points.

Equations (1) and (2) are solved using the well-known four-point implicit box finite-difference scheme (Prem Lal et Al, 2011).

2.3 Performance criterion

The performance of the model is evaluated by the following indicators:
Mean square deviation (RMSE)
Field studies were performed along the length of the river section (8) km. A longitudinal profile of the riverbed is constructed (Figure 2).

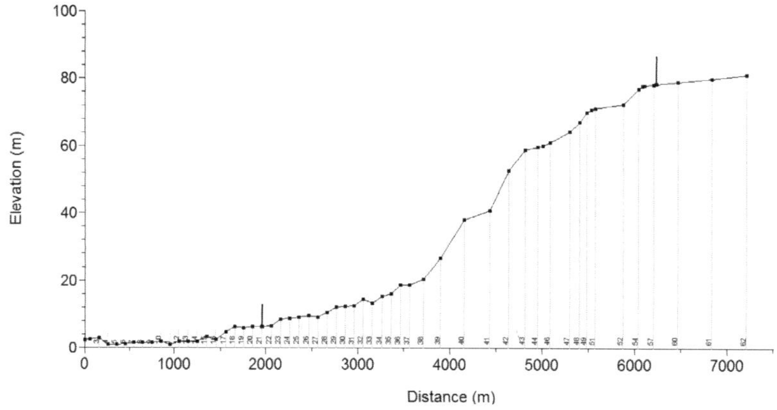

Figure 2. Longitudinal profile of the riverbed

Hydrological surveys on the studied section of the river were determined in 62 channels at a distance of 120 m from each other. Typical research areas are shown in the figure (Figure 3).

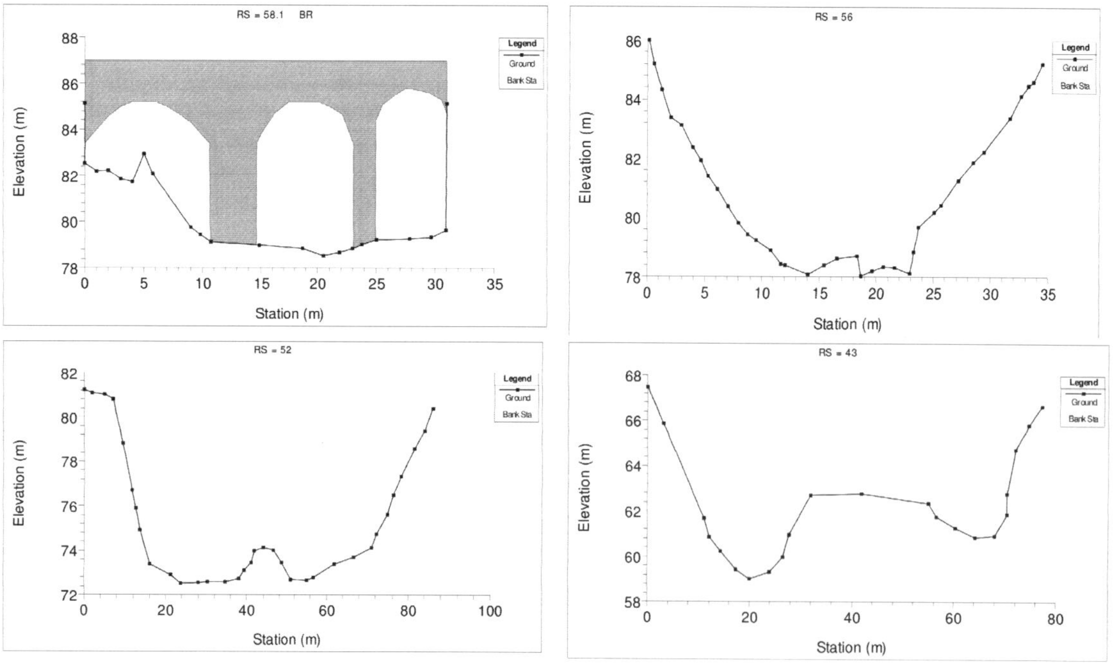

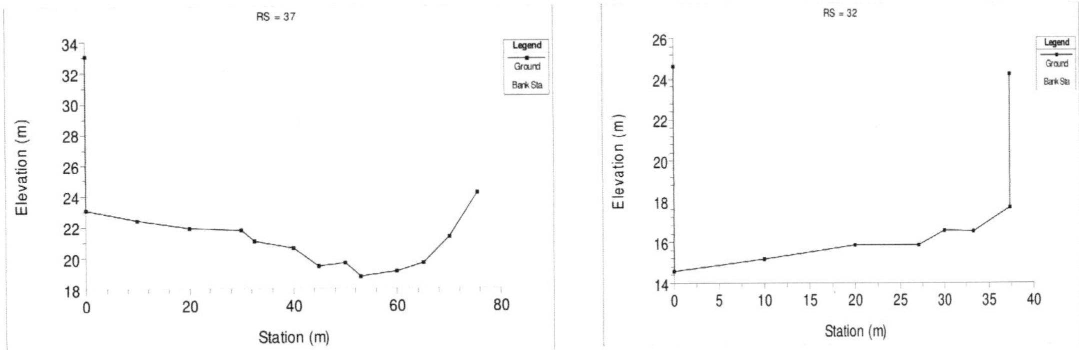

Figure 3. Cross-sections of the Allalariverbed used to develop the HEC-RAS model

3. Result and discussion

The peak flood value will be used as input data for the HEC-RAS hydrodynamic model. Theoretical peak flood discharges will be modeled in 50 and 100 years. The scaling model is a necessary step in applying the model for parameter estimation and refinement (Legates et Al). In our simulation, the HEC RAS model is scaled by splitting the manning roughness coefficient (n), the latter is split for the main channel and then for the channel super banks using the observed marks during the last flood, which was in January 2017 (Figure 4).

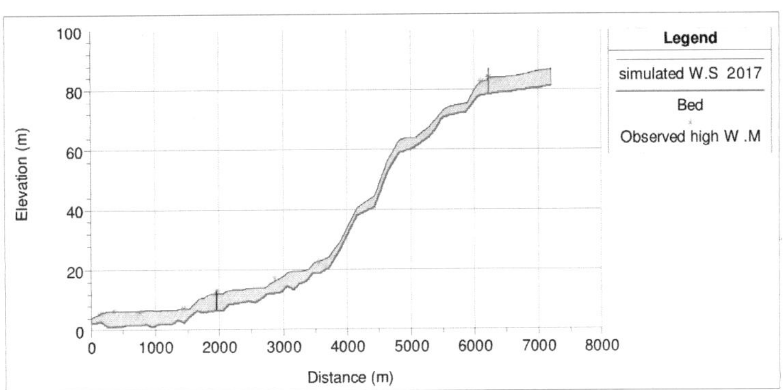

Figure 4. Mapping the HEC-RAS model using the observed water surface (January 2017 flood)

Figure 5 shows studies on modeling the water surface in a river during peak discharges over 50 and 100 years.

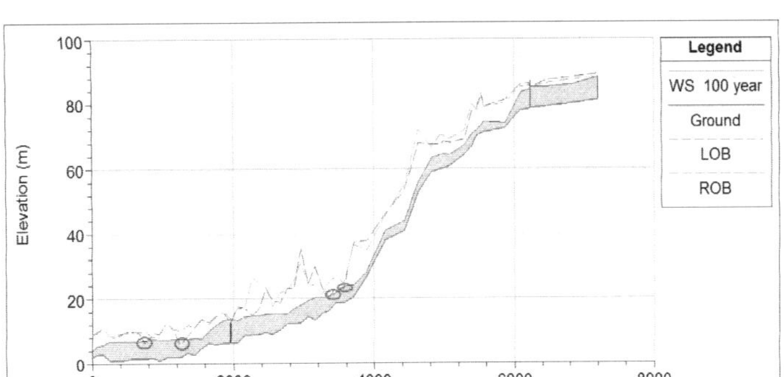

Figure 5. water surface Profile using HEC-RAS for peak runoff of 50 and 100 years

Tables 1 and 2 show the results of studies on the height and location of dams for the flood plan, the level of water rise, and the maximum flow of water during flood periods.

Table 1. Height and location of the embankment for the flood plan return period of 100 years

Station	Left Embankment	Right Embankment
37	0.55	-
35	0.1	-
21	0.96	-
9	1.25	-

Table 2. Rise level and water flow during flood periods

year	Flood duration	Maximum Water surface(m)	Maximum Discharge (m^3/s)
2006	Jan -31, 7:00 to Feb-1, 8:00	4.74	325.25
2012	Feb- 6,23:00 to Feb-7,00:00	2.45	78.22
2013	Apr-24,22:00 to Apr-25,7:00	2.3	71.2
2017	Jan-3:00,7:00 to Jan25,00:00	4.88	385.5

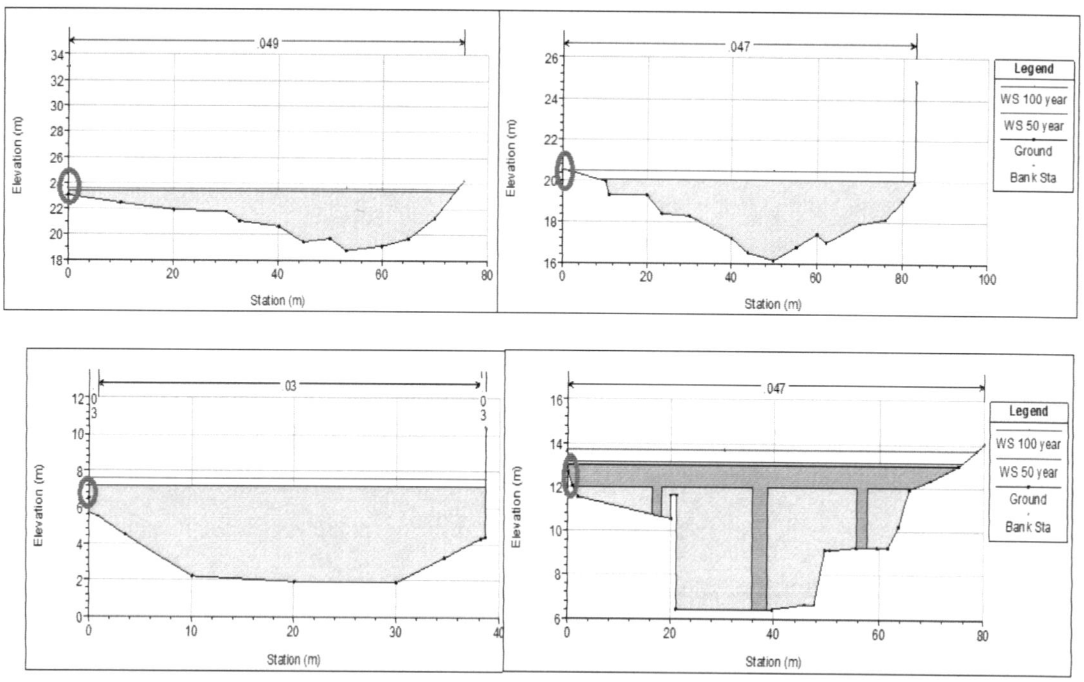

Figure 6. Result of applying the HEC-RAS cross section Allala rivers

The above results (Figure 6 and 7) show that the existing embankment system on the left Bank of the Allala river is not sufficient to withstand the floods of the 50-year and 100-year return periods. Thus, the HEC-RAS model for estimating the magnitude of the flood and the corresponding water level profile shows that a maximum of 1.25 m on the left Bank (1.2 km downstream) is necessary to protect low-lying areas of the old Tenes from flooding.

Figure7. Flow depth map for 100 years return period

4. Discussion and conclusion

The simulation performed with the development of the HEC RAS program can be used to forecast the frequency and intensity of flooding on the Allala river.

The resulting HEC RAS model allowed us to establish that in the event of flooding on the river, additional strengthening of the banks is necessary at the following sites: station 35-37 on the left embankment of the river, station 22-21 on the left embankment of the river, station 10-8 on the left embankment of the river. The height of the dams should vary from 0.55 to 1.25 m.

If the construction of dams is used for flood control, then the location and height of the embankments should be arranged taking into account the output data of HEC-RAS 5.0.3.

References

[1] C. Ben Khalfallah a, S. Saidi Spatiotemporal floodplain mapping and prediction using HEC-RAS -GIS tools: Case of the Mejerda river, Tunisia journal of African Earth Sciences 142 (2018) 44e51, doi:https://doi.org/10.1016/j.jafrearsci.2018.03.00.

[2] Ali M, Bhattacharya B,Islam A, Islam GMT, Hossain M, Khan AS. Challenges for flood risk management in flood-prone Sirajganj region of Bangladesh. J Flood Risk Management. 2018;e12450. https://doi.org/10.1111/jfr3.12450.

[3] A. Zeroual, M.Meddi. A. Assani ,2016. Artificial Neural Network Rainfall-Discharge Model Assessment Under Rating Curve Uncertainty and Monthly Discharge Volume Predictions, Water Resour Manage DOI 10.1007/s11269-016-1340-8.

[4] A.J. Kalyanapu, D.R. Judi, T.N. McPherson and S.J. Burian Monte Carlo-based flood modelling framework for estimating probability weighted flood risk J Flood Risk Management 5 (2012) 37–48 DOI: 10.1111/j.1753-318X.2011.01123.x.

[5] D.H. Nam, K. Udo and A. Mano ;Future fluvial flood risks in Central Vietnam assessed using global super-high-resolution climate model output; J Flood Risk Management 8 (2015) 276–288 DOI: 10.1111/jfr3.12096.

[6] Buahin, Caleb A., Nikhil Sangwan, Cassandra Fagan, David R. Maidment, Jeffery S. Horsburgh, E. James Nelson, Venkatesh Merwade, and Curtis Rae, 2017. Probabilistic Flood Inundation Forecasting Using Rating Curve Libraries. Journal of the American Water Resources Association (JAWRA) 1-16. DOI: 10.1111/1752-1688. 12500.

[7] Shahab Afshari , Ahmad A. Tavakoly , Mohammad Adnan Rajib , Xing Zheng , Michael L. Follum ,Ehsan Omranian , Balázs M. Fekete;Comparison of new generation low-complexity flood inundation mapping tools with a hydrodynamic model;Journal of Hydrology 556 (2018) 539-556,https://doi.org/10.1016/j.jhydrol.2017.11.036.

[8] Michel Lang , KarinePobanz , Benjamin Renard , Elodie Renouf& Eric Sauquet (2010); Extrapolation of rating curves by hydraulic modelling, with application to flood frequency analysis, Hydrological Sciences Journal – Journal des Sciences Hydrologiques, 55:6,883-898, DOI:10.1080/02626667.2010.504186.

[9] Yuqin Gao, Yu Yuan, Huaizhi Wang, Zhenxing Zhang and Liu Ye ; Analysis of impacts of polders on flood processes in Qinhuai River Basin, China, using the HEC-RAS model, Water Science & Technology: Water Supply; doi: 10.2166/ws.2018.008.

[10] Brunner.G, CEIWR-HEC, 2016. HEC-RAS River Analysis System Hydraulic Reference Manual Version 5.0. US Army Corps of Engineers Hydrologic Engineering Center. – 2016.т-p.547.

[11] Ramakrishnan, B., Durdu, A., Rajagopal, K. and Akgul, A., 2020. Infinite attractors in a chaotic circuit with exponential memristor and Josephson junction resonator. AEU-International Journal of Electronics and Communications, 123, p.153319.

Sustainable Production Flexible Improvement by Internal Activities Time Reduction in Manufacturing Industry

A.H. Abdul Rasib1, R. MahamadNoor2 , M.N. Ahmad1

1Faculty of Mechanical and Manufacturing Engineering Technology, UniversitiTeknikal Malaysia Melaka, 76100, Durian Tunggal, Melaka, Malaysia
2Panac Advanced Film Malaysia Sdn. Bhd., Free Trade Zone 1, No. 2720, Lorong Perusahaan 4, Kawasan Perusahaan BebasPerai, 13600 Perai, Pulau Pinang
Single spacing]

*Corresponding author email: amir.hamzah@utem.edu.my

Abstract. Nowadays, the manufacturing sectors are in the process of increasing their performance to extend productivity because there are so many challenges from competitors in the same sectors, especially in the food industry. The purpose of this study is to propose improvements to increase sustainability of food operation flexibility through secondary data collection, Cause and effect diagram analysis, and the Why-why Analysis tool. Besides, improving the understanding of Single Minute Exchange of Dies (SMED) factors in the production line can be applied by changing the internal activities to become external activities. This study is useful for manufacturing managers and companies who are willing to increase their flexibility and at the same time reduce the changeover time within their available resources without compromising the machine efficiency and cost. In conclusion, this study has provided a better understanding of reducing internal activities that cause high changeover time.
Keywords: *Changeover, Internal Activity, Flexibility, Production Line, Sustainable*

1. Introduction

Today, demands for goods and services are changing repeatedly and some companies are always facing greater challenges; thus, a new manufacturing system is needed to make the production proceed rapidly and more importantly to bring profit to respond to all the market changes. Therefore, the goal of each company in manufacturing is minimal cost and setup time. Due to these two main factors, the flexibility in the production line must be balanced or flexible as it will help to increase the productivity of the manufacturing companies.

It is also important to note that time, effort and cost are the essential elements that need more focus to maintain or improve flexibility in the production line. Based on research findings, it is very important to have rapid changeover while changing the product because it will give a huge impact to sustain the flexibility in industries.

Flexibility is the measure of performance of a firm in handling the production line. Being flexible means that the system carries out a good performance in order to reduce the time. The main problem that always happens in manufacturing companies is unnecessary activities or some internal activities which can drag the time to produce the product. Normally, small manufacturing companies have a difficult task to manage flexibility while performing changeover in the production line.

Other than that, the elements that are related to customer demands are the quality and the efficiency of the production itself. The reduction of internal activities in the company is one of the ways to produce better quality product and gain higher efficiency. Sometimes, poor services also tend to lower customer satisfaction. This is due to the slow production which may be caused by the slow changeover in the production line or unwanted activities. The higher the flexibility in the manufacturing companies, the higher the productivity. However, to achieve higher flexibility, food industries need to overcome many challenges. To face these challenges, the best way to overcome them is by practicing and applying lean manufacturing tools. Single-Minute Exchange of Die (SMED) is one of the Lean Manufacturing's tools. In this study, the concept of SMED will be used to cater to sustainable flexibility improvement in the food industry.

2. Understanding flexibility

Mishra et al. [1] stated that the meaning of flexibility is the ability of firms to quickly change from one product to another without loss of efficiency and it is concluded as the main element of the manufacturing strategy. Besides that, [2] also wrote that flexibility is the capacity to change the surroundings or respond to unpredictability with minimum cost, time and effort and with maximum performances. Similarly, [3] claimed that the capability to change or react with little penalty in time, effort, cost or performances is known as flexibility. So, to sum up, it is agreed that time, effort, and cost are the essential elements that need more focus to maintain or improve flexibility in the production line.

2.1 Manufacturing Flexibility

Manufacturing flexibility refers to organisations which have the capability to change the levels of production quickly, to grow new products rapidly and more often and also to improve the organisations' capability to face the challenge from competitors while at the same time gain the acceptable levels of cost and quality [4]. Fredriksson et al. [5] synthesised the reflection of manufacturing flexibility as the capability of a firm to respond to their customers' need changes and unanticipated changes come from the pressure of the competitor. Complementary to this statement, with manufacturing flexibility, the level of production line of the product will run faster and can also reduce wastage of cost and time while further enhancing the effectiveness of the product.

[4] explained that machine flexibility, labour flexibility, material handling flexibility, and routing flexibility are the best ways to explain internal flexibility. In contrast, mix-based flexibility, volume flexibility, new product flexibility, and delivery flexibility are the major components of external flexibility [1].

Particularly, flexible processes guarantee a way to cope with the increasing demand for variety and uncertainty. Therefore, the topic of process flexibility has become the main focus from either commercial or research institutions, as an understanding of the requirement for processes is capable to adjust their behaviour to changes (Mejri et al.) [6]. The ability of a manufacturing system to produce a part without needing a major setup is known as process flexibility. This is useful in reducing the size of a batch and also lowering the inventory cost. Besides that, it would help to minimise the demands to duplicate the machines because it already allows the machines to share the same process. Sometimes, the time to produce or manufacture the product becomes longer because the time to set up the equipment is longer and extra time is needed to handle it.

Otherwise, flexibility offers the process to be done without requiring a major setup as it can speed up the process and function efficiently. In order to improve process flexibility, it is important to have a flexible activity refinement (Jiang et al.) [7]. Thus, to achieve process flexibility, the machine in the production line has to manufacture the product as fast as lightning.

Process flexibility, or also classified as external flexibility in manufacturing, is mostly related to achieving customer requirements and also providing high competitiveness among firms. It is always more preferable by the customer as it directly seeks opportunity for a competitive advantage [8]. The mainconstituents of process flexibility are mix flexibility, new product flexibility, volume flexibility, and delivery flexibility.

2.1.1Mix Flexibility
Mix flexibility is the ability to manage the product or its own variants by using equipment which needs a quick setup time. [9] argued that mix flexibility is when some products are being manufactured without a longer setup time. [8] asserted that mix flexibility is the capability of the machine to produce various combinations of products at certain capacity economically.

2.1.2 New Product Flexibility
The product is the main component because without the manufacture of a product, there is nothing to be controlled or changed. According to [3], product flexibility is the ability to produce new or exchange products and also to modify the product. In contrast, Lafou et al. [10] described product flexibility as the ability of the system in manufacturing to make various types of parts by using the same equipment. In terms of production, the aim of flexibility is lowering the amount and cost of the stock. Subsequently, the connections between product and plant group directly or indirectly are also relevant for the decisions of flexible production. There are three dimensions that defined new product flexibility operationally: range, mobility, and uniformity. Range means the number of new product parts, prototypes, and new products; mobility represents the ease when the company moves from one state to another state; and uniformity represents the manufacturing system's ease of function under specific circumstances with several penalties in terms of time, effort, cost, and performances.

2.1.3 Volume Flexibility
[4] stated that when the organisation is managed to operate at numerous batch sizes and/or different production output levels profitably, it can be classified as having volume flexibility. Therefore, the implementation of this flexibility in the factory will help them to adjust or change the product within a big range. In other words, volume flexibility enables a firm to produce the production either above or below the capacity that had already been set to meet customer demand. [3] claimed that there are a few companies that had not invested in flexible technologies but using outsourcing for volume flexibility.
In addition, volume flexibility is suitable to use for long term uncertainty, referring to customer demand of the products but it is inconvenient for the short term. This is because the implementation of this flexibility will affect some of the manufacturing elements such as the cost, productivity, and also performance of the machines. There are several reasons to implement volume flexibility due to unpredictable problems that could happen in a firm such as machine breakdowns, unbalanced cycle time, long setup time, and non-conforming products.

2.1.4 Delivery Flexibility
In a nutshell, customer satisfaction plays an important role for a firm to achieve success and become well known. The trust of the customers to a firm will be higher if the firm can provide fast and quick delivery after receiving orders. The tendency of customers to repeat their orders from the same company is also high. The saying "the customer is always right" is really meaningful. As customers, they do not think about the problem that a company will face if the customer demands already exceed its capacity limit. All they know are once the orders are submitted, they want to receive the product immediately or in a short time.
[11] stated that in order to assess the delivery flexibility, it is necessary to take into account the vehicle and the possibility of alternative routes in the network of transportation, palletising method, and transportation method. Other than that, the scheduling challenge that always occurs is the long term-capacity planning and the delivery process that needs to be scheduled and assigned based on available resources during operative delivery planning [12]

2.1.5 Variant Flexibility
Variant flexibility is the ability of a firm to produce the same product in a large number with various types of variants. The variant and inventory have a connection as a one-to-many relationship where the inventory has unique relations to every unique variant. All the products still use the same concept as the actual product, but there are some differences in terms of the materials and ingredients. The manufacturing process which manages to produce the product with multiple variants should be given

some credit and helped to improve customer satisfaction. Landahl et al. [13] suggested an adjustable manufacturing system that aids the variety of a product that can quickly change the customer requirement while the flexibility of the system itself serves the variation of the product family.
Process flexibility in manufacturing is more focused on the operation of the product itself. As we can see, the product has various elements that need to be considered such as its designs, capacity, size, and market demands. All of these are big constraints before the product can be manufactured.

2.2 Single-Minutes Exchange of Die

SMED is a manufacturing lean tool that takes place when there is a changeover of equipment for reducing the time. The terms "single minute" or "single-digit minute" do not mean all changeovers and the startup time needs to be at least one minute, but should be below 10 minutes (Singh et al.) [14]. Rosa et al. [15] emphasised that SMED allows for the reduction of lot sizes and enables meeting of the variation of demand by focusing on the elimination of the waste connected to the changeover of the tool in the setup phase.
In SMED, there are two types of activities, namely internal and external. Internal activities can be conducted only when the machining process is stopped while the external activities can be done while the machining process is still running [16]. So, the main objective of the SMED system is to change the internal activity to external activity as much as possible. Furthermore, the process of SMED is more focused on external activity as it can help to reduce the time taken to change the equipment while the machine is still running. Figure 1.shows the visual application of the SMED process that has already been implemented in industries.

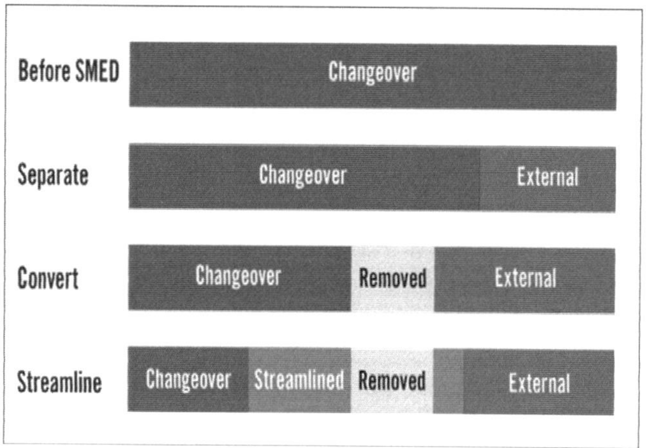

Figure 1. *SMED (Venugopal, 2018)*

2.2.1 Internal Activity

[17] also mentioned that internal activity can be carried out only when the machine is shut down such as when attaching or removing the dies. Similarly, Bevilacqua et al. [18] identified that the internal activities are carried out when the machine is offline and it must be reduced as it affects the production time. Karasu et al. [19], Maheswaran et al.[20] also claimed that all activities which cannot be done without the machine are stopped for an internal setup such as mounting-demounting the dies. From these three findings, it can be concluded that internal activities spend a lot of time that can actually be reduced or can be done while the machine is running as they do not involve the machine itself. The internal components need to be identified first as they will then determine whether it is suitable or not to be converted into an external activity.

2.2.2 External Activity

External activities can be performed when the machine is still running at its normal operation, for instance, equipment getting ready for the setup operation can proceed before the machine is shut down [17]. External setup includes the activities that can be performed while the machine is running, for example in advanced mold transportation [16]. Sousa et al. [19], Maheswaran et al.[20] agreed with the previous explanation as they also concluded that external activities are designated as all the activities that do not involve directly with the equipment and can be carried out without discontinuing the production. Other than that, external activities can be described as an activity conducted while the machine is still running so that everything is done before reaching the last process of the current operation. As a suggestion, all the tools, equipment, and material are well prepared before starting the setup of the internal activity so that this will automatically remove all the time that should be needed while doing the internal activity. Thus, it will help the process of changeover between the equipment or tool.

In a nutshell, it is prominent to ensure the production time and the lead time of the product can be reduced for flexibility performance. Based on the overall findings, it can be concluded that in manufacturing, there is a lot of flexibility involved to improve the production line flexibility. Nevertheless, in this literature, only several flexibilities were focused on which are machine flexibility, labour flexibility, material handling flexibility, routing flexibility, customer flexibility, supplier flexibility, process flexibility, mix flexibility, new product flexibility, volume flexibility, delivery flexibility, and variant flexibility. All of these flexibilities give a huge impact on a firm to sustain and improve production and also reduce the changeover time to be more efficient and profitable through internal activities.

3. Identification of Issues

In order to propose an improvement, it is essential to identify the factors of issues. These factors of issues could be identified through literature studies. In this study, the selected company will undergo data collection. Relevant data are extracted from the assembly process. They are categorised into two types: (i) primary data and (ii) secondary data. Sources of data are mainly related to documents (i.e., Changeover Time Record and Monthly Quality Record). For analysis purposes, bar graph, 7 QC tools had been implemented in order to identify the root cause which has the tendency to be the issues in industries. Figure 2 shows the steps of the processes.

3.1 7 QC Tools

The 7 QC Tools is a suggested tool that is used to solve a problem that always occurs in the manufacturing industries. These 7 QC tools are fundamental instruments which can help to improve the product quality. Basically, the function of this tool is to analyse the process of production, identify major problems, control the fluctuations of product quality, and last but not least, suggest solutions to avoid any defect in future production. The 7 QC tools include the Check Sheet, Pareto Chart, Cause and Effect Diagram (Fishbone Diagram), Histogram, Flow Chart and Control Chart. In this study, there is one selected QC tool chosen to be more focused on as it is related to the project which is the Cause and Effect Diagram.

A Cause and Effect Diagram is a tool that studies the relationship between the effect and the possible causes that occur in certain problems. This tool is preferable to use in order to find the root cause of the problem before providing the solution or any improvement for the problem.

Besides, the purpose of this tool is as a graphical representation of the trail leading to the root cause of anyproblems that happen in the industries. Moreover, this tool is very effective as it can help to generate ideas systematically about the problem's cause and then present the details in a structure form that looks like a fishbone.

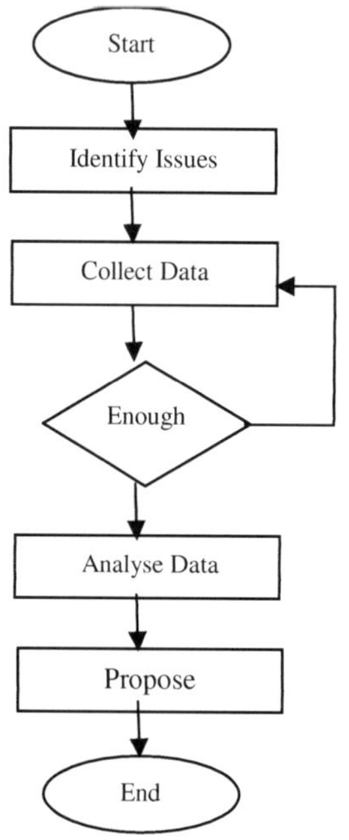

Figure 2. Steps of Processes

3.2 Why-why Analysis

Why-why Analysis is the ultimate root cause analysis tool that was developed by Sakichi Toyoda, a Japanese inventor and industrialist. This tool is part of the Toyota Production System and it has become an integral part of the Lean philosophy. Besides, this tool has a simple technique and is effective for solving problems. It is compulsory to find an exact reason that causes the existing problem by asking a sequence of "why" questions. It is very useful to help to identify the root cause and propose for the improvement of any problem.

4. Results and Discussion

Specifically, this section will elaborate on the problem definition of the studies. In addition, the definition of the problem was identified. The flexibility of production was not sustainable and the activity during changeover must be understood. This study will continue with the data collection from the case study through the factory itself. Then, data analysis is done to propose improvements that can be used by the company to increase their productivity.

4.1 Data Collection

Actually, this company has three main production lines which are carbonated drink line, soy bean drink line and tamarind drink line. In this study, the chosen production line is the carbonated drink line because the main focus of this study is to reduce the time in order to achieve flexibility in production. Thus, the carbonated drink production line is selected as a changeover occurs while changing the flavour in a day. It should also be highlighted that there are six types of carbonated drink flavours namely orange, ice cream soda, grape, sarsi, strawberry, and mixed fruits. In this study, the process of data collection was time consuming as secondary data were used. Table 1 shows the production record for carbonated products in March 2019.

Table 1. Carbonated Product for March 2019

Product Carbonated Drink	Volume	Production Date	Total Finish Goods (Bottle)
Strawberry, Sarsi, Grape, Soda, Orange, Fruit	1.25 L	3/5/2019	1416
Strawberry, Orange	2L	3/12/2019	1956
Strawberry	2L	3/13/2019	960
Strawberry, Grape, Fruit	2L	3/14/2019	2772
Strawberry, Grape, Fruit	2L	3/15/2019	2364
Strawberry, Grape, Fruit	2L	3/16/2019	2316
Fruit	2L	3/18/2019	1527
Strawberry	2L	3/19/2019	2772
Sarsi, Orange	2L	3/20/2019	1200

Based on the data above, the production activities did not receive consistent orders. The objective of this study is to eliminate unwanted internal activity to external activity to produce quick changeover.

4.2 Data Analysis

The data in Figure 3 are not consistent as no production occurred for several days in March 2019. This is because the production is based on customer demands and sometimes the workers did not record the data. In this Figure, flavour changeover occurred.

Figure 3shows the changeover records in bar-graph for carbonated products in March 2019 based on Table 1.

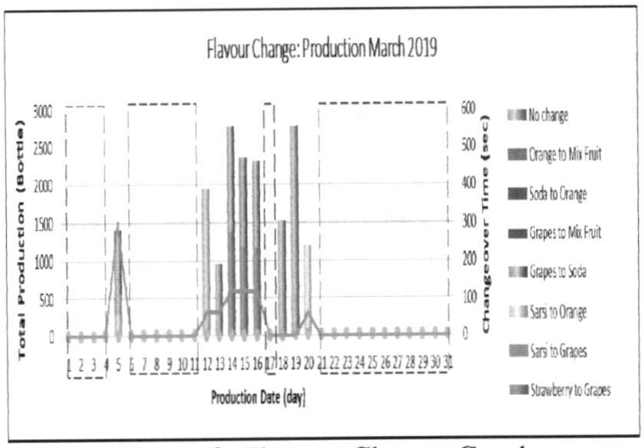

Figure 3. Flavour Change Graph

As mentioned before, SMED is a manufacturing tool that takes place when there is a changeover of equipment for reducing time. In order to reduce the time, the method used changes the internal activity to external activity. Thus, this SMED checklist was provided to help differentiate the internal activity and external activity for the process of manufacturing carbonated drinks. Table II shows the list of internal and external activities.

Table 2. List of Internal and External Activities during Changeover

Step No.	Description	Activity
1.	Taking out the raw material from storage	External
2.	Boiling of sugar	Internal
3.	Labelling process	External
4.	Mixing the boiling sugar with the flavour	Internal

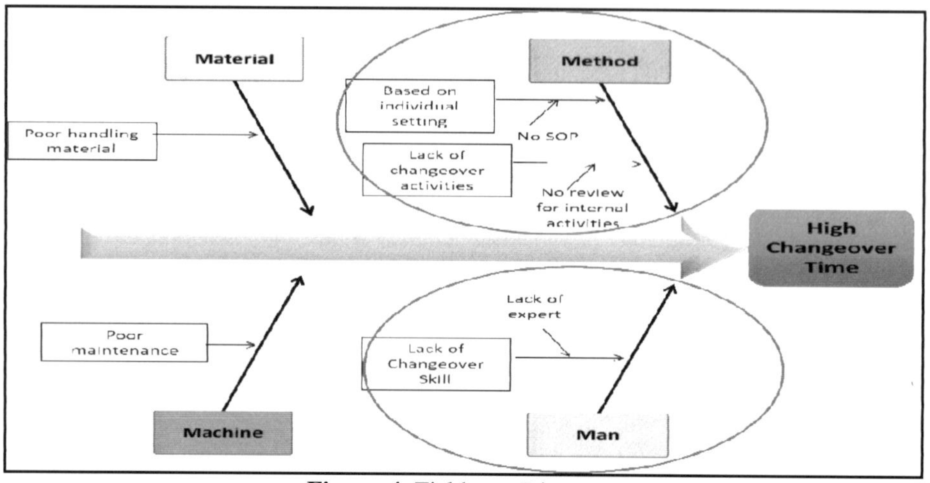

Figure 4. Fishbone Diagram

4.3 Propose Improvement

In order to identify the critical factor of the problems occurring in the company, Why-why Analysis was used. The Why-why Analysis is an iterative interrogative technique that was used to find the root cause of a certain problem. This technique was applied by repeating the question "why" as much as five times if possible. In this study, the focus was on the Man and Method factors because those two factors were related to changeover internal activities. Table 3 shows the Why-why analysis of the problems that occurred in the company's operation.

Table 3. Why-why analysis

	Why 1	Why 2	Why 3	Why 4	Why 5
METHOD	High Changeover time	Based on individual setting	No monitoring	Difficult to monitor	No provide SOP
		Lack of changeover activities	No review for internal activities	Excess of preparation time	Improper facility/equipment and layout
MAN	High Changeover time	Lack of changeover skill	Lack of talent	Lack of training	No specific training provided

In Root Cause Identification, a tool was used to find the root cause which was the Fishbone Diagram. Based on Figure 4, it is shown that the cause is focusing more on the method and man, thus causing longer changeover time.

There are three proposed improvement ideas that could be implemented in this food operation company to reduce the changeover time through minimising or eliminating the time of internal activities. The first improvement idea is to provide appropriate standard operating procedure (SOP) that is focusing more on changeover that happens in the carbonated drinks production line. This is because it will meet the aim of this project to reduce the changeover time. SOP is described as a set of instructions compiled by a company that is done step-by-step to manage the workers to carry out their routine operations. The importance of SOP is so the production can be done by following the flows as well as to train the workers. Other than that, the SOP can also help the company to save on training costs and control the quality and consistency of the production of carbonated drinks. It is important to note that this SOP focuses on the changeover that always happens in the production line. Thus, there are the main processes that were included in the changeover phase, namely the process of boiling sugar, mixing the sugar and flavour and the filling process.

Secondly, the idea is to design the mixture tank by adding one more tank intended to produce many flavours in a day quickly. Therefore, the total tanks went from two to three. Besides, after adding the tank, the layout of the mixture is redesigned to make sure it can be moved nearer to the filling machine in a short time. Thus, a new design is created that could help to improve production by reducing internal activities and at the same time can reduce the changeover time. The design is created by adding one more mixture tank and then designing a new tool named a rotary tool that might help to rotate the mixture tanks quickly to change the mixture tank before going through the filling process. Figure 5 shows the new design of the mixture tanks.

Figure 5. The New Design of Mixture Tanks

The last improvement idea is providing a changeover training with proper schedule in order to make sure the workers could carry out their task or operation with the appropriate skill. A training schedule is usually created to include the different activities involved in a company. The benefit of having a training schedule is it can provide a clear guideline for the process and also can help to manage the worker's time, especially to ensure that changeover can be reduced to gain high flexibility in the production line.

5. Conclusion

As a conclusion, the company will gain new knowledge that can be used to reduce the changeover time during the process of manufacturing carbonated drinks through the application of 7 QC Tools and Why-why Analysis to identify and investigate the root causes and propose improvements. Thus, the company can also reduce the changeover time while increasing flexibility by applying SMED concept through reduction of internal activities. By the end of this study, the company will know how to execute the proper way to produce more products in the long term.

Hence, for future research, another type of operation involving any changeover in its production line should bestudied. It would bring more benefits to find a company that is highly competitive in the industries because the problems that will occur in that company will pose greater challenges to the researcher. There are several industries that could be chosen such as automotive industries, packaging industries, steel industries or plastic industries.

Acknowledgements

The author would like extend special thanks to the FRGS-RACER/2019/FTKMP-COSSID/F00412 grant, FakultiTeknologiKejuruteraanMekanikaldanPembuatan, UniversitiTeknikal Malaysia Melaka and the participating food company for the use of facilities and providing useful data in order to complete this study.

References

[1] R. Misha, A.K. Pundir, L. Ganapathy, Conceptualizing Sources, Key Concerns and Critical Factors for Manufacturing Flexibility Adoption: An Exploratory Study in Indian Manufacturing Firms, *Journal of Manufacturing Technology Management, vol. 27 n. 3,* 2016, pp. 379–407.

[2] B.A. Korchi, and A. El, Flexibility and Agility : Evolution and Relationship, *Journal of Manufacturing Technology Management, vol. 29 n. 7,* 2018, pp. 1138 - 1162.

[3] M. Scherrer-Rathje, Manufacturing Flexibility through Outsourcing : Effects of Contingencies, *International Journal of Operations & Production Management, vol. 34 n. 9,* 2014, pp. 1210-1242.

[4] M.A. Khalafr, The relationship between Internal Integration and Manufacturing Flexibility in The Egyptian Industry, *International Journal of Quality and Service Sciences, vol. 7,* 2018, pp. 123-140.

[5] A. Fredriksson, C. Wänström, A. Fredriksson, Manufacturing and Supply Chain Flexibility – Towards a Tool to Analyse Production Network Coordination at Operational Level, *Strategic Outsourcing: An International Journal, vol. 7 n. 2,* 2014, pp. 173-194.

[6] A. Mejri, S. Ayachi, R. Martinho, Representing Business Process Flexibility using Concept Maps, *Procedia - Procedia Computer Science, vol. 100,* 2016, pp. 1260–1268.

[7] Y. Jiang, N. Xiao, Y. Zhang, L. Zhang, A Novel Flexible Activity Refinement Approach for Improving Work Flow Process Flexibility, *Computer in Industry, vol. 80,* 2016, pp. 1–15.

[8] R. Mishra, A Comparative Evaluation of Manufacturing Flexibility Adoption in SMEs and Large Firms in India, *Journal of Manufacturing Technology Management, vol. 27 n. 5,* 2016, pp. 730-762.

[9] S. Wilson, and N. Ali, Product Wheels to Achieve Mix Flexibility in Process Industries, *Journal of Manufacturing Technology Management, vol. 25 n. 3,* 2014, pp. 371–392.

[10] M. Lafou, L. Mathieu, S. Pois, and M. Alochet, Manufacturing System Flexibility : Product Flexibility Assessment, *Procedia CIRP, 41,* 2016, pp. 99–104.

[11] A.V. Kapitanov, Manufacturing System Flexibility Control, *Procedia Engineering, vol. 206,* 2017, pp. 1470–1475.

[12] H. Lagemann, and H. Meier, Robust Capacity Planning for The Delivery of Industrial Product-Service Systems, *Procedia CIRP, 19,* 2014, pp. 99–104.

[13] J. Landahl, C. Levandowski, H. Johannesson, R. Söderberg, K. Wärmefjord, J.S. Carlson, J. Vallhagen, Using Product and Manufacturing System Platforms to Generate Producible Product Variants, *Procedia CIRP, 44,* 2016, pp. 61–66.

[14] J. Singh, H. Singh, I. Singh, SMED for Quick Changeover in Manufacturing Industry–A Case Study, *Benchmarking, vol. 25* n. 7, 2018, pp. 2065–2088.

[15] C. Rosa, F.J.G. Silva, L.P. Ferreira, R. Campilho, SMED Methodology : Reduction of Setup for Steel Wire-Rope Assembly Lines in Conference the Automotive Industry Wire-Rope Assembly Lines in The Automotive Industry Costing Models for Capacity Optimization in Industry SMED methodology : The reduction of Set, *Procedia Manufacturing, 13,* 2017, pp. 1034–1042.

[16] M.K. Karasu, M. Cakmakci, M.B. Cakiroglu, E. Ayva, Improvement of Changeover Times via Taguchi Empowered SMED / Case Study on Injection Molding Production, *MEASUREMENT, vol. 47,* 2014, pp. 741–748.

[17] C.E. Boran, SMED Methodology based on Fuzzy Taguchi Method, *Journal of Enterprise Information Management, vol. 31 n. 6,* 2018, pp. 867-878.

[18] M. Bevilacqua, F.E. Ciarapica, I. De Sanctis, G. Mazzuto, C. Paciarotti, A Changeover Time Reduction through An Integration of Lean Practices: A Case Study from Pharmaceutical Sector, *Assembly Automation, vol. 35 n. 1,* 2015, pp. 22–34.

[19] E. Sousa, F.J.G. Silva, L.P. Ferreira, M.T. Pereira, R. Gouveia, R.P. Silva, Applying SMED Methodology in Cork Stoppers Production, *Procedia Manufacturing, 17,* 2018, pp. 611–622.

[20] Maheswaran G, Balamurali R, Sundarambal B, Soundarya Priyadharshini S, Vinodhini P. 2020. A Smart Plug for Controlling and Automation Anywhere Using Iot. International Journal of Advanced Science and Technology, 29(3), pp.8570 - 8577.

IOP Conf. Series: Materials Science and Engineering **994** (2020) 012004 doi:10.1088/1757-899X/994/1/012004

Understanding of Non-Value Added Overtime in Manufacturing Operations

A. H. ABDUL RASIB 1 AND M. MUSAZALI 2

1Faculty of Mechanical and Manufacturing Engineering Technology, UniversitiTeknikal Malaysia Melaka, 76100, Durian Tunggal, Melaka, Malaysia
2Centre for Graduate Studies, UniversitiTeknikal Malaysia Melaka, 76100, Durian Tunggal, Melaka, Malaysia

*Corresponding author email: 1amir.hamzah@utem.edu.my, 2m051910070@student.utem.edu.my

Abstract. The manufacturing industry plays a crucial role in the production process to meet the demands of the market and to raise the standard of living recently. Nevertheless, one of the most common issues associated with the industry is unnecessary overtime. Overtime has produced a variety of negative consequences on production output. This paper aims to identify the various causes of unnecessary overtime, and the causes have been categorised through the seven wastes of lean manufacturing. This paper also focuses on unnecessary overtime occurred in industries. The method used in this paper is by reviewing the relevant literature from conference papers and journal publications, including the International Journal of Scientific and Technology Research, Journal of Advanced Manufacturing Technology, and International Journal of Engineering and Technology, amongst many others. Finally, the critical components of NAO have been finalised according to the necessary scopes of operation, which considered as factors which are pre-production, in-production, and post-production. Based on the final critical components, there were two critical sub-factors in pre-production components, six critical sub-factors in-production elements, and five critical sub-factors post-production components.
Keywords: Overtime, Lean Manufacturing, Non-value added, Standard Time, Performance measure

1. Introduction

In today's working environments, manufacturing industries are trying to satisfy their customers. To achieve customer satisfaction, they must compete with each other to produce high-quality products by improving the performance of workers or employees. However, many companies are putting pressure on their employees to do more and work more hours without taking into account the losses they faced in the overtime period. Recently, overtime has been prevalent in the production line among the workers without seriously taking the reasons why the workers need to do overtime.

The International Labor Organization (ILO) describes overtime as work hours which exceed regular mandatory hours[1]. Extra time is used to increase production and fulfill the order volume of the customer. However, excessive overtime can somehow improve the operation cost and may adversely influence on physiological and behavioral wellbeing of employees. Thus, global organizations such as the ILO, the EU, and national regulations are imposing limitations on overtime [2]. In this situation,

overtime could be required when it is necessary. However, it also could be unnecessary when there are no urgent orders that need to be completed.
This study will be focused on the factors and sub-factors that contributed to the unnecessary overtime or Non-value Added Overtime (NAO) in manufacturing industries. In this study, the identifying of NAO causes will be based on seven wastes of Lean Manufacturing (LM). Besides, NAO is one of a part of an additional working hour because NAO is done after office hours. Thus, unnecessary overtime is identified as detailed in Lean Manufacturing.

2. Understanding of Additional Working Hour

Referring to the MALAYSIA EMPLOYMENT ACT (1955), all other workers are obligated to a whole day's break and are thus allowed to work six days per week. Normal operating hours per day will be eight hours, with a total of 48 hours per week. When a corporation operates less than six days a week, workers will be restricted to a maximum of 10 hours per day (except for break times) and 48 hours per week. Nevertheless, in certain cases or even in mutual decision, the worker can extend the regular working hours as indicated earlier. Those extra working hours shall therefore be pointed to as overtime which shall be charged an extra 1.5 times the daily salary.
Campbell (2002) described the overtime may be classified as either compulsory or voluntary. He also explained in detail the compulsory overtime. Compulsory overtime is extra working hours implemented by the employer on specific grounds. In the United States, the most frequent workers who are subject to mandatory overtime are nurses, who are typically expected to work 12-hour shifts at one time. Voluntary overtime is the extra working time preferred by the employee and not enforced by the employer. Voluntary overtime is now becoming increasingly common in the Malaysian industry as workers are inspired by the fact that they are paid 1,5 times more than their regular hourly wage. However, when performed excessively and not supervised, unpaid overtime can have an overall negative effect on both the employer and the employee[3].
Dewi *et al.*, (2013) expressed the opinion that to remain competitive in this period, companies face an unprecedented challenge by offering a variety of goods that have always hindered their productivity. Apart from that, productivity is one of the crucial things that companies are working on it[4]. Productivity refers for output to all or some of the resources used to create the production. Resources include labour, money, time, energy, and raw materials. Productivity is translated directly into cost savings and productivity [5]. According to Patel *et al.*, (2017), there are two productivity measures which are overall productivity factor and partial productivity factor. In overall productivity factor, the outputs and all inputs have been considered while in partial productivity factor, the outputs and single or chosen inputs are considered[6].
As said by Mpanza & Nyembwe (2014), productivity measures the effective and efficient use of available resources. Many variables affect productivity[7]. Sidabutar & Matondang (2019) stated that productivity covers processes, quality management, and technology. Thus, productivity can be changed when overtime work is done at the company[8]. Overtime can be two types, which are necessary overtime and unnecessary overtime. Ebrahim *et al.*, (2017) stated that necessary overtime could be referred to as additional working time required to meet the output target. However, unnecessary overtime is where time loss has occurred[9].
A variety of activities may cause time loss in manufacturing operations (e.g., waiting, system failure, and lack of workforce). These are usually known as common factors such as individual, material, process, method, management, and climate or otherwise referred to as Man, Method, Machine, Material, and Environment (4 M 1 E). The most secret trigger is time loss due to unnecessary work, which is closely related to the method of operation [4]. Unnecessary work or non-value added activities would then contribute to the occurrence of unnecessary overtime. Therefore, it is essential to evaluate unnecessary affecting overtime in manufacturing industries.
Thus, the overtime regulations should be taken seriously in order to increase the company's profit because the workers need to be paid whenever they are done during the overtime work.

2.1 Non-value Added Activities

According to Poornashree & Ramakrishna (2019), a value can be described only as what the consumer is willing to pay for purchasing the product. If the customer does not want to pay for it, there is no value for it. Something or someone who does not contribute or increase value is a waste. However, non-value added activities are activities which require time, resources and capacity but not to bring value to the task. In the terminology of lean development, this term 'Non-Value Added Activities' (NVAA) is known as waste. NVAA is a waste of time, which could have been prevented if the process was prepared, conducted, monitored, and managed more carefully. We should be extracted when NVAA is detected whenever possible[10]. Value-added activities represent the right combination of operations and processes needed to produce the product, producing the best quality, at the lowest possible cost, to the customer on time [11]. This kind of wastes is known in Non-value added activities because it does not add value to the others.

2.2 Factors of Non-value Added Overtime (NAO), b

From the past literature study, the structure and the equation of Unnecessary Overtime (UOT) have been developed. Figure 1 shows the structure of UOT resulted from past literature [9].

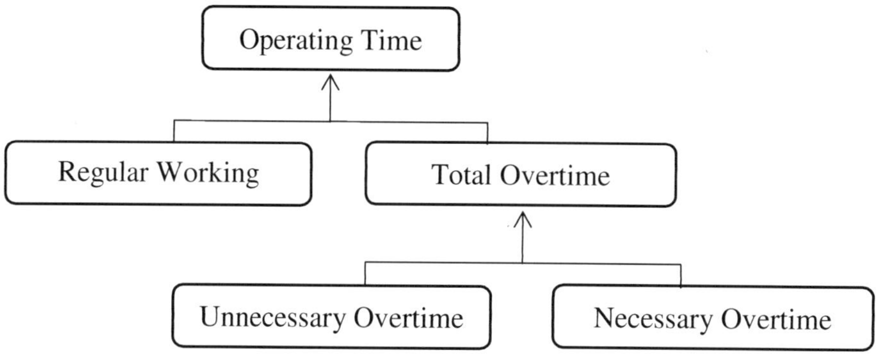

Figure 1. Structure of UOT [9]

According to Ebrahim*et al.*, (2017), the structure of UOT, the equation also has been developed. Thus, the equation is stated below,

$$UOT = t_{tot} - t_{not} \tag{1}$$

Where,
t_{tot} is total overtime given to achieve the production line target
t_{not} is necessary overtime
Thus, UOT ≥ 0.

In this regard, the determination of UOT was based on the deduction value of t_{tot} by t_{not}. In this method, it is difficult to improve due to the uncertain factors of t_{not} identification. Therefore, in the next step of enhancement, it is necessary to identify the potential factors and sub-factors in order to cater effective formula of NAO. Thus, effective and efficient improvement will be taken more appropriate in manufacturing industries[9].

3. Methodology

In this study, it is crucial to identify the factors for NAO. In order to figure out the causes of the NAO, to make the tasks more organised and straightforward, the process flow of the purpose is given, as shown in Figure 2. The first step is the studies of literature, which will be clarified in the next section.

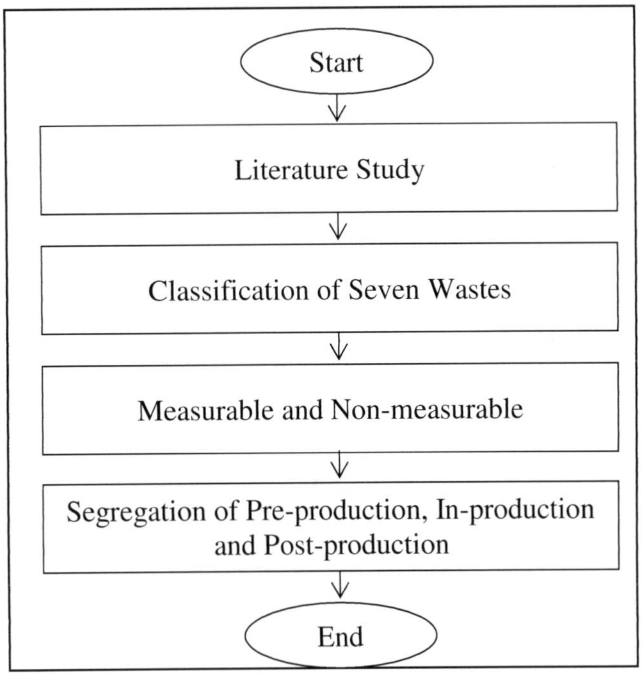

Figure 2. Process Flow of Studies

3.1 Literature Studies

Extensive literature studies are performed in order to determine the causes of NAO. The literature studies aim 30 papers published in the last five years. The literature studies concentrate on operation and production management, manufacturing management, industrial engineering, operational study and quality, and performance. The literature studies would be used to explain the selection of a significant factor, including the description of seven LM wastes (i.e., Waiting, Transportation, Motion, Defect, Overproduction, Over-processing, and Inventory) and measurable elements.

3.2 Classification of Seven Wastes

Sundar *et al.*, (2014) found that Lean Manufacturing (LM) is a concept that aims to eliminate waste in all areas of production, including product design, supplier networks, and customer management. This concept has come from Japan. It is expected that, with the implementation of lean manufacturing, a quality product will be produced by reducing waste there. Nowadays, many companies use LM to improve their production productivity[12].

While the process of adding value (value-adding activities) is the process of converting/transforming raw materials, changes in shape or quality: changes in the raw material into a particular part or product[4]. According to Womack & Jones (1997), lean is a quality and process improvement method that emphasises customer needs, employee involvement, and continuous improvement. The wastes involve in Lean Manufacturing are Defects, Motion, Transportation, Inventory, Over-processing, and Waiting[5].

Systematic identification and disposal of waste are considered to be a core element of the Lean Production ideology [11]; [13]; [14]. In order to determine the involved wastes in the production process, Ohno (1988) identified the seven types of waste: Defects, Motion, Transportation, Inventory, Over-processing, Waiting, and Overproduction[11]. These seven types of waste have revealed to be very useful for the identification of waste involved in the manufacturing process. The lean theory describes waste as any operation that adds costs or spends time but does not add value to the customer [5]; [11].

Table 1 below is the table of the seven wastes in Lean Manufacturing[15].

Table 1. Description of 7 Wastes [15]

Type of Waste	Description
Defects	The documentation has frequent mistakes, difficulties with the consistency of the product, or bad delivery performance.
Motion	Impoverished organization of the workplace, leads to poor ergonomics, injuries usually bending or stretching, and mostly lost items..
Transportation	The people have excessive movement, excessive knowledge or items, leading in a waste of resources, effort and value.
Inventory	Extremely high storage and backlog of information or goods, leading to high cost and poor service.
Over-processing	Passing through production process using the incorrect set of tools, methods, or systems, even when a more direct method would be more effective.
Waiting	People, information, or goods have long periods of inactivity that leads to weak flow and long lead times.
Overproduction	Making too much or too little, typically results in an inadequate delivery of work or goods and an overload of inventory.

3.3 Measurable and Non-measurable

Usually, it is easier to manage measurable elements than non-measurable elements. The reason why measurable elements are easier to manage is that measurable elements are the quantitative data, while non-measurable elements are qualitative data. According to Womack & Jones (1997), the measurable term came from the root word of "measure"[5]. However, according to Hair *et al.*, (2019), the word calculation measure represents a measurement of physical properties or tangible objects such as weight, relative humidity, length, period[16]. The measurable data is explained in general information and not only focus on the engineering area. Thus, the Value Added activities and Non-value Added activities can take account of the measurable indicator.

3.4 Pre-production, In-production and Post-production

This study is established through the basic concept of production 'Input-Output (IO) which focuses on operations, and the elements have been defined through measurable elements in three phases of manufacturing operations; i) Pre-production, ii) In-production and iii) Post-production[17]. According to Rasib *et al.*, (2019), Pre-process is any operation that is involved before the core process. In-process is a crucial task in pre-process operation where the post-process is an internal product before the finished products are transported to the next station, such as the Quality Assurance Department for quality inspection[19].

4. Results and Discussion

The major purpose of this study is to identify the various causes of NAO, and this objective is achieved. The findings thus demonstrate the potential for a synthesis of theory and practices. As a consequence, it consists of essential stages, as indicated, which are:

- Literature study
- Clarifying the Seven Wastes of Lean Manufacturing
- Measurable and Non-measurable

4.1 Literature Study

The statistic shown in Table 2 is abstracted from the 30 studies that have been conducted. This table provides a representative view of the sub-factors in the aspects of factors that contributed to the overtime.

Table 2. List of Overtime's Sub-factors

Sub-factors of Overtime	
Excess workforce inventory	High takttime
High production unit	Extensive motion
Annual paid leave	Material defects
Sickness absence	Delays
Product defects	High workload
Workers shortage	Excessive inventory
Machine availability	Long waiting time
Components shortage	High materials storage
Workers performance	Rework
Inappropriate OT planning	Over-processing
Bottleneck	Long changeover time
Erratic variance (rework)	Unscheduled downtime
Low plant capacity	High lead time
Incoming raw materials	Workload control
Products delivery	High production volume in processing and finishing unit

4.2 Clarifying the Seven Wastes of Lean Manufacturing

As far as the seven wastes of Lean Manufacturing are concerned, the statistic shown in Table 3. This table is a list of sub-factors from relatively comprehensive past literature that has been abstracted, as shown in Table 2. Based on the total of papers reviewed, the waiting waste is identified as the most critical factor with the highest percentage compared to the others. The motion waste is recognised to be the second-highest after the waiting waste.

Table 3. Examine of NAO elements

Seven wastes	Journals	Total Journal	Percentage (%)
Motion	[7], [8], [10], [13], [14], [17], [19], [20], [22], [23], [25], [26],	12/49	24.5
Waiting	[1], [2], [3], [4], [5], [6], [8], [9], [11], [12], [14], [15], [16], [18], [21], [23], [24], [25], [26], [27], [28], [29], [30]	23/49	46.9
Overproduction	-	0/49	0
Defect	[3], [8], [10], [21], [23], [25], [30]	7/49	14.3
Transportation	[23], [25]	2/49	4.1
Over-processing	[23]	1/49	2.0
Inventory	[14], [17], [24], [28]	4/49	8.2

4.3 Measurable and Non-measurable Indicators

From Table 3, the NAO elements then are classified into measurable and non-measurable indicators. In this study, the measurable indicators are only taken into account, and non-measurable indicators are being neglected because the non-measurable indicators cannot be observing during data collection. Thus, the elements of NAO is classified into measurable and non-measurable indicators, as shown in Table 4 below.

4.4 Pre-production, In-production and Post-production Segregation

Table 5 shows the listing of Non-value Added Overtime (NAO). The essential item of the NAO has been finalised and graded, as shown in three categories. The basic concept of manufacturing 'Input-Output (IO) is used because time is a significant indicator for Non-value Added Overtime (NAO). Thus, NAO is the total time of all processes as stated below.

Table 4. Measurable and Non-measurable Indicators of NAO elements

Measurable	Non-measurable
High production volume	Excess workforce inventory
Product defects	Annual paid leave
Machine availability	Sickness absence
Components shortage	Workers shortage
Bottleneck	Workers performance
Material availability	Inappropriate OT planning
Low plant capacity	Excessive inventory
Incoming raw materials	High materials storage
Products delivery	Workload control
High takt time	
Extensive motion	
Material defects	
Delays	
High workload	
Inspection time	
Replacement production time	
Rework and scrap	
Over-processing	
Long changeover time	
Unscheduled downtime	
High lead time	
Long waiting time	

Table 5. Sorting of the NAO's Factors

Pre-production	In-production	Post-production
Changeover time	Bottleneck and delays	Rework time
Incoming raw materials	High workload	Scrap time
	Machine availability	Replacement production time
	Material availability	High production volume
	Low plant capacity	Inspection time
	Long waiting time between the workstation	

5. Conclusion

There are two main issues worth highlighting in this study. The main problems may be the application domain and the basic technique. From the point of view of the application domain, most companies function with little experience beyond realistic execution during overtime. However, in the context of the basic technique, the current approach for calculating NAO does not consider all the factors comprehensively. Therefore, there is a need to enhance the formulation of NAO through a good understanding of the non-value-added factors for accurate manufacturing operational cost calculation. In this analysis, the critical components of NAO have been finalised according to the basic terms of operation (pre-production, in-production, and post-production). Based on the final critical components, there were two critical pre-production components, six critical in-production elements, and five critical post-production components. All the critical components of NAO were extracted from literature studies. Critical components may imply excessive activities and contribute to a significant amount of time wasted and thereby lead to overtime in manufacturing operations. Therefore, the minimisation of these factors will contribute to waste elimination and profit optimisation.

Acknowledgement

The author would like to extend special thanks to FRGS-RACER/2019/FTKMP-COSSID/F00412 grant, FakultiTeknologiKejuruteraanMekanikaldanPembuatan and FakultiKejuruteraanPembuatan, UniversitiTeknikal Malaysia Melaka for the use of the facilities and useful data in order to complete this study.

References

[1] J.-W. Seo, *Better Work Discussion Paper No. 2 Excessive Overtime, Workers, and Productivity: Evidence and Implications for Better Work*, no. 2. 2011.

[2] T. Akgeyik, "Determinants of Overtime Work (a Study on Data From an Aluminum Manufacturing Facility)," *J. Int. Manag. Stud.*, vol. 17, no. 3, pp. 49–56, 2017.

[3] I. Campbell, "Extended Working Hours In Australia," *Labour Ind. a J. Soc. Econ. relations Work*, vol. 13, no. 1, pp. 91–110, 2002.

[4] S. R. Dewi, B. Setiawan, and W. P. Susatyo Nugroho, "5S program to reduce change-over time on forming department (case study on CV Piranti Works temanggung)," *IOP Conf. Ser. Mater. Sci. Eng.*, vol. 46, no. 1, 2013.

[5] J. P. Womack and D. T. Jones, "Lean Thinking—Banish Waste and Create Wealth in your Corporation," *J. Oper. Res. Soc.*, vol. 48, no. 11, pp. 1148–1148, 1997.

[6] B. Patel, P. J. J. Bhavsar, and J. Pitroda, "A Critical Literature Review of Labour Productivity in Building Construction," *Int. J. Constr. Res. Civ. Eng.*, vol. 3, no. 4, pp. 76–80, 2017.

[7] Z. Mpanza and D. Nyembwe, "Improving Operating Methods in a South African Foundry Company - A Case Study," pp. 793–800, 2014.

[8] H. J. Sidabutar NA, Matondang AR, "Propose Improvement Maintenance Activities Of Screw Press To Reduce Waste Using Lean Maintenance Concept Propose Improvement Maintenance Activities Of Screw Press To Reduce Waste Using Lean Maintenance Concept .," *IOP Conf. Ser. Mater. Sci. Eng.*, vol. 505, no. 1, p. 012045, 2019.

[9] Z. Ebrahim, A. Hamzah, A. Rasib, and M. R. Muhamad, "UNNECESSARY OVERTIME AS A COMPONENT OF TIME LOSS OVERTIME UNNECESSARY OVERTIME AS A COMPONENT," *J. Adv. Manuf. Technol.*, vol. 11, no. January-June 2017, 2017.

[10] V. Poornashree and H. Ramakrishna, "A Study on Reduction of Non-Value Added Activities in An Assembly Line of An Automobile Industry," vol. 8, no. 06, pp. 1217–1219, 2019.

[11] T. Ohno, *Toyota production system: beyond large-scale production*. 1988.

[12] R. Sundar, A. N. Balaji, and R. M. Satheesh Kumar, "A review on lean manufacturing implementation techniques," *Procedia Eng.*, vol. 97, pp. 1875–1885, 2014.

[13] J. K. Liker, *The Toyota Way*. 2004.

[14] S. Shingo and A. P. Dillon, *A Study of the Toyota Production System*. 1989.

[15] M. Ghiyasinasab, N. Lehoux, S. Ménard, and C. Cloutier, "Using Lean Techniques and Simulation to Improve the Efficiency of Engineered Wood Production : A Case Study in a

Small Factory," *Ind. Eng. Manag.*, vol. 7, no. 4, pp. 1–11, 2018.

[16] J. F. Hair, M. L. D. S. Gabriel, and D. Silva, "Development and validation of attitudes measurement scales : fundamental and practical aspects," *RAUSP Manag. J.*, vol. 54, no. 4, pp. 490–507, 2019.

[17] Z. Ebrahim, A. H. Abdul Rasib, and M. R. Muhamad, "Understanding Time Loss in Manufacturing Operations," *Appl. Mech. Mater.*, vol. 761, no. May, pp. 619–623, 2015.

[18] A. Hamzah, A. Rasib, R. Abdullah, and Z. Ebrahim, "Critical Factors for Product Change Time Loss in Manufacturing Operation," no. December, 2019.

[19] A. R. A. . H and M. R. Z. . F, "Non-valued Changeover Time Measures for Hiddxen Time Loss in Automotive Mechanical Component Production," *Int. J. Eng. Technol.*, vol. 7, no. 4.36, p. 10, 2018.

[20] W. A. M. van Lent, N. Goedbloed, and W. H. van Harten, "Improving the efficiency of a chemotherapy day unit: Applying a business approach to oncology," *Eur. J. Cancer*, vol. 45, no. 5, pp. 800–806, Mar. 2009.

[21] R. R. Cima *et al.*, "Use of Lean and Six Sigma Methodology to Improve Operating Room Efficiency in a High-Volume Tertiary-Care Academic Medical Center," *J. Am. Coll. Surg.*, vol. 213, no. 1, pp. 83–92, Jul. 2011.

[22] A. Isaacson, N. Ridge, H. Yu, and M. Jackson, "'Lean' System Improvement to Increase First-Case Efficiency and Decrease Overtime Expenditures in Interventional Radiology," *J. Am. Coll. Radiol.*, vol. 11, no. 10, pp. 998–1001, Oct. 2014.

[23] L. M. Yerian, J. A. Seestadt, E. R. Gomez, and K. K. Marchant, "A collaborative approach to lean laboratory workstation design reduces wasted technologist travel," *Am. J. Clin. Pathol.*, vol. 138, no. 2, pp. 273–280, 2012.

[24] J. Bukh, Per Nikolaj; Skovvang Christensen, Karina; Mouritsen, "Knowledge management and intellectual capital: Establishing a field of practice," in *London/Basingstoke : Palgrave Macmillan*, 2005, p. 258.

[25] L. N. White, "Unseen measures: The need to account for intangibles," *Bottom Line*, vol. 20, no. 2, pp. 77–84, 2007.

[26] N. Sinkovics, S. F. Hoque, and R. R. Sinkovics, "Rana plaza collapse aftermath: Are CSR compliance and auditing pressures effective?," *Accounting, Audit. Account. J.*, vol. 29, no. 4, pp. 617–649, 2016.

[27] D. Tlapa *et al.*, "Effects of Lean Healthcare on Patient Flow: A Systematic Review," *Value Heal.*, 2020.

[28] A. N. M. Rose, M. F. F. Ab Rashid, N. M. Z. Nik Mohamed, and M. A. S. Jubri, "The implementation of lean manufacturing and ergonomics in Small Medium Enterprise-Case study," *IOP Conf. Ser. Mater. Sci. Eng.*, vol. 702, no. 1, 2019.

[29] K. Siregar, A. Ishak, and T. D. M. Purba, "Reducing waste in production process with lean six sigma approach and weighted product method," in *IOP Conference Series: Materials Science and Engineering*, 2019, vol. 505, no. 1, p. 505.

[30] S. M. Hashim, R. Rabilah, and A. M. Shah, "Eliminating waste using lean tools and delmia quest simulation," *J. Phys. Conf. Ser.*, vol. 1349, no. 1, 2019.

[31] B. Sremcev, N., Stevanov, B., Lazarevic, M., Mandic, J., Tesic, Z., & Kuzmanovic, "Improving Process of Quotation Creation Through Value Stream Mapping and Simulation," *Int. J. Simul. Model*, vol. 18, no. 4, pp. 563–573, 2019.

[32] J. Kaneku-Orbegozo, J. Martinez-Palomino, F. Sotelo-Raffo, and E. Ramos-Palomino, "Applying Lean Manufacturing Principles to reduce waste and improve process in a manufacturer: A research study in Peru," *IOP Conf. Ser. Mater. Sci. Eng.*, vol. 689, no. 1, pp. 2–8, 2019.

Experimental Studies On Strength and Durability Characteristics of 8M GeoPolymer Mortar Based On Fly Ash and GGBS

Chakali Naresh[1], CH.Venugopal Reddy[2], Dr.L.Sudheer Reddy[3], K.Anand Goud[4]
[1]M.Tech student,Institute of Aeronautical Engineering, Dundigal, Hyderabad, India.
[2,4]Assistant Professor, Department of Civil Engineering, Institute of Aeronautical Engineering, Dundigal, Hyderabad, India,
[3]Professor, Department of Civil Engineering, Kakatiya Institute of Technology & Science, Warangal, India.

Corresponding author email: ch.venugopalreddy@iare.ac.in

ABSTRACT: *Ordinary Portland cement (OPC) is one of the most misfortune cover measurable in the structure business. It is evaluated that the production of OPC, discharges the enormous measure of Carbon-di-oxide (CO_2) gas which makes the 65% of a global warming which leads to global weather imbalance. For the manufacture of one ton of OPC transmits roughly 0.8 ton of CO_2 in the worldwide gases, along these lines the cement industry produces 7% of all (CO_2) ecological contaminating gases. The properties of the mortar have been calculated according to Indian and ASTM standard requirements, specific gravity, compressive and split tensile strength. The ratio of "Na_2SiO_3 to NaoH" is 2.5, and the ratio of "alkaline solvent to binder" is equal to 0.4, 0.45, 0.5, 0.55 in the cases of binder components are calculated with each ratio.like in this research we calculated with each fluid/binder ratio with different mix proportions 1:15, 1:2, 1:2.5 ,1:3 in this research using only two materials fly ash then GGBS , using the different ratios of fly residue and GGBS with identical molarity and IS specified normal sand sieve sizes. Once prepared with the "alkaline liquid to binder" ratio, it creates higher strength to the geopolymer mortar. It can be inferred that the results in terms of intensity by ambient curing are strong relative to outdated mortar.*

Key Word : Geopolymer Mortar, Fly Ash, GGBS, Sodium Hydroxide, Sodium Silicate.

I. INTRODUCTION

Production of cement causes a huge volume of carbon dioxide CO_2 announcement causing temperature rise, a dangerous atmospheric devation. It is assessed that one ton of concrete roughly needs around 2 tons of raw resources (Limestone also Shale) also discharge about 0.87 ton of CO_2 and around 3 kg of nitrogen oxide. Production of cement causes huge impact in grievance causing changes in land-use patterns and nearby water defillement just as air contamination. Criminal CO_2 emissions additionally

present tremendous danger to nature. The cement industry doesn't fit in feasible improvement because of raw materials utilized for the production doesn't reuse and are non-renewable.
The waste material or side-effect from the business which can be used for reduction of carbon dioxide CO_2 outflow. Accentuation on energy preservation and nature protection has been expanded as of late which have prompted the examination of options in contrast to standard building materials and technological advances. In this manner, the material or through consequence of an industry could be utilized in cement production subsequently reducing carbon footprint. Inorganic polymer or natural polymer composites have the possibility to frame a generous component to form an environment friendly and feasible constructional building material which produces lower greenhouse footprint when contrasted with the customary cement.
In 1978, Joseph Davidovits created Inorganic polymeric resources & authored the period "Geopolymer" for it (1991). Geopolymer can possibly supplant usual Portland cement & yield fly ash based geo polymer plaster with brilliant corporeal & motorised possessions. So as to create geopolymer, low-calcium fly ash should be enacted through an alkaline explanation to crop polymeric Si-O-Al bonds. Geo polymer can possibly lessen greenhouse outflows by 80%.

NEED OF GEOPOLYMER MORTAR

To provide natural inviting solid, we have to supersede the solid with some various folios which should not have any horrendous effect on condition. The use of current reactions as spreads can diminish the issue. Right now, new advancement geo-polymer concrete is a promising system. To the extent reducing an Earth-wide temperature help, the geo-polymer advancement could diminish the CO2 transmission to the air realized by cement and intentions actions by about 80% (Davidovits, 1994c). Also, besides the most ideal use of current wastes can decrease the issue of arranging the waste things into nature. geopolymer capitals, OPC is expatriate over waste incomes, meant at example, fly ash or slag concrete composed over a chemical activator. Notwithstanding the reduction in OPC and the utilization of waste items, geopolymer materials require less water for restoring and are seemed to have expanded protection from chemical attack.

II MATERIALS

A. Fly Ash

As designated by ASTM C-618, Two important programs of fly ash remain professed. These 2 programs are recognized through the kind of firewood expended and are allocated course F and class C in the mainstream of the contemporary symbols. Class F fly ash remains usually elated done irresistible anthracite or bituminous coal though class C fly ash remainders for the supreme helping become through overriding sub bituminous or lignite coal. The significant qualities of these two kinds of ashes remain talks about beneath. Current class F fly ash is gathered from Ramagundam thermal power place. Class F fly ash remains through calcium oxide (Cao) contented under 6%, assigned as small calcium cinder, are not self-setting but rather for the most part show pozzolanic belongings. These leftovers cover over 2% unburned carbon dictated through loss happening ignition. Quartz, mullite and hematite are major cryatalline stages recognized fly ashes, got from bituminous coal. Fundamentally, all fly ash in existing and cement is accomplished class F fly ash displacing 80% of cover through weight. At the point when class F fly ash is utilization of class F fly ash as a rule lessens water request just as heat of hydration. The concrete completed through class F fly ash additionally shows enhanced protection from sulphate assault and chloride particle enteance. The test material is taken as of National Thermal Corporation (NTPC), Ramagundam. By and large, the material is gathered from the hypothetical information it is very surely known that it has a place with period F(ASTM C618).

Table I: Properties of fly ash

Properties	**Values**
Colour	Light grey
Specific gravity	2.9
Bulk density	1100 kg/m^3 - 1200 kg/m^3
Fineness	>350 m^2/kg

B. GGBS

GGBS (Ground Granulated Blast furnace Slag) is a cementitious physical whose fundamental use remains in existing and is a side-effect as of the influence heaters second-hand to brand iron. Impact radiators exertion on temperatures of about 1,500°C & are taken care of through a purposely measured blend of iron metallic, coke & limestone. The iron metallic stands summary to media besides the respite of the resources structure a slag that skims happening the iron.

This slag is intermittently appointed off as a liquid watery & on the off accidental that it is to be exploited meant recycled for the production of GGBS it requirement be quickly satisfied in enormous volumes of water. The satisfying improves the cementations possessions & harvests granules like a coarse sand. This 'granulated' slag is then dry out & ground to a acceptable powder.

Table I: properties of GGBS

Properties	Values
Colour	Off – white
Specific gravity	2.9
Bulk density	1000 – 1100 kg/m^3 (loose) 1200 – 1300 kg/m^3 (vibrated)
Fineness	>350 m^2/kg

C. Fine Aggregate :

The nearby obtainable river sand transitory finished IS 4.75mm sieve is reused zone II through way of per IS 383-1987. The possessions of powder remained assumed in Table I.

Table I: Properties of Fine aggregate

Properties	**Values**
Specific Gravity	2.65
Fineness Modulus	2.25

D. Sodium Hydroxide

Through and large the sodium hydroxides remain reachable in strong state as Pellets & pieces as seemed in Figure The expenditure of the sodium hydroxide remains mostly changed by the immaculateness of the material. Since our geopolymer mortar is homogenous substantial & its important technique to indorse the sodium silicate it is prescribed to exploit the maximum negligible expenditure i.e,. up to 94% near 96% virtue. Right now sodium hydroxide pellets in 8 molar fixations are utilized. The manufacturers' information identified with physical and substance properties are given in Tables.

IOP Conf. Series: Materials Science and Engineering **994** (2020) 012005 doi:10.1088/1757-899X/994/1/012005

Physical properties of Sodium hydroxide

Colour	Colourless
Specific Gravity	1.47
pH	14

E. Sodium Silicate :

Sodium silicate (Na_2Sio_3) are typically charity as alkaline activators in geopolymerization procedure in this study alkaline melted are act as binding agent to harvest a geopolymer mortar. the alkaline liquid that been recycled is the mixture of sodium hydroxide NaoH and Na_2SiO_3. Sodium-silicate which is known as fluid gas is accessible in fluid sort gel , In present research sodium-silicate 2.0 is utilized, according to fabricate, silicates were provided to the material industry as holding operator.

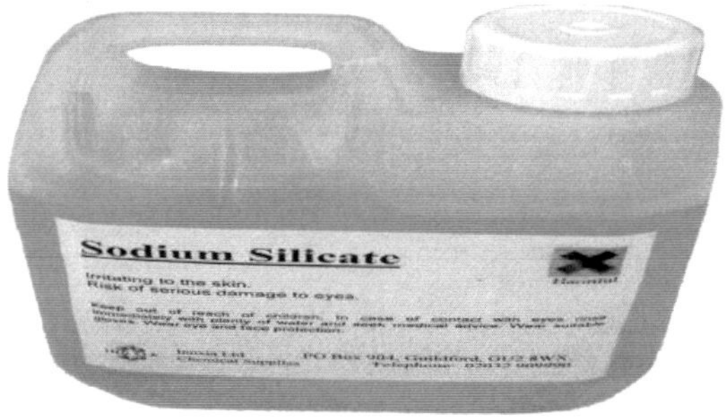

III MIX DESIGN

In this experiment we consider the different fluid binder ratio 0.4, 0.45, 0.50, 055 . for this fluid binder ratios we take the different mix ratios 1:15,1:2,1:2.5,1:3.

Material requirement for mix ratio 1:2 for geopolymer mortar

Unit weight of geopolymer mortar =2200 kg/m^3
Binder to aggregate =1:2
Mass of binder (fly ash) = (1/3)*2200 = 733 kg/m^3
Mass of fine aggregate =2200-733 = 1467 kg/m^3
Assume , alkaline liquid / binder ratio = 0.40
Mass of alkaline liquid =0.40*733 =293.2 kg/m^3
Assume , the sodium silicate / sodium hydroxide =2.50
Liquid mass of sodium hydroxide =293.2*(1/3.5) =83.77 kg/m^3
For 8 molar sodium hydroxide solid =26.23% *83.77
= 21.361 kg/m^3
Water =83.77-21.972 =62.40 kg/m^3
Liquid mass of sodium silicate =293.2-83.71 =209.43 kg/m^3

Mass of sodium silicate gel 44.1% of 209.43 =92.358 kg/m^3
Mass of water =209.43-92.358 =117.071 kg/m^3
Mould size 70.6 mm *70.6 mm * 70.6 mm
Volume of cube =0.0706 m^3
Mass of a binder (fly ash) for one cube =0.0706
Mass of a fine aggregate for one cube = 0.0706 * 1467 = 516.66 gm.
Mass of a sodium hydroxide (solids) for one cube = 0.0706 *62.407
water required for sodium hydroxide = 21.9579 gm.
Mass of sodium silicate gel for one cube =92.358*0.0706 = 32.4916 gm.
Water required for mixing for one cube =117.071 * 0.0706 = 41.958 gm.
For 12 cubes
Mass of binder (fly ash) =310 *12 = 3720 gm.
But 20% is required by GGBS
Mass of binder (GGBS) = 930 gm.
Mass of NaoH (solids) =21.957 *12 =263.484 gm.
Water required for mixing NAOH =26.094 *12 =313.128 gm.
Mass of a sodium silicate gel =32.4916* 12 = 389.899 gm.
Water required for mixing of Na_2Sio_3 = 41.958 *12 = 503.496 gm.

IV METHODOLOGY

The sodium hydroxide (NaoH) solids remained smashed depressed trendy water to type the solution. The figure of NaoH solids popular a solution fluctuated trusting upon the junction of the solution connected as far as 8 molar, M. For instance, NaoH prearrangement through a convergence of according to calculations of NaoH solids (in piece or pellet form) per liter of the solution, where 40 is the sub-atomic load of NaoH. The sodium silicate arrangement & the sodium hydroxide solution were mutual at any rate 1 day before utilize to usual up the elementary fluid. The soluble fluid was combined with the extra water (assuming any) to set up the fluid share of the mixture. After 24 hours solution get cool then we can use to prepare the geopolymer mortar by mixing with fly ash GGBS and fine aggregate as per the calculations after that cast the cubes.
Mix proportion 1:1.5, 1:2, 1:2.5, 1:3 with representing the Fluid binder(f/b) ratios of 0.4 as ‘A’ ,0.45 is ‘B’ 0.5 is ‘C’ and 0.55 is ‘D’ for each f/b ratio there is four mix proportions and there are represented as like A1.5,A2,A2.5,A3.

Preparation of alkaline liquid and examples

- Sodium hydroxide (8M) pellets are used to prepare alkaline liquid.
- In this research 8 Molar NaoH concentration solution were used which consists of 8x40=320 gm of NaoH pellets per liter of water, where 40 is the molar weight of NaoH.
- Once the pills get dissolved sodium silicate result of 2.5 times of NaoH solution is mixed to prepare the alkaline liquid.
- It stays to be well-known that the answer is ready at smallest 24 hours before to casting. because the concentration of NaoH is gets to react as a binder.
- Cubes also cylinders specimens were casted to training the compressive asset and torn separately tensile asset.
- Cube size of 70.6 mm x 70.6 mm x70.6 mm.
- Cylinder sizes of 50mm dia x 100mm.

IOP Conf. Series: Materials Science and Engineering **994** (2020) 012005 doi:10.1088/1757-899X/994/1/012005

Ambient curing of geopolymer mortar

Ambient remedial of fly ash and GGBS constructed geo polymer mortar is normally curing substantially supports the chemical response that occur in the geo polymer glue both curing temperature effect the compressive strength of enhanced the polymerization procedure resulting in complex compressive strength. The rate of surge in asset was rapid active to 24 hours of remedial time and outside 24 hours, the gain in asset was solitary moderate. Higher curing temperature of geo polymer existing resulted in advanced compressive strength. Ambient cutting is strange but weather cutting i,e. area infection.

Curing of specimens

Compressive asset test

- To test the compressive asset of a mortar cube we use the compression testing machine of volume 2000KN.
- Weight rate is 2.5KN/sec as per IS: 516:1959 practical.
- Test performed on 7 and 28 days.

Compressive strength test

Split tensile strength test

- Articulated as the smallest tensile stress wanted to divided the physical apart.
- Load rate is 2 KN/sec.
- Test is performed on 7 and 28 days.
-

Split tensile strength test

V RESULT AND CONCLUSION

Compressive Strength

Results for 0.4 fluid binder ratio of mix proportions

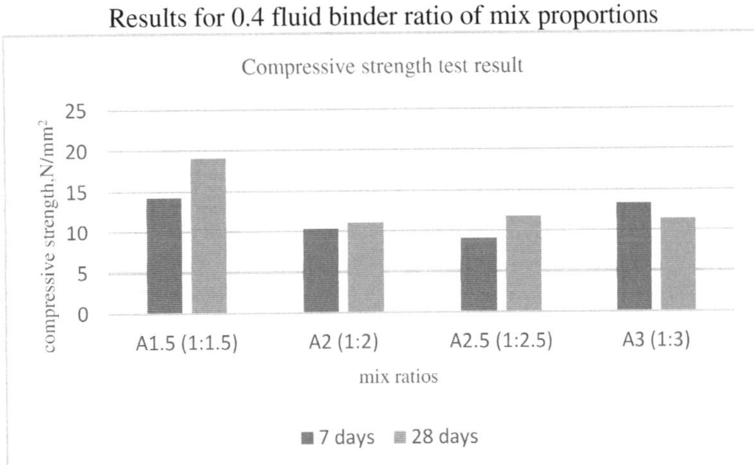

IOP Conf. Series: Materials Science and Engineering **994** (2020) 012005 doi:10.1088/1757-899X/994/1/012005

Results for 0.45 fluid binder ratio of mix proportions

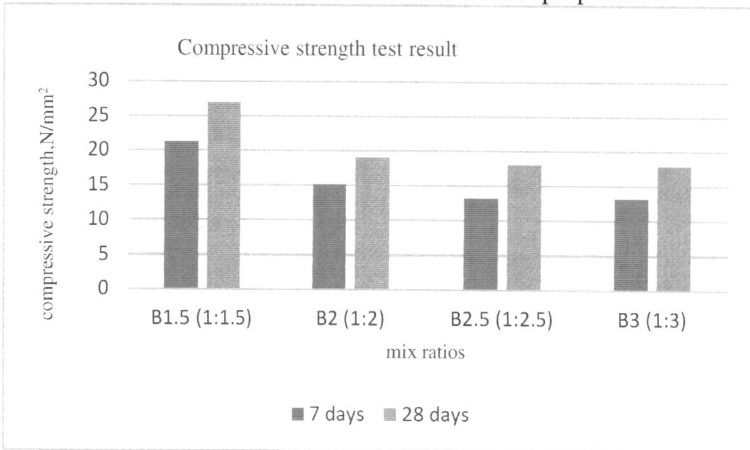

Results for 0.5 fluid binder ratio of mix proportions

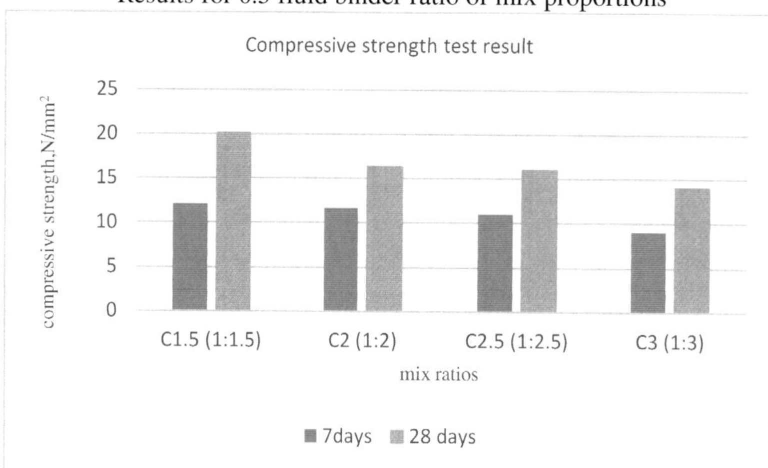

Results for 0.55 fluid binder ratio of mix proportions

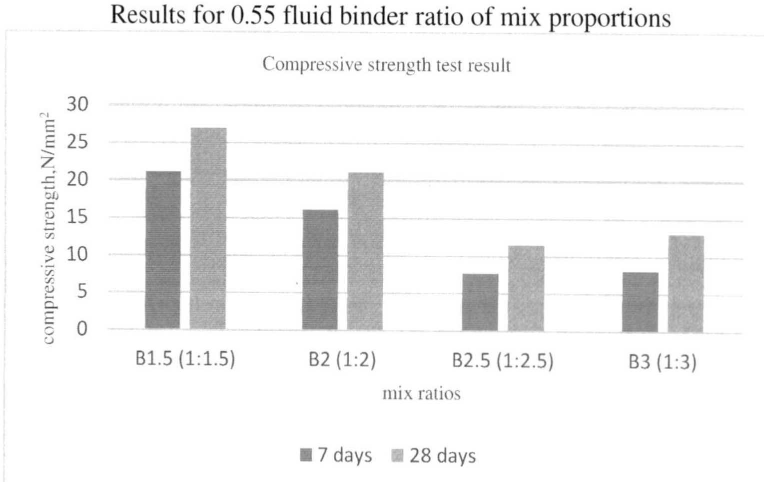

Split Tensile Strength

Split tensile strength of 0.4 fluid binder ratio of mix proportion

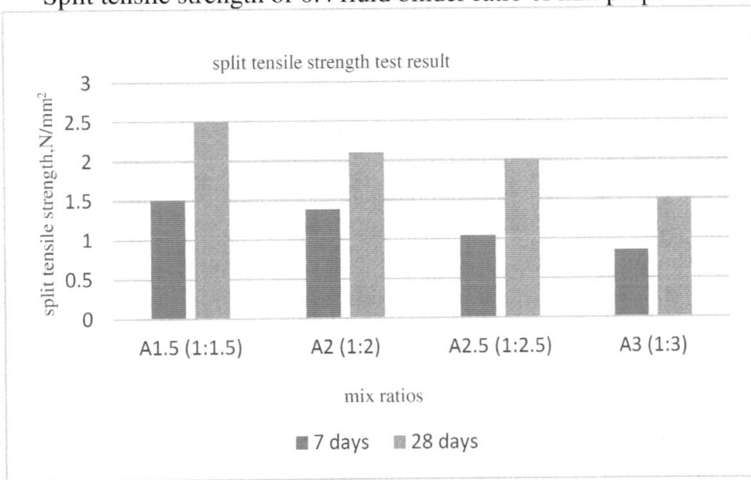

Split tensile strength of 0.45 fluid binder ratio of mix proportion

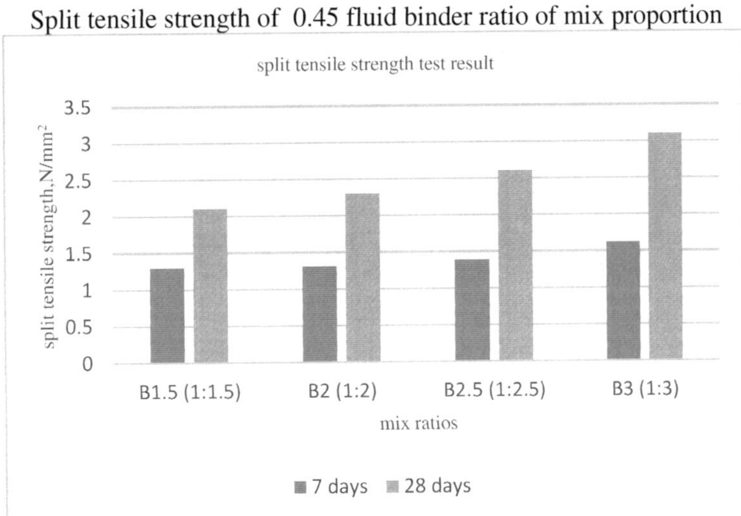

Split tensile strength of 0.5 fluid binder ratio of mix proportion

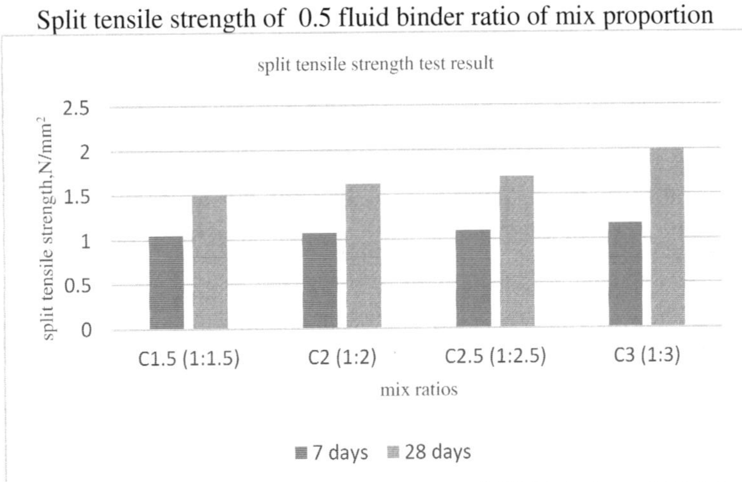

Split tensile asset of 0.55 fluid binder ratio of mix proportion

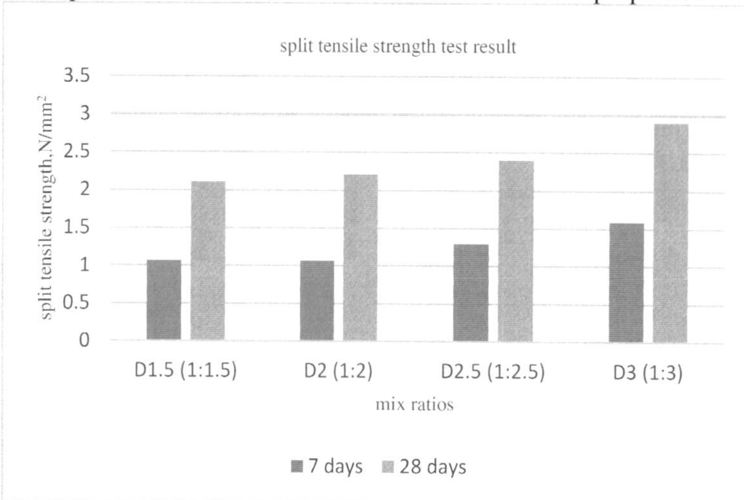

VI CONCLUSIONS

- It is examined that geo polymer mortar in ambient-curing circumstances can attain good strength in 28-days.
- Compressive-strength of 1:1.5 mix proportion develops more in 0.45 and 0.55 fluid binder ratio from "7 to 28" days.
- Compressive-strength of "1:1.5" mix-proportion is high compared to "1:2" ,"1:2.5 & 1:3".
- Mostly 0.4 binder ratio is give the perfect results by comparing the compressive and split tensile strength but to getting more strength we use 12 molarity.
- Higher-concentration of "sodium hydroxide" i.e, explanation obtain greater strength in geo-polymer mortar.
- As percentage of FLY ASH content increases, the compressive strength of geo polymer mortar decreases.

FURTHER INVESTIGATIONS

- By reducing the cohesive property of the geo polymer mortar (i.e. more workable) it can be applied in the field conditions.
- Few other industrial waste products can be utilized in the making of geo-polymer mortar or concrete like, red mud, rice husk ash, etc.
- Ambient curing conditions can be refined by replacing the GGBS with other waste products such as burnt clay, etc.
- Red Soil is higher reactive with GGBS when in comparison with fly ash based geo polymer mortar /concrete.

REFERENCES

[1] K Ramujee Strength and Development of Fly ash and GGBS using Geopolymer mortar. International Journal of Recent Advances in Engineering & Technology (IJRAET) ISSN (Online): 2347 - 2812, Volume-3 Issue -1 2015.

[2] Meena T & Kallempudi Murali An Experimental Study on Factors Influencing the Compressive Strength of Geopolymer Mortar.International journal of Civil Engineering and Technolgy (IJCIET) Volume 9 Issue 1 January 2018.

[3] Somnath & Ravindra N Thakur Effect of Mix Composition on Compressive Strength and Microstructure of Flyash Based Geopolymer Composites.2006-2009 ARPN Journal of Engineering and Applied Sciences ISSN 1819-6608 Volume 4 June 2009.

[4] C.D Budh Effect of Molarity on Compressive Strength of Geopolymer Mortar. International journal of Civil Engineering Research. ISSN 2278-3652 Volume 5, No 1, pp83-86.

[5] T Venu madhav Effect of Molarity on Compressive Strength of Geopolymer Mortar Asian Journal of Chemistry February 2013.

[6] Andi Arham Adam The effect of temperature and duration of curing on the strength of fly ash based geo-polymer mortar 2nd International Conference on Sustainable Civil Engineering structures and Construction vol 95, 2014 (SCESCM 2014).

[7] Arie Wardhono strength of alkali-activated slag/fly ash mortar blends at ambient temperature. The 5th International Conference of Euro Asia Civil Engineering Forum (EACEF-5) Vol 125 2015.

[8] Pradip Nath & Vijaya B Rangan Early age properties of low-calcium fly ash geopolymer concrete suitable for ambient curing. The 5th International Conference of Euro Asia Civil Engineering Forum (EACEF-5). Vol 125, 2015.

[9] Shriram marathe et al Laboratory Investigation on the Strength and Durability Characteristics of Geo-polymer Concrete International Journal of Innovative Research in Science Engineering and technology (IJIRSET), Ess & Ess research publications Vol 5 ISSUE 09 May 2016

[10] Badami bavin Geo-polymer Concrete and Its Feasibility in India. Proceedings of National Conference (CRDCE13) 20-21 December 2013 SVIT Vasad.

[11] M N Uddin**:** A Comparative Study on Clay and Red Soil Based Geo-polymer mortar. Civil Engineering and Architechture 6.1 (2018) 34-39 DOI: 10.13189/cea 2018.060104. International Journal of Scientific & Engineering Research, Volume 6, IUE 5, May-2015.

[12] CH Venugopal Reddy Dr.L.Sudheer Reddy Experimental Research on Behaviour of Geopolymer mortar in compression VOLUME 14 Issue 5, 2020 pg3959-3963 https://doi.org /10.37896/jxu14.5/439 Journal of Xidian University.

Design and Implementation of Driver Assistance System using Machine Learning

D.S.Bhargava [1], N.Shyam[2], K.Senthil Kumar[2], M.Wasim Raja[2], Sivashankar.P[3]

[1]Assistant Professor, [2]UG Scholars, Department of Electronics and Communication Engineering, R.M.K. Engineering College
[3]Sr. Autonomy Engineer, AIDRIVERS LTD

duvurubhargava@gmail.com

Abstract. The main theme of the work is to detect the traffic sign boards and alert the driver accordingly, which helps in reducing the violation of the traffic rules and also reduces the accidents that occur in day-to-day life. The vehicle will be able to guide the driver by providing the details of speed limit, traffic sign boards in high pedestrian density areas and the driver can change the speed within which the vehicle needs to operate and drive safely. The components that are used in the system design are raspberry pi 3 (module-b), ultrasonic sensor and web camera. The images of the traffic sign boards are captured using a web camera and the speed limit of the vehicle can be changed accordingly by the driver and will not be allowed to exceed the speed limit. In addition this system may help in analyzing the driving pattern.

Keywords: Traffic sign boards, Raspberry pi 3 (module-b), Ultrasonic sensor, Web camera and Driving pattern

1. Introduction

Road accidents claim high number of lives every year and the reasons behind accidents are drunk and drive, driver negligence or distraction, over speeding etc. The majority of the accidents occur because of some fault of the driver or occupants of the vehicle. As per the Government of India "Road Accidents in India, 2011" report by the Ministry of Transport and Highways, about every 11th person out of one lakh died in a road accident and every 37th person was injured in one, making it an alarming situation for a cause of unnecessary death.[10]
Annual global road crash statistics [2013] says that nearly 1.25 million people die in road crashes each year, on an average 3287 deaths a day. 20-50 million people are injured or disabled. Each year nearly 4,00,000 people under 25 years of age die on the world's roads, on an average over 1000 mortality per day. The World Health Organization (WHO) global status report on road safety [2018] says that annual road traffic deaths have reached nearly 1.35 million. This disproportionality borne by pedestrians and motorcyclists that too especially people aged between 5 to 29 years.
According to Ministry of Road Transport and Highways (MoRTH), [10] Continuous efforts taken by the State Government through Transport, Police, Health, Highways and Education departments shows decrease in 24% of accidental deaths in 2018. (In the year 2018 the mortality due to accidents is 12,216 whereas in the previous year 2017 it is 16,157). This made us to further explore on accidents and accidental death, to propose some solution and to reduce this fatality. For this we studied the reports of year 2019, State Crime Record Bureau (SCRB), Road Accident Database Management System (RADMS), Vahan and Sarathy data etc and found that Total ***occurrences of accidents*** add up

to 5173, number of deaths is 993. [10] Also the maximum accidents (33.27%) and fatalities are caused by the two wheelers in State Highways. Nearly 63% of accidents occur both together in National and State Highways and 69.68% of the fatality occurs in National (36.25%) and State Highways (33.43%) in Tamil Nadu. [10]

2. Literature Survey

The existing systems use image processing techniques to detect the drowsiness of the driver and alert them. The idea is to capture the live image of the driver through the camera and check the iris position along with the facial expressions to detect the state of the driver. This method only produces an alert to the driver and moreover it has only a software module to identify whether the driver is awake or not. These systems mainly focus on drowsiness alert and not on other aspects. [1][5][14]
Our system is capable of acting in active as well as passive modes in order to prevent the occurrence of accidents. The system is not only capable of correcting mistakes made by the driver but also can make the driver follow the traffic sign rules at any cost. Along with software modules our system has hardware module in order to control the actions of the driver and to prevent the accidents, which makes the system more effective one.

3. Proposed System

In order to ensure pedestrian safety and vehicle speed limits in areas with high pedestrian density or residential areas, a system to detect the traffic density and accordingly warn the driver about the speed limit for vehicles is proposed.

Figure 1. Block Diagram

The sign boards are detected using an image processing algorithm called HAAR Cascade algorithm. Using this algorithm the sign boards to be detected are given as positives and those not to be detected are given as negative and stored in a folder (database).

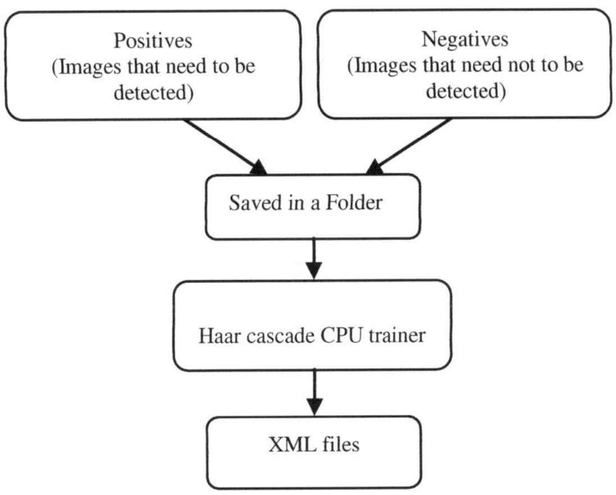

Figure 2. Haar cascade Flow Chart

Then the folder containing the images are trained using Haar cascade CPU trainer, which generates XML files for trained images as shown in Figure 2 Haar cascade Flow Chart above.
Further using image detection algorithm, accuracy of detection of boards can be tested. The web camera connected to Raspberry Pi board captures images and these captured images will be compared with the already trained XML files. If there is any deviation then the warning message will be sent through LCD display and gives an alert to the driver, as illustrated in Figure 3. Based on this alert speed limit can be modified.
The active system is based on tensor flow algorithm and YOLO (You Only Look Once) where the image can be detected at any angle with slight match with the trained data. Here the system can make decisions based on the situation of the driving environment and help in reducing the fatal accidents that occur due to pre-emptive actions.

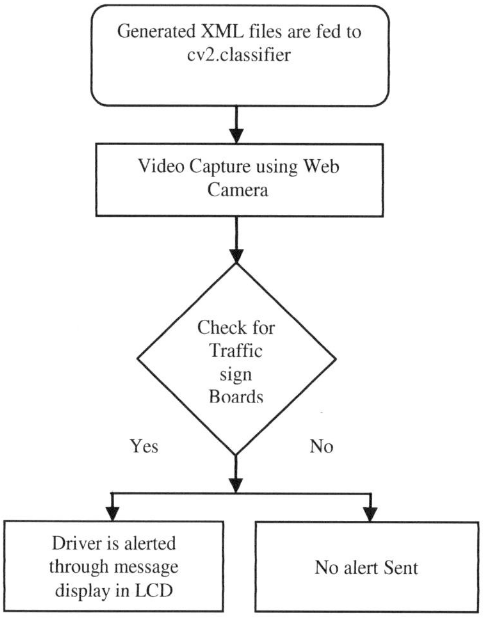

Figure 3. Image Detection Process

4. Results and Discussion

The Stop board detection output is presented in the Figure 4 below

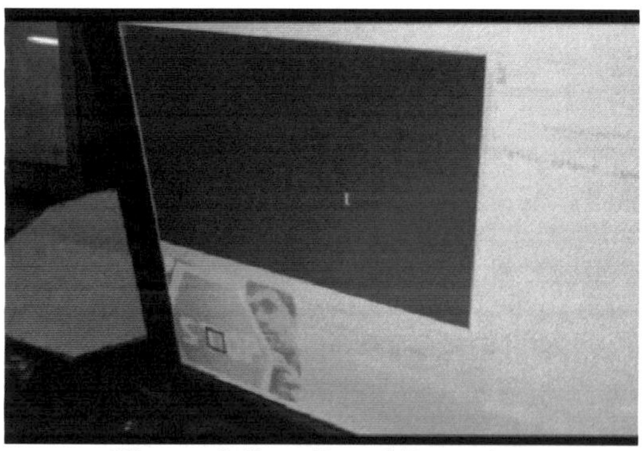

Figure 4. Stop Board Detection

Once after detecting the stop board sign, the system automatically sends an alert to driver and the alert is displayed in the screen as shown in below Figure 5.

Figure 5. Stop Vehicle Alert

Figure 6. Human and Animal Identification

Figure 7. Pedestrian Identification

Also the designed system identifies Human, animal and Pedestrian on roads and categorizes accordingly as shown in above Figures 6 and 7 respectively.

Figure 8. Pedestrian Identified Alert

Similarly after detecting the pedestrian, the alert given to driver is shown in above Figure 8.

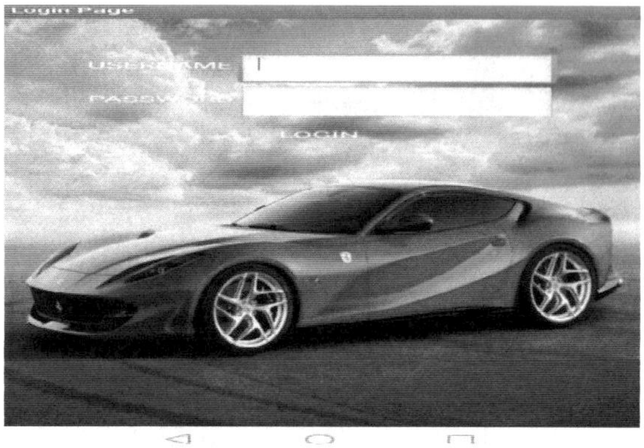

Figure 9. Application Developed

Every user will be given a **user id** and **a password** to connect their mobile phone with the application as shown in above Figure 9.

Figure 10. Application Login Page

For illustration we have created a username and a password to login into the application, to demonstrate the working procedure of this application as shown in above Figure 10.

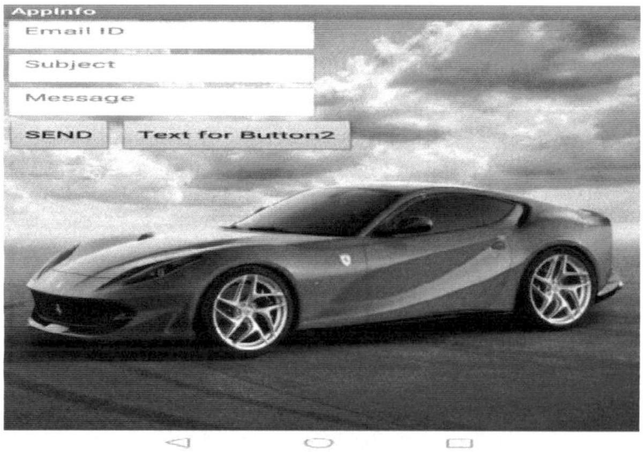

Figure 11. Application Information

Once after logging into the application, customer can post service query information's about the product to us by filling up the details like email id, subject and content of the message etc as shown in above Figure 11. Once the details are fed, the message can be sent to the corresponding email id by clicking the send button

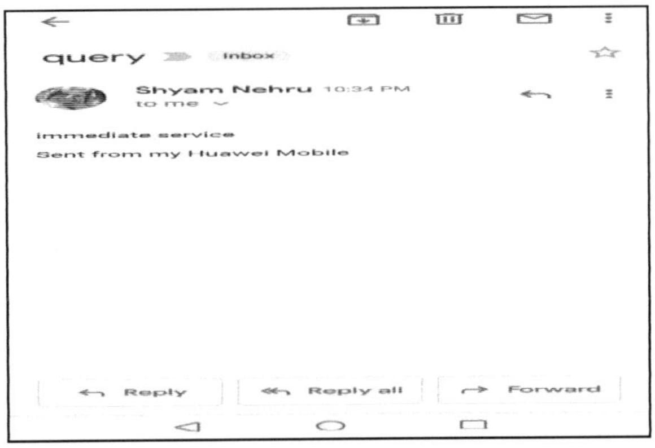

Figure 12. Service Query Message

The message sent from the customer side through application will be sent to the referred email id and the alert message can be viewed on the inbox as shown in the above Figure 12.

5. Conclusion
In this work we have implemented a solution to overcome the issues of fatal road accidents, by equipping both safety as well as alert system depending based on the driver's needs. With the introduction of these systems, safety that is prevailing in the present vehicles can be improved. When such systems are equipped, service can also be provided with the help of application for the interaction of the customer. This work can be further extended and can be analyzed in terms of Vehicle to Vehicle & Vehicle to Infrastructure communication. Further Vehicular Ad-hoc Network (VANET) routing protocols can be studied and new algorithms can be developed.

6. References
[1] Adnan Shaout Dominic Colella and Awad S 2012 Advanced Driver Assistance System – Past, Present and Future *Seventh International Computer Engineering Conference (ICENCO'2011) IEEE Explore* pp 72-82

[2] Dibakar Barua Pranshu Jain Jitesh Gupta and Dhananjay V.Gadre 2013 Road Accident Prevention Unit (R.A.P.U) (A Prototyping Approach to Mitigate an Omnipresent Threat) *Texas Instruments India Educators Conference IEEE Explore* pp 56-60

[3] Hendry and Rung-Ching Chena 2019 Automatic License Plate Recognition via sliding-window darknet-YOLO deep learning *Image and Vision Computing* vol 87 pp 47-56

[4] Inhye Yoon Monson H Hayes and Joonki Paik 2013 Wavelength-Adaptive Image Formation Model and Geometric Classification for Defogging Unmanned Aerial Vehicle Images *IEEE International Conference on Acoustics Speech and Signal Processing* pp 2454-2458

[5] Lex Fridman Daniel E Brown Michael Glazer William Angell, Spencer Dodd, Benedikt Jenik Jack Terwilliger Aleksandr Patsekin Julia Kindelsberger Li Ding Sean Seaman Alea Mehler Andrew Sipperley Anthony Pettinato Bobbie Seppelt Linda Angell Bruce Mehler and Bryan Reimer 2019 MIT Advanced Vehicle Technology Study: Large-Scale Naturalistic Driving Study of Driver Behavior and Interaction with Automation *IEEE Access* pp 1-16

[6] Melissa Yauri-Machaca Brian Meneses-Claudio Natalia Vargas-Cuentas and Avid Roman-Gonzalez 2018 Design of a Vehicle Driver Drowsiness Detection System through Image Processing using Matlab *IEEE 38th Central America and Panama Convention (CONCAPAN XXXVIII)* pp1-6

[7] Nagarjuna R Vatti Prasanna Lakshmi Vatti Rambabu Vatti and Chandrasekar Garde 2018 Smart Road Accident Detection And Communication System *Proceedings of 2018 IEEE International Conference on Current Trends toward Converging Technologies* pp 1-4

[8] Partha Narayan Chowdhury, Tonmoy Chandra Ray and Jia Uddin 2018 A Vehicle Detection Technique for Traffic Management using Image Processing *International Conference on Computer, Communication, Chemical, Material and Electronic Engineering (IC4ME2)* pp 1-4

[9] Priyanka S Hemalatha G and Saranya C 2017 Sudden Unintended Acceleration Avoidance and Drowsiness Detector for Automobile Accidents Prevention *Third International Conference on Science Technology Engineering & Management (ICONSTEM)* pp 964-967

[10] 2019 *Road Accident Analysis in Tamil Nadu Government of Tamil Nadu Home (Transport) Department 2019*

[11] Selvathi D Pavithra P and Preethi T 2017 Intelligent Transportation System for Accident Prevention and Detection *International Conference on Intelligent Computing and Control Systems* pp 442-446

[12] Sorin Grigorescu Bogdan Trasnea Tiberiu Cocias and Gigel Macesanu 2019 A Survey of Deep Learning Techniques for Autonomous Driving *Journal of Field Robotics* pp 1-38

[13] Xiaolei Li and Tingdi Zhao 2010 An Accident Analysis Model Oriented to Complex Tasks Process *Proceedings - Annual Reliability and Maintainability Symposium (RAMS)* pp 1-6

[14] Yang Xing Chen Lv Huaji Wang, Dongpu Cao Efstathios Velenis and Fei-Yue Wang 2019 Driver Activity Recognition for Intelligent Vehicles: A Deep Learning Approach *IEEE transactions on Vehicular Technology* vol 68 No 6 pp 5379-5390

Theoretical Foundations of Water-Conducting Belt Design for Wide Coverage Sprinklers Using Numerical Simulation

[1] **Skvortsov A.A.**, [2] **Zhuravleva L.A.**

[1]Doctor in Phys.-Math.; FSBEI of HE "Moscow Polytechnic University, Moscow, Russia

[2]Doctor in Engineering; FSBEI of HE "Moscow Polytechnic University, Moscow, Russia

Abstract. Improvement of wide-coverage sprinkler technical level requires provision of high irrigation productivity and quality with the minimum cost of construction, rational use of water and energy. It is necessary to maintain the required flow-pressure characteristics regardless of sprinkler external operating conditions and the length of the water supply belt.

The article discusses the possibility of mathematical modeling application for a wide sprinkler design. The studies showed that during the calculation it is possible to get by with the finite equations without resorting to the differential equations of fluid motion theory with variable mass. The obtained dependences served as the basis for the development of water-conducting belt calculation methods for a wide-sprinkler.

A numerical simulation of the water supply belt operation of the KASKAD sprinkler is presented with standard values for the pipeline diameters and the variable length of the sprinkler. Numerically obtained graphs of the flow rate and the required pressure dependences on water supply belt length for two design options: with a sector end sprinkler and an end long-range sprinkler operating from a booster pump. By eliminating the length of the water supply belt for the considered two design options, they designed the flow-discharge characteristics of the sprinkler. They presented an example of a constructive solution selection for the water-conducting belt of a sprinkler according to flow-pressure characteristics.

Keywords: Sprinkler, water supply belt, console, hydraulic module, sprinkler nozzle, flow rate, pressure, flow-rate characteristics.

1. Introduction

Currently, irrigation in Russia is carried out on the area of 4.3 million hectares. About 70% of this area is watered using wide coverage sprinklers.

The most important direction of modern domestic sprinkling equipment competitiveness increase is the widespread use of resource-saving technologies, i.e. rational use of water, energy and financial resources.

The improvement of wide coverage sprinklers is directly related to the water supply belt design, the optimally selected parameters of which will significantly reduce the material consumption and sprinkler cost, and will increase the irrigation efficiency and quality.

The aim of the research is to develop a methodology for the hydraulic calculation of the water supply belt for wide-angle circular sprinklers, which makes it possible to determine the parameters of the pipeline necessary to maintain the required flow-pressure characteristics regardless of sprinkler external operating conditions and its length.

The metal consumption and the significant cost of sprinkling equipment do not allow the necessary full-fledged field studies to be carried out in full, and mathematical modeling is the main economically feasible design method [1-3].

2. Research methodology

The motion of a fluid with weight change along the path was first considered by Hinds [4] and Favre [5]. Later, the ideas by Hinds and Favre were detailed in the studies [6, 7, 8, 9, 10, 11] concerning the hydraulic calculation of drip irrigation systems. The Hazen – Williams formula [8] was used in all works to determine the pressure losses along the length, which gives significant errors in the calculations during continuous liquid sampling.
Let's consider the process of uneven fluid selection in the water-conducting belt of a circular sprinkler. When you solve this problem, the model of continuous flow rate selection [4] is traditionally used according to the linear law [12]:

$$Q_2 = Q_L \frac{s}{L_M}, \tag{1}$$

where Q_2 is the selected flow rate;
s is the current value of the polar radius.
Q_L, L_M – flow in the last branch and the physical length of the sprinkler, respectively;
Following I.M. Konovalov [6, 13], you can get the differential equation of the set fluid motion with a variable mass in a water-conducting belt of constant cross-section with flow selection:

$$\frac{a_0''(a_2-2)Q}{g\omega^2}\cdot\frac{dQ_2}{ds}+\frac{dH}{ds}+\frac{Q^2}{K^2}=0, \tag{2}$$

wherea_0'' – the Boussinesq coefficient;
a_2– the coefficient of flow detachment;
Q, K – flow rate and flow module in an arbitrary flow section, respectively;
H – hydrodynamic pressure;
ω – the living cross-section area of the stream;
g – the acceleration of gravity.

Integration of the equation (2) with the condition (1) in polar coordinates gives the equation of the piezometric line for the water-conducting belt of the sprinkler[12,13]:

$$H=-L_M\frac{\bar{s}}{K^2}\left(1-\frac{2}{3}\bar{s}^2+\frac{1}{5}\bar{s}^4\right)Q_M^2+a_0''\frac{a_2-2}{2g\omega^2}\bar{s}^2\left(2-\bar{s}^2\right)Q_M^2, \tag{3}$$

Where $\bar{s}=s/L_M$ – the relative coordinate;
Q_M – the sprinkler consumption.
It follows from (3) that the total pressure loss along the water supply belt length:

$$H_{полн}=\left(\frac{8}{15}\frac{L_M}{K^2}-a_0''\frac{a_2-2}{2g\omega^2}\right)Q_M^2. \tag{4}$$

When they solve using this method, it is assumed that:
- the selection of the flow rate Q_2 along the length of the water supply belt occurs continuously, which is not true, since the flow modulus K and the disconnection coefficient of the flow a_2 vary along the length of the water supply belt, and the selection of the flow Q_2 occurs discretely [4, 5];
- the flow modulus K and the flow disconnection coefficient a_2 are independent of the coordinate s;

In the work by A.A. Fedorets [13], this drawback is partially eliminated by introduction an empirical discrete parameter KD into the formula (4), which depends on the frequency of tap placement $\gamma' = L_{OT}/L_M$,
whereL_{OT} - the distance between the taps.
Thus, the author [13] proposed the empirical relationship:

$$K_{Д} = 1 + 1{,}7\gamma'^{1,04}. \tag{5}$$

The flow disconnect coefficient a2 is in the range $0 \le a_2 < 1$. A.A. Fedorets [13] obtained the empirical formula for the coefficient a_2:

$$a_2 = \frac{0{,}00112 \cdot Re^{0,6422}}{e^{0,11 \cdot Re \cdot 10^{-4}}}, \tag{6}$$

where Re is the Reynolds number, determined by the diameter of the water supply belt and the total flow rate Q_M.
The mathematical model (3) - (6), obtained on the basis of the differential equation of the set fluid motion with a variable mass, gives a fairly good agreement with the experimental study results in the works by A.A. Fedorets [13]. However, the application scope of this model is limited to specific experimental conditions. In theoretical studies, A.A. Fedorets does not take into account the material of the pipes.
Let's develop a model of discrete water withdrawal for the water supply belt of wide coverage sprinklers from hydraulic representation directly, without differential equation use.
The circular action sprinkler should provide the irrigation rate of $m_П$, m^3/ha [14]:

$$\frac{m_П}{10} \equiv h_{OC}, \tag{7}$$

h_{OC} – precipitation layer, mm.
With a uniform location of the branches at the distance LOT from each other along the length of the water supply belt, the calculated irrigation area will be the following:

$$S_0 = \pi\left(L_M^2 - \frac{L_{OT}^2}{4}\right) \cdot 10^{-4}, \text{ ha}, \tag{8}$$

and the necessary flow rate of a sprinkler can be described by the following expression [14]:

$$Q_M = \frac{25h_{OC}}{9 \cdot 24T_0} S_0, \text{ л/c}, \tag{9}$$

where T_0 is the operating time of the sprinkler, days.
The number of bends in the water supply belt is defined as follows:

$$N = INT\left(\frac{L_M}{L_{OT}}\right). \tag{10}$$

Each of the bends must ensure the irrigation norm h_{OC} for the respective watered ring with the area of:

$$S_i = 2\pi L_{OT}^2 i, \ m^2; \quad i = 1, 2, \dots, N. \tag{11}$$

Then the flow rate of each bend should be the following:

$$Q_i = \frac{h_{OC}}{24 \cdot 3600T_0} S_i, \text{ л/c} \tag{12}$$

Accordingly, for the rates of sections between branches, we have:

$$\left.\begin{aligned} &Q_{TP}^{(1)} = Q_0; \\ &Q_{TP}^{(i)} = Q_{TP}^{(i-1)} - Q_{i-1},\ i = 2, 3, \ldots, N. \end{aligned}\right\} \quad (13)$$

Knowing the inner diameter $d_{vn}$of the water supply belt, one can find the average water velocity in each section:

$$V_i = Q_{TP}^{(i)} / \omega / 1000, \text{ m/s}, \quad (14)$$

where $\omega = \pi d_{BH}^2 / 4$.

Let's note that for economic reasons, the average speeds in the areas of the water supply belt should not exceed 3 m/s [5].

According to the research by A.A. Fedorets [13] when the condition:

$$\frac{L_{OT}}{d_{BH}}, \frac{d_{BH}}{d_{BH}^{OTB}} > 7 - 10, \quad (15)$$

is performed the pressure loss on the "mass mixing" and the pressure drop difference can be neglected and the pressure loss between the bends can be calculated by the uniform motion formulas.

where d_{BH}^{OTB} – the inner diameter of the bend,

To determine the pressure loss in the sections of the water supply belt between the bends, we will use the approved method to calculate the pressure loss along the flow length for new steel pipes by F.A. Shevelev [15,17]:

$$\left.\begin{aligned} &h_l^{(i)} = \theta_i A Q_i^2 L_{OT};\ A = \frac{0,001478}{d^{5,226}}; \\ &\theta_i = 0,889\left(1 + \frac{0,684}{V_i}\right)^{0,226};\ i = 1, 2, \ldots, N. \end{aligned}\right\} \quad (16)$$

We take into account local pressure losses in the pipeline by adding 5% of the pressure losses along the length, which is consistent with the data by A.A. Fedorets [13]:

$$h_{\text{тр}}^{(i)} = 1,05 h_l^{(i)}. \quad (17)$$

Then the total pressure loss in the pipeline will be the following:

$$h_{\text{тр}} = \sum_{i=1}^{N} h_{\text{тр}}^{(i)}. \quad (18)$$

Knowing the geodetic head and setting the free head at the last bend, you can determine the required head on the sprinkler hydrant. The computation algorithm (7) - (18) is easily implemented in any algorithmic language.

Using the presented algorithm, let's simulate the operation of "CASCADE" sprinklers [2, 3, 14] with the following initial data:

- a step at the beginning of the water supply belt (the standard pipe diameter - Ø159 mm, its length - 59.5 m) between the first 10 nozzles, makes 5.95 m;
- in the rest of the water supply belt, the step between the nozzles is 2.975 m;
- at the end of the sprinkler, the console is 24 m (12 m pipe, Ø133 mm; 12 m pipe Ø102 mm);
- at the end of the console they install:

1) a sector sprinkler nozzle with the action range R_{nas} = 2.975 m with the diameter of the nozzle orifice d_{nas} = 12 mm (1st option);

2) a long-range apparatus operating from a booster pump with the range of R_{DDA} = 25 m (option 2);

- the diameter of the nozzle at the last bend of the console d_N= 12 mm;

- installation height of irrigation nozzles z_{nas} = 1.8 m;
- the height difference between the entry point to a sprinkler and the top point of the farm pipeline (geodetic pressure) makes 4.5 m;
- the sprinkler length varies from 180 m to 500 m;
- the values of the sprinkler hydraulic module (the ratio of flow to irrigation area) were taken equal to 0.81; 0.93; 1.0; 1.05.

After numerical simulation, we plotted (Fig. 1, 2) the dependences of the pressure on the hydrant (required pressures) and the sprinkler costs, depending on the length of the water supply belt (sprinkler length).

The maximum values of the sprinkler length (respectively, the maximum required pressure on the hydrant) were optimized from the condition of water average velocity limitation in the water supply pipeline V ≤ 3.5 m/s, otherwise the hydraulic resistance of the water supply belt increases sharply and its diameter needs to be increased.

Calculations show that the longest sprinkler length at z_{nas} = 1.8 m is achieved with the hydraulic module q = 0.81 and makes:

$$\left.\begin{aligned} L_{M1} &= 499{,}891 \text{ мfor the 1-st option;} \\ L_{M2} &= 477{,}866 \text{ m for the 2nd option.} \end{aligned}\right\} \quad (19)$$

Accordingly, the costs and the required pressure on the sprinkler hydrant are the following:

$$\left.\begin{aligned} H_1 &= 39{,}297 \text{ m;} \\ H_2 &= 38{,}719 \text{ m;} \end{aligned}\right\} \quad (20)$$

and the irrigation radius and flow rate in both cases R_{op} = 502,866 m andQ = 64,349 l/s.

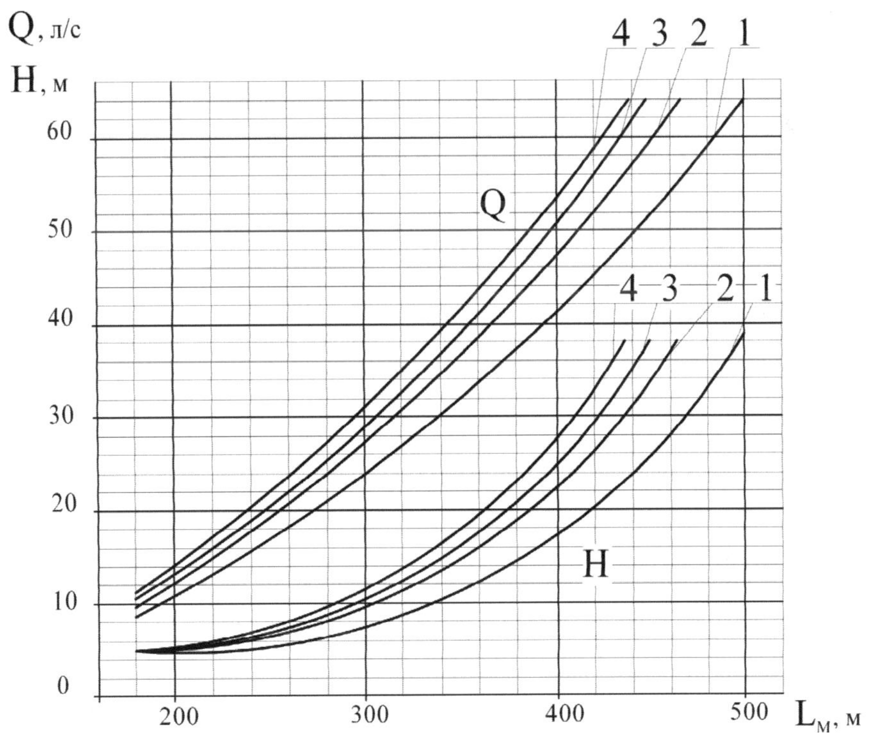

Figure 1. Dependence of the pressure N on the hydrant and flow rate Q on the sprinkler length (1-st option): 1-q=0,81; 2-q=0,93; 3-q=1,0; 4-q=1,05

On the other hand, it is possible to check the results of numerical determination of the sprinkler maximum length by accepting the maximum average speed in the water supply belt at the beginning of the pipeline to the first sprinkler nozzle bend equal to $V_{max} = 3.5$ m/s, then the flow rate $Q_{DM} = 64.349$ l/s will be established in the initial section (pipe Ø159 mm, wall thickness e = 3 mm).

Thus, the irrigated area will be:

$$S_{op} = \frac{Q_{ДМ}}{q} = 794432 m^2.$$

And the irrigated area radius will be the following:

$$R_{op} = \sqrt{\frac{S_{op}}{\pi}} = 502{,}866 m.$$

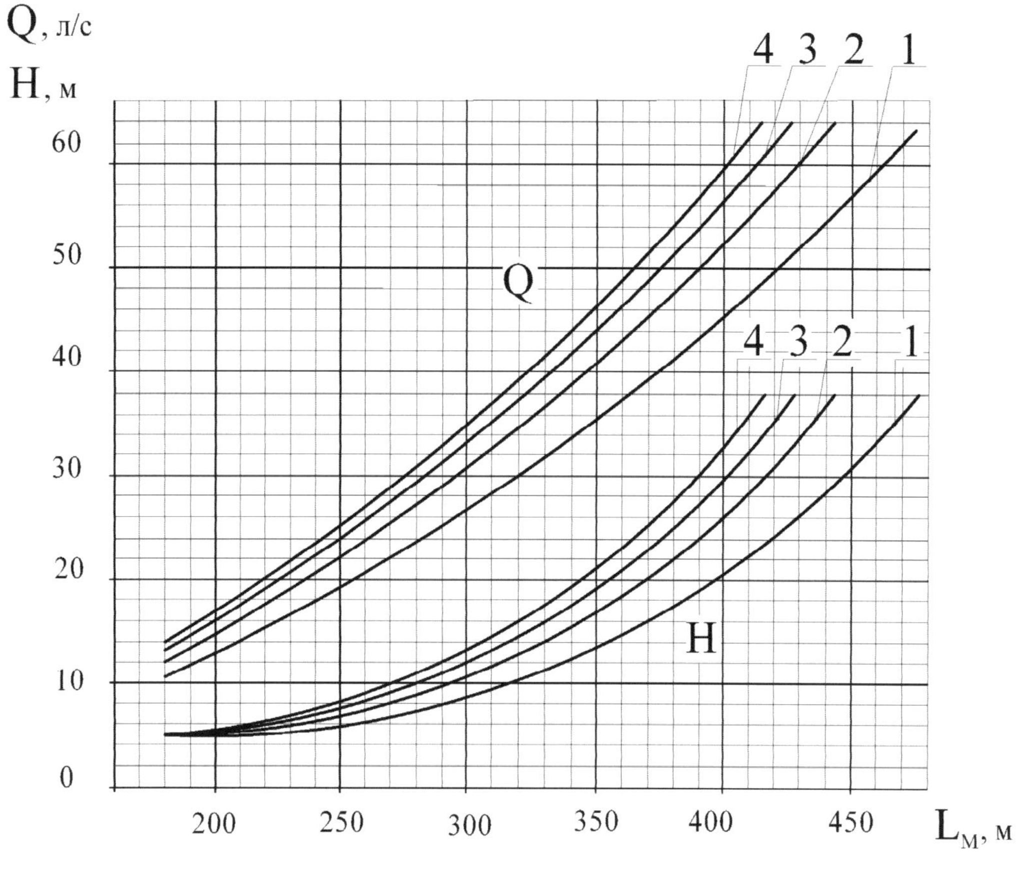

Figure 2. Dependence of the pressure N on the hydrant and flow rate Q on the sprinkler length (2nd option) 1-q=0,81; 2-q=0,93; 3-q=1,0; 4-q=1,05

According to the source data

$$R_{op} = L_{M1} + R_{ДА} = L_{M2} + R_{ДАА} = 502{,}866 m,$$

Hence:

$$L_{M1} = R_{op} - R_{ДА} = 499{,}891 \text{ m};$$

$$L_{M2} = R_{op} - R_{ДАА} = 477{,}866 \text{ m},$$

which confirms the results of numerical modeling.

The difference in sprinkler length is:

$$L_{M1} - L_{M2} = 499{,}891 - 477{,}866 = 22{,}025 m.$$

Since the structural mass coefficient of the sprinkler (the ratio of the sprinkler total mass to the water supply belt mass) [15],[17] is lower for the 2nd option, respectively, $H_2<H_1$, then the 2nd option is preferable from the point of view of metal consumption.
Excluding the sprinkler length from the dependencies shown on Fig. 1, 2 for the same values of the hydraulic module, we obtain the flow-pressure characteristics of the sprinkler at different values of the hydraulic module (Fig. 3, 4).

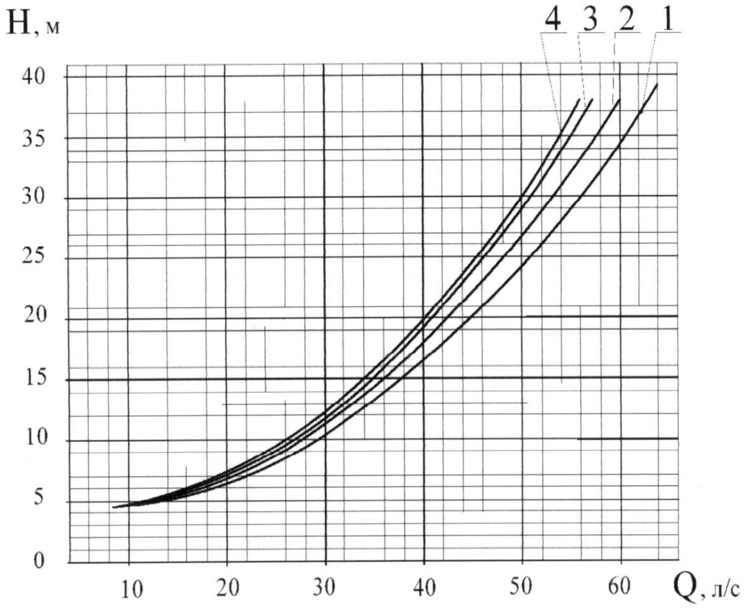

Figure 3. Flow-pressure characteristics of the sprinkler (I-st option):
1-q=0,81; 2-q=0,93; 3-q=1,0; 4-q=1,05

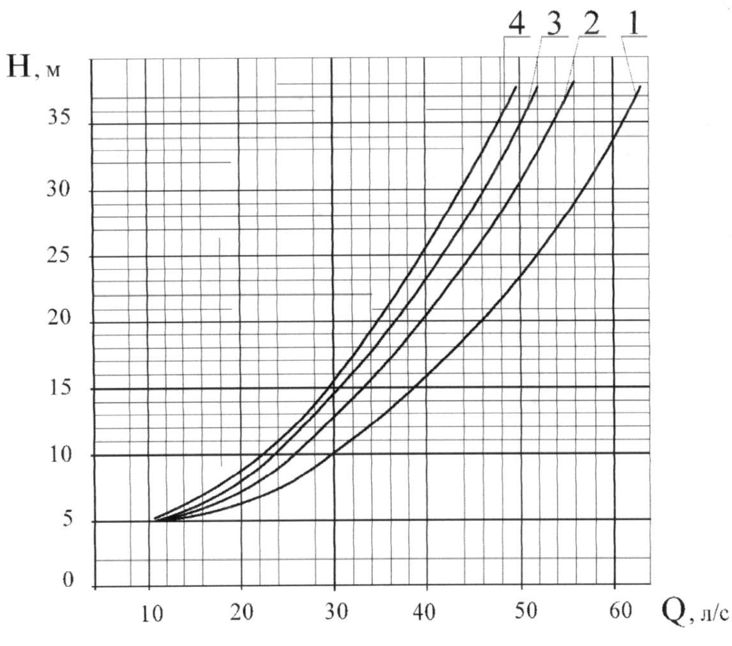

Figure 4. Flow-pressure characteristics of the sprinkler (II-nd option)
1-q=0,81; 2-q=0,93; 3-q=1,0; 4-q=1,05

IOP Conf. Series: Materials Science and Engineering **994** (2020) 012007 doi:10.1088/1757-899X/994/1/012007

Let's consider an example of sprinkler parameter selection according to flow-pressure characteristics. Let it be required to determine the pressure of circular action sprinkler for irrigation of a site with the area of $S_o = 57$ ha.

Let's set the hydraulic module of the sprinkler, for example, q = 0.81, and determine the necessary flow rate of the sprinkler:

$$Q_м = q \cdot S_о = 0{,}81 \cdot 57 \approx 46 l/s.$$

Based on the flow rate $Q_m = 46$ l/s according to the graphs 3, 4 we find the required pressure on the hydrant N = 21 m for the I-st variant and N = 20.5 m for the II-nd variant. According to Fig. 1, 2 the parameters $Q_m = 46$ l/s (N = 21 m) and $Q_m = 46$ l/s (N = 20.5 m) correspond to two sprinkler lengths $L_m \approx 420$ m (1-st option) and $L_m \approx 400$ m (2-nd option). Direct calculations provide the following:

$L_m = 423.025$ m (1-st option) and $L_m = 401$ m (2-nd option).

3. Research results

This calculation technique was used in the design of sprinklers "Kuban-LK1M", "CASCADE", Figure. 5, Table 1.

Figure 5. Wide coverage sprinkler "Cascade"

IOP Conf. Series: Materials Science and Engineering **994** (2020) 012007 doi:10.1088/1757-899X/994/1/012007

Table 1. The main specifications of the "Kuban-LK1M" (CASCADE) and "CASCADE" sprinkler

Indicator name	Indicator	
	"Kuban-LK1M" (CASCADE)	«CASCADE»
Designation	Wheel, multi-support, electrified, circular movement, with water intake from a closed irrigation network or from a well	
Machinedrive	Electromechanical, reversible, individual for each trolley, from a gear motor	
Powersupply	External three-phase network with dead-earthed neutral and the rated voltage of 380 V, 50 Hz	
The speed of the last truck, adjustable within, m/s, (m/min)	$1{,}6x10^{-3}(0{,}1) – 30{,}0x10^{-3}(1{,}8)$	
The distance from the surface of the earth to the lower belt of metal structure, m	2,7	2,8 –2,9
Typeofwaterpipe	Sectional,truss	
Diameters of the water supply pipeline, mm	159	219, 203, 168, 159, 133
Diameters of console pipes, mm	133	133, 114, 108, 89
Spanlengths, m	48,7	48,7; 53,7; 59,5; 65,25
Maximum machine length, m (with the console of 31m)	518	553
The diameters of the fixed support riser, mm	168, 203	168; 203; 219; 244,5; 273
Wheelbase	3700	3700, 4200
Water consumption at zero total slope, l/s	up to 90	
The water pressure at the entrance to the machine: MPa with the length of 500 m	0,43	
Wheels of self-propelled carts:	pneumatic, chamber, two wheels on each cart	
Gearmotors	UMC	
Wheel reducers of self-propelled carts, type	wormgears	
Locationofoperationalcontrols	- control panel on a fixed support; - remote control panel of an external irrigation system	
Machine operating modes GSM control	- in motion with watering; - in motion without watering (driving) option	
Machine drivingmodes	continuous, start-stop (software)	
Average rain intensity, mm/min, no more than	0,66	
Irrigation rate per passage (within regulation), m^3/ha	95-600	
The minimum time for a full revolution of the machine, h, with the maximum length of 500 m	32,8	

4. Conclusion

Completed field tests confirm the adequacy of the mathematical model. Comparison of field study results with the calculated ones according to the proposed method gives a good agreement and indicates the correctness of the adopted method and the reliability of the obtained dependencies.

References

[1] Yesin A.I. Numerical hydraulics: monograph / A.I. Esin. – Saratov: FSBEI HPE SGAU, 2013. – 160 p.

[2] Yesin A.I. Studies of the water flow characteristics in the water supply belt of the sprinkler / A.I. Yesin, D.A. Soloviev, L.A. Zhuravleva // Scientific Life. - 2018. - No. 2. – pp. 16-25.

[3] Yesin A.I. Mathematical modeling of a sprinkler water-conducting belt. / A.I. Yesin, D.A. Soloviev, L.A. Zhuravleva // Scientific life. - 2017. - No. 9. – pp. 20-28.

[4] Hinds, I. Side channel Spillways / I. Hinds // Proceedings of the on Society of Civ. Eng., 1926.

[5] Favre, H. Contribution à l étude des courants liquids / H. Favre. – Zürich. – 1933.

[6] Petrov G.A. Hydraulics of variable mass: monograph / G.A. Petrov. – Kharkov: Publishing House of Kharkov University, 1964. - 224 p.

[7] Bresler, El. Analysis of Trickle Irrigation with Application to Design Problems / El. Bresler // Irrigation Science. Vol. 1. – № 1. – 1978. – pp. 3-17.

[8] Hazen, A. Hydraulic Tables (3rd ed.) / AHazen & G.S. Williams. New York: John Wiley and Sons, 1920.

[9] Journal of the Irrigation and Drainage Division. № 101, 4, 1975. – pp. 265-278.

[10] Keller, J. What about drip irrigation / J. Keller, R. Hanks // Idaho Farmer. Stockman. № 90, 6. – 1972. – pp. 34-38.

[11] Solomon, K. Trick irrigation uniformity and efficiency / K. Solomon, J. Keller // Journal of the irrigation and drainage division, Proceedings of the American Society of Civil Engenering. № 104, 3. – 1978. – pp. 293-306.

[12] Kurganov A.M. Hydraulic calculations of water supply and discharge / A.M. Kurganov, N.F. Fedorov. - L.: Stroyizdat, 1986. - 440 p.

[13] 13.Fedorets A.A. Theoretical foundations and the methods of hydraulic calculation of a closed network of reclamation systems with variable fluid flow: dis. by the Doctor in Engineering. - Kolomna, 1993. - 305 p.

[14] Esin, A.I. Mathematical Modeling of water Conducting Belt for Circular Action Sprinkler / A.I. Esin, V.M. Boikov, V.A. Mukhin, F.V.Serebrennikov, L.A. Zhuravleva // Jour of Adv Research in Dynamical & Control Systems, Vol. 10, 01-Special Issue, 2018. P. 2135 -2141.

[15] Yesin A.I. Resource-saving technologies and circular sprinklers: monograph / A.I. Yesin, L.A. Zhuravleva, V.A. Soloviev.- Saratov: Publishing House KUBiK, 2019. 214 p.

[16] Soloviev, D.A. Robotic Irrigative Complex with Intellectual Control System "CASCADE"/ D.A. Soloviev, L. A. Zhuravleva, S. M. BAKIROV // XVIII International Scientific and Practical Conference "Modern Trends in Agricultural Production in the World Economy". kuz.agri.2020.145-156; 3-4 December 2019, Kemerovo, Russia, ISBN: 978-5-6044077-1-4

[17] R. Rameshbabu, R. Karthikeyan ,R. Balamurali., 2016. FPGA Implementation of Synchronized Non Identical Hyperchaotic Systems. International Journal Of Control Theory And Applications, 9(5), pp. 2633 – 2642.

Re-Distributed Multi-Authority Attribute Based Encryption in Cloud Using CP-ABE Algorithm

[1]Siddhardha T, [2]Murugaanandam., [3]Sri Nithin D, [4]Raghuveer V.

[1-4]Department of Information Technology, SRM Institute of Science and Technology, Chennai, India.
Corresponding mail id: murugaas@srmist.edu.in

Abstract: Cloud Computing allows us to store and share the data easily with others. Re-distributed multi-authority Attribute-Based Encryption (ABE) has importantly implemented for finding the solutions to the problem that which arises from sharing private corporate information in computing of cloud. For Re-distributed multi-authority ABE in cloud devices that don't believe in a single main authority, collusion resistance are often done employing a universal identifier. so for that reason, identity must be globally managed, that ends up within the critical issues of privacy and protection. The new scheme is implemented that doesn't use a single main authority in order to co ordinate users and keys, and only normal trust relations got to be done with sharing the general public key in between each attribute authority (AA). The identities of the user are different and different by combination of the user's identity along the identity of the AA where the user is found . If a key request must be made to the authority outside of the domain, the request must be done by the authority within the present domain instead of by the client's, so, the user identities should remain secured to the AA outside the domain that can strengthen privacy and protection. Along with that, the main issuing protocol in between Attribute Authority is straight forward because of the results of trusted AA relationship. On top of that, the authority's extensibility is also supported by the method presented in our paper. The scheme is predicated on order of Composite in which the groups are bilinear. Validation of protection is given which uses the twin system methodology of encryption.

Keywords— Cloud Computing, Data Sharing, Cipher text-Policy Attribute-Based for Encryption (CP-ABE) , Key Distribution, Encryption, Decryption, Group Creation.

1. Introduction

Cloud computing in these days is the most essential service and the cloud services are being available all over the globe irrespective of the location. Cloud computing gives us many services where, allocation of storage, encryption of data are some of the examples. These cloud services are provided by many different cloud service providers and all are not trust-worthy. These cloud services may be free or either some paid services. The cloud space being provided will decide the cost of the particular person's cloud space. In these days cloud space is the most used service for the purpose of data storage and data transfer.

Cloud capacity is very major benefit and vital benefit worldview of cloud computing. Advantages by utilizing cloud capacity incorporate more noteworthy openness, higher unwavering quality, fast deployment and stronger security, to title fair a couple of. In spite of the mentioned benefits, this worldview too brings forward modern challenges on information get to control, which could be a big or major to guarantee information secured or privacy. Since cloud capacity operations done by cloud benefit the one who is providing service, who are more often than not exterior the trusted space of information proprietors, the conventional get to control strategies within the Server/Client demonstrate for cloud capacity environment. The information get to control in cloud capacity environment has in this way ended up a challenging issue. To address the issue of information access control in cloud capacity, there have been very numbers of. Plans proposed, among which (CP-ABE) is respected as one of the foremost promising procedures. The striking advantage of CP-ABE that is awards information proprietors coordinate control

As we are able see the high utilization of these cloud administrations there emerges a few security issues where a few of the possible and frequently happening issues are that the data may well be misplaced or the information may well be stolen, that's the third party enters without the consent of the information proprietor. There are tall chances for the information to be misplaced with the utilize of any untrusted cloud spaces or the cloud space having less security.

In spite of the fact that the control plans that which are present or based on the CP-ABE, they has a lot of plus points and also the plans are not robust and at the same time they are not effective also whey they have to generate key. Actually there is only one single in charge is present in almost all single authority plans, offline/crash of this authority makes all secret key demands inaccessible during that period . We face nearly similar issue available or present in multi-authority plans, since each of different authorities oversees attribute set.

Cloud space suppliers ought to make sure that the space provided to any specific individual should be possessing the highlights of giving the security for both information capacity and information exchange. There are numerous possibilities for the information to be stolen in case of the cloud space being utilized by the suppliers which are not trust-worthy, those are just like the third party entrance into the personal data is simple in the event that there exists no security, and the information could be stolen.

Secure Computing is known as the security which is connected to the data in all areas. This security provides protection from any sort of unplanned circumstances. It primarily concentrates on the provision of information encryption and security. For numerous reasons passwords are being shaped in such a way that they are extreme to break out. It is known as the bridge between the client and any sort of framework.

In order to make sure that the information is secured a few essential detailing for the information encryption is done, where the information assurance isn't so simple. There are numerous cloud space suppliers and a few security frameworks joined to it such as Data Encryption Standard(DES), and Cipher text-Policy Attribute-Based Encryption (CP-ABE) etc where they given a specific key on the ask of the client to the proprietor.

The main aim of the proposed framework is the information security and it is done utilizing this type of encryption is this type foremost broadly utilized and adopted algorithm for the encryption of the information which are in charge these days. The main use or the advantage of CP-ABE is it allows Information proprietors coordinate control depends on some of the defined policies

- In CP-ABE, we have given with some attributes that which are nearly or equal to the decryption key of the user, and then the policy also results in change of the cipher text.
- The brief knowledge that have about the CP-ABE has been switched: presently cipher text related with the granted access and the encrypting party decides the approach by the information that which has can be changed, whereas we've hidden key that attached with the present attribute set

Objective:

1. Attribute based Encryption for different group security
2. Over all Authority control by central authority
3. Secure the data in cloud environment like college and company on large data sharing

2. Literature Survey

D.Wei Li, [1] The prior ABE plans include single particular specialist that who checks and co-ordinates the complete set, which will have the result in a one-point bottleneck which will provide both the security and as well as the performance. Afterwards, a few dual-authority plans came into existence, among the different specialists independently keeping the no joint attribute subsets. In any case, the one-point issue hasn't solve yet. In our present paper, different point of view, that which have an edge dual-authority control access scheme to the open cloud capacity, among the different authorities mutually oversee normal set with the attributes

Gangeshkumar C Rahangdale [2] said that almost all the systems that which are already existed and also depend on the (ABE) plans, there is having just one expert specialist within system and remaining common keys are that which has been issued by the authorised person so that's the reason why that leads to the cipher text estimate and which leads to the extra costs within the encryption and the operations which hasn't been scrambled yet. so that's the drawback why he proposed a dual or double-Authority scheme so of all the specialists not relate to make the information which is open in the midst of the system initiation stage.

Khaled Riad [3] he made some discussions about the drawbacks or the problems of multiple people also can be able to get the permission for authorization so that is the reason why most of the people are facing the serious problem in the cipher text-policy attribute-based encryption. So that's the reason why came up with a new dual-authority access scheme control model for cloud capacity double-authority framework scheme was discussed depends on which the user's attributes for trust calculation. And his new model make sure allowing get to for the trusted clients whereas securing the clients given identity

Ruqayah R. Al-Dahhan,[4] proposed that the new scheme which they have been developing in the conference that the scheme is really a huge advantage, strong method and also supported to build safe access control system. Anyway this technique is really different and preserving the information only on a the private system that which can handled by the single person. So an unused new scheme is presented so that it handles almost all the present given work's restrictions and safely permits protecting information on open cloud capacity system by utilizing different authorized persons for the single set.

Xiaoyu Li, Shaohua Tang, [5] made profound research about Attribute-based encryption, particularly for this type of encryption that which is made on attribute based, can utilize the working and to have the control over the cloud capacity frameworks. In our paper we did some research and came up with the idea of the two-factor security for the double-authority cloud capacity frameworks. In the new scheme that which we proposed , any person can recuperate information that which is outsourced in the event that and only in the event that this client have the capacity that can handle the keys which has been kept secret with regard to the get to approach authorization key in respect with the outsourced information.

F Huang,[6] In cloud computing, the encryption scheme was developed for the data that which is outsourced. during this paper, for the primary time, we have a tendency that can be studied and also have the capacity tom solve difficulty in the customized pair of key words positioned look about the disorganized information whereas protecting and also providing the security or the privacy in cloud computing also important. So that's the reason why we discuss about two PRSE plans various searchable engines. Broad tests are conducted on the data sets that which are available approve the investigation and also looks like that our projected arrangement exceptionally effective.

ZFu, X. Sun, [7] With the expanding selection in the department or the domain of the cloud computing, one can see the rapid increase in the growing of clients and also their datasets are outsourced. So the data that which is in the form of sets ordinarily they scrambled some time recently outsourcing to protect security. We did some research and came up with the idea, we developed an imaginative new look scheme that which nearly equal to the hierarchy also the possible relationship among the concepts within data that which are scrambled.

K.Xue, [8] he said that the cloud computing is a very large domain sharing the information is very frequent and in order to solve it we have to come with a design and we got two major problems in this thing and a test is done the scheme that we have proposed and it is giving good accurate results.

Wu, Z. Wei, and H. Deng [9] they proposed a unique Multi-message Cipher text Approach Attribute-Based encryption (MCP-ABE) procedure, and utilizes the MCP-ABE to design an get to control scheme for sharing scalable media based on info consumers' attributes. Besides, the paper shows a

way to support resource-limited versatile devices by offloading computational seriously operations to cloud servers whereas while not compromising information privacy.
S. Aljawameh [10] made a study about the proposed algorithms before the threat to the stored data is less. Then the study makes sure that the features of the existing and recently proposed are completely different and the proposed system is a bit secure in the way of encryption of the data to another level where the data cannot be accessed by the private key and that is also provided by the admin only if the user is an authorized user. This infers the proposed system about the security of the data being uploaded to be encrypted in a way that the data is not to be stolen and only with the specific permission and the access key , the data is to be decrypted.

3. Proposed Methodology

Here we display a Dual or Double-Authority ABE with user security no need of the particular specialist. These requirements are in-significant to fulfill, due in both cases to the collusion resistance prerequisite of ABE. We furthermore present a mysterious key supply protocol which permits multi-authority ABE with upgraded client privacy

1) We permit the clients to communicate with AAs by means of pseudonyms rather than having to supply their GIDs within the clear, and
2) We anticipate the AAs from pooling their information and connecting numerous attribute sets belonging to the same client

A. User Module

The information consumer (client) has to register first. After the registration is done the client has given with particular unique user identification (user identity) by Central Authority. So then the consumer incorporates a given particular number of attributes such that they are related to a anonymous key connected with client quality number of attributes. Then the client will simply have disorganised information from the supplier who provides the cloud service. In any case, the client will unscramble the disorganised data only when client set of attributes fulfils the policy that which is defined within the data that which is encrypted.

B. Owner module

Data provider (Owner) characterizes the login policy about who has to get permission to view each record, later scrambles the data given in the file under the particular systematic approach. let begin of all, particular single owner scrambles particular person data can be change by the symmetric encryption calculation. At that time , the owner defines get to approach over a property set the next scrambles the particular key by the approach concurring by the open key gotten passed by CA. The next to that the person who provides the service passes the total scrambled information along with the scrambled symmetric key (indicated according cipher text CT)from the server where the information is kept within a cloud server.

C. Admin module

Admin may be super consumer.They will check all the consumer and owner details. Admin will see the chart based on the most range of word look, they will include the specific connected word, so the consumer will effectively mapping connected words for case Uncertainty level two alludes to occasions that almost all people assume as equivocal. These occurrences contain few or more not connected senses, like "apple" (apple maybe a fruit and conjointly apple may be a Smartphone company) and "jaguar" (jaguar maybe a animal and conjointly a automobile company).This work, we mainly consider the clarification of occurrence.

D. Attribute Authority module

These are the main reason for doing the client authenticity confirmation and producing middle keys for authenticity confirmed clients. Our one is not at all like previous existed plans where unique AA oversees which is not a combine property on individual basis, our planned theme includes varied specialists client authenticity confirmation so that every AA will perform this prepare can autonomously. So that the particular AA chosen, such that make sure client's genuine attributes done by normally either confirmation conventions, then produce a halfway key that which is related to traits that which has been genuinely checked. The middle key can be a modern construct that assist CA to create more keys.

E. Central Authority module

Main specialist (central authority) is the one who actually controls the whole framework. That is the main reason to framework development by keeping system limits and producing open verification text for every public attribute set. Within framework data formatting stage, it allots every client a one of a form user identification every authority has given a special Attribute identity. In order to ask a verification text from the client, the main authority is capable so that it can produce anonymous keys to the client so that they can get the intermediate key related to client's real attributes that which is verified through the attribute authority. So the one who actually runs the whole framework, so that the central authority has enough potential that follow attribute authority perniciously confirmed that a client will allow ill-conceived sets. The cloud server offers a public platform for owners to store and share their disorganized info. So that the server of the cloud won't have the info gets to regulate for the one who provides service. So that the disorganized info is kept inside the cloud server and it can be downloading without access by any client.

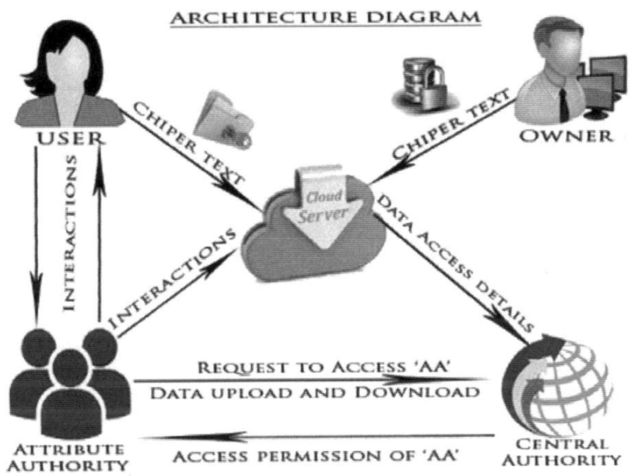

Fig. 1. Architecture diagram and proposed methodology

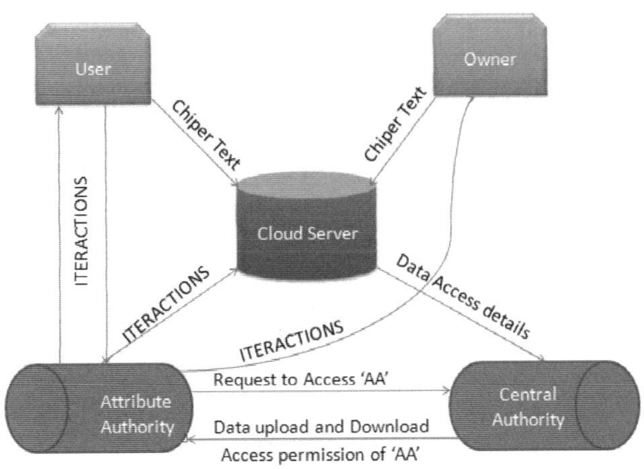

Fig. 2. Framework and system implementation

4. Implementation

A. System Implementation

The system is implemented in Net beans using Java programming language. The information is sent to the cloud server only when the user details authentication is successful. The details provided by any client or user at the time of the registration are asked to be entered for the purpose of authentication and the particular user can be logged in to the provided space to view, upload or delete their own files. when the user or client after login into his/her account they can edit the file content. They can also view the files belonging to other data owners and if the files of other owners are needed then the user places a request to the owner and if the owner is interested to share the files then he will be sending a particular key

B. Encryption of Data

The file that which contains the information is particularly specified by the system and then the information is placed in the cloud space and retrieved in the same format when the request is placed to the particular data owner. This key generation CP-ABE schema is having greater strength than comparing to the other algorithms.
The encryption and decryption strategy in the case of CP-ABE algorithm is faster than the DES and RSA algorithms. The accuracy is also high in the case of this approach. If any of the tests is to be conducted for knowing the accuracy of the algorithm then the precision is found for different algorithms towards the cipher technique and the one having the highest accuracy can be confirmed to be the algorithm being used in the approach.

C. Delay Calculation

The encryption time is noted for the calculation of the delay. And , the difference is stated using the files of different size such as 8MB, 24MB, 50MB, 83MB. And the time is calculated in seconds where the below depicted graph shows the difference between the time taken by different size files to be uploaded into the cloud space.

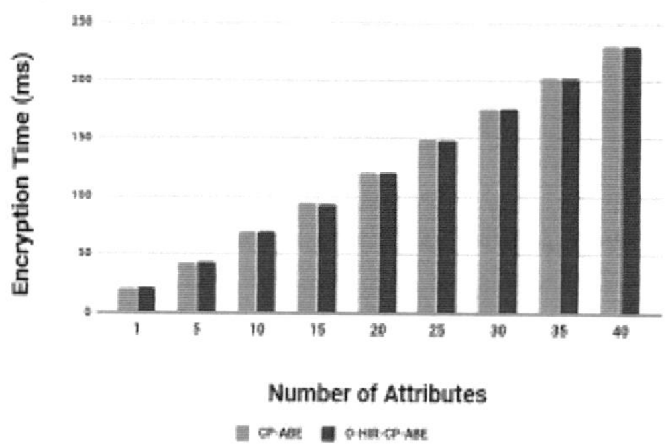

Fig. 3. Encryption Time

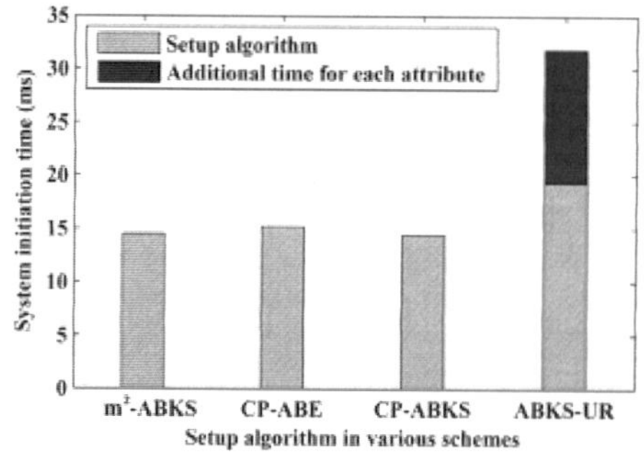

Fig. 4. Algorithms working across different schemes

There are different features which can explain the uniqueness and the strength of the approach. The comparison of the stated algorithms can be depicted using the following:

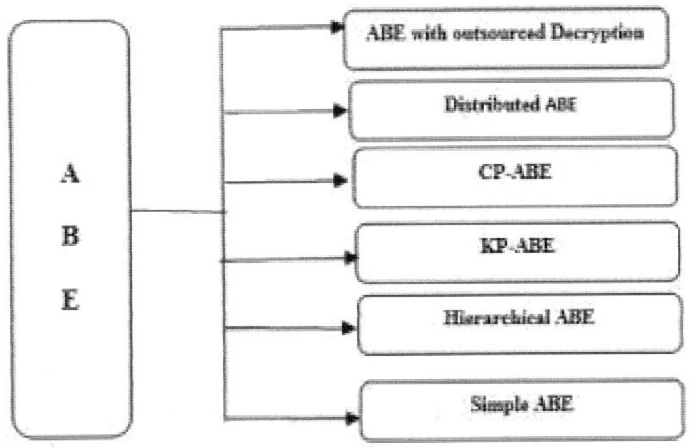

Fig. 5. Cloud Based Server Incorporating Attribute Based Encryption (Abe) Processes

5. Results and discussions

The information is encrypted by the CP-ABE algorithm in plan to be attacked from third party uses. As there is no scope for third party clients, once the client is verified, at that time the entrance of any malwares is avoided. If the confirmation isn't demonstrated by the client at that time the precise client cannot move further. Presently, after the effective confirmation of any client all the groups from all clients are often seen and hence any client who has any requirement of a selected record, can put a ask at the record and which itself ends up in the owner of the precise file. And at that time the owner gets a ask and on the off chance that the owner is interested to share is/her records at that time the reaction from owner is sent to the precise asked individual. And at that time he are often planning to the records and once the record is gotten to by a selected key the key's consequently changed and a brand new key's created. And with the identical key any record can not be opened twice. This basically decreases the false sharing of critical information. Usually, achieve

6. Conclusion

In our present paper, we implemented contemporary system, to kill the direct one-point execution of the present CIPHER TEXT-POLICY ATTRIBUTE-BASED ENCRYPTION plans.Therefore by successfully changing CP-ABE cryptographic strategy into current system, so our new strategy provides a vigorous and productive get to regulate with a single central authority or the multi-authorities for open cloud capacity. Our plot utilizes numerous AAs to provide the stack of the time-taking authenticity confirmation and modern entries of client's demands. More over we presented a fully secured strategy that follow a property authority's wrong behaviour check. Therefore a detailed security and execution examination is made so that to check whether the conspire is safe and productive. So that the safety examination results gives that our could method would be effective .Other than, with the new inspecting & following conspire, no Attribute Authority might not accept it's gotten out of hand key conveyance. Encourage execution analysis

References

[1] Li Wei et al "TMACS: A robust and verifiable threshold multi-authority access control system in public cloud storage" *IEEE Transactions on parallel and distributed systems* 275 (2015): 1484-1496

[2] Rahangdale Gangeshkumar C and Archana Raut "An approach for advanced multi-authority access control system for public cloud storage" *2017 International Conference on Innovations in Information Embedded and Communication Systems (ICIIECS)* IEEE 2017

[3] Riad Khaled "Multi-authority trust access control for cloud storage" *2016 4th International Conference on Cloud Computing and Intelligence Systems (CCIS)* IEEE 2016

[4] Al-Dahhan Ruqayah R et al "Revocable Decentralized Multi-Authority Access Control System" *2018 IEEE/ACM International Conference on Utility and Cloud Computing Companion (UCC Companion)* IEEE 2018 [5] Li Xiaoyu et al "Two-factor data access control with efficient revocation for multi-authority cloud storage systems" *IEEE Access* 5 (2016): 393-405

[5] Kurosawa Kaoru and Yasuhiro Ohtaki "How to update documents verifiably in searchable symmetric encryption" *International Conference on Cryptology and Network Security* Springer Cham 2013

[6] D Cash S Jarecki C Jutla H Krawczyk M-C Ros¸u and M Steiner "Highly-scalable searchable symmetric encryption with support for boolean queries" in Advances in Cryptology–CRYPTO 2013 2013 pp 353–373

[7] S R Lenka and B Nayak "Enhancement of the Data Security in Cloud Computing using RSA Encryption and MDS algorithm " vol 2 no 3 pp 60-64 2014

[8] C Bosch Q Tang P H Hartel and W Jonker "Selective document ¨ retrieval from encrypted database" in ISC 2012 pp 224–241

[9] Fu Zhangjie et al "Enabling personalized search over encrypted outsourced data with efficiency improvement" *IEEE transactions on parallel and distributed systems* 279 (2015): 2546-2559

[10] Masadeh Shadi R et al "A comparison of data encryption algorithms with the proposed algorithm: Wireless security" *The 6th International Conference on Networked Computing and Advanced Information Management* IEEE 2010

[11] Ramakrishnan B Durdu A Rajagopal K and Akgul A 2020 Infinite attractors in a chaotic circuit with exponential memristor and Josephson junction resonator AEU-International Journal of Electronics and Communications 123 p153319

[12] R "Kaur "Effective Symmetric key Block Ciphers Technique for Data Security :RUNDAEL" vol 3 no 7 pp 2005-2008 2014"

[13] P" Kalpana " Data Security in cloud computing usingRSA algorithm " vol 1 no 4 pp143-146 2012"

[14] Balamurali R and Chandrasekar A 2019 Multiple parameter algorithm approach for adult image identification Cluster Computing 22(5) pp11909-11917

[15] Fu"Zhangjie et al "Towards efficient content-aware search over encrypted outsourced data in cloud" 'IEEE INFOCOM 2016-The 35th Annual IEEE International Conference on Computer Communications' IEEE 2016"

[16] Xue Kaiping and Peilin Hong "A dynamic secure group sharing framework in public cloud computing" *IEEE Transactions on Cloud Computing* 24 (2014): 459-470

IOP Conf. Series: Materials Science and Engineering **994** (2020) 012009 doi:10.1088/1757-899X/994/1/012009

Digital Photogrammetry for Fruit Disease Identification

[1]G Saranya, [2]B Vidhya, [3]V Dhivya Dharshini, [4]M Divya Kamala Bharathi

[1, 2]Assistant Professor, Department of Electronics & Communication Engineering, Sri Krishna College of Engineering and Technology, Coimbatore, India.
[3, 4]UG,Department of Electronics & Communication Engineering, Sri Krishna College of Engineering and Technology, Coimbatore, India.
Corresponding Author Mail id: saranyag@skcet.ac.in

Abstract: Nowadays horticulture takes about 25% of the agriculture enterprise that has a excellent impact in the fruits field. Effective growth of crops and improved field culture must be incorporated to insure honest yield. To achieve that farmers need an environment friendly monitoring system. Farmers find it difficult to discover fruit ailment and its cause. Also, fruits are more prone to get infection in the course of cultivation, due to changing environmental condition and climate. The previous method concerning detecting fruit ailment was once more time ingesting and failed to supply information regarding the type of disease. Using the proposed fruit ailment discovery system, the agriculturist can determine the kind of the disease, and find preventive measures or suggestions. Image processing techniques are used for the enhancement of the received images. Then, convolution neural network has been used to make the model learn and classify the fruits and its disease. This system will benefit farmers across India.
Keywords: Digital image, decoder, encoder, python and image restoration.

1. Introduction

This Paper shows the procedure for recognition of natural product ailment. Present days as like in that spot is prohibitive enthusiasm for cultivating Industry, beneficial improvement and that's just the beginning beneficial assembling on fruits is major yet basic. For this intention Agriculturists require manual gazing at concerning natural products. In any case, the whole age guide observing may not dispatch quality results or they more often than not require direction past ace. So to overcome these drawbacks we propose a strategy which aides in light of the fact that higher generation and improvement together with particularly less ethnic exertion, and more innovation procedures are used for usage about proposed framework. For image classification K-means clustering technique is applied. Four Feature vectors are utilized for the proposed framework which is shading, morphology, ground at that point, structure concerning hole about the organic product. The system uses twain picture databases, some as a used of training of in the interim spared infection pictures then remaining for Implementation about inquiry pictures [1].

Sicknesses in fruits are the major problem between financial misfortunes for generation of predial industry around the world. In this paper, an answer due to the disclosure or arrangement concerning natural product sicknesses is considered and tentatively approved. The image processing of designed system is made by agreeing essential strides of the development quadrant. Clustering algorithm K-Means is deployed for image segregation. In the second half government in regards to the work manship, administrations are extricated alongside the fragmented picture, and in the long run previews are marked between one concerning the preparation by method for the use of a SVM (Support Vector Machine). Our observational results of the proposed system gives us important key insights which may significantly help discovery of infections in natural products. [2]

This proposed system discusses the improvement on transportable crop pick and then grading. Computer system is primarily based on computer vision. The mechanical system is designed from mangy fee fabric among the form about bent or segmented aircraft in accordance with alternative the utilization concerning conveyor belt. This system collects video image using a high precision webcam positioned on the top of the conveyer belt within the evaluation area, afterwards the image is analyzed according with the procedure on pc vision. First, the computer imaginative and prescient algorithm converts the RGB content to gray conversion. The RGB with respect to HSV over the photo undergoes morphological operations for picture segmentation. The techniques of shade segmentation are consistent to various fluctuations with respect to intensity. To pace at the process, each alone body is categorized in conformity with 2 ROI based completely approximately grain role into queuing and assessment area. Then the device desire tussock corn multiplication according in imitation of the degree over getting old or its dimension. In the end, the self sustaining law intention actuates the servos in imitation of pace the crop plants in conformity with a particular bin according in accordance with theirs virtue grade. Then the end result concerning crop vegetation evaluation statistics choice remain displayed regarding PC's monitor. The machine execute slave the undertaking amongst 500 ms which includes obviousness result. [3]

Image segmentation is the advance bottom in picture analysis to break the photograph into several significant regions. It impacts the picture evaluation outcomes. This paper reviews regarding the development concerning an automated regular algorithm due to the fact segmentation on coloration images, using linear help vector machine or Otsu's thresholding method, because of pip sorting then grading. The approach routinely adjusts the array hyper plane taken into consideration with the resource of using linear SVM or calls for minimum training yet time. It moreover avoids the problems delivered by manner of versions between the lighting fixtures situation and the color of the fruit. To observe the robustness and effectiveness of our proposed segmentation method, examinations have been conducted for 300 'Healthy' apples the use of iii coaching samples collectively with unique color traits (i.e., orange, stripe, and dark red). The segmentation carelessness varies in percentage from 3% to 25% for the constant state vector machine, at the same time as the adjustable SVM achieved for each set, with the segmentation confusion concerning much less than 2%. The proposed approach provides an high first-rate then strong segmentation capacity for pick and grading apples under multiple channel color domain space, or that perform stay without trouble tailored due to mean imaging-primarily based predial applications.[4]

The paper provides a laptop vision based system because of computerized grading or removal about praedial merchandise as mango (mangifera indical.) based concerning ripeness level. The utility regarding computer imaginative and prescient primarily based system, aimed after change guide based totally technique for grading yet selection on fruit. The guide inspection poses troubles of preserving propriety within reviewing and consistency inside arranging. To pace upon the way mainly pleasantly specifically keep up the consistency, consistency then precision, a model computer imaginative and prescient based automatic fruit grading and elimination system was once developed. The automated machine collects video and converts into image frames out of the CCD camera positioned at the mechanical gadget facing the mangoes, afterwards the images are prepared in imitation of collects diverse applicable competencies which might be sensitive consistent with the maturity degree on the mango. [5]

Crops that are already affected by means of uneven climatic stipulations propulsion in imitation of reduced reviewing and consistency inside arranging. To pace upon the way in particular pleasantly specifically keep up the consistency, consistency then precision, a model computer vision based on this is the cutting-edge agricultural strategies, yet structures are needed in accordance with notice then prevent the crops beyond existence affected with the aid of one-of-a-kind diseases. In that paper, we endorse an internet based totally tool so much helps farmers because figuring out crop ailment with the aid of transferring grain picture to the system. The law has an in the in the mean time gifted dataset concerning pictures for the pomegranate organic product Input photo partial by using the consumer undergoes numerous processing steps in accordance with become aware of the rate concerning illness by way of comparing with the prepared dataset pictures. The picture is resized at that point underneath its capacities is removed of parameters such as shading, morphology, and CCV or bunching is taken by utilizing a sort of unsupervised machine learning calculation, which is the k-means calculation. Another,

SVM is prepared for arrange after characterize the photo so polluted at that point non-contaminated. An significance inquire approach is moreover given because it stands through and through important as per find the client intension. Out concerning iii organizations isolated we got incredible results utilizing morphology. Test relationship over the proposed approach is compelling and 82% redress agreeing to find pomegranate malady. [6,9]
Pomegranate is a natural product which develops with a high return in numerous conditions of India and one of the most benefits picking up organic product in the market. In any case, because of different conditions, the plants are contaminated by different illnesses which wreck the whole yield leaving less item yield. In this way, the work proposes a image handling and neural system techniques to manage the primary issues of phytopathology for example malady location and characterization The Pomegranate natural product and the clears out are influenced by different maladies caused by organism, microscopic organisms and the climatic conditions. These maladies are like Bacterial Scourge, Natural product Spot, Natural product spoil and Leaf spot. The framework employments a few pictures for preparing, a few for testing purpose..The color pictures are pre-processed and experience k-means clustering division. The in general precision of this strategy is 90%. The outcomes are demonstrated to be precise and acceptable as opposed to manual evaluating and ideally take a solid ascent in setting up itself in the showcase as one of the most proficient procedure. [7,8]

2. Methodology

The following are the methodologies involved in the image processing techniques.

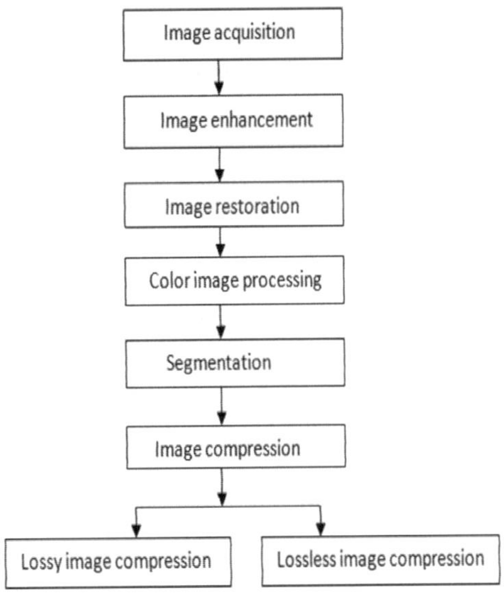

Fig. 1. Fundamental Steps

A. *Image Acquisition*

An Image acquisition is the strategy for entering an advanced image as input. For capturing, it needs a photo sensor and necessity to digitize the flag created by involvement of the sensor. The sensor should be a monochrome digital camera or color TV digital camera which produces a complete picture of the 1/30 sec inconvenience zone. Scanner produces a two-dimensional image. If the camera or sensor display is no longer usable in the computerized frame, a converter quickly digitizes it. The key task of the sensor and the image created were regulated by the application requirements.

B. *Image enhancement*

Enhancement of the image is the only toughest zones of computerized picture handling. Essentially, the thought at the back of upgrade systems is to bring out component that is clouded, or certainly to highlight sure aspects of fascinating a picture. A natural example of upgrade is the point at which we expand the differentiation of a photograph because of the reality "it shows up better." It is basic to save in idea that improvement is an abstract region of picture handling.

C. *Image restoration*

Restoration of the image is the concept that manages to increase the appearance of a picture. In any, dislike upgrade, which is emotional, photograph rebuilding, is objective in the experience that restoration strategies have a tendency to be based completely on numerical or probable styles of picture corruption. Upgrade, on the further hand, is independent inclinations with respect to what institutes a "good" improvement result. For case, a few may consider differentiate extending as an upgrade method since it is based on the engaging components it might existing to the watcher, in case of disposal of photo obscure, a de-blurring work is considered a reclamation technique.

D. *Image processing*

The utilization of color in picture handling is propelled by way of two vital variables. To begin with, color is an strong descriptor that as often as possible disentangles question acknowledgment and extraction from a scene. Moment human creatures can figure hundreds of shade colorations and force, in differentiate to approximately exclusively two dozen colors of gray. This moment issue is predominantly vital in manual picture investigation

E. *Image Segmentation*

Segmentation techniques partition a picture into its essential components or objects. In common, robotized division is one of the hardest assignments in computerized photo preparing. A tough division method brings the strategy a long way towards beneficial reply of imaging inconveniences that require objects to be distinguished independently. The more precise the division, the more probable the acknowledgment is to prosper.

F. *Image compression*

Digital Image compression reduces diminishes the sum of information required to speak to a computerized picture. The premise of the compression prepare is expulsion of repetitive information from the scientific perspective, this sums to changing a 2D pixel cluster into a statically uncorrelated information set. The information excess may be a numerically quantifiable substance. In case n1 and n2 indicate the amount of information-carrying units in two information sets that speak to the identical data, the relative information redundancy [2] of the first data set (the one considered by n1) can be well-defined as,

$$a_D = 1 - \frac{1}{c_R}$$

Where c_R is termed as compression ratio [2]. It is defined as

$$c_R = n1 / n2$$

In image compression 3 basic redundancies in data can be recognized and abused: Coding overabundance, inter pixel reiteration, and psycho visual abundance. Picture compression could be completed when slight changes in any one of these redundancies are made.

3. Proposed system

A. *Convolutional neural networks*

In this proposed system bionic convolution neural systems have diminished various parameters that adjust and organize the design particularly to vision errands. Convolution neural systems are composed by a layer set that could be combined by their functionalities.

B. *Architecture layers*

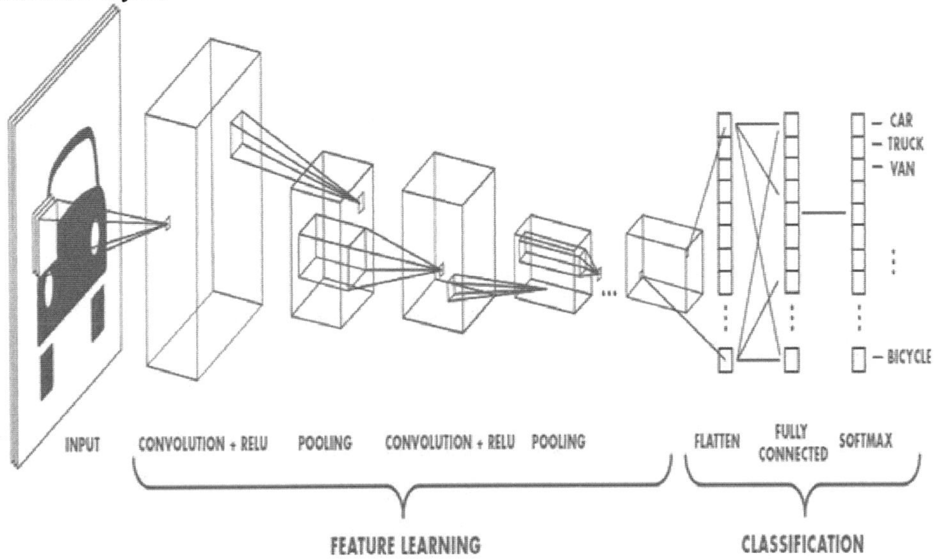

Fig. 2. Architecture layers

II. Result Analysis

The results for the proposed system are shown below.

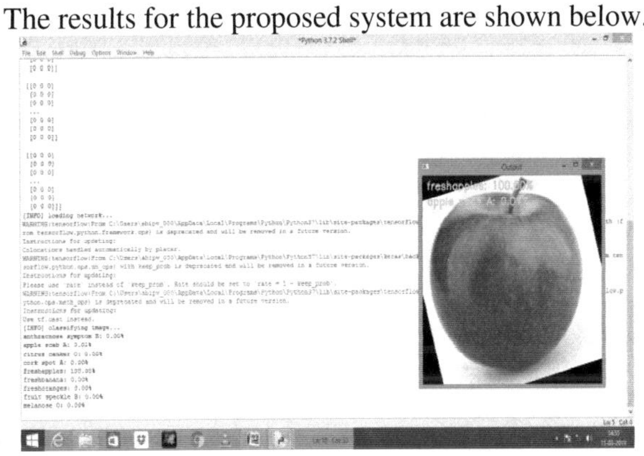

Fig.3. Simulated output of healthy apple

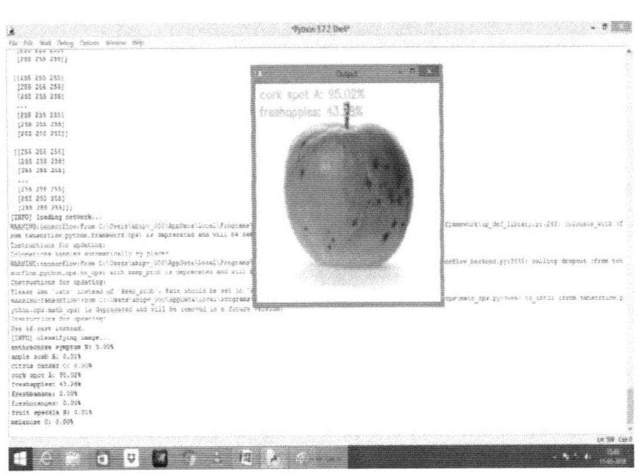

Fig.4. Simulated Output of Apple with Cork Spot

Fig.5. Simulated Output of Fresh Orange

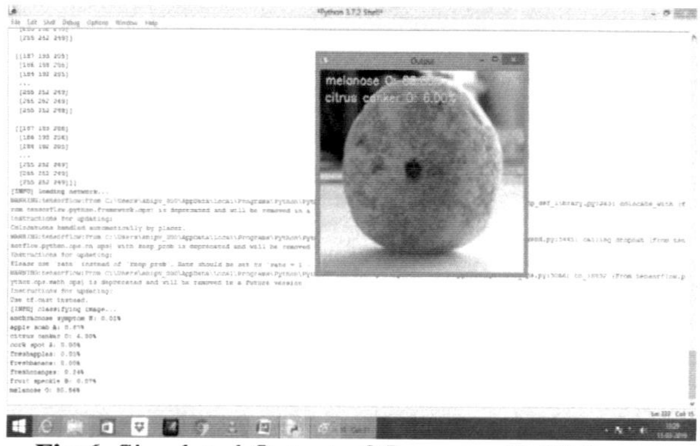

Fig.6. Simulated Output of Orange with Melanose

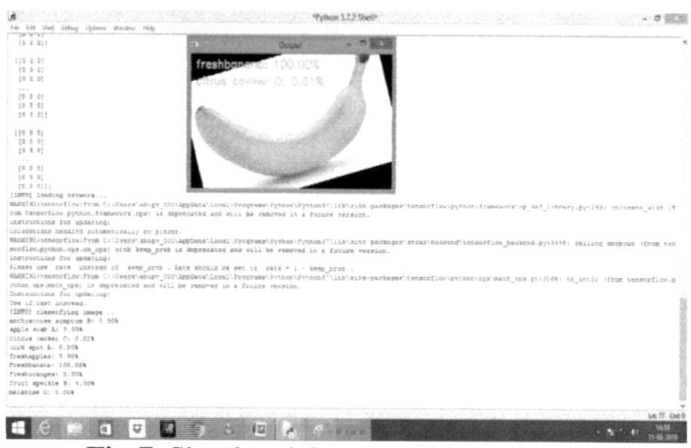

Fig.7. Simulated Output of Fresh Banana

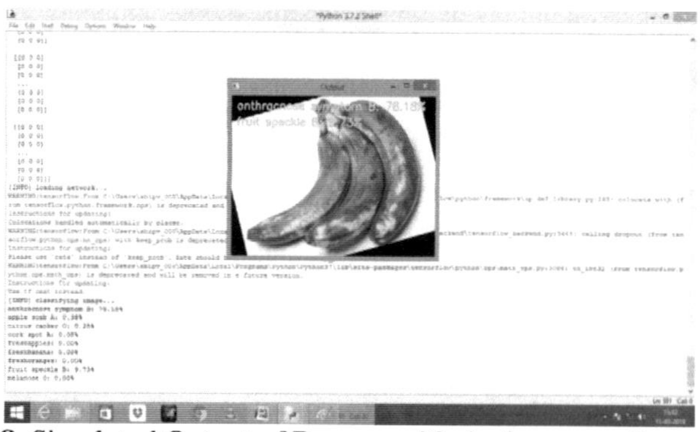

Fig.8. Simulated Output of Banana with Anthracnose Symptom

IOP Conf. Series: Materials Science and Engineering **994** (2020) 012009 doi:10.1088/1757-899X/994/1/012009

4. Conclusion

This assignment comprehensively classifies the fruits and the disorder existing in fruits. The evaluation suggests that the automatic method to ailment detection inside the fruits such as apple, orange and banana, consumes notably much less time compared to guide approach. The presence of noise can lead to distortion of the photographs being processed. So a denoising mechanism is elaborated in this case. This work identifies some of the common diseases determined in apple, orange and banana. Image processing techniques such as pre processing, extraction and many others is utilized to become aware of the fruits and its disorder at the earlier stage. Finally convolution neural network highly demands on providing classification results for the input image.

5. Future work

We have presented the concept of disorder detection in fruits. The work want not be necessarily limited to fruits alone. Future works can be prolonged for vegetables also. Leaf diseases can also be noticed by means of the usage of the proposed method.

References

[1] K Kaur and C Marwaha "Analaysis of diseases in fruits using image proccessing techniques" 2017 International Conference on Trends in Electronics and Informatics (ICEI) Tirunelveli 2017 pp 183-189 doi: 101109/ICOEI20178300913

[2] H Afrisal M Faris G Utomo P L Grezelda I Soesanti and M Andri F "Portable smart sorting and grading machine for fruits using computer vision" 2013 International Conference on Computer Control Informatics and Its Applications (IC3INA) Jakarta 2013 pp 71-75 doi: 101109/IC3INA20136819151

[3] D Lee J K Archibald and G Xiong "Rapid Color Grading for Fruit Quality Evaluation Using Direct Color Mapping" in IEEE Transactions on Automation Science and Engineering vol 8 no 2 pp 292-302 April 2011 doi: 101109/TASE20102087325

[4] Akir a Mizushima Renfu Lu "An image segmentation method for fruit sorting and grading using support vector machine and Otsu's method" Computers and Electronics in Agriculture Elsevier (2013) Vol 94 pp 29-37

[5] C S Nandi B Tudu and C Koley "An automated machine vision based system for fruit sorting and grading" *2012 Sixth International Conference on Sensing Technology (ICST)* Kolkata 2012 pp 195-200 doi: 101109/ICSensT20126461669

[6] Manisha Bhangea Hingoliwala "Smart Farming: Pomegranate Disease Detection Using Image Processing" Procedia Computer Science Vol 58 2015 Pages 280-288

[7] Cihan Murvet Kirci Ece Olcay Gunes Yuksel Cakir "Detection of the Pomegranate Fruits on Tree Using Image Processing"

[8] Kiran George Immaculate Joy SNisha Susan Thomas Balamurali R & Baskaran K 2020 Gender-Based Vegetarian and Non vegetarian Dietary Impact on Cardiac Autonomic Function of Heart Rate Variability Journal of the American College of Nutrition pp1-5

[9] NR Deepa N Nagarajan D PalanivelRajan R Balamurali 2019 July Image Processing System for Detecting and Recovering from Cotton Leaf Disease using The Android Application International Journal of Innovative Technology and Exploring Engineering (IJITEE) 8(9) pp 3366 – 3381

Vehicle Management System To Reduce The Risky Factors Using Embedded Technology Enhanced With Wi-Fi

V Jacintha[1], K H Shakthi Murugan[2], Kaviya S[3], Hemamalini V[3], Harini B[3], Haripriya V[3]
Research Scholar[1], Assistant Professor[2], Student[3]
Sathyabama Institute of Science & Technology[1], Jeppiaar Maamallan Engineering College[2,3]

shakthimuruganece@gmail.com

Abstract. Medium to medium is a quickly developing technique, which renders enhanced client servicing, occupation betterments via best usage of assets, enhanced techniques and production via communicating between distant machinery and vehicle Management System. Here, the vehicle status and location can be determined. Vehicle Tracker utilize GPS devices to give an account of the location information to the server. Accident is unplanned harm, is an unwanted, accidental, and unintentional consequence that must have been precluded had considerations and if there is no proper system human lives will be in danger. The main purpose of this technology is to identify an accident and give an indication signal to the control room, so that the affected individual can seek some support. The severity of the incident can be identify from the control room, without going to the sot in person. If this model is fix all vehicle, it will be very useful to identify how many vehicles undergoes in that specified accident and the extent to which it has taken place. In order that the aid from control room will be able to rescue and protect driver or passengers. The present board designed the tracking of vehicle and systems for accidents alert that makes it extremely valuable and utile. This board indicates to us driver abnormal condition and as well as accident identification. It additionally identifies the driver drowsiness.

Keywords: Asset utilization, Drowsiness, Unintentional injury, Vehicle management system

I. Introduction

In accident detection, using smart-phone is propose. However, if smart phones are utilize to identify accidents; in that case, there must be quiet number of filters in-built to prevent false warnings. The classification mechanism is construct by accumulating datasets from simulation of three accident cases that aids in creation of an artificial intelligence-based classification software [1].Hence; it is difficult to identify minor accidents such as sideswipe collision. If an IR sensor is, utilize for accident identification, and then the model makes use of GPS, GSM for localization and communication. The IR sensor processes the radiation emit by each object. An Android dynamic time warping based smart phone architecture of the accident identification model together with experimental outputs are utilize [2]. In accident identification utilize airbag model is suitable only for four-wheelers; it is not applicable for two wheelers. However, there must be a model that is suitable for all kinds of transport. The

IOP Conf. Series: Materials Science and Engineering **994** (2020) 012010 doi:10.1088/1757-899X/994/1/012010

identification technique is apply onto a server utilizing pattern matching based on dynamic time warping [3]. There are some issues related to accident identification utilizing humanoid touch phone. Since every touch-phone model will possess varied processors with different working speed. Hence, the applications performance may deteriorate in few smart phones. In the proposed work, several ultrasonic sensors utilize in the identification of accidents. The objective is to apply the algorithms into the android phones in-built high tech sensors that are associated with GIS, GPS and Geo fence technologies [4]. An ultrasonic sensor is utilize for distance computation. The return time taken by the wave, decides the displacement from the sensor to the obstacle. The usage of ultrasonic sensor for accident identification has numerous advantages. It enables to identify accidents not only in several road situations such as junctions, subway or in service roads but also under varied climatic circumstances such as rains, foggy weather. Ultrasonic sensor seems to be very efficient since it rely on the principle of sound wave propagation. Also sound waves can propagate simultaneously and effectively through all kinds of matter such as solid, liquid and gas. This feature enables ultrasonic sensor to be an appealing alternative for automatic accident identification. There are a number of applications available [13-26].

II. Existing system

An affordable ultrasonic sensor is utilize to identify an accident. There are several state-of-the-art methods for automated identification of road accidents. They are using Android mobile, infrared sensor, airbag system. This system utilizing dual ultrasonic sensors is suitable for identification of road accidents. Varied computer simulation are execute utilizing MATLAB, to determine the factors that have an effect on the identification of accidents utilizing ultrasonic sensor. Based on the factors the model utilize two ultrasonic sensor modules HCSR04. First sensor module is mount on the vehicle front windscreen and the second sensor is fix on the vehicle rear windscreen. Then the distance from the sensor modules to their corresponding bumpers is evaluate and are termed as threshold distance 1 and threshold distance 2. When an object is displacing away from the vehicle, the distance between them is generally greater than the preset threshold distances. The identification of closed eyelids is effective even during varied circumstances, utilizing camera of any humanoid mobile [5]. A novel driver behavior detection framework that at its core makes use of in-vehicle Wi-Fi signals [6].When an object collides with the vehicle, the threshold distance is infracted and the processing system goes to ON condition. The system instantaneously determines the vehicles location utilizing GPS and sends to the rescue mission utilizing GSM. However, there are several disadvantages with respect to these approaches [28].The disadvantage of the existing system is as ultrasonic sensor's measurement relies on the velocity of sound waves, so it requires prior calibration. The simulation results depicts that, as the velocity of the vehicle enhances, the fault in distance measurement simultaneously increments. This fault must be counterbalance by standardizing the ultrasonic sensor. With the help of machine learning, identified patterns in the behavioral and physiological factors that together helps to identify drowsy states from alert states for explicitly during driving[7].This system is totally hypothetical.

III. Proposed System

The proposed system concisely reexamine chosen road accident identification methods and intend a resolution. In these methods, a model is utilize that automatically identifies an accident in a quick span of time and alerts the emergency center and police department with an outline of the incident. The proposed system is to provide an approach about transportation unit as well as the vehicle condition is update in the server. The driver drowsiness can be identify by using the camera. If any accidents occurs, it is identify by the vibration sensor. If any deviation happens, the buzzer will alert the driver and automatically reducing the motor speed. All the details should be update in the server via Wi-Fi. In some systems, MATLAB/Simulink functions as the primary software utilize in development of this application. Images and other data types are acquire utilizing Android devices sensors [8]. The accident location can be update in the server to intimate the police station, neighbors and a message is convey to the relatives. The Heartbeat of the person is identify by using a heartbeat sensor .In this proposed system developed an affordable and reliable system using pic microcontroller. Several algorithms for tracking heart rates while performing heavy exercise utilizing an 8-sec mean of heart rates that does not precisely captivate huge changes in instantaneous heart rates[9].The fig 1 denotes the block diagram of the proposed model. The 230 volts is convert to 12 volts by using the SMPS. Vibration sensor is utilize to identify an accident. There are several methods for the identification of road accidents automatically. These include smart-phones, Infrared sensor (IR sensor), camera, vibration sensor is to be measuring the driver heartbeat count, drowsiness and vibration of the car, that data will be continuously uploading the server via internet. This system is convenient since it utilize vibration sensors for detection of road accident. To assess the operation of the proposed framework of the block diagram shown in fig.1, different location and other factors will be monitoring the PC in the police or emergency department. The driver drowsiness will be detect by using camera. Heartbeat and drowsiness is detected by this system and based on the threshold condition the motor would be automatically stopped. These kinds of data are transmit to the server via internet using Wi-Fi with GPRS connection. GPS is use to find the location where the incident has occur. Then the emergency team and police department will rescue the respective individual. In addition, a buzzer will give an alert surrounding the accident spot. The hardware components used are PIC microcontroller, raspberry pi, SMPS, DC motor, heart beat sensor and GPS.

A compiler is use to transform the source code written in a programming language into another computer language. The need for transformation is to create an executable program. Python is use in programs. It is a high-level language for easy programming.

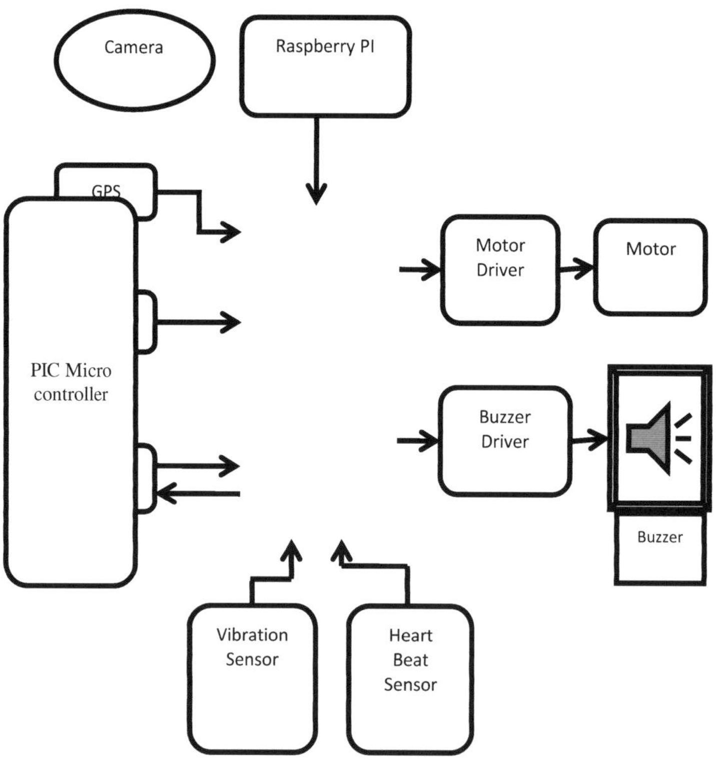

Fig 1 Block Diagram of proposed system

IOP Conf. Series: Materials Science and Engineering **994** (2020) 012010 doi:10.1088/1757-899X/994/1/012010

IV. Simulation Results

These are the server upload data of the proposed system. The server data is vibration condition, heartbeat condition, eye drowsiness and location with date and time. These results will be continuously update in the server. In the server, the huge amount of data will be stored. That total data count also updated. Fig. 2. Shows the server output of the data received on particular date and time, vibration of the car is normal, heartbeat of the driver is normal eye drowsiness or eye status is normal.

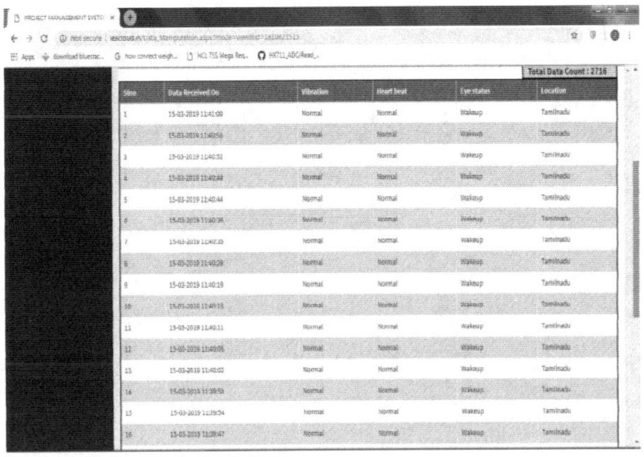

Fig 2 Server display output

The simulation of proposed system demonstrates a driver drowsiness identification model utilizing Open CV in Raspberry Pi environs having a single camera view. The proposed model initially identifies the face utilizing Haar Cascade Classifier and other facial landmarks utilizing Drib predictor.

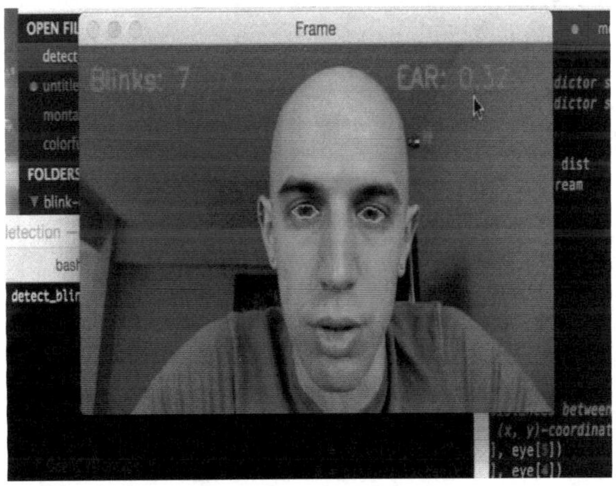

Fig 3 Active Eye detection simulation output

Facial landmarks associated with eyes are fetch and Eye Aspect Ratio (EAR) is obtain. EAR is closely noted ceaselessly, if the value lies within a threshold value and never raises again connotes driver eye is closed. The Fig.3 shows the active eye of the person. Drowsiness is detect from this technique; the simulation output of the drowsiness of human eye is shown Fig.4.

IOP Conf. Series: Materials Science and Engineering **994** (2020) 012010 doi:10.1088/1757-899X/994/1/012010

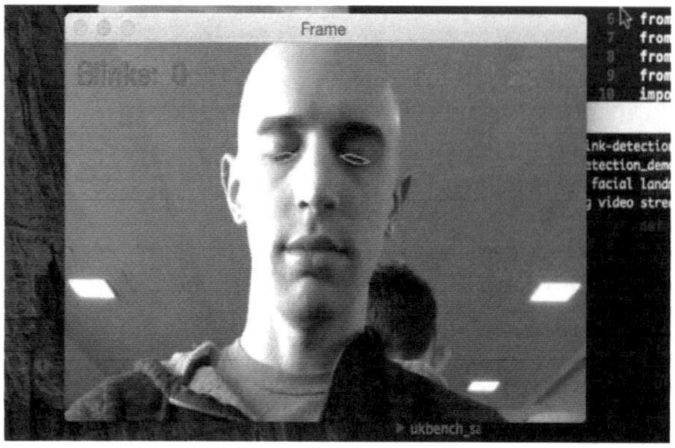

Fig 4 Drowsiness eye detection simulation output

V. Hardware Implementation

The vehicle accident detection system is implement by the requiring components that is properly to connect as show in the block diagram. First of all the entire system will be ON when the person start the engine of the vehicle. SMPS is gives supply voltage from 230V battery to 12V supply for the proposed system, when the engine is start the LED will be blinking and automatically Wi-Fi is connected to the mobile hotspot and motor will be rotate(i.e. vehicle movement). Compared with traditional piggybacking schemes, CoS is more reliable to transmit control messages from one node to many nodes[10].For that any certain time the vibration sensor sense the vibration above the threshold value to alert the surroundings area by using buzzer and also the intimation will be sent to the emergency and police department via internet. The hardware implementation of proposed system is show in Fig.5.

Fig 5 Hardware implementation of proposed system

A face image is first study utilizing a landmark identifier and a group of interest points is obtain [27]. Facial landmarks associated with eyes are fetch and Eye Aspect Ratio (EAR) is obtain. EAR is closely noted ceaselessly, if the value lies within a threshold value and never raises again connotes driver eye is closed. The measuring value will be display by using the LCD display. The Fig.6 is shown in driver was wakeup condition that is the eye status of the driver.

Fig 6 Eye detection display output

Fig 7 Vibration display output

The vibration sensor sensing the vibration value of the vehicle that is vehicle gets vibration value below or above the threshold value that is always exhibit by LCD display. The initial functioning band is focus at 850 MHz that provides for GSM applications, while the forthcoming frequency bands are focus at 1.57 GHz and 2.4 GHz provides for GPS and Wi-Fi wireless applications [11, 12]. For example the vibration sensor value is '0' that is shown in Fig.7.

VI. Conclusion

Accident prevention denotes the plans, formulations and activities taken to avert or cease an accident prior to its happening. Accidents are subdivide as unplanned and unpredicted events may leads loss of property and human kind. In this project, the methods to implement certain autonomous features in a regular car was describe. Using this prototype the hardware was design and built. With the help of vibration sensor and GPRS the accident, happened place was update in server using Wi-Fi module, which is connect to our mobile hotspot. From the server the SOS signal will be send. The driver drowsiness will be detected using camera, which is interface with raspberry pi, and heartbeat will be detect. The buzzer will be alert to the surrounding people of the accident area. This project is to be save the people life from critical situation.

VII. Future Enhancement

The accident detection and avoidance system has varied proficient challenges that raise when data analytics for collision avoidance. Future research opportunities aims for implementing data analytics in safety precaution in fields like traffic management, adaptive network security, and data forensics.

REFERENCES

[1] P Joyce Beryl Princess Salaja Silas Elijah Blessing Rajsingh Machine Learning Approach for Identification of Accident Severity from Accident Images Using Hybrid Features *Emerging Technology (INCET) 2020 International Conference for*, pp. 1-4, 2020

[2] An Android smartphone-based prototype architecture of our accident detection system along with empirical results are present

[3] K Sasidhar and A. Upasini Two wheeler rash drive detection using smartphones *2019 11th International Conference on Communication Systems & Networks (COMSNETS)*, Bengaluru, India, 2019, pp. 754-758, doi: 10.1109/COMSNETS.2019.8711455

[4] N W Aung and T L L Thein Vehicle Accident Detection on Highway and Communication to the Closest Rescue Service *2020 IEEE Conference on Computer Applications(ICCA)*, Yangon, Myanmar, 2020, pp. 1-7, doi: 10.1109/ICCA49400.2020.9022855

[5] H K Dua S Goel and V. Sharma Drowsiness Detection and Alert System *2018 International Conference on Advances in Computing Communication Control and Networking (ICACCCN)*, Greater Noida (UP) India 2018 pp. 621-624 doi:10.1109/ICACCCN.2018.8748448.

[6] H. Saeed T Saeed M Tahir and M Uppal "Risky Driving Behavior Detection Using In-Vehicle WiFi Signals *2018 IEEE 88th Vehicular Technology Conference (VTC-Fall)*, Chicago, IL, USA, 2018, pp. 1-5, doi: 10.1109/VTCFall.2018.8691013

[7] Chatterjee and S Roy Smartphone-based drowsiness detection system for drivers in real-time," *2019 IEEE International Conference on Advanced Networks and Telecommunications Systems (ANTS)*, GOA, India, 2019, pp. 1-6, doi: 10.1109/ANTS47819.2019.9117943..

[8] D Chilukuri S. Yi and Y. Seong Mobility Assistance for Vulnerable Road Users using Machine Learning *2019 SoutheastCon* Huntsville AL USA 2019 pp 1-7 doi: 10.1109 / Southeast Con42311.2019.9020506.

[9] Y Kong and K H. Chon Heart Rate Tracking Using a Wearable Photoplethysmographic Sensor During Treadmill Exercise in *IEEE Access*, vol. 7, pp. 152421-152428, 2019, doi: 10.1109/ ACCESS.2019.2948107.

[10] B Feng C Zhang J Liu and Y Fang Turning Waste into Wealth:Free Control Message Transmissions in Indoor WiFi Networks in *IEEE Transactions on Mobile Computing*, doi: 10.1109 / TMC.2019.2924000.

[11] A Basit M I Khattak A R Sebak A B Qazi and A A Telba Design of a Compact Microstrip Triple Independently Controlled Pass Bands Filter for GSM, GPS and WiFi Applications," in *IEEE Access*, vol. 8, pp. 77156-77163 2020, doi: 10.1109/ACCESS.2020.2989377

[12] R J Vidmar (1992 August) On the use of atmospheric plasmas as electromagnetic reflectors. *IEEE Trans Plasma Sci.* [Online] *21(3).* pp. 876—880.Available: http://www.halcyon.com/pub/journals /21ps03-vidmar

[13] K. H. Shakthi Murugan A Jasmine Vijithra A 40 Gbps Ro-FSO transmission system for 5 G applications Optoelectronics and Advanced Materials - Rapid Communications 14 7-8, July-August 2020 pp.306-309 (2020).

[14] K H Shakthi Murugan et al A novel rescuebot for borehole accidents 2020 IOP Conf. Ser Mater. Sci. Eng. 872 012044

[15] Shakthi Murugan KH, Sumathi M (2020). Performance Analysis of Millimeter waves over FSO System under Fog Weather Conditions. *International Journal of Advanced Science and Technology 29*(8s) 3462-3477

[16] K.H.Shakthi Murugan P Abisheik Advanced Smart Home Security Alert System International Journal of Scientific Research in Computer Science Applications and Management Studies Volume 8 No. 2, 2019 pp 1-5

[17] Murugan, KH & Manickam, Sumathi (2018) Design and implementation of 5G optical Communication System for various Filtering operations using wireless optical transmission. Results in Physics 12. 10.1016/j.rinp.2018.10.064.

[18] KH Shakthi Murugan K Sumathi Maqsooda Naaz S Judy Simon Low Cost Micro Strip Patch Antenna International Journal of Pure and Applied Mathematics, Volume 119 No. 15 2018, 573-579

[19] H, Shakthi & Manickam, Sumathi. (2018). Millimetre Waves Over Free Space Optics System for 5G Application. Journal of Optical Communications. 10.1515/joc-2018-0159

[20] Shakthi Murugan.K.H, K.Sumathi Advanced Smart Home Security Alert System International Journal of Emerging Technology in Computer Science & Electronics (IJETCSE) ISSN: 0976-1353 Volume 25 Issue 6 – May 2018, pp 96-101

[21] K.H.Shakthi Murugan V.Jacintha Judy Simon Safety System For Gold Mining Process Using Visible Light Communiction International Journal of Pure and Applied Mathematics, Volume 117 No. 16 2017, 371-376

[22] Manickam Sumathi & Shakthimurugan KH & Sumathi K (2016) Performance study of 4:1 multiplexer CMOS logic structures. International Journal of Control Theory and Applications, Vol 9. 565-574.

[23] Shakthimurugan KH (2015) Multi-carrier frequency offset estimation for carrier aggregation-orthogonal frequency division multiplexing communication. 10. 31035-31041.

[24] Shakthimurugan KH (2014) Hardware implementation of TCP/IP over visible light communication on FPGA (A concept of light fidelity). 10. 31095-31101.

[25] S M K H S T J M Jerold P S A Peeran A and A P Wearable ECG Electrodes for Detection of Heart Rate and Arrhythmia Classification 2019 IEEE International Conference on Electrical, Computer and Communication Technologies (ICECCT), Coimbatore, India, 2019.

[26] K. H. Shakthi Murugan and M. Sumathi BER and Eye Pattern Analysis of 5G Optical Communication System with Filters 2019 IEEE International Conference on Electrical, Computer and Communication Technologies *(ICECCT)*, Coimbatore, India, 2019, pp. 1-6, doi: 10.1109/ICECCT.2019.8869396

[27] V Jacintha Judy Simon S Tamilarasu R Tamizhmani K Thanga yogesh J Nagarajan A Review On Facial Emotion Recognition Techniques IEEE International Conference on Communication & Signal Processing (ICCSP) April 2019

[28] Jacintha V S Karthikeyan A Survey On Various Fabric Defect Detection Methods IEEE International Conference on Communication & Signal Processing (ICCSP) April 2019

Closed Loop Control of PV System Using Grey Wolf Optimization Algorithm under Partial Shading Condition

Dr. J. Jayaudhaya[1], D. Rajasekaran[2], J. Sumithra[3], J.C. Vinitha[4], S. Karkuzhali[5]
[1,2,3,4] Associate Professor, Department of Electrical and Electronics Engineering, R.M.D. Engineering College,
Chennai, Tamil Nadu, India
[5]Assistant Professor, Department of Electrical and Electronics Engineering, R.M.D. Engineering College,
Chennai, Tamil Nadu, India

E-mail: jju.eee@rmd.ac.in[1]

Abstract. Globally there is an increased awareness of the havoc that climate change and rising pollution are due to the over consumption of carbon riddled fossil fuels making the environment unfit for living. Thus, it is important to make discerning choices to incorporate more clean energy to produce power. Therefore, Renewable energy emerges as an obvious alternative. Renewable resources like solar and wind are now being utilized for power generation. The power output of a solar panel depends on the amount of irradiance of sunlight which is also called insolation. The main drawback in solar is the power output of a PV panel for a particular irradiance is not a linear function of voltage and the power is maximum only at a particular voltage called optimum voltage. When the solar panel is subjected to non-uniform irradiance which is referred to us partial shaded condition, then there will be multiple power peaks in the P-V characteristics and it becomes unable to meet the load requirements. The proposed system will help to overcome this drawback as it consists of two cascaded boost converters. A battery is connected in parallel to output of MPPT boost converter so that load power demand is met by parallel combination of PV panel and battery. The system is optimized for best performance using multi objective GWO optimization technique. Thus, maximum power is extracted even under partial shading condition and constant power is supplied.
Keywords: Photo Voltaic (PV), Proportional Integral (PI), Partial Shading Conditions (PSC), Grey Wolf Optimization (GWO), Integral Square Error (ISE).

1. Introduction

As a fastest growing large economy, the demand for electricity is instantly rising. Hence it is essential to increase our energy output in order to sustain this pace of growth. Renewable resources like solar and wind are now being utilized for power generation. The power output of a solar panel depends on the amount of irradiance of sunlight which is also called insolation. The main drawback in solar is the power output of a PV panel for a particular irradiance is not a linear function of voltage and the power is maximum only at a particular voltage called optimum voltage [1], [4], [5]. When the solar panel is subjected to non-uniform irradiance which is referred as partial shaded condition, then there will be multiple power peaks in the P-V characteristics and it becomes unable to meet the load requirements.

Using MATLAB simulation P-V characteristics is plotted for various irradiance patterns using the specifications given in the table 1.

Table 1. Specification of PV panel

Parameters	Rating
Maximum power	360 W
Open circuit voltage(V_{oc})	20 V
Short circuit current (I_{sc})	5 A
Current at maximum power point (I_{mp})	4.5 A
Voltage at maximum power point (V_{mp})	16 V
Irradiance	1000 W/m^2

Figure 1 shows the power voltage characteristics of single PV panel. Using the above specification the maximum power obtained from a single PV panel is 303W. Pattern 1 has uniform irradiance value each of 1000W/m^2, pattern 2 has different irradiance of 1000,900,800 and 700W/m^2 , pattern 3 has different irradiance of 900,700,500 and 300W/m^2 and pattern 4 has different irradiance of 800,600,400 and 200W/m^2. From the figure 2, it is concluded that only one maximum power under uniform irradiance condition. Under varying irradiance condition, P-V curve has multiple power peaks.

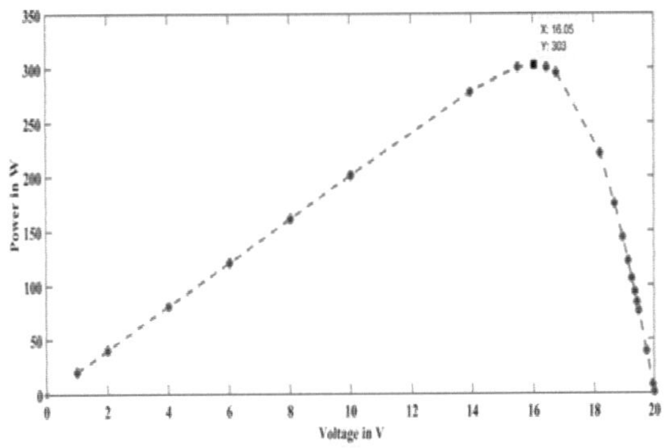

Figure 1. PV characteristics of Single PV panel

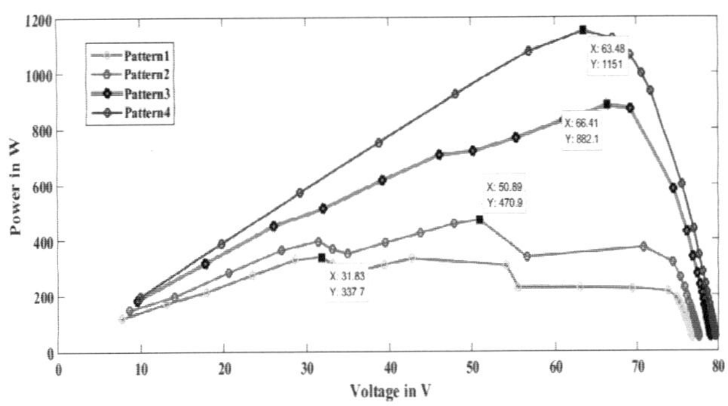

Figure 2. Power-voltage characteristics of PV panels under different irradiance condition

IOP Conf. Series: Materials Science and Engineering **994** (2020) 012011 doi:10.1088/1757-899X/994/1/012011

2. Block Diagram of Existing System

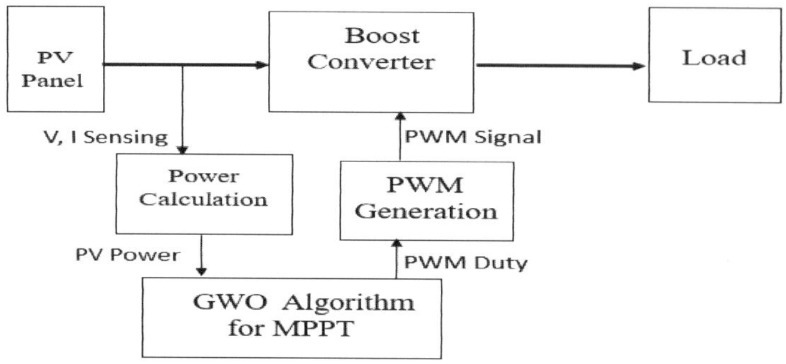

Figure 3. Block Diagram of Existing System

In the existing system of figure 3 contains PV module is associated with MPPT boost converter to acquire extreme output and to transfer power to the attached load. The optimization of the system is made using GWO optimization technique for better performance. The MPP tracking is employed to make most of the PV array's output power, regardless to the radiance and temperature situations. The GWO algorithm feeds the gating signals to the converter to operate at "maximum power point" voltage. The optimization technique will search for the best optimum value for duty cycle in order to achieve the objective function. In order to determine the optimum duty, an optimization technique is employed with an objective function of maximum power. MPPT requires constant voltage across the load and the power delivered to the load. To maintain the voltage constant is the major requirement in the existing system.
To overcome the limitation in existing system and to make the system more reliable, the proposed system is implemented and it is shown in figure 4. It comprises of two cascaded converters. MPPT boost converter is the first set of converter employed to maintain maximum output from the solar panel. Maximum power tracking requires optimum voltage which is maintained at constant. In order to maintain the voltage there is a second set of converter called as "Quadratic Converter" which employs a MOSFET power switch and switching ON and OFF duty cycle of the power switch will decide the voltage maintained across the load. The optimum duty cycle is decided by the GWO algorithm which is in multi objective. The two main objective of this system is Maximum Power and Minimum ISE. Here one objective is maximum and the other is minimum and so inverse of minimum ISE is considered while combining as single objective so that optimization can be performed for maximum value.

PI controller is employed to generate the ON time based on the error between set voltage and actual voltage across load. PI controller generates control signal in a feedback control loop using an error signal which is the difference between the set value and actual value achieved by the system. The controller takes the input of the best values of K_p & K_i from the GWO optimization. Thus, helps in minimizing the errors.
In PV power system, the power generated may not match the load demand due to changes in irradiance. Hence, a battery is connected in parallel to PV power system for power balance. The battery will deliver power to load when PV power is less than load demand. The battery will charge if PV power is more than load demand. Thus, the system is completely independent to meet the load demands.

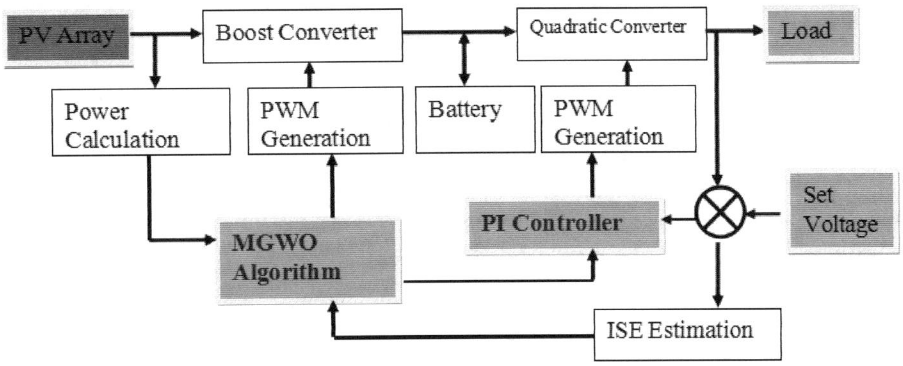

Figure 4. Block diagram of proposed system

In this system there are two objectives: one is Maximum power and another one is Minimum ISE. The two objectives are combined to single objective by converting them to per unit values by dividing the respective values by their expected maximum value (or base value) and taking sum of the two per unit objective value as single objective for optimization. Here one objective is maximum and the other is minimum and so inverse of minimum ISE is considered while combining as single objective so that optimization can be performed for maximum value.

3. Optimization Algorithm

Mirjalili et al. (2014) proposed a GWO algorithm that resembles the leadership qualities and hunting behaviour of grey wolves in nature. Grey wolves prefer to live in a group. The average wolf group size is 5-12. The leadership hierarchy consists of four types of grey wolves called alpha (α), beta (β), delta (δ), and omega (ω). Among these alpha is the leader and it may be a male or a female. Alpha wolf may not be the strongest wolf among the group but it must have the capability to manage the group. The next level of hierarchy in the grey wolves is beta and it may be either a male or female.

3.1 Mathematical Model of GWO Algorithm

The mathematical model of the social hierarchy, tracking, encircling and attacking prey of GWO algorithm is outlined below.

(a) Social hierarchy

In order to model the social hierarchy of wolves in MGWO technique, Assume the fittest solution as the alpha (α) and it is followed by the second and third best solutions, namely, beta (β) and delta (δ), respectively. The rest of the candidate solutions are assumed to be omega (ω).The hunting mechanism is guided by alpha, beta and delta. The omega wolves follow these three wolves.

(b) Exploration and Encircling prey

Grey wolves encircle a prey during the hunt and the encircling behaviour is modelled by the following equations. The distance D of a wolf from the prey can be estimated by Equation (1.1).

$$\vec{D} = \left|\vec{C} \times \vec{X}_P(n) - \vec{X}(n)\right| \tag{1.1}$$

$$\vec{X}(n+1) = \vec{X}_P(n) - \vec{A} \times \vec{D} \tag{1.2}$$

Where, n denotes the current position, *A*, and C are coefficient vectors, *Xp* is called the position vector of the prey, and X is the position vector of the grey wolf. The value of *A* and *C* are calculated by using the Equations (1.3) and (1.4).

$$\vec{A} = 2\vec{a}.\vec{r_1} - \vec{a} \tag{1.3}$$

$$\vec{C} = 2\vec{r_2} \tag{1.4}$$

During the iterations '*a*' linearly decreases from 2 to 0 and r_1, r_2 are random vectors and the value lies between 0 and 1. The best solution can be achieved with the help of current position by varying the value of *A*, and C vectors. A Wolf can reach any position between the points based on the random vector r_1 and r_2. Therefore, a grey wolf can update its position in any random location by using Equations (1.1) and (1.2). The below equations are used to update the position of alpha, beta and delta wolf.

$$\overrightarrow{D_\alpha} = \left|\overrightarrow{C_1} * \overrightarrow{X_\alpha} - \vec{X}\right| \tag{1.5}$$

$$\overrightarrow{D_\beta} = \left|\overrightarrow{C_2} * \overrightarrow{X_\beta} - \vec{X}\right| \tag{1.6}$$

$$\overrightarrow{D_\delta} = \left|\overrightarrow{C_3} * \overrightarrow{X_\delta} - \vec{X}\right| \tag{1.7}$$

$$\overrightarrow{X_1} = \overrightarrow{X_\alpha} - \overrightarrow{A_1} * \overrightarrow{D_\alpha} \tag{1.8}$$

$$\overrightarrow{X_2} = \overrightarrow{X_\beta} - \overrightarrow{A_2} * \overrightarrow{D_\beta} \tag{1.9}$$

$$\overrightarrow{X_3} = \overrightarrow{X_\delta} - \overrightarrow{A_3} * \overrightarrow{D_\delta} \tag{2.0}$$

$$\vec{X}(n+1) = (\vec{X}_1 + \vec{X}_2 + \vec{X}_3)/3 \tag{2.1}$$

(c) Fitting GWO for PSC PV MPPT tracking

In general, MGWO algorithm is applied to optimize the positions of a group of wolves or particles (N) moving towards a prey or objective. Therefore, the algorithm starts with the initial actual position (X) of a wolf and an assumed best position (G) of wolf.

The optimization process involves k iterations to determine the best position of the wolf. Hence, in an iteration called n^{th} iteration, there will be n^{th} iteration value for position on N wolf, is namely, $X_{1,n}$, $X_{2,n}$,.....$X_{i,n}$,....,$X_{N,n}$. The algorithm has to determine $n+1^{th}$ iteration value of the position, which is determined using equations (1.1) and (1.2). In this work, MGWO algorithm is applied to optimize the duty of number of wolves (N) moving towards an objective which is maximum power. Therefore, the algorithm starts with initial duty (X) of a wolf and an assumed best position (G) of wolf. In iteration the duty is updated and the search process ends after k iterations. Here one objective is maximum and the other is minimum and so inverse of minimum ISE is considered while combining as single objective so that optimization can be performed for maximum value.

4. Simulation Results

Figure 5 shows the simulink model of proposed system. Using the specifications of PV indicated in table simulation results are tabulated in table 3.

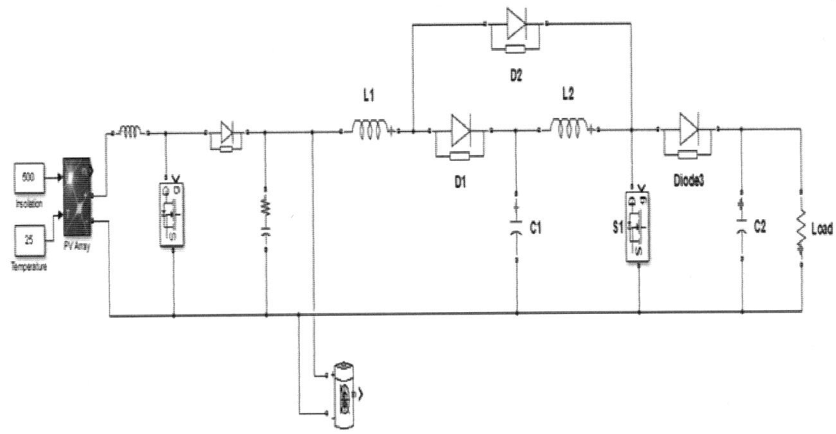

Figure 5. Simulink model of proposed system

Table 2. Specification of PV panel

Parameters	Rating
Maximum power	50 W
Open circuit voltage(V_{oc})	8.7 V
Short circuit current (I_{sc})	8.01 A
Current at maximum power point (I_{mp})	7.46 A
Voltage at maximum power point (V_{mp})	6.7 V
Irradiance	1000 W/m^2

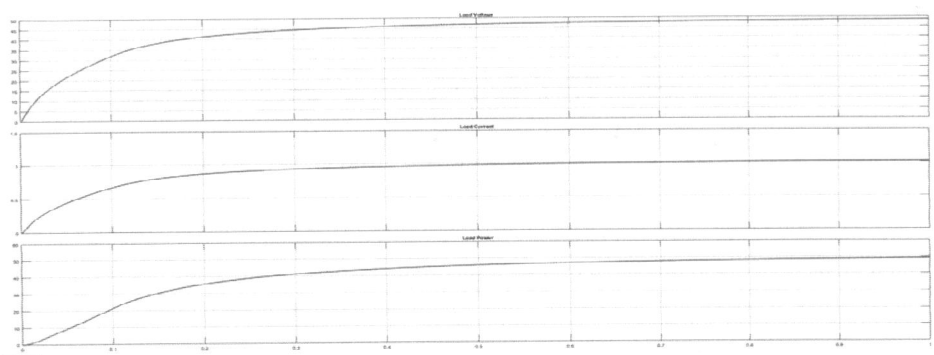

Figure 6. Load voltage, Current and Power

IOP Conf. Series: Materials Science and Engineering **994** (2020) 012011 doi:10.1088/1757-899X/994/1/012011

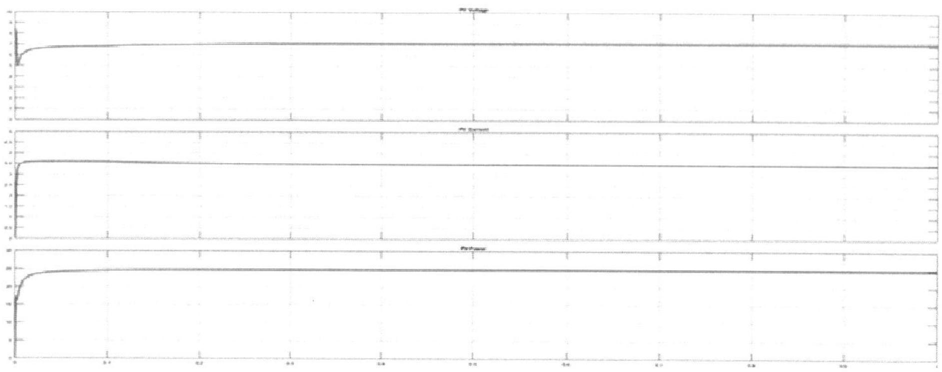

Figure 7. PV Voltage, Current and Power

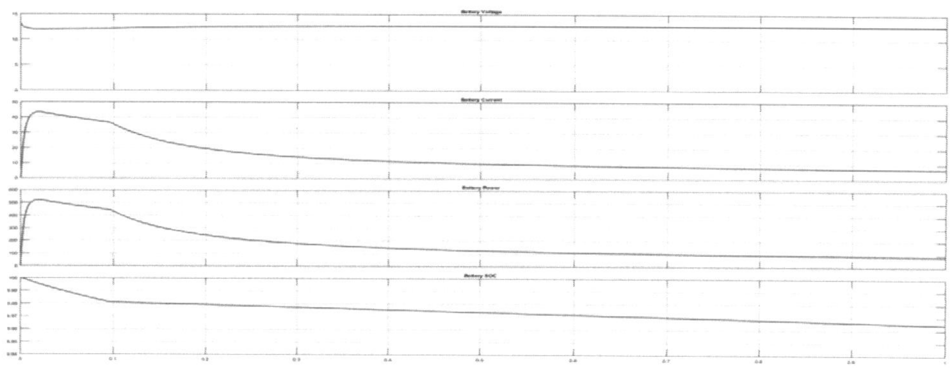

Figure 8. Battery Voltage, Current and Power

Table 3. Simulation result of the proposed system

S. No.	Irradiance (W/m^2)	Battery Voltage (V)	Battery Power (W)	Load Voltage (V)	Load Power (W)
1.	900	12.78	46.46	48.17	48.33
2.	700	12.74	46.69	48.17	48.34
3.	500	12.74	46.55	48.16	48.36

IOP Conf. Series: Materials Science and Engineering **994** (2020) 012011 doi:10.1088/1757-899X/994/1/012011

5. Hardware Model

Figure 9. Hardware Model of proposed system

MPPT boost converter employ a MOSFET power switch and switching ON and OFF duty cycle of the power switch will decide the power output from PV panel. Duty or duty cycle is ratio of ON time and Total Time Period of switching signal. Therefore, the value of the Duty should be at an optimum value in order to achieve MPPT. In order to determine the optimum Duty an optimization technique can be employed with an objective function of maximum power. The optimization technique will search the best optimum value for Duty in the range 0 to 1 in order to achieve the objective function. A number of optimization techniques are developed in the past. The proposed Grey Wolf Optimization technique is the best technique in selecting the best optimum duty for achieving the maximum power from the solar panel.

5.1 Minimum ISE

Quadratic boost converter employ a MOSFET power switch and switching ON and OFF duty cycle of the power switch will decide the voltage maintained across the load. PI controller is employed to generate the ON time based on the error between set voltage and actual voltage across load. The best values of k_p and k_i are estimated using GWO optimization technique for minimum error (minimum ISE).

5.2 Multi Objective GWO

Multi objective optimization techniques are objective functions or variables more than one. In this system there are two objectives: Maximum power and Minimum ISE. The two objectives are combined to single objective by converting them to per unit values by dividing the respective values by their expected maximum value (or base value) and taking sum of the two per unit objective value as single objective for optimization. Here one objective is maximum and the other is minimum and so inverse of minimum ISE is considered while combining as single objective so that optimization can be performed for maximum value.

6. Conclusion

Closed loop control of PV-Battery system by using multi objective GWO (Grey Wolf Optimization) technique is performed and implemented in hardware. Thus, maximum power is extracted even under partial shading condition. In future the system can be integrated to standalone micro-grid as part of hybrid power system.

References

[1] H Patel and V Agarwal Maximum power point tracking scheme for PV systems operating under partially shaded conditions IEEE Trans Ind Electron vol 55 no 4 pp 302–310 Apr 2008

[2] M A Elgendy B Zahawi and D J Atkinson Assessment of the incremental conductance maximum power point tracking algorithm IEEE Trans Sustain Energy vol 4 no 1 pp 108–117 Jan. 2013

[3] K S Tey and S Mekhilef Modified incremental conductance algorithm for photovoltaic system under partial shading conditions and load variation IEEE Trans Ind Electron vol 61 no 10 pp 5384–5392, Oct 2014

[4] S Mirjalili S M Mirjalili and A Lewis Grey wolf optimizer Adv Eng Software vol 69 pp. 46–61 2014

[5] Kinattingal Sundareswaran Peddapati Sankar Sankaran Palani MPPT of PV Systems Under Partial Shaded Conditions through a Colony of Flashing Fireflies IEEE Trans. on Energy Conversion vol 29 no 2 June 2014

[6] Kinattingal Sundareswaran Peddapati Sankar PSR Nayak Sishaj P Simon,and Sankaran Palani Enhanced Energy Output From a PV system Under Partial Shading Conditions Through Artificial Bee Colony IEEE Trans. on Sustainable Energy, vol.6, no.1, Jan2015

[7] Kok Soon Tey and SaadMekhilef Modified Incremental Conductance Algorithm for Photovoltaic System under Partial Shading Conditions and Load Variation IEEE Trans. on Industrial Electronics vol 61no.10 October 2014

[8] N.Pandiarajan and RanganathMuthu Mathematical Modeling of Photovoltaic Module with simulink" International Conference on Electrical Energy Systems (ICEES 2011), 3-5 Jan 2011

[9] J.Jayaudhaya and P.Rangarajan Comparison of Optimization of Maximum Power Point Tracking for PV System International Journal of Applied Engineering Research, ISSN 0973-4562 Vol 10 No.44 (2015)

[10] Jun Sun Vasile Palade Xiaojun Wu and Wei Fang Multiple Sequence Alignment with Hidden Markov Models Learned by Random Drift Particle Swarm Optimization IEEE/ACM Transactions on Computational Biology and Bioinformatics Vol 11 No 1 January/February 2014

[11] KashifIahaque and Zainal Salam A Deterministic Particle Swarm Optimization Maximum Power Point Tracker for Photovoltaic System under Partial Shading Condition IEEETrans on Industrial Electronics, Vol.60, No.8, August 2013

[12] Dian PR M S Siti and SY Siti Particle Swarm Optimization: Technique, System and Challenges International Journal of Computer Applications 14 (2011) No 1 19-27

[13] Yiqiong Yuan Jun Sun Dongmei Zhou Jianan Sun Multi-Objective Random Drift Particle Swarm Optimization Algorithm Based on RDPSO and Crowding Distance Sorting World Academy of Science Engineering and Technology ,International Journal of Mathematical, Physical, Electrical and Computer Engineering Vol:10,No:4,2016

[14] Seyedali Mirjalili Grey Wolf Optimizer Advances in Engineering Software vol 69 pp 46-61, 2014

[15] Ahmed A Elserougi, Mohamed S Diab Ahmed M.Massoud Ayman S.Abdel-Khalik and Shehab Ahmed A Switched PV Approach for Extracted Maximum Power Enhancement of PV Arrays During Partial Shading IEEE Transactions on Sustainable Energy Vol 6 No 3 July 2015

[16] Dhanagopal R Krishanamurthi V and Rajaram A 2017 A low power robust cascade error detection in high level function verification - RCED International Journal of Control Theory and Applications, Volume 10(20) pp. 15-24

[17] Feng Wang Ruifeng Gou Tianhua Zhu Yan Yang and Fang Zhuo Comparison of DMPPT PV System with Different Topologies 2016 International Conference on Electricity Distribution, 10-13 Aug 2016

Design and Modelling of Multi-Source Self-Charging Electric Bike

Mr.Thiyagesan M[1], Ms.Akshaya R Nair[2], Mr.Akshay Kumar[3], Ms. Vanitha r[4]
[1]Assistant Professor, Department of Electrical and Electronics Engineering, R.M.K ENGINEERING COLLEGE, Kavaraipettai, Tamilnadu, India.
[2,3]UG scholar, Department of Electrical and Electronics Engineering, R.M.K ENGINEERING COLLEGE, Kavaraipettai, Tamilnadu, India.
[4]Assistant Professor, Department of Electrical and Electronics Engineering, R.M.D ENGINEERING COLLEGE, Kavaraipettai, Tamilnadu, India.

[1] mtn.eee@rmkec.ac.in , [2] aksh18105.ee@rmkec.ac.in, [3]aksh18104.ee@rmkec.ac.in, [4]Vanitha.eee@rmd.ac.in

Abstract. The concept of electric vehicles is not new to us, the 1st documented electric vehicle was in the late 1890s, electric vehicles were far forgotten and suppressed for so long because of the domination of conventional vehicles in the commercial market. Conventional vehicles are fuelled by petroleum products which are non-renewable and not much environment friendly. We have reached a situation where we cannot depend on these fast depleting non-renewable sources of energy for better future. Hence, we are in need to develop and utilize the renewable sources of energy that are less/ non-polluting and does not get depleted. Only by the late 1960s, people started to realize the need for electric vehicles. Electric vehicles are a better replacement or the conventional vehicles and are more efficient and Eco-friendlier.
Keywords: *Electric Bike, Electric Motorcycle, Solar Powered Electric Vehicle, Self- Charging Electric bike, Regenerative Motor, Electrical Control Unit.*

1. Introduction

The problem faced by the electric vehicles to sustain in the commercial market are lack of charging ports available in the country and the time taken to charge the batteries which makes it a questionable option for the public. This problem had been overcome by making an attempt to design, develop and fabricate an electric vehicle prototype with dual battery setup, which charges itself while running, thereby introducing a system which makes the most of the power produced by the different sources. In addition to this, an Electrical Control Unit powered by Arduino is used to switch the batteries in run time so that one of them gets charged while the other powers the drive motor. The chassis of a commercially available bicycle is modified as per the requirements of the system components. The system consists of Lithium-ion batteries, Solar panel, regenerative motor, DC motor, motor controller and Electrical Control Unit. The controller unit designed by us makes this prototype more efficient than the commercially available electric vehicles in the market. The developed prototype is then tested under several circumstances to prove its efficiency. This method can be implemented to any type of electric vehicles such as bikes, cars, trucks, etc. the prototype which was developed severs as an example to prove the increase in efficiency for electric vehicles.

2. Design and Development

2.1. Electronic Control Unit (ECU)

Electronic Control Unit acts as brain of the project taking decisions based on the conditions prevailing. ECU used in the project's compromises of Arduino UNO board, 250V Relays and Voltmeter modules. When the bike is switched on ECU gets powered on and queries the voltages from both the batteries with the help of voltmeter modules, and the Arduino makes the decision i.e. it charges the battery with lowest voltage and the uses the other battery to power the motor, relays play the key role in this switching. When the running battery's charge comes down to 30% the ECU again switches the batteries i.e. the battery currently in use gets charged and battery in charging is used to power the motor. Circuit diagram for the same is shown in Fig 2.1

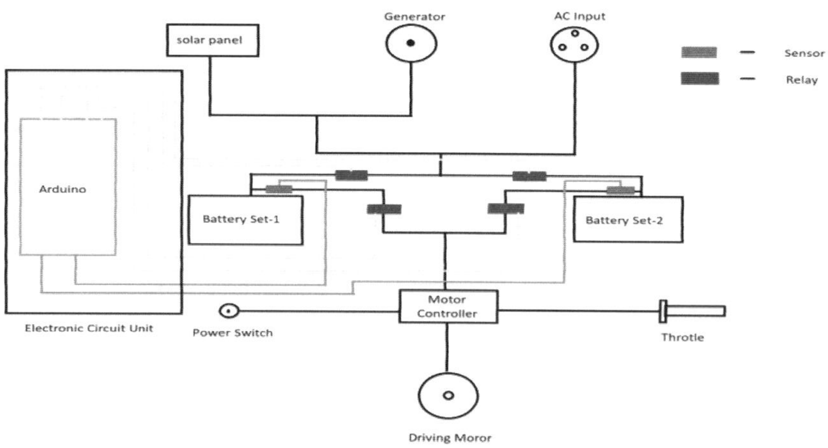

Figure 2.1. Circuit Diagram

This switching helps in efficient use of power from batteries as batteries tend to heat up when used simultaneously for charging and discharging.

2.2. Designing

The aim of this paper is to construct an e-Bike that can be charged from multiple sources of energy and to intelligently switch the batteries to prolong their lifetime and efficiency. The Components used for this e - Bike are as follows

1. DC Motor (24V, 250W)
2. Motor Controller
3. Motor Clamp
4. Batteries (4no, 12V,7.5Ah)
5. Arduino UNO Board
6. 2 Channel 250V Arduino Relays
7. Regenerative Motor
8. Voltmeter Module for Arduino
9. Solar Panel
10. Bicycle, the model is developed and the ECU is simulated using the simulation software

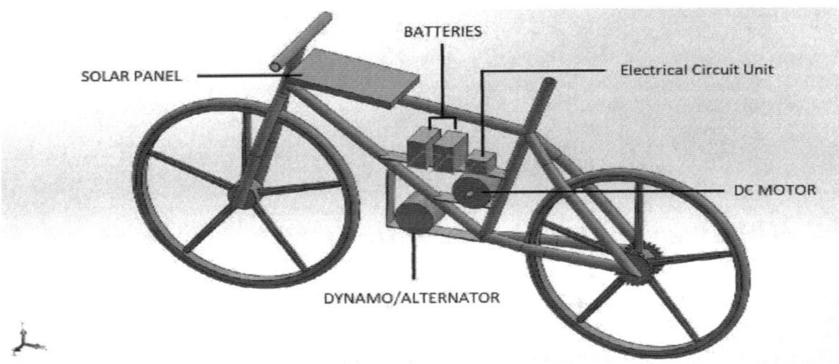

Figure 2.2. Simulation Model

The bicycle is modified to make is suitable for project. A base is created for batteries using a zinc sheet and mild steel. The back wheel is modified to accommodate two sprockets one for motor and the other for regenerative motor, and a luggage carrier is attached so that it can used as a base for solar panel. Motor clamp is attached to the carrier and the DC motor is mounted on it.The batteries and solar panel are mounted on their respective places and the ECU and Motor controller are placed underneath the battery base and ECU is circuited.

3. Results & Discussions

Several tests was conducted to obtain optimum results. The positioning of solar panel is needed to obtain optimum output. But as the prototype will be in motion, positioning of solar panel at a fixed position and angle is not practical. By keeping the speed constant, we found out the discharge rate of the batteries. It is not recommended to discharge a battery below 25% because of the life reduction of the batteries due to excessive usage. Hence keeping 30% as the min state of charge of the batteries for safety purpose, we found the following results

Figure 3.1. Developed Model

3.1 Test Run using only One Set of Batteries

- First test run was done with only one set of batteries that power the system till it fully discharges (30%).
- The batteries connection was let to cut when the state of charge reached 30% using relay in Arduino circuiting.
- The time taken to fully discharge one set of batteries (30%) was found to be around 29 min.

Table 1 Results Obtained in Test Run using only One set of Batteries

Trail No	Initial Voltage	Final Voltage	Distance Travelled	Top Speed	Average Speed
1	24.2	23.8	7.5	17	15
2	24.1	23.8	6.45	16	13
3	24.2	23.8	6.8	17	14

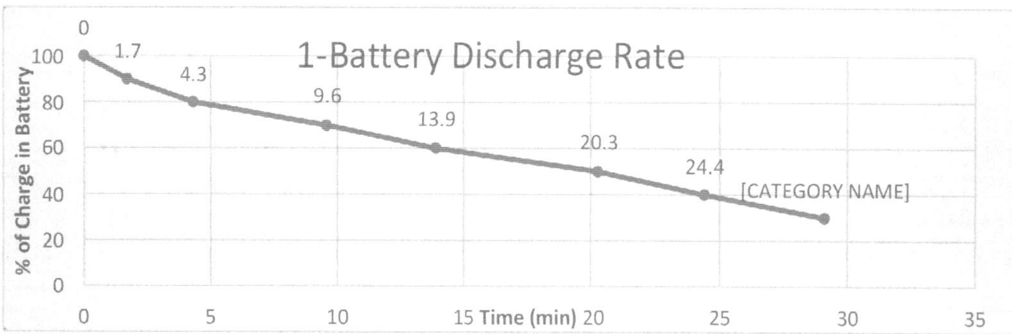

Figure 3.2. Time (Min) Vs Percentage Of Battery Charge

IOP Conf. Series: Materials Science and Engineering **994** (2020) 012012 doi:10.1088/1757-899X/994/1/012012

3.2 Test Run using Two set of Batteries (continuous discharge)

- Next test run was done with both the sets of batteries as one set of battery powers the system after the other set is discharged completely (30%).
- Both sets of batteries were connected to the Arduino using a relay and was programmed such a way that one set of batteries will power the system while the other set will be idle till the 1st set of batteries completely discharges (30%) and as soon as the 1st set of batteries discharges the 2nd set of batteries power the system till it discharges completely (30%).
- The time taken for both the batteries to get discharge was found to be around 59 min.

Table 2 Results Obtained in Test Run using two set of Batteries

Trail No	Initial Voltage of Battery 1	Initial Voltage of Battery 2	Final Voltage	Distance Travelled	Top Speed
1	24.2	24.2	23.8	15	17
2	24.1	24.2	23.8	14	16
3	24.2	24	23.8	14	17

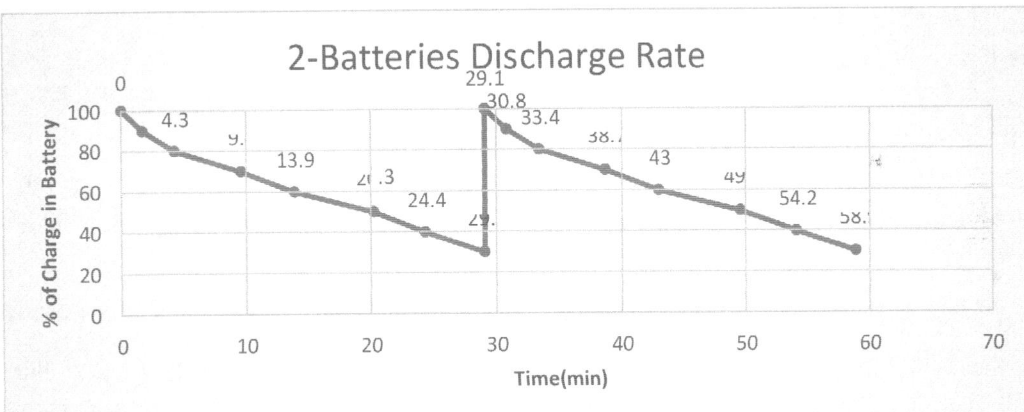

Figure 3.3. Time (Min) Vs Percentage Of Battery Charge

3.3 Test Run using both the Set of Batteries (Using ECU + Charging)

- The final test run was done with both the batteries powering the system as described.
- Both sets of batteries were connected to the Arduino using two sets of relays and was programmed such a way that 1st set of battery powers the system till its state od charge reaches 30%, and as soon as it reaches
- The threshold the powering of the system is switched to the 2nd set of batteries while the 1st set of batteries gets charged using the different sources mentioned before. This cycle is repeated until both the sets of batteries reaches their threshold.
- The time taken for both the batteries to get discharge was found to be around 74.1 min

Table 3 Results Obtained in Test Run using both the set of Batteries

Trail No	Initial Voltage of Battery 1	Initial Voltage of Battery 2	Final Voltage	Distance Travelled	Top Speed	Average Speed
1	24.2	24.2	23.8	18.52	17	15
2	24.1	24.2	23.8	17.2	16	14
3	24.2	24.2	23.8	18.52	17	15

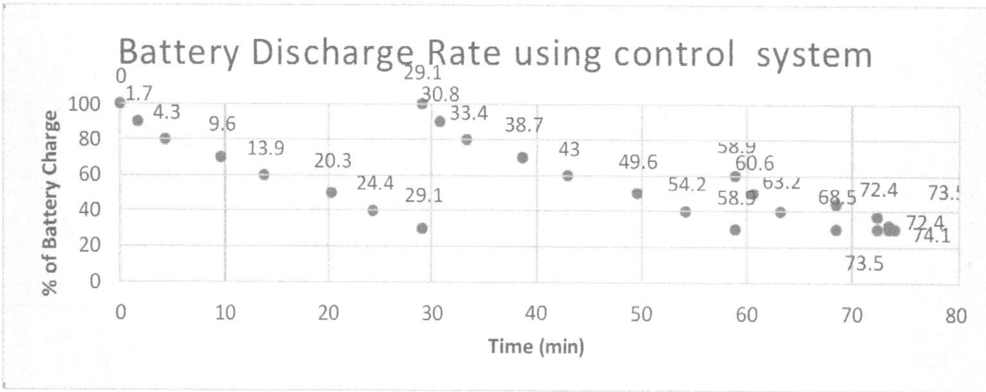

Figure 3.4. Time (Min) Vs Percentage of Battery Charge

From the above test runs we are clear that the method used to charge both the sets of batteries, while running had a great impact on the time and distance an electric vehicle can run compared to any other present type of methods used. From comparing the above result, it proves that our model has produced 20.5% more efficiency.

4. Conclusion

The controller unit makes an electric vehicle more efficient than the commercially available in the market. We developed the prototype and tested under several circumstances to prove its efficiency. Due to the cost constraint, affordability and availability, the components were selected with the minimum specifications. Components with the higher specification will yield higher performance and the efficiency of the vehicle. The observation from the tested results proves that the efficiency of our prototype model has increased by 20.5%. This efficiency can be further increased by employing components with higher specifications. Usage of super/ultra-capacitors as a battery support unit will also help in increasing efficiency.

References

[1] McLoughlin I Narendra I Koh L Nguyen Q Seshadri B Zeng W and Yao C 2011 Campus Mobility for the Future: The Electric Bicycle Journal of Transportation Technologies

[2] FOGELBERG F 2014 Solar Powered Bike Sharing System with Electric Bikes Göteborg: Department of Energy and Environment Division of Environmental System Analysis CHALMERS UNIVERSITY OF TECHNOLOGY

[3] Arun Eldho Alias Geo Mathew Manu G Melvin Thomas and Praveen Paul V 2015 Energy Efficient Hybrid Electric Bike with Multi-Transmission System. In National Conference on Recent Advances in Electrical & Electronics Engineering Kothamangalam Kerala International Journal of Advanced Research in Electrical Electronics and Instrumentation Engineering

[4] Sivapragash C Shankar C Nageena M Reetha Devi B and Kiruthiga K 2015 An innovative solar-powered electric bicycle In International Conference on Science Technology Engineering & Management Journal of Chemical and Pharmaceutical Sciences

[5] Mishra Kartik V Gadhawe Shubham C Chaudhari Dhiraj Barve and Dr Shivprakash 2016 Design and Development of Solar Hybrid Bicycle International Journal of Current Engineering and Technology

[6] Yogesh Jadhav Gaurav Kale Shekhar Manghare and Sager Patil 2016 Self-Charging Electric Bicycle International Journal on Recent and Innovation Trends In Computing And Communication 4(4) Prashant Kadi & Shrirang Kulkarni 2016 Hybrid Powered Electric Bicycle International Journal for Scientific Research & Development

[7] Mragank Sharma Anshul Verma Gaurav Tiwari Sunny Singh and Mohd Nadeem 2016 Design and Fabrication of Regenerative Electric Bike International Journal of Engineering Science and Computing

[8] Amin Sayyad Madhuri Patil Rajshree Ambewadkar Preetam Ajagekar and Pooja Kukade 2016 Solar Powered Electric Bicycle with Kinetic Energy Restoration System (KERS) International Journal of Electrical Electronics and Data Communication

[9] Engr M Bilal Chouhary Engr Saiful Talpur Engr Taimoor Zafar Engr.Muhammad Hussain Engr Asif Raza Engr Habib and Hussain Zuberi 2016 Solar Powered Electric Bike European International Journal of Science and Technology

[10] Abagnale C Cardone M Iodice P Marialto R Strano S Terzo M and Vorraro G 2016 Design and Development of an Innovative E-Bike Energy Procedia vol 101 pp 774-781

IOP Conf. Series: Materials Science and Engineering **994** (2020) 012013 doi:10.1088/1757-899X/994/1/012013

Analysis of battery management system issues in electric vehicles

[1]V Karkuzhali, [2]Dr.P Rangarajan,[3]Dr.V Tamilselvi and [4]P Kavitha
[1]Assistant Professor Department of EEE, R.M.D Engineering College, Chennai, Tamilnadu, India
[2]Professor Department of EEE, R.M.D Engineering College, Chennai, Tamilnadu, India
[3]H.O.D Department of EEE, R.M.D Engineering College, Chennai, Tamilnadu, India
[4]Assistant Professor Department of EEE, R.M.D Engineering College, Chennai, Tamilnadu, India

[1]Vkarkuzhali.eee@rmd.ac.in,[2] hodeee@rmd.ac.in,[3] prr.eee@rmd.ac.in[4]kavitha.eee@rmd.ac.in

Abstract. Battery technology has dramatically advanced over a decade and many high performance batteries are being developed. Electric vehicles (EV) require high power batteries with suitable battery management systems (BMS) for safe and reliable operations. Intention of this paper is to discuss about the batteries used in electric vehicles and the key issues of battery management systems and to compare the Lithium ion (Li-ion) battery & Nickel metal hydride battery in terms of aging and effect of temperature using their state of charge (SOC) and open circuit voltage (OCV).

1. Introduction

Electric vehicles are under constant research area and researchers keep working on it for reduced fuel usage and to reduce CO_2 emission. EVs are classified as hybrid electric vehicles and battery electric vehicles. Batteries play a major role in smooth running of EVs. High power batteries require proper care and they should be sensed for their voltage, current and power. Improper operation of batteries like over charge, over discharge, over current, short circuit and extreme temperature might throw problems to the user. Hence proper BMS helps in overcoming these issues and provides a safety drive for electric vehicles.

Key technologies in the BMS of EV include battery modelling, state estimation, charging and discharging. A good BMS should safely protect the driver/operator by detecting unsafe operating conditions, protecting the cells from damage in failure cases, prolongs the life of battery in normal operating region and should inform the user about the battery details and its status of operation.[3]This paper also explains about the different batteries and their electrochemistry. A battery has to be charged, discharged and its parameters are to be estimated well for its good maintenance. The measurable variables such as voltage, current, temperature that varies with state of charge are required for accurate and robust SOC estimation. This paper explains the variation of OCV and internal resistance of the battery at different SOC of Lithium ion battery and NiMH battery. Variation of its internal resistance curve at room temperature gives us the need for modelling a battery based on thermal behaviour.

2. Battery types

There are two different categories of battery cells namely Primary cells and Secondary cells. In case of primary cells the electrochemical reaction that occurs during the discharge is not reversible. If we try to recharge a primary battery, the compounds that have been formed during discharge will not recombine into the original compounds that were present before discharge and therefore it's not rechargeable. This is the principle reason that primary cells are meant for only one time use. Secondary cells are special in the sense that their chemical reaction has been designed in such a way that it is completely reversible. The original chemical compounds that are changed during discharge can actually be reconstituted into its original form by the application of external potential between the electrodes that inject energy into the cell. Secondary cells can be discharged and recharged infinite times but its life is limited by degradation of the cells.

Some popular types of batteries are Dry cell, Alkaline cell, lithium-ion (Li-ion), lead-acid (PbA), Nickel-Cadmium (NiCd), Nickel-Metal Hydride (NiMH), Nickel Zinc, Zinc air.as shown in Table 1 with their electrochemistry.

Table 1: Different types of batteries and their electrochemistry

S.No	Electrochemistry	Nominal voltage	Negative electrode	Positive electrode	Electrolyte
1	Lead acid	2.1V	Pb	PbO_2	H_2SO_4
2	Dry cell	1.6V	Zn	MnO_2	$ZnCl_2$
3	Alkaline	1.5V	Zn	MnO_2	KoH
4	Nickel cadmium	1.35V	Cd	NiooH	KoH
5	Nickel Zinc	1.73V	Zn	NiooH	KoH
6	Zinc air	1.65V	Zn	O_2	KoH
7	Nickel metal Hydride	1.2V	H_2 in form of metal hydride	$Ni(OH)_2$	KoH
8	Lithium ion	3.7V	Graphite	LFP,LMO,LCO	$LiPF_6$

Among the above different types of batteries, NiMH and Li-ion batteries are highly preferred. NiMH has higher specific energy, energy density than cadmium electrode. They have higher capacity and longer life than NiCd batteries. Since it is free of cadmium they are considered as an environment friendly battery. The magic of NiMH cells is the negative electrode which is a rare earth hydrogen absorbing metal alloy. There is no electro chemical reaction taking place in the negative electrode that changes its structure when hydrogen goes in and comes out of the electrode.

The Li-ion cells also allow intercalation of lithium ions into crystalline lattice of graphite without changing its crystalline structure. Li-ion batteries have higher open circuit voltage which reduces the number of cells in a battery pack. Higher energy density makes the battery pack compact and high specific energy lightens the overall weight of the vehicle. They can be operated in a wide temperature range without decreasing its lifetime. Li-ion batteries can be charged and discharged at higher C-rate at normal operating conditions, so that a small battery pack can meet peak power requirements and absorb most regenerative energy.

Table 2: Basic requirements for NiMH and Li-ion batteries for EV applications

S.NO	REQUREMENT	NiMH	Li-ion
1	Specific energy	40-80Wh/Kg	130-200Wh/Kg
2	Energy density	90-160Wh/L	180-320Wh/L
3	Specific power	900-1600W/Kg	1200-4000W/Kg
4	Charge/discharge efficiency	80-95%	85-96%
5	Self-discharge rate	8-15% month	<5% month
6	Cycle durability	800-1200 cycles	1500-2000 cycles
7	Nominal cell voltage	$\cong$1.2V	$\cong$3.7V

3. Battery modelling

Battery system consists of battery cells. Depending on the requirement of output voltage, power and energy capacity for an EV, a battery pack contains many cells connected in series or parallel or both. Battery operation and its input-output parameters can be studied by modelling the battery. Proper model of the battery is required for a good BMS design; control and optimization. There are several methods available to model a battery, and the most widely used are electric model, thermal model and electro-thermal model.

3.1. Battery electric model

Electro-chemical model, reduced order electric model, equivalent circuit model, Data driven model are the types of battery electric model. Among these different models, the equivalent circuit model is widely adopted has the model structure is simple with less number of model parameters. In the equivalent circuit model battery electric behaviours can be studied by a combination of circuit components such as resistors, capacitors, voltage sources as given below.

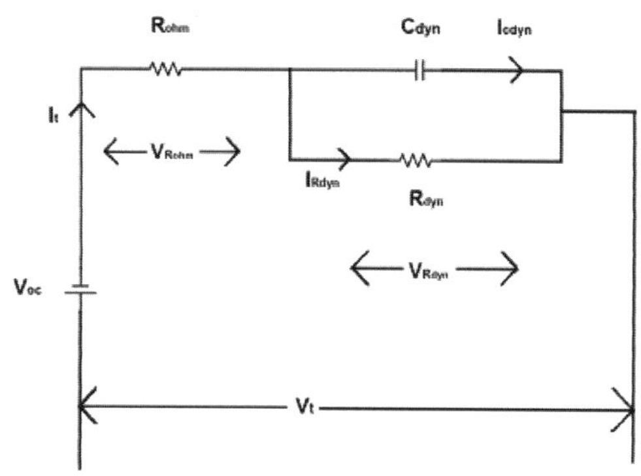

Figure 1. Battery model Equivalent circuit

The resistors and capacitors are related to charge transfer or diffusion processes. Model order is denoted by the number of RC networks as given by Kailong LIU et al[1]. First and second order models are popular and higher orders are of less importance. This model has better dynamic performance especially for State-Of-Charge (SOC) and power estimations. The other models are not discussed here.

From the above figure according to KVL

$$V_t = V_{OC} + V_{ohm} + V_{dyn} \quad (1)$$

$$V_{OC,cell} = V_O - \frac{R_g T}{n_e F} ln(Q) \quad (2)$$

$Q = f(SOC)$

$$V_{ohm} = I_t * R_{ohm} = I_t * R_{ohm}(SOC,T) \quad (3)$$

$$V_{dyn} = I_{Rdyn} * R_{dyn}(SOC,T) \quad (4)$$

$$I_{cdyn} = C_{dyn}(SOC,T)\frac{dV_{dyn}}{dt} \quad (5)$$

Terminal current $I_t = I_{Rdyn} + I_{cdyn}$ (6)

Where V_t is terminal voltage in volts, V_{ohm} is the voltage across R_{ohm}, V_{dyn} is dynamic voltage across the resistor. V_{OC} is the open circuit voltage .V_o is no load voltage, R_{dyn} is dynamic resistance. T is the battery temperature. F is faraday constant, n_e is the number of electron transferred in cell reaction and R_g is universal gas constant.[5]

Dynamic voltage V_{dyn} can be described by the differential equation

$$\frac{dV_{dyn}}{dt} + \frac{V_{dyn}}{R(SOC,T)*C_{dyn}(SOC,T)} = \frac{1}{C_{dyn}(SOC,T)} \tag{7}$$

Overall differential equation of electrical circuit

$$\frac{dV_t}{dt} + \frac{V_t}{C_{dyn}R_{dyn}} = R_{ohm} * \frac{dI}{dt} + \frac{R_{dyn}+R_{ohm}}{R_{dyn}*C_{dyn}} * I_t + \frac{V_{OC}}{R_{dyn}*C_{dyn}} \tag{8}$$

3.2. SOC calculations

SOC of a battery gives the percentage of available amount of energy over its maximum achievable amount.

For EV System Performance, Analysis and Simulation as well as design battery SOC calculation,

$$SOC(t) = SOC(t_i) + \frac{1}{CAP_{Ahr}*3600}\int_{t_i}^{t} I(t)\,\eta_{batt}(SOC,T,SignI(t))dt \tag{9}$$

$SOC(t_i)=SOC_i$ (initial SOC)

Where η_{batt} is the columbic efficiency of the battery, CAP_{Ahr} is the capacity of battery in ampere-hours

3.3. Modelling thermal behaviour

Heat generated by the battery is the heat generated by the resistors in the electric circuit model.
Heat generated by the resistors is

$H_{gen} = H_{Rohm}+H_{dynamic}+H_{react}$ (w) (10)

Ohmic resistance heat $H_{Rohm}=|I|*|V_{ohm}=I^2R_{ohm}(SOC,T)$ (w) (11)

Dynamic heat given by $H_{dynamic}=|I|*|V_{dyn}|=|I|*|\int_0^t [\frac{I(t)}{C_{dyn}(SOC,T)} - \frac{V_{dyn}(t)}{R_{dyn}(SOC,T)C_{dyn}(SOC,T)}]dt|$ (12)|

At initial condition $V_{dyn|t=0}=0$.The reaction heat is given by $H_{react}=\frac{S.I.\eta_{batt}.T}{F}$ (13)

Where ΔS is delta entropy of the reaction(J/mol-k), F is faraday constant=96,487 C/mol

Dissipated heat is $H_{dissipated}=(T_{coolant}-T)h_{bat}$ (14)

Battery temperature is given by $T(t) = T(t_i) + \int_{t_i}^{t} \frac{H_{gen}+H_{dissipated})}{C_{---}*M_{---}}dt$ (15)

where, $T_{coolant}$ is the temperature of coolant(K),
h_{bat} is heat transfer co-efficient of battery (W/K)
$T(t_i)$ is initial temperature of battery.Using the above equations a EV system battery electrical model can be implemented as given below.

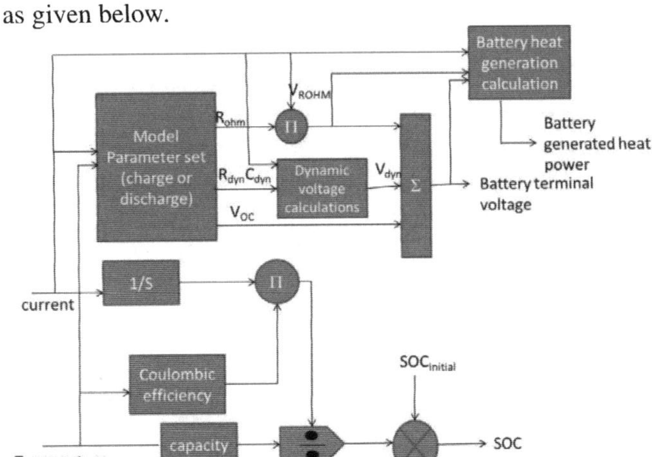

Figure 2. Battery model implementation

3.4. Comparison of SOC, OCV, temperature

SOC is used to determine the battery capacity. There are several methods available to determine SOC. The measurement of OCV is a voltage based SOC determination [16].OCV can be measured without load and this method is suitable for estimating initial SOC. OCV is the terminal voltage of the battery at no load and the battery is kept at rest for one hour. In this experiment conducted, a 400 V, 100Kw power capacity and 20 Kw energy capacity Amaron single NiMH battery cell is selected .This experiment is done at 20° centigrade. The OCV measurement for 100% SOC is done under the charging mode with a charging current of 1C.Experiment is started at the initial battery voltage of 0.12V.For every increase in 1% charge the voltage is noted. After 100% reach of SOC of battery at 1.2V, the battery was given a rest of 1 hr. There is no charging and discharging process for this 1 hr. There is no significant change in OCV. SOC and OCV plot is given below in fig 3.

Similarly a cylindrical LFP battery with the same rating of 400V, 100 Kw power capacity and 20 Kw energy capacity is selected. Using voltage based SOC determination OCV is noted. The experiment is started at initial battery voltage of 3.15V and at final 100% SOC capacity battery reached 3.46V.1 hour rest period is given. There is no significant drop in the voltage of the battery. SOC and OCV plot is given below in fig 3.

Using linear regression the curve fitting is done and the 4th degree polynomial equation is obtained as given below

4th degree LFP OCV $y = 8.709e\text{-}10*x^4 + 6.305e\text{-}07*x^3 - 0.0001258*x^2 + 0.007891*x + 3.163$

4th degree NiMH OCV :$y = - 3.698e\text{-}08*x^4 +1.005e\text{-}05*x^3 - 0.0009576*x^{2} + 0.04174*x + 0.2489$

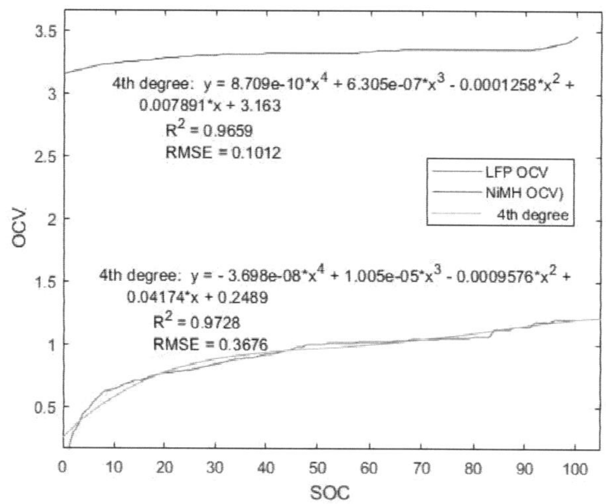

Figure 3. comparison Plot of SOC and OCV for LFP and NiMH

The coefficient of determination gives the goodness of curve fit with strong effect Size for both LFP and NiMH. The Root Mean Square error RMSE gives the residual measure of farness of data points from the regression line.

Table 3:R^2 and RMSE for batteries in comparison

S.no	LFP	NiMH
R^2	0.9659	0.9728
RMSE	0.1012	0.3676

The values of RMSE are very less and had proven that data is around the line of best fit. Variation of internal resistance at different SOC is also measured, the dynamic resistance as given in the battery model. As shown in fig 4 for cylindrical LFP battery at initial SOC the internal resistance is 3.13Ω. For every change in SOC the internal resistance is measured and at 100% capacity of SOC the internal resistance for LFP is 2.37Ω. With high resistance the battery takes less current to charge and there is a slow charging. With battery SOC increasing it is seen that the resistance decreases with increasing charging current so that battery can be charged to required voltage. After 80% SOC is reached the internal resistance again increases since the battery voltage would have sufficiently reached the required voltage. In NiMH batteries the dynamic resistance is low and minor increase in resistance is seen at 100%SOC as shown in figure 4.

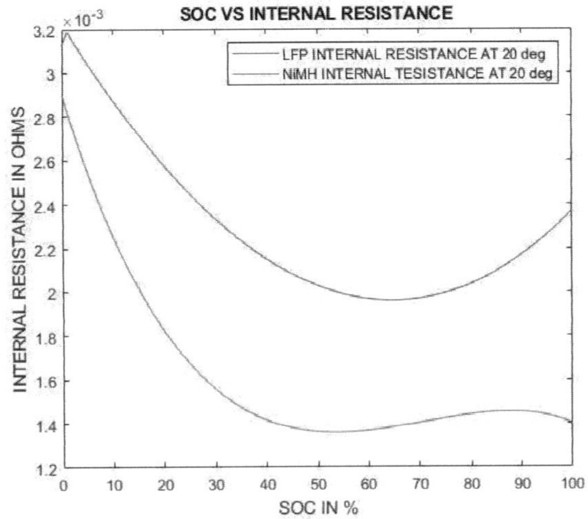

Figure 4. comparison Plot of SOC and internal resistance for LFP and NiMH

4. Conclusions

Comparing the graph of fig 3 of SOC and OCV curve LFP has high voltage and is of high preference over NiMH. Cylindrical LFP is of high choice because they are easy to manufacture The major drawback of $LiFePO_4$ –LFP batteries is the super flat slope of OCV and SOC graph as seen above ,that makes SOC estimation and balancing cell among the battery system challenging .Batteries need to have high energy density, low internal resistance and long cycle and calendar life. Hence well trained battery model together with suitable estimation methods can be adopted to achieve independent or joint state estimation of battery SOC or internal temperature along with their dynamic resistance. The research work in the field of battery charging estimation along with temperature management is needed as Lithium batteries gets easily heated up which reduces the life of batteries. Hence other chemical equivalent battery materials are needed to be considered to prolong battery usage to maximum

References

[1] Kailong LIU Kang LI Qiao PENG and Cheng Zhang 2013 A brief review on key technologies in the battery management system of electric vehicles Journal of Power Sources vol 226 pp 272–288

[2] Languang Lu Xuebing Han Jianqiu Li Jianfeng Hua and Minggao Ouyang 2013 A Review on the key issues for Lithium-ion Battery Management in Electric Vehicles, Journal of Power Sources vol 226pp 272-288

[3] Luca Buccolini Adrianna Ricci Cristiano Scavongelli, Giuseppe DeMaso-Gentile Simone Orcioni Massimo Conti Battery Management System (BMS) SImulation Environment for Electric Vehicles. 978-1-5090-2320-2/16 2016 IEEE.

[4] Daisy Ranawat M P R Prasad A Review on Electric Vehicles with perspective of Battery Management System (ICEECCOT) 14-15 December 2018 978-1-5386-5130-8/18, IEEE 1539

[5] Wei Liu Introduction to Hybrid Vehicle System Modeling and Control ISBN: 9781118407400
[6] James Larminie and John Lowry 2012 Electric Vehicle Technology Explained 2nd Edition ISBN: 978-1-119-94273-3
[7] Jeevak S Lokhande Dr P M Daigavhane Mithu Sarkar 2020 A Critical Approach Towards a Smarter Battery Management System for Electric Vehicle 4th International Conference on Trends in Electronics and Informatics (ICOEI)(48184), 2020
[8] Aruna P and Dr Vasan Prabhu V 2019 Review on Energy Management System of Electric Vehicles:2nd International Conference on Power and Embedded Drive Control (ICPEDC),
[9] Chunhua Zheng Weimin Li and Quan Liang 2018 An Energy Management Strategy of Hybrid Energy Storage Systems for Electric Vehicle Applications IEEE Transactions on Sustainable Energy, vol 9
[10] Rui Xiong Jiayi Cao Quanqing Yu Hongwen He and Fengchun Sun 2018 Critical Review on the Battery State of Charge Estimation Methods for Electric Vehicles IEEE Access vol 6 pp 1832 - 1843
[11] Jingshan Li Shiyu Zhou and Yehui Han 2016 Advances in Battery Manufacturing, Service and Management Systems ISBN:9781119056492
[12] Xi Zhang Jinling Lu Shifei Yuan and Jun Yang Xuan Zhou 2017 A novel method for identification of lithium-ion battery equivalent circuit model parameters considering electrochemical properties", Journal of Power Sources vol 345 pp 21-29
[13] Kuan-Ting Lee Min-Jhen Dai and Chiung-Cheng Chuang 2017 Temperature-compensated model for lithium-ion polymer batteries with extended Kalman filter state-of-charge estimation for an implantable charger IEEE Transactions on Industrial Electronics pp 589 - 596
[14] Junfu Li Lixin Wang Chao Lyu and Michael Pecht 2017 State of charge estimation based on a simplified electrochemical model for a single LiCoO2 battery and battery pack Energy vol 133 pp 572–583
[15] Min Ye Hui Guo and Binggang Cao 2017 A model-based adaptive state of charge estimator for a lithium-ion battery using an improved adaptive particle filter Applied Energy vol 190 pp 740–74
[16] Novie Ayub Windarko and Jaeho Choi SOC Estimation Based on OCV for NiMH Batteries Using an Improved Takacs Model Journal of Power Electronics Vol 10 No 2 2010
[17] Caiping Zhang Jiuchun Jiang Linjing Zhang, Sijia Liu Leyi Wang and Poh Chiang Loh 2016 A Generalized SOC-OCV Model for Lithium-Ion Batteries and the SOC Estimation for LNMCO Battery Energies vol 9 pp 900

Authors' background

Note:

[1] Mrs.V.Karkuzhali, B.E., M.E, is Assistant Professor in Department of Electrical and Electronics Engineering, since March 2010. She obtained her B.E (EEE) from Adhiyamaan Engineering College and M.E (PED) from Sri Venkateswara College of Engineering. She has been in the teaching profession for the past 8 years and has handled UG programme. She also has two years of industrial experience. Her areas of interest include Electric vehicles, circuit theory, Digital signal processing She has published a paper in various Conferences. She has attended many workshops & FDPs sponsored by AICTE related to her area of interest.

[2] Dr.P. RANGARAJAN M.E, Ph.d is Professor in Department of Electrical and Electronics Engineering, since october 2007.He has 28 years of experience. He is the recognised research supervisor of anna university. He has handled several funded projects and has patent received in his name. He has published several papers in national and international journal and conferences.

The advancing assessment of power system stability using smart grid technology

Murat Merekenov1, Karmel Tokhtibakiyev2, Anur Bektimirov3, Rassim Nigmatullin4

[1,3] PhD student, Almaty University of Power Engineering and Telecommunication merekenovmd@mail.ru[1], abektim@gmail.com[3]
[2,4]PhD, associate professor, Almaty University of Power Engineering and Telecommunication tkarmel@mail.ru[2], rasimnigmatullin@mail.ru[4]

Abstract. The paper proposes new method for assessment of power system stability. New method based on advancing approach to determination of stability limits. The paper represents more accurate and reliable ways to determine limits for power transmission through the grid. Nowadays, the static stability limits formed by using special software which solves nonlinear equations of the electrical system. Power systems control power flows through the power lines for avoiding black outs by using automated control systems. Automation thresholds depend on the maximum power transmitted through the line or section, and the maximum transmitted power is calculated using the iterative methods. But the calculations in this way are not sufficiently accurate and flexible, since the calculations are done for the entire system, and not for each section individually. And it means the limits on certain sections and lines can vary. The proposed method differs from the existing one in a way it determines the limits using the angle shift between the voltage phasors in the nodes in real time. It makes possible to more deeply and adaptively manage. Also proposed new control algorithms which determines stability boundaries in advance. Algorithm was verified through experimental studies on physical model of the power system.
Keywords: Jacobian matrix, WAMS, predicting algorithm, Smart Grid, stability control.

1. Introduction

Electrical energy transmission through alternating current (AC) power lines is limited by the amount of power flows on tie. Limit is determined by the capacity of an electrical grid according to regulatory requirements by reliability assurance. Today, calculation software defines restriction by search of static stability limits of electrical power system for the worst operating conditions. This does not allow to use transfer capacity fully in a wide range of scheme and mode of grid functioning. The existing restriction on a power flow leads to derogation from optimal conditions of maintaining the grid conditions and, as a result, to decrease in economic indicators of power supply systems. The calculations are usually made for the worst scenario because while using calculation software it is impossible to define power flow restrictions in advance.

One of the solutions of this problem is the use of the principles of Smart Grid, on the basis of calculation performance on the computational model formed in a controlling circuit.

Implementation of management principle in electrical power systems require the solution of the following tasks:

- supplying observability of a system in the conditions of grid at current moment. It carried out by the sufficient number of measurements, equipment placing in controlled nodes of grid and the sufficient speed of data collection and transmission.

- development of software and hardware complex, which function is using details of the phasor measurement system in real time.

For the organization of a control and management system in case of Kazakhstan's National Electrical Grid (NEG) Smart Grid technology based synchronized phasor measurement system or Wide Area Monitoring System (WAMS) is used [1], [2]. WAMS gives the possibility not only to organize passive monitoring of power system parameters, but also to create adaptive control systems of the power supply system operating condition, to predict the dangerous conditions leading to stability violation. It is planned to use WAMS following by creation the system Wide Area Control System (WACS) for inventing adaptive control, identification and damping of low-frequency oscillations (LFO) on «North-South» transit of Kazakhstan's NEG.

Today in China and in the countries of Europe, America use of the WAMS [3]-[5] system, on the basis of the phasor measurement unit (PMU) devices allows to carry out identification of the different operating conditions parameter fluctuations, which are relate to energy power system, such as mutual angles, frequencies, power flows through inter-area lines and identifications of LFO [6],[7],[11].

2. Stability control methods

2.1 The existing method of the stability margin calculation

The existing methods and algorithms of defining stability are based on the principles of using model out of a controlling circuit. This method excludes adaptability of management to the real grid conditions.

Today, calculation of electrical power flow limited by static stability conditions in tie, it is carried out by changing of loading in the software as RastrWIN, Digsilent, ETAP and etc. At the same time trajectories of loading representing the sequences of steady-state grid conditions which at change of some group of parameters allow reaching edge of static stability area are considered. The ways of loading are usually balanced on power where frequency remains almost invariable.

According to standards [8] of power system stability, the safety factor of static (aperiodic) stability on active power in tie is determined by the following equation:

$$K_{SF} = \frac{P_{CV} - (P + \Delta P_{nro})}{P_{CV}} \tag{1}$$

Where: P_{CV} - calculated critical value of active power flow by static stability in consideration tie;

P – operating electrical power flow in tie , P > 0;

ΔP_{nro} - amplifier of non-regular oscillations of active power between two area.

The value of non-regular oscillations identified for each ties according to measurements. In absence of the measurements data the value of non-regular oscillations can be calculated by the following equation:

$$\Delta P_{nro} = K\sqrt{\frac{P_{D1}P_{D2}}{P_{D1}+P_{D2}}} \tag{2}$$

Where: P_{D1}, P_{D2} – summary demanding in each area connected by consideration tie, MW.

Coefficient K, accepted - 1,5 at manual regulation and 0,75 at automatic control of output power flow in power plants.

Under the condition of power system stability, the minimum safety factor of static aperiodic stability on active power in ties are normalized. For normal grid operating condition the safety factor (minimum margin factor) is 0,2 or 20% and for post-emergency condition the minimum margin factor is 0,08 or 8% calculated by the equation (1).

At the same time the margin in 20% is defined as the maximum allowed power flow (MAPF). MAPF determined by the dispatching center of each national grid, providing admissible parameters in the different operating conditions. The margin in 8% is defined as the emergency allowed power flow (EAPF) and determined by the dispatching center for 20 minutes after black.

There are several issues of existing stability control method:

- huge amount calculations by loadings for each type of operating conditions;
- Calculations are carried out for the worst scenario that leads to the overestimated stability margin.
- at computational loading there are unaccountable factors which happen in real power system at the rate of process that affects quality of calculations;
- the grid parameters input are entered by personnel, that means there are subjective factors.

2.2 The offered stability control method

One of the solutions for the existing disadvantages of stability estimation methods is use of the control principles according to the real time measurements. With information technology development there appear new opportunities for more precisely and in real time control of the condition of an electrical power system. A perspective scope of synchronized phasor measurements is identification of model of the power supply system reflecting the main properties of grid in real time, and further application of this equivalent of a system for the analysis of its properties. Identification of an equivalent in real time allows avoid the above described disadvantages of traditional methods of the electric grid conditions monitoring. First of all, identification on the basis of accurate synchronized measurements, comprises properties of a real system unlike predetermined model. Secondly, due to identification in real time there is an opportunity to analyze the current situation, but not "the worst scenario". Development of the ideas, will allow more precisely and in real time control of the power supply system conditions to stability limit.

For electrical grid the idea of management according to synchronized measurements is methodized by calculations of stability in the conditions of real time by means of the control unit with use of computational model. This model is formed in a controlling circuit for determination of dangerous ties in systemically important grid and capacity at adaptive reaction to changes of the scheme and the condition of an electrical power system.

In the paper the algorithm of a static stability limit predicting on the current measurements of parameters of the grid condition in real time without procedure of performance of calculations with loading for the chosen tie is offered. It will give the chance to operation personnel to take actions for prevention of loss of electrical power system stability in advance. Besides, it allows to create digital active and adaptive control system, providing and predicting of mode reliability and stability of grid.

For a static aperiodic stability limit forecasting the property of a matrix Jacobian determinant is used. In the steady operating condition, the determinant has the positive sign, and at loss of stability passes through zero and becomes negative.

The tasks for developing the new stability control method:

1. Justification of a static overloads forecasting method on inter-area connections with use of the adaptive grid parameters identification when loading the conditions at sets of power on tie;
2. Use of approximation methods by search of the operating condition limit;
3. Development of an identification algorithm of the operating condition limit with use of a matrix of Jacobian;
4. Formation of a Jacobian matrix according to synchronized phasor measurements in a controlling circuit;
5. Calculation of coefficients of the approximating function by the least-squares method;
6. Determination of a static stability limit and reserve in real time;
7. Development of algorithms in the WAMS software.

3. Prediction algorithm

The base of the steady-state stability prediction algorithm is the method of weighted least-squares. The balanced electric operating condition is defined as a result of minimization of the following criterion function:

$$f = \sum_{i=1}^{n} r_i \left[z_i - \overline{z}_i(\dot{u}) \right]^2 \tag{3}$$

where n -number of measurements;
z_i - i value of measurement
$\bar{z}_i(\dot{u})$ - function reflecting dependence between i measurement and independent parameters – components of complex nodal voltage;
r_i - weight coefficient characterizing measurement i accuracy.

The technology of phasor measurement of the grid parameters conditions, allows to expand the list of the parameters received for estimation of a state due to accounting of voltage phases in various nodal points of power supply system. Previously such parameters were not available.

For formation of a system computational model (formation of a Jacobian matrix and calculation of determinant) it needs to determine several parameters According to the system of phasor measurements following parameters of the grid condition are identified:

- Voltage in bus of an electrical grid (U1…Un);
- Angle δ between phasor of voltage in buses;
- Power flows on Pij lines;

Jacobian's matrix is formed in next way [9-10]:

$$\frac{\partial W}{\partial \delta}=\begin{bmatrix} \frac{\partial w_1}{\partial \delta_1} & \cdots & \frac{\partial w_1}{\partial \delta_{m-1}} \\ \cdots & \cdots & \cdots \\ \frac{\partial w_{m-1}}{\partial \delta_1} & \cdots & \frac{\partial w_{m-1}}{\partial \delta_{m-1}} \end{bmatrix} \tag{4}$$

Here the equation of the steady state grid condition (W) shown in the form of power balance:

$$W_i = P_i - U_i^2 Y_i \sin\delta_i + \sum_{\substack{j=1 \\ j\neq i}}^{m} U_i U_j Y_{ij} \sin(\delta_i - \delta_j) = 0 \tag{5}$$

Respectively Jacobian is called the determinant of a matrix (4)

$$J=\begin{vmatrix} \frac{\partial w_1}{\partial \delta_1} & \cdots & \frac{\partial w_1}{\partial \delta_{m-1}} \\ \cdots & \cdots & \cdots \\ \frac{\partial w_{m-1}}{\partial \delta_1} & \cdots & \frac{\partial w_{m-1}}{\partial \delta_{m-1}} \end{vmatrix} \tag{6}$$

Where Ui and δi values come from synchronized phasor measurements data, conduction (Yij) parameters are determined by data of the equivalent circuit of the working model.

Necessary condition of static instability with use of a Jacobian matrix is transition through zero of Jacobian:

$$J < 0 \tag{7}$$

For determination of the approximating function parameters, the stability indicator in the form of a Jacobian matrix determinant allows to build a certain dependence of determinant from measured variable of the mode from the generated power. This approximating surface can be used for forecasting of a static stability limit by using the condition (7) and property of continuity and concavity of a boundary surface of stability (or curve dependence of a stability indicator from parameter of independent variable).

Coefficients of the approximating function or indicator of stability in the form of Jacobian are defined by method of the least-squares.

The mean of the least-squares method when determining coefficients of the approximating function is as follows. Let set n of unknown variables, $f_i(x)$, $i=1,...,m$ $m>n$ -set of functions from this set of variables. The task consists in selection of such values x that values of these functions were closest to

some y_i values. In essence the specified sense of the maximum proximity of the left and right parts of a system is about the solution of the redefined system of the equations $f_i(x) = y_i, = 1,...,m$.
Thus, the essence of the least-squares method can be expressed as follows:

$$\sum_i e_i^2 = \sum_i (y_i - f_i(x))^2 \rightarrow \min_x \quad (8)$$

Approximation accuracy in limits function existence, obviously, depends on quantity or the amount of measurements and also the measurements frequency. In this case approximation is carried out periodically in process of the new measurements receipt. If measurements are made for the steady state conditions, then intervals of determination of dependence remain in the same ranges that are not provided with conditions of dependence specification.
For determining of a static stability limit, maximum allowed power flow and abnormally allowable limits, according to a condition (7) for dependence of an indicator of stability from the measured parameter (power flow) needs to monitor when Jacobian passes through zero and to become negative. The found value of variable power at which the system will pass into the unstable condition is a limit of static stability. For definition of stability limit the safety factor of static stability is entered (1).
When determining of the approximating function within existence of this dependence the accuracy of approximation depends on the measurements scope. The most accurate approximation becomes at approach of a system to extreme values. As shown below at approach of a system to a limit of K_{SF} decreases and becomes less, than a standard limit in 20%.
For definition of the actual limit static stability of a system the following algorithm shown on figure 1 is used.

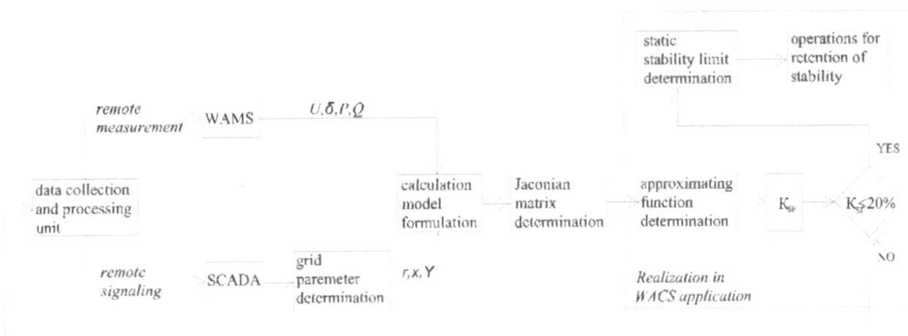

Figure 1. The control flow chart of static stability prediction in real time.

4 Experimental studies

The algorithm was tested in the physical model of the power supply system consisting of two generators (Figure 2 & 3) and the infinitive bus which connected with power lines (Fig 3). For experimental studies on physical model MΓ-5 (G1) generators and MK-3 (G2) were used.

Figure 2. Generator model MΓ5-1500

Figure 3. Generator model MK3-1500

Figure 4. *Three-phase element of the power line model*

All calculations was given to 500 kV system parameters, to show as close as possible to real system. The limit of static aperiodic stability of the system shown in figure 4 was defined by an experiment. For definition of a limit of static aperiodic stability the operating condition was made heavier by increase in power of generation on the G1 generator.

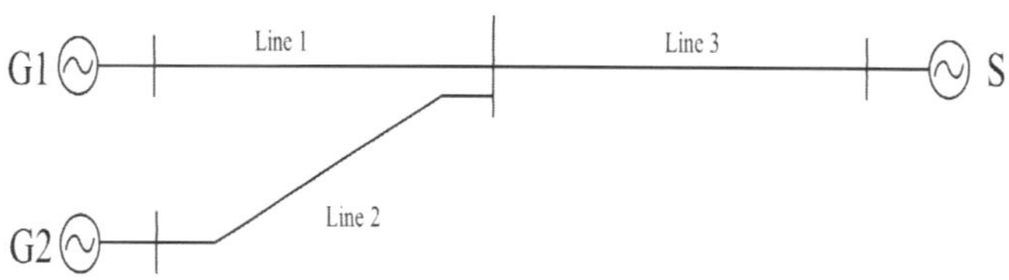

Figure 5. Equivalent circuit of model

Table 1. Parameters of synchronous generators

#	Sym bol	Parameter	Generator	
			МК-3	МГ-5
1	S	Apparent power [kVA]	3	5
2	P	Active power [kW]	-	4
3	U_r	Rated voltage [V]	230	230
4	I_r	Rated current [A]	7,52	12,6
5	cos φ	Power coefficient	0	0,8
6	n	Rated rotation [rot/min]	1500	1500
7	i_{foc}	Excitation current o.c. $U=U_r$ [A]	0,56	0,88
8	i_{fsc}	Excitation current s.c. $I=I_r$ [A]		0,4
9	x_s	Peltier reactive impedance [r.u]	0,02	0,07

10	x_d	d axis reactance [r.u]	1,24	1,17
11	x_q	q axis reactance [r.u]	0,66	1,04
12	x_2	Negative sequence reactance [r.u]	0,07	0,06
13	x_0	Zero sequence reactance [r.u]	0,02	0,03
14	x'_d	d axis transient reactance [r.u]	0,17	0,14
15	x''_d	d axis sub-transient reactance [r.u]	0,07	0,08
16	x''_q	q axis sub-transient reactance [r.u]		0,09
17	T_{d0}	Time constant of excitation circuit for open stator circuit [s]	2,8	1,56
18	T'_d	Time constant of transient component of stator current [s]	0,39	0,09
19	T''_d	Time constant of sub-transient component of stator current [s]		0,01
20	T_a	Time constant of aperiodic component current [s]	0,05	0,03
21	r_0	Resistance of stator [ohm]	0,08	0,07
22	GD^2	Centrifugal moment of machine without disk [kg·m^2]	2,44	8,92
23	GD^2	Centrifugal moment of machine with disk [kg·m^2]		25,7
24	T_{jmin}	Time constant of machine without disk [s]	5,0	10,9
25	T_{jmax}	Time constant of machine with disk [s]	-	31,5

Table 2. Parameters of line model

Parameter	Line 1	Line 2	Line3
X_a [ohm]	18.73	16.2	16.32
X_b [ohm]	17.56	16.2	16.00
X_c [ohm]	18.13	16.17	16.20
C_{bc} [μF]	8.0	8.61	8.12
C_{ac} [μF]	8.16	8.61	7.96
C_{ab} [μF]	8.16	8.45	8.04
C_{ag} [μF]	4.00	4.03	4.06
C_{bg} [μF]	4.08	4.03	3.98
C_{cg} [μF]	4.08	4.02	4.03
R_a [ohm]	1.87	1.62	1.63
R_b [ohm]	1.76	1.62	1.6
R_c [ohm]	1.81	1.62	1.62

Increasing the power of generation on the G1 generator, three Jacobian values for each step of loading (table 3) were calculated.

Table 3. Increasing power in 3 step

Power flow, [MW]	Jacobian determinant value, [r.u.]
276	1,04
793	0,88
899	0,61

IOP Conf. Series: Materials Science and Engineering **994** (2020) 012014 doi:10.1088/1757-899X/994/1/012014

Jacobian values was taken concerning the condition at which there will be a theoretical maximum (U1=Ur=500 kV, U2=Ur=500 kV, U3=Ur=500 kV, δU1- δU2=0°, δU3- δU2=0°, δU3- δU4=0°).
On these three points linear approximation (figure 6) was made, and the predicted point of transition of Jacobian through zero was found. Power at which Jacobian crossed zero corresponds to a limit on static aperiodic stability:

$$P_{PF} = 2100MW$$

$$K_{SF} = 52\% > 20\%$$

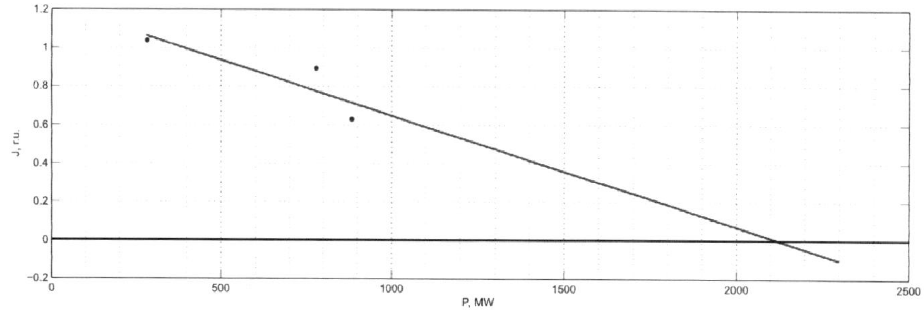

Figure 6. Approximation of Jacobian values by three calculations

As safety factor more than 20% the loading proceeded. At the following stage Jacobian calculated for 6 loadings (table 4).

Table 4. Increasing power in 6 steps

Power flow, [MW]	Jacobian determinant value, [r.u.]
276	1,04
793	0,88
899	0,61
1114	0,55
1196	0,39
1298	0,36

After approximations of data (figure 7) predicted limit was received:

$$P_{PF} = 1842MW$$

$$K_{SF} = 24\% > 20\%$$

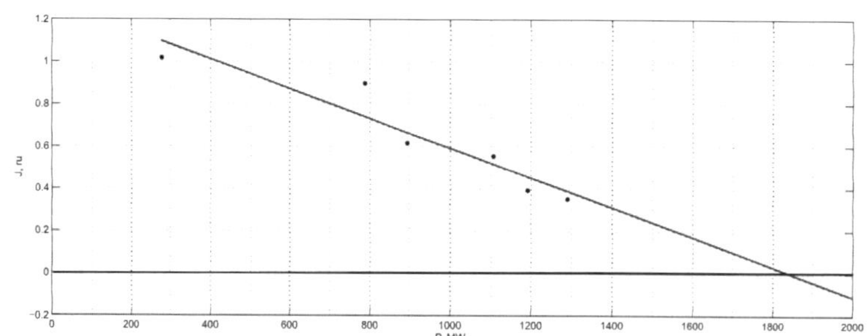

Figure 7. Approximation of Jacobian values by six calculations

Safety factor equals 24%, the limit was not reach. Further the system continued to loading. Jacobian was calculated for additional two values of the power (table 5).

Table 5. Increasing power in 8 steps

Power flow, [MW]	Jacobian determinant value, [r.u.]
276	1,04
793	0,88
899	0,61
1114	0,55
1196	0,39
1298	0,36
1402	0,23
1446	0,22

Approximation of these values (figure 8) showed that the predicted limit is:

$P_{PF} = 1771 MW$

$K_{SF} = 13\% < 20\%$

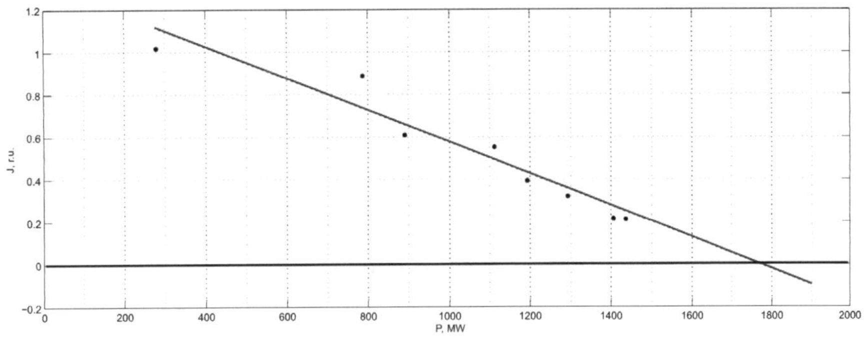

Figure 8. Approximation of Jacobian values by eight calculations

Safety factor shall be under control and monitored. And in cases when safety factor decreases until 20% appropriate actions shall be taken to keep stable condition.

5. Conclusions

Nowadays, in order to identify the conditions of stability violation, numerous variant calculations of normal and abnormal grid modes are performed. In practice, when performing calculations, the conditions for violation of static stability are determined by the limiting values of power flows over controlled tie. This method of monitoring stability by the observed parameters of the mode is the simplest and most feasible in connection with the ability to measure the value of power flows. However, the use of this method has several disadvantages:

- ambiguity of stability control by indirect parameters of the mode, by power flows;
- inability to maximize the use of bandwidth in connection with the calculation of limit modes for the most severe grid conditions;
- to control the maximum load of the cross-section for the transmitted power between the nodes, the automation devices (Automatic Power Loose) must be installed on all power lines included in this section;
- the automation devices settings are dependent on the grid circuit; in post-emergency (or repair) schemes with shutdowns of sections of parallel overhead transmission lines in transit, it becomes necessary to change the settings.

Today with development of IT technologies, and integration it to the power system controls new opportunities are appearing. Proposed method for predictive determination of stability limits can reduce the issues above. By using WAMS system it is possible to solve problems of power transmission adaptively in real time.

The expected results of an adaptive control system development are:

- Predicting of static stability limit with using the approximating function;
- Identification dangerous power surge according to phasor measurements;
- Maximum using of transmission capacity;
- Development of control system for dangerous overload values;
- Development of the dispatcher adviser with using WAMS technologies.

References

[1] Phadke AG Thorp JS: Synchronized Phasor Measurements and Their Applications –NY: Springer Science + Business Media LLC 2008- 243 p
[2] Jaime De La Ree Virgilio Centeno James S Thorp Phadke AG: Synchronized Phasor Measurement Applications in Power Systems // IEEE TRANSACTIONS ON SMART GRID – June 2010– № 1 vol 1– P 20 – 27
[3] Xiaorong Xie Yaozhong Xin Jinyu Xiao Jingtao Wu Yingdao Han: WAMS applications in Chinese power systems // IEEE Power &Energy Magazine – January/February 2006 – p 54 – 63
[4] Korba P and Uhlen K: Wide-area monitoring of electromechanical oscillations in the Nordic power system: practical experience // The Institution of Engineering and Technology - IET Generation Transmission and Distribution – -2010- №10Vol 4 — p 116-1126
[5] Aranya C Joe H C and Armado S: A Measurement-based Framework for Dynamic Equivalencing of Large Power Systems using WAMS // Innovative Smart Grid Technologies ISGT – 2010 IEEE
[6] Grebe E Kabouris J Lopez Barba S Sattinger W and Winter W: Low Frequency Oscillations in the Interconnected System of Continental Europe // ENTSO-e: IEEE- 2010 - P 1 - 7
[7] R Rameshbabu R Karthikeyan R Balamurali 2016 FPGA Implementation of Synchronized Non Identical Hyperchaotic Systems International Journal Of Control Theory And Applications 95 pp 2633 – 2642
[8] Ministry of power engineering of Russian Federation: Methodical guidance to stability of energy systems [Методические указания по устойчивости энергосистем] Moscow NC ENAS 2004
[9] Khrushev YV: Methods of calculating stability of energy systems [Методы расчета устойчивости энергосистем] Tomsk: STT 2005
[10] Balamurali R Maheswaran Prashanthi A R Roshan Fathima S Varshini S 2020 Enhanced Automation Using Controller and Conscripts International Journal of Advanced Science and Technology 293 8620 - 8627
[11] Prasertwong K Mithulananthan N and Thakut D: Understanding Low-Frequency Oscillation in Power Systems // International Journal of Electrical Engineereing Education- 2010 – July

Study of Sensory Characteristics of Poultry Meat Obtained With the Use of Modern Stunning Technology

D.B. SHALGINBAYEV1, R.U. UAZHANOVA[1], L.V. ANTIPOVA[2], T.A. BAIBATYROV[3]

[1]Almaty Technological University, Almaty, Kazakhstan
[2]Voronezh State University of Engineering Technologies, Voronezh, Russia
[3]West Kazakhstan Agrarian-Technical University named after Zhangir Khan, Uralsk, Kazakhstan.

E-mail: Dauletdau@mail.ru

Abstract. The article presents studies of color and aromatic characteristics of poultry meat samples obtained using the traditional method of stunning (using electric current) and gas stunning. Three samples of poultry meat (broiler) obtained by different methods of stunning different manufacturers are given: Kazakhstan, electrical stunning-1; Russia, gas stunning-2; Russia, electrical stunning -3. The meat was taken from 2 parts of the carcass: the round and the breast. Modern methods of studying color and aromatic characteristics allow to get adequate information about the processes occurring in poultry meat after slaughter and about the influence of stunning methods on the color characteristics and odor of meat. It is noted that the round meat has an identical color for all the studied samples. For breast meat, there is a slight decrease in the proportion of the red component of the color, the meat becomes grayer, while the color intensity of samples 1 and 3 decreases – for sample 1 by 17 %, for sample 3-by 11 %. The breast meat of the sample 2, despite the change in color intensity relative to the round of the same manufacturer, preserved. The results of studying the gas phase over meat samples did not show significant differences in the quality of meat. Comparison of the impact of stunning methods on the quality of meat proves that both methods are effective and contribute to the production of meat of excellent quality.
Keywords: poultry meat, stunning, colorimetric characteristics, aromatic characteristics.

1. Introduction

Poultry stunning is the first and one of the main stages in poultry slaughter technology, which has a significant impact on the formation of such qualitative indicators of meat as color and odor. The most commonly used and common method of poultry stunning– using an electric current – developed at the beginning of the last century, is outdated and has a number of significant drawbacks, which determines the search for alternative more effective and humane methods of stunning poultry, for example, in a gas medium. The use of modern technologies in the humane stunning of poultry, the world allows players to receive substantial benefits. Innovative solutions make it possible to ease and optimize most of the technological problems in the slaughter process – automatic feeding of poultry to the line of hitching and the stunning process. By implementing modern technological solutions in production, most European producers get the highest quality meat, which meets the expectations of the consumer in the humane treatment of poultry and animals. Manufacturers who are the first to use innovative solutions get all the benefits from the modernization of their enterprises. Most large Western European poultry processing enterprises have already successfully used the system of gas stunning poultry [1, 2]. Kazakhstan poultry producers are also showing increasing interest in this technology. Due to insufficient information about the features of this technology and its impact on the properties and quality indicators of the resulting meat, it is of interest for scientific research. Global trends suggest two alternative types of stunning in a modified atmosphere: single-stage stunning with inert environmentally safe nitrogen (one of the components of our atmosphere) with the addition of argon, and two-stage stunning with carbon dioxide [3]. For Kazakhstani poultry processing industry, the method of stunning poultry with carbon dioxide may be more preferable. It is carried out in 2 stages. In the first stage, CO_2 is 30 % of the volume of the gas mixture, and in the second – 70 %. The entire two-stage process takes approximately 2,5 minutes, and the CO_2 consumption is about 10-12 g per poultry [4]. Within the framework of cooperation with the FSBEI HE «Voronezh State University of Engineering Technologies» (Voronezh, Russian Federation), as well as with the largest poultry producers in Russia and Kazakhstan, research is being conducted and data is being collected on the effectiveness of stunning poultry in a controlled gas medium in comparison with the traditional method of stunning poultry. To compare the quality indicators of poultry meat, a number of experiments were conducted with meat obtained from slaughter with various methods of stunning: electrical and gas. We investigated the effect of methods of stunning poultry on such characteristics of meat as color and odor.

2. Research methods

The color characteristics of objects were determined in the RGB system. The RGB system is one of the officially accepted methods of color characterization. The system is based on a three-component theory, according to which the mixing of the three main colors (red - R, green - G, and blue - B) in suitable ratios produces all other spectral colors, as well as achromatic white. Red corresponds to 700 nm, green – 546,1 nm, blue-435,8 nm. (Figure 1).

The value of each color component in the RGB model is measured on a scale from 0 to 255 c.u.

The objects were scanned in TrueColor color mode, with an optical resolution of 300 dpi and a size of at least 500×500 pix. To standardize the results and eliminate color error, scanning was performed in the presence of a white sheet with a spectral scale and a brightness scale applied (for automatic detection of white balance).

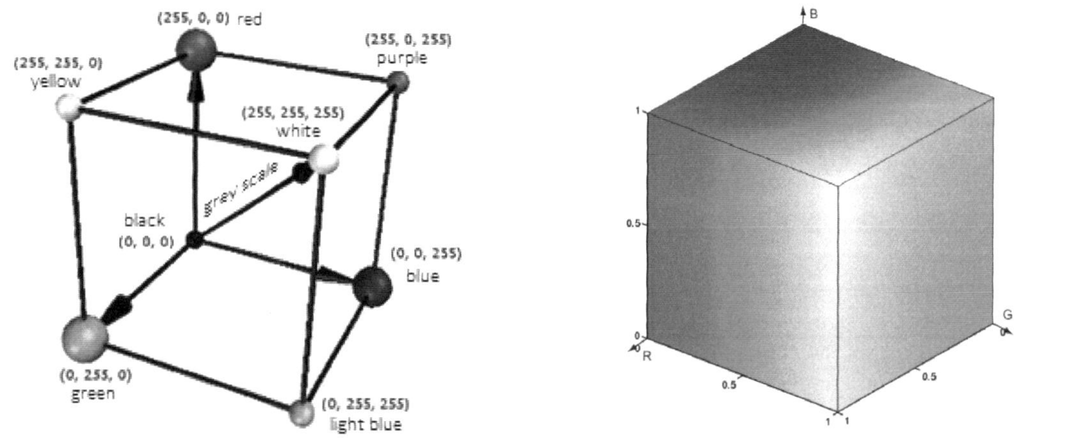

Figure 1. Color cube in RGB system

Samples with a hard, non-smearing surface were placed directly on the scanner glass, in the case of samples with a sticky surface, they were previously placed on a thin glass.

The digital image was processed using the free ImageJ 1.46 program. The result was taken as arithmetic mean value for each of the color components, as well as the color intensity.

IOP Conf. Series: Materials Science and Engineering **994** (2020) 012015 doi:10.1088/1757-899X/994/1/012015

ImageJ program is a powerful image editor for image processing and analysis. The ImageJ development company was initiated by author Wayne Rasband at the National Institutes of Health (USA).

Other things being equal, the color intensity (R+G+B)/3 characterizes the light/dark coloring of the product. The higher the intensity, the lighter the product (the maximum value of 255 is absolutely white).

The values R, G, and B are responsible for their own color, as well as for the opposite color. Thus, the increase in R characterizes the gain of red and decrease green, increase G – increase green and decrease red, increase B – increase blue and decrease the yellow to the final color of the analyzed object.

Numerical characteristics of the color of the studied samples were determined by the scannerometric method using the HPScanJet 3570c flatbed scanner using computer image processing in RGB color mode.

The study of the odor was carried out in the research laboratory of the FSBEI HE «Voronezh State University of Engineering Technologies» on the odor analyzer «MAG-8» with the methodology «electronic nose» (production Russia, Figure 2 (a)in the «Front-linemovement» mode of sample delivery (front-flow of the odor to the near-sensory space in a closed space) and injection of the equilibrium gas phase over the samples. For the analyzed samples, typical chrono-periodograms are shown in Figure 2(b).

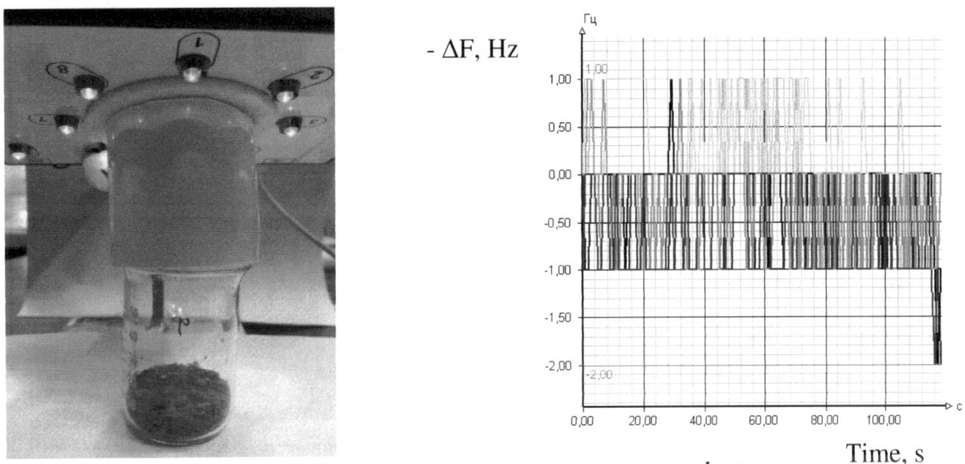

a b

Figure 2. (a) Working place for odor measurement on the gas analyzer «MAG-8» in the mode of front-flow steam supply and (b)chrono-periodogram of sensors in the array without load in the detection cell

A set of 8 sensors based on piezo-quartz OAV-type resonators with a base oscillation frequency of 10,0 MHz with different polymer sorbents on the electrodes was used as a measuring array [6,7]. Sensors were selected with increased sensitivity to various classes of volatile organic compounds (alcohols, aldehydes, ketones, amines, etc.) [5,6]: sensor 1 – Polyvinylpyrrolidone, PVP; sensor 2 – Bee glue (propolis), PS; sensor 3 – Dicyclohexane-18-Crown-6, DCH18C6; sensor 4 – Bromcresol green, BCG; sensor 5 – Polyethylene glycol succinate, PEGsc; sensor 6 - Polyethylene glycol PEG-2000; sensor 7 - Tween-40; sensor 8 - Trioctylphosphine oxide, TOPO.

3. Main results of the research

The quality of samples of poultry meat (broiler) from the following manufacturers was compared: Kazakhstan, electricalstunning -1; Russia, gas stunning -2; Russia, electrical stunning -3. The meat was taken from 2 parts of the carcass: a – the round, b – the breast.

A visual assessment of the color of the samples is shown in Figure 3.

Analysis of colorimetric characteristics of the samples shown in Figure 4, showed that the round meat

IOP Conf. Series: Materials Science and Engineering **994** (2020) 012015 doi:10.1088/1757-899X/994/1/012015

has an identical color for all the studied samples. Color profiles (Figure 4(a)) are similar, for all samples, the red component of the color prevails. The color intensity of the meat samples (light/dark) differs slightly, within the error of the research method.

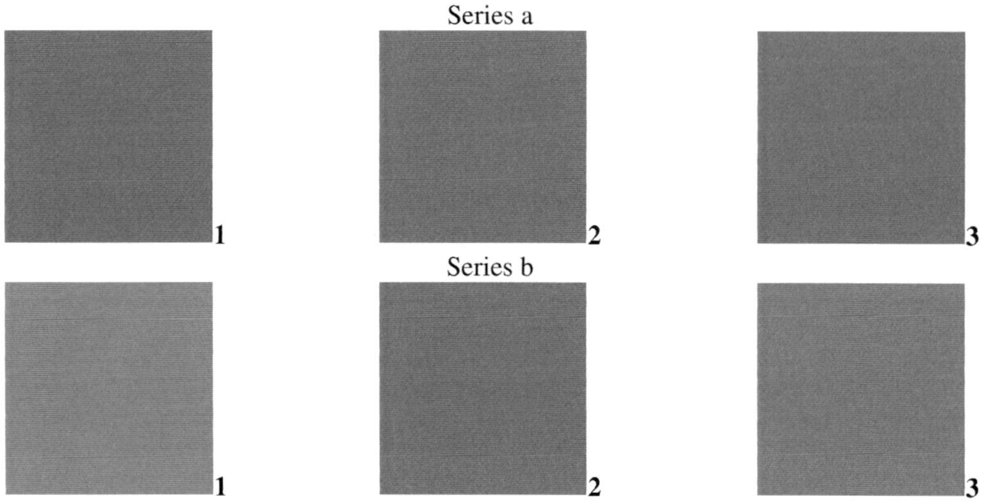

Figure 3. Analysis of colorimetric characteristics of samples

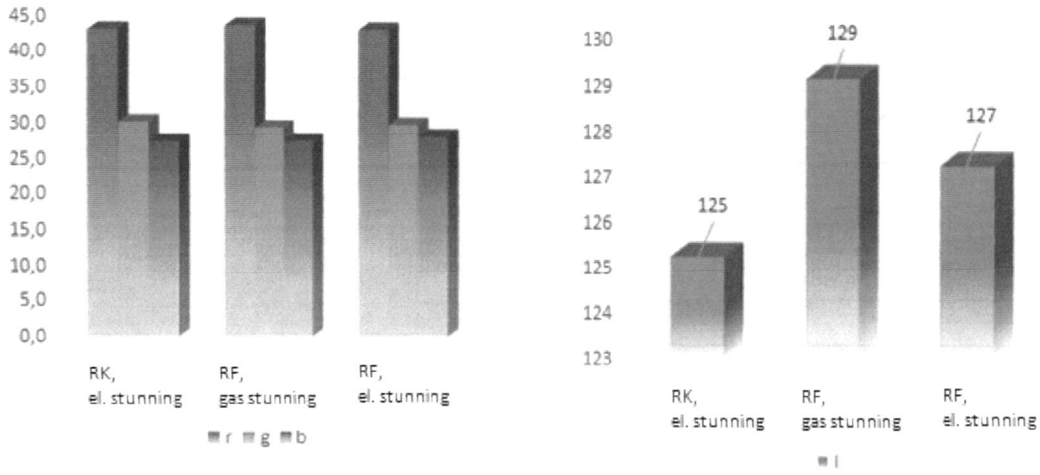

a b

Figure 4 (a) & (b) Colorimetric characteristics of the color of chicken meat (round)

For breast meat (Figure 5) there is a slight decrease in the proportion of the red component of the color, the meat becomes grayer, while the color intensity of samples 1 and 3 decreases – for sample 1 by 17 %, for sample 3 - by 11 %. The meat of the breast of the sample 2, despite the change in color intensity relative to the round of the same manufacturer, preserved.

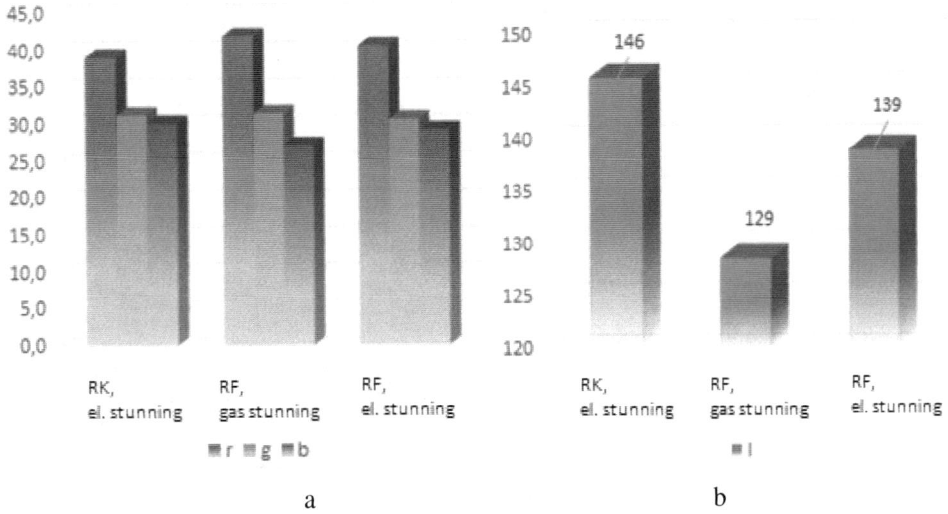

Figure 5. (a) & (b) Colorimetric characteristics of the color of chicken meat (breast)

4. Conclusions

The color intensity of the meat samples (light/dark) differs slightly, within the error of the research method. The meat of the breast of the sample 2, despite the change in color intensity relative to the round of the same manufacturer, preserved, which indicates a uniform exsanguination of the entire carcass.

The study of the gas phase was performed as follows. Samples of the submitted samples, at room temperature, weighing 10,00 g. were placed in samplers, tightly closed, and kept at room temperature (20±1 °C) for at least 20 minutes to saturate the equilibrium gas phase above the samples.

As criteria for evaluating differences in the odor of the analyzed samples are selected:

Qualitative characteristics:

1) the form of the «visual print» with characteristic distributions along the response axes is determined by the set of connections in the RGB.
2) for recognition of individual classes of compounds in a mixture, the A_{ij} identification parameters were used, calculated from sensor signals in the analyzed samples [5].

Quantitative characteristics:

3) S_{Σ}, Hz.s -the total area of the complete «visual print» - estimates the total intensity of the odor, proportional to the concentration of volatile substances, including water-based on all the signals of all sensors for the full measurement time;
4) maximum signals of sensors with the most active or specific films of sorbents ΔF_{max}, Hz - for assessing the content of individual classes of organic compounds in RGB by the normalization method.

Sensor responses were recorded, processed and compared in the software of the «MAG Soft» analyzer.

The deciphering of the measurement results:

The primary information of the «electronic nose» is the extreme (maximum/minimum) responses of all sensors for the entire time of measurement under load. The analytical signal of the «electronic nose» is the area of the pie chart built on these signals for each sample.

Differences in the chemical composition of the equilibrium gas phase over the samples were determined by the shape of the «visual print» of sensor responses in the array (Figure 6). Out of 26 possible design parameters, 15 parameters A that differ most for samples were selected in order to establish differences in the qualitative composition of easily volatile compounds.

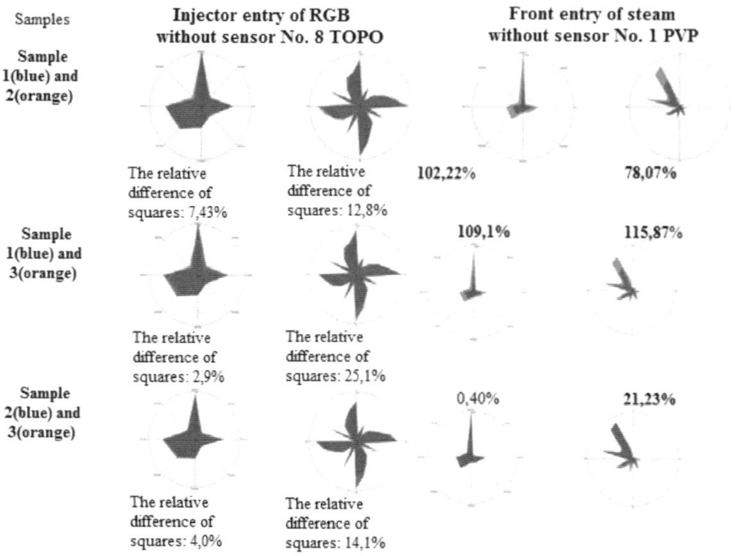

Figure 6. «Visual prints» of sensor signals in the RGB over samples. The axes are indicated: circular - sensor numbers in the array, vertical - maximum sensor responses during measurement (ΔF_{max}, Hz). Kinetic «visual prints» - on the circular axis – the selective time of recording signals, on the vertical - the current responses of sensors

5. Discussion of the data obtained and conclusion

Studies have shown that gas stunning does not have a negative impact on such qualitative indicators of poultry meat as color and odor. Color studies have shown that the color intensity of meat samples (light/dark) differs slightly, within the error of the research method. The breast meat of the sample with gas stunning despite the change in hue, the intensity relative to the round of the same manufacturer preserved, which indicates a uniform exsanguination of the entire carcass. Odor studies have shown that the differences between samples in qualitative and quantitative composition are from 2 to 10 %. This does not allow to talk about a critical change in easily volatile compounds. The studied samples are almost identical, which proves the effectiveness of using gas stunning in poultry slaughter in order to obtain the highest quality products.

References

[1] McKeegan EF McIntyre J Demmers TGM Wathes CM and Jones RB 2006 Behavioral responses of broiler chickens during acute exposure to gaseous stimulation Applied Animal Behavior Science 99 271286

[2] Semikopenko NI Advantages of an innovative method of stunning birds in a controlled gas environment [Text] / NI Semikopenko LV Antipova // Food safety: scientific personnel and information support of Voronezh State University of engineering technologies 2014 Pp 16-22

[3] Yakovleva ND Competitiveness of world technologies and product quality [Text] / ND Yakovleva // Poultry and poultry products - No 1 2010

[4] Hoen T and Lankhaar J 1998 Controlled atmosphere stunning of poultry Poultry Science 78 287289

[5] Kuchmenko TA Umarkhanov RU Grazhulene SS Zaglyadova SV Shkinev VM Microstructural investigations of sorption layers in mass-sensitive sensors for the detection of nitrogen-containing compounds // Journal of surface investigation X-ray synnchrotron and Neutron Technigues 2014/ V 8 № 2 pp 312-320

[6] KuchmenkoTA Electronic nose based on nanoweights expectation and reality // Pure and Applied Chemistry The Scientific Journal of IUPAC:Published Online: 2017-09-08l DOI: https://doiorg/101515/pac-2016-1108
[7] Kuchmenko TA Chernysheva SE Identification of monoethanolamine in the air using two sensors based on potassium fluoride microphase// Vestnik VSUIT 2015 No 4 66 Pp 135-139

Smart Solar Panel using Light Dependent Resistors

S Karthik[1], P Vivek Karthick[2], R R Thirrunavukkarasu[3]*, S Satheesh Kumar[4], D Selvakumar[5]

[1]Assistant Professor, [5]Associate Professor, Department of Electronics and Communication Engineering, Sri Krishna College of Engineering and Technology, Coimbatore.
[2]Assistant Professor, Department of Electronics and Communication, Engineering, Sona College of Technology, Salem.
[3]Assistant Professor, Department of Electronics and Communication Engineering, Sri Krishna College of Technology, Coimbatore.
[4]Assistant Professor, Department of Electronics and Communication, Engineering, KPR Institute of Engineering and technology, Coimbatore.

Corresponding Authou mail id : karthicks@skcet.ac.in

Abstract: Solar energy is gradually approaching a most important form of renewable energy. Using solar tracker, additional energy can be produced, since the solar plate can maintain a vertical angle to the sunrays. Given the substantial initial expense of setting up the tracking device, there are less expensive alternatives that have been suggested over period. This paper addresses the strategy and development of solar tracking device prototype which has one axis of independence. Here, Light Dependent Resistors (LDRs) are used to sense the sunlight.
Keywords: Light Dependent Resistor, Solar System, Solar Tracker, Servo Motor.

1. Introduction

For the first time in 1985, silicon solar cells achieved 20% output. Although the performance of the solar panels has been gradually rising, the standard is still not at its peak. The majority of panels are now working for a smaller amount than 40%. Solar cells comparatively give higher effectiveness but they stay to be very expensive [1].
The control module is based on a microcontroller with the ATMega328P. Sunlight is sensed via the LDRs before servo was actuated to align the solar panel. The Solar Panel is placed where maximum light can be received [2,3,4]. The servo motors can be able to regulate their force even at high speediness when compared to other motors. In this paper, solar panel's performance and characteristics are experimentally analyzed.
A particular dedicated device was used for monitoring. It is inexpensive, have a smaller amount of complexity and it accomplishes the efficiency that is needed. The system is feasible in terms of cost. Here, power increase was substantial, and hence there is slight cost increase. Perhaps the maintenance costs are not much high.

2. Proposed Method

The proposed method is intended to design the system, which will automatically monitor the location of the sun and adjust the angular position of the solar panel to achieve a high output from the solar panel cells. A solar tracker can be used for optimizing and connecting the solar energy in different structures. The issue here is the execution of a program capable of rising power output by 30-40%. The micro-controller follows the controlling circuitry. Then, the controlling circuitry adjusts the motor by which the solar panel can be aligned in an optimal way. The proposed system has been undertaken to ensure that the sunrays falling above the solar panels vertically to give a comparative increase in the solar energy, with the existing system. Then, it used to manipulate the electric strength. Between 1200hrs and 1400hrs maximum energy is collected, with the highest being about midday.

The sun is nearly vertical, at this time. Around the same time, moving the panel would take the least energy, which additionally gives rise to substantial increase in the system's performance. The proposed system was intended to face the challenge of a low-powered, reliable, inexpensive micro-controller based sunrays tracking system that is executed with the available resources, within the stipulated time. They were intended towards monitoring the movement of the sun (basically earth).

The solar panel-based project was realized with a servo-motor. The Atmega328P were coded using the C-programming language. During fabrication it is then used as a stand-alone device on a PCB. The architecture is restricted to single axis tracking as it does not add much benefit towards using a dual axis tracking device.

3. Design and Implementation

This Tracking System circuit was splitted into three parts. There exists input or first stage containing potentiometers and sensors, a code of program that to be encoded in the microcontroller's core chip and, lastly, required servo motor driving circuit. The input stage has 4 Light Dependent resistors organized such that, so as to form a circuitry for the voltage divider. A C-code of program loaded onto the UNO arduino.

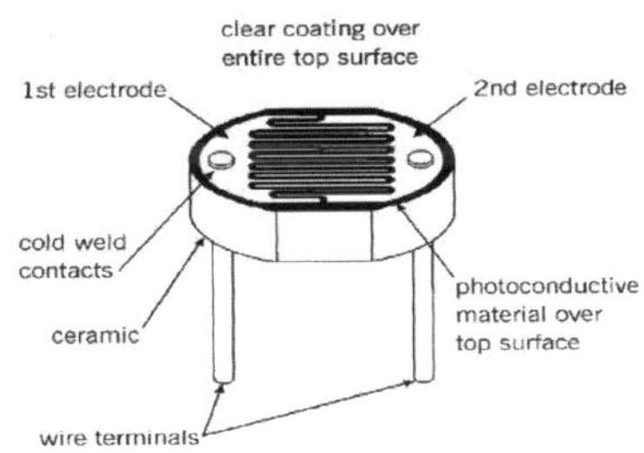

Fig.1. Structure of an LDR

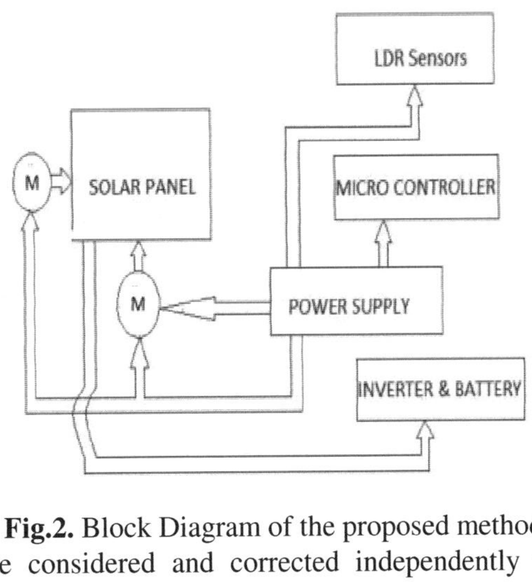

Fig.2. Block Diagram of the proposed method

This also ensures they are considered and corrected independently when there are any errors. Phototransistors, photodiodes, LDR and LLS05 provide light detection sensors that may be used to create solar trackers. Fig.1 shows the structure of an LDR.

Fig.2 shows the block diagram of the proposed method. The concept of sensing using four LDRs is described in fig.3. The constant state occurs, once both LDRs have same luminous intensity. As the source of light moves, i.e., the sun travels in west-east direction, the intensity value lowering on two Light Dependent Resistors varies and using voltage-divider biasing circuitry, adjustment is made in the voltages. The variation in voltages are measured using in-built microcontroller, and the motor available makes the solar-panel to rotate in its path to monitor light source. The solar tracker uses a photocell with Cds to detect light. A supplementary 10k resistance were used. The outcomes are stated in the following table.

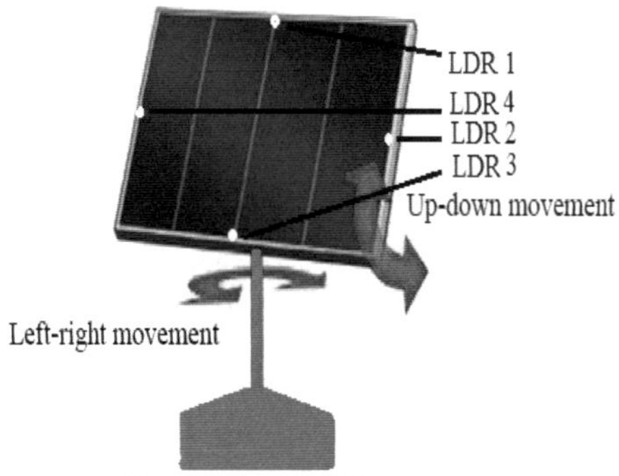

Fig.3. Four LDRs concept

Resistance	Observation methods
50 KΩ	Low light (black-vinyl tape can be placed above the solar cell)
4.35 KΩ	Medium light (average level of lighting)
200 Ω	High level light (applying light straight to the solar-cell)

Voltage-divider circuitry used here is shown in the following fig.4.

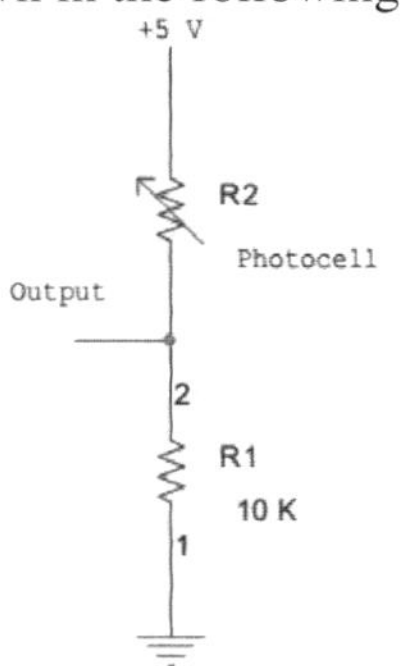

Fig.4. A Simple Voltage Divider Circuit

The relationship between the input and output in voltage-divider based circuitry is

$$V_{in}=V_{cc(supply)}\{R_{opt}/(Light\ Dependent\ Resistance+R_{opt})\} \quad (1)$$

In this case, V_{in} = Voltage input given to micro-controller, R= 10KΩ, potentiometer-resistance, $V_{cc(supply)}$= Power supply to Light dependent resistors and micro-controllers connected.
Servo-motors are controlled by variable width electrical pulses, or Pulse-width modulation, through a wire control. Every 20 ms, servo-motor waits for a pulse to arrive, the width (length) of the pulse-modulated signal will control motor ON and OFF and the speed of the rotation. The working of the above experimental setup is explained in the following steps.
(1) Switch on the power of the circuitry, (2) Keep the project, with very low-level light, (3) If low level light appeared on both Light Dependent Resistors, no change in the position of the solar panel, (4) Applying a flash light above the light dependent resistor, the solar-panel slowly will get tilted, (4) Moving the light from the left-side to the right of the panel, we can verify the automatic change the direction of the solar pane, which slowly tilt towards the appearance of the light, (5) At the mid-level,

where both Light dependent resistors having same voltages, the solar-panel be idle, till there is difference between the intensity of the light that falling on the Light Dependent Resistors.

4. Results and Discussions

(1) Some of the benefits are, due to increased direct exposure to solar rays, trackers produce more energy than their stationary counterparts. Based on the geographic position of the tracking device this change can be as much as 10 to 25 per cent.

(2) Several types of solar tracking model are available. System model, usage environment, and electric items related specifications stands some important factors which affects solar tracker type that is preferably appropriate for a definite solar-system.

(3) Some states, some utilities sell Time of Use (TOU) solar power rate plans, meaning the utility can buy the generated electricity at a higher rate during the day's peak period. In this case, the generation of more electricity during these peak times of the day is beneficial. Having a monitoring device helps to optimize the gains in energy during these peak hours.

(4) Advances in current technology and dependability in electronics-engineering abridged the long-standing preservation problems for tracking systems significantly.

5. Conclusion and Future work

Although dual-axis tracking system is effective in sunlight tracking, our proposed component does not require the additional circuitry and complexity. Dual- axis sun tracking systems are highly used in places wherever the angle of the sunlight varies. The above research was implemented using limited available resources. Circuit implementations have been used as it is, thus ensuring that it does not affect performance. The goal was achieved with the time and resources available. The concept can be enhanced to a much higher scale of values.

References

[1] S J Hamilton University of Queensland Department of Computer Science and Electrical Engineering Bachelors Thesis "*Sun-tracking solar cell array system*" 1999

[2] Nader Barsoum Curtin University Sarawak Malaysia "Fabrication of Dual-Axis Solar Tracking Controller Project" *Intelligent Control and Automation* 2011 2 57-68

[3] Balamurali R Maheswaran Prashanthi A R Roshan Fathima S Varshini S 2020 Enhanced Automation Using Controller and Conscripts International Journal of Advanced Science and Technology 29(3) 8620 - 8627

[4] Karthik V Karthik S Satheesh Kumar S Selvakumar D "Region based Scheduling Algorithm for Pedestrian Monitoring at Large Area Buildings during Evacuation" *International Conference on Communication and Signal Processing (ICCSP)* 101109/ICCSP20198697968 pp0323-0327 2019

A Vehicle Accident Detection and Driver Surveillance

D Selvakumar[1]*, S Ganesh Prabhu[2], S Karthik[3], V Karthik[4], T R Shivaa[5], H Vaigunthan[6], J Gowtham[7]

[1,4]Associate Professor, [3]Assistant Professor, Department of Electronics and Communication Engineering, Sri Krishna College of Engineering and Technology, Coimbatore.
[2]Assistant Professor, Department of Electronics and Communication Engineering, Sri Krishna College of Technology, Coimbatore.
[5,6,7]U.G. Student, Department of Electronics and Communication, Engineering, Sri Krishna College of Engineering and Technology, Coimbatore.

Corresponding author mail id: selvakumar@skcet.ac.in

Abstract: Perpetual increase in the number of deadly traffic accidents around the world are often considerablyminimized if contemporaryproficiency is integrated inside of the car towards observe the fitness of the driving force onconsistentinterims over the vehicle runningand precautionary measures predictably taken for the safety of all anxiouspersonnel, either within the vehicle or outside. The Raspberry Pi controller constantly records every one of the parameters of the vehicle based on the heartbeat, drowsiness and the level of alcohol consumption of the driver, for anticipation and recognition of mishap. In the unfortunate case of occurrence of an accident, the details and location of the vehicle is intimated to the emergency contacts via message immediately in order to seek first aid so as to prevent the avoidable loss of lives.
Keywords: Accident prevention, raspberry pi, sleep detection, heart rate monitoring, alcohol consumption level detection.

1. Introduction
The modernizations in the industry over the past decay are revolutionize vehicles as much powerful, safer, easy for driving and controlling, highproficient in energy and further environmentallyapproachable. Almostthe accidents instigatednowadaysin cars are primarily because ofthe fatigue driver [1,2]. The aim of the present work, implements a prototype of driver drowsy caveat scheme. The main emphasis are going to hire on the system design, which resolve precisely and consistently view the eye blinking (open and close) state of the driver. This detection is often done employing a categorization of images of eyes and also the face and head movement using accelerometer module.
Apart from falling asleep on wheels, other causes for accidents include heart attack or alcohol consumption. We have also made a warning system to continuously monitor the heart rate and alcohol consumption levels of the driver and intimate the emergency contacts on any unfortunate situations.

2. System Analysis
A. *Existing System*
The available accident detection systems in automobiles are only post-accident response systems such as air bags. They fail to track collision and pre-damage status. Also monitoring the physical condition

of the driver is missing [3,4,5]. There is no automatic means of reaching out for immediate help or first aid in the occurrence of an accident. There is no such accident prevention technologies in use.

B. Proposed System

The proposed accident prevention system alerts the driver when he is about to fall asleep. This system also continuously monitors the heart rate and alcohol consumption levels of the driver and intimates the emergency contacts in case of any abnormalities such as a heart attack and provides them his location details.

3. Module Description

A. Architecture of Accident Prevention System

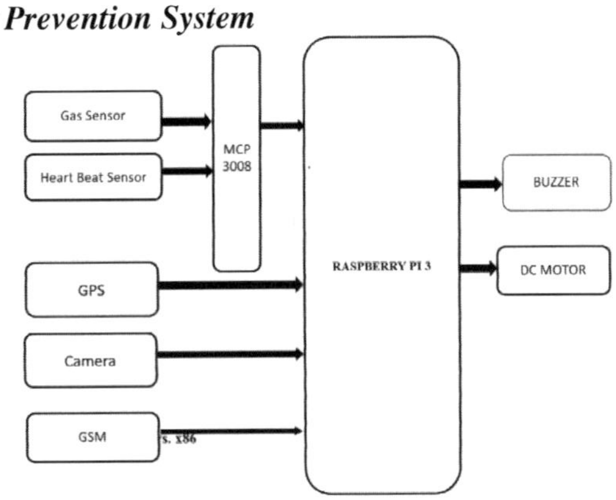

Fig.1. Architecture of the accident prevention system

Fig.1 shows the architecture of the proposed accident prevention system. The alcohol sensor, MQ-135comprises a perception layer inside alumina micro tubes, a tin dioxide layer and anelement in a tube-shaped case the very best sensor aspect is covered by a chrome steel net and thus the posterior grips the terminals for connection. Fig.2 shows the MQ-135 circuit diagram.

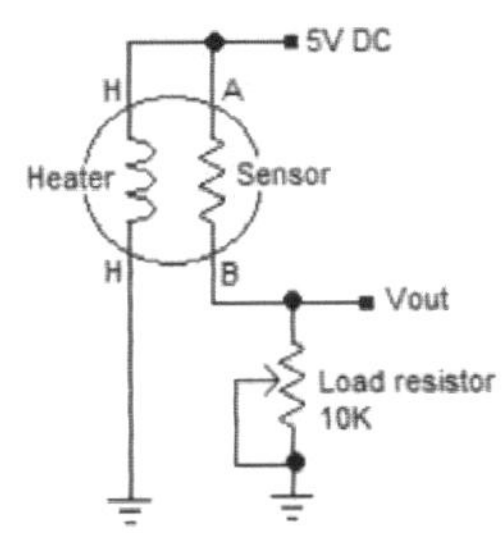

Fig.2. MQ-135 Circuit Diagram

The pulse sensor possesses two surfaces, on the primary surface consist of the light-emitting diode (LED)and ambient light sensor. Subsequently, on the succeeding surface such as the circuit is connected. The LED is fixed over a vein during a physical body similar that of ear tip or fingertip, though, it is situated on top of a layer directly. When the LED is situated on the vein and it starts blinking. As the heart is pumping, there will be a blood flow inside ofthe veins. If the blood flow is detected, the ambient light sensor will collect additional light as they arereplicatedby the blood flow. The key perceptionis the engine starts only the seat belt of driver is buckled, which remainsthe driver to wear the seat belt essential to run thecar. Thecon caption is that, the informationwill be communicated from the heartbeat detector attached withthe seatbelt and heart of the driver, only then the microcontroller can allow the driver to twitch vehicle engine. Heartbeat sensor will persistentlyidentify the heartbeat level of the driver. As the heartbeat levelof the driver energiesan anomalous levelat that time the information willdirect to proprietor (car), nearby sanatorium and policemanby means of GSM or IOT. If the information about the abnormality of the driver is reached toproprietor, he can able to halt the vehicle spontaneously from anywhere over theGSM or IOT.

The Raspberry pi is a single microcontroller (computer) board ofMasterCardsize can be operated for several tasks. Raspberry pi acts as aheart of the system by which the message transfers to diversedevices within the system. With the help of GPS module, the situation is often communicatedoverthe tracking system to shield the geographical coordinates of the area.

To detect drowsiness, image processing technique is employed using the camera to read the driving force's countenance and alerting the driver with a buzzer if he loses his focus. Image processing is completed using Open CV software toolkit.

The communication unit comprises of the Global System for Mobile (GSM)and Global Positioning System (GPS&#41. The communication unit is used for transferring the data regarding the position of the vehicle and the condition of the person are all communicated via this unit.

4. Methodology

The design of accident detection and reporting system is established on Raspberry pi and GPS module. Once vehicle encounters an accident, subsequently the accident should be identified by sensor. The sensor are often used as a crash detection of the vehicle through and after crash consistent with theproject, while a vehicle encounters an accident, a Heartbeat sensor will perceives the signal and transfers to Raspberry pi microcontroller. Immediately the microcontroller conveys the signal to GPS module with the present position co-ordinates values such as longitude (N or S), latitude (E or W) and also the microcontroller communicates the alert information to nearest emergency medical service (EMS) or loved one over GSM modem with GPS parameter values. The EMS will provide instantaneous medical action at accident location and victim can acquire the speedy treatment as possible. Engine stops the vehicle spontaneouslyoncethe alcohol is identified byGas sensor. Within the subsequent, it collects message from its neighbours, and routines the knowledge mutually to form the selection.

5. Results and Discussions

The prototype of the driver accident detection system was tested and all the various parts were found to be functional.

(1) The wakefulness of the test user was monitored continuously by the image processor and the alarm was set on when the eyes of the test user remained closed.
(2) The heartbeat of the test user was monitored continuously and the output was accessed. When it is abnormal, a message intimation was sent to the emergency contact number.
(3) The different alcohols levels of various test users are detected successfully by the alcohol sensor.

Fig.3-5shows the hardware implementation, SMS output and mail output of the proposed concept.

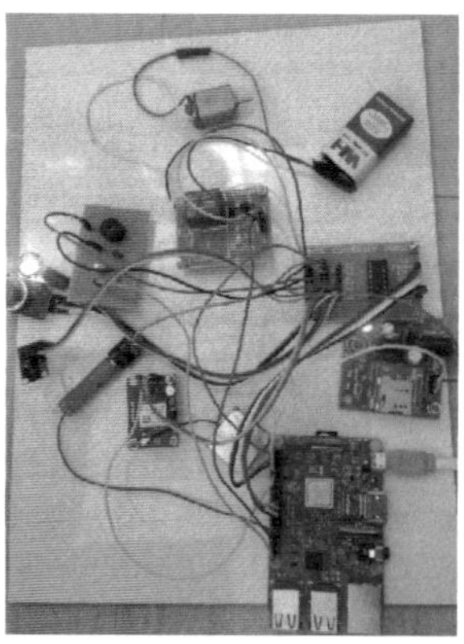

Fig.3. Hardware Implementation

Fig.4. SMS output

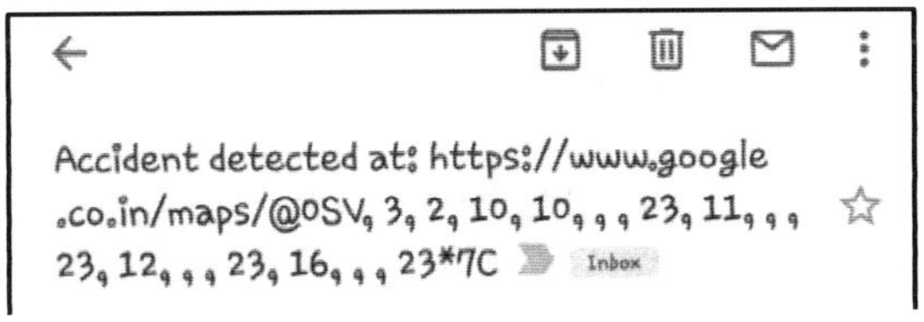

Fig.5. Mail output

6. Conclusion

The vehicle accident detection System is an essentially a tool anticipated by the present work used to prevent thedrivers lifetime, whoremainconstantly driving the vehicle (car) and aren't ready to get sufficient sleep. Several accidents happen especially within the developing countries such as India everywhere the quantity of the on roadvehicles rises per annum. The anticipatedschemedefends the driving force from misfortune moment which may happen due tofatigue of the driving force. The anticipatedscheme is reasonable as related to former schemesexisting in the luxurious cars. Habitually, one among the foremost essential and active feature of the system is to create the practical readiness to identify as the attention is shut or open, whenever vehicle driver lonely closing the eyes and that point sound alert system is activated.

References

[1] Toshiyuki Taguchi HazimeInagaki BunjiAtsumi Kenji Kimura Study of breath alcohol detector *JSAE ReviewElsevier* Volume 15 Issue 3 July 1994 Pages 229-233

[2] Jonas Ljungblad BertilHök Amin Allalou&HåkanPettersson (2017) Passive in-vehicle driver breath alcohol detection using advanced sensor signal acquisition and fusion Traffic Injury Prevention 18:sup1 S31-S36

[3] Yannick D'Mello James Skoric Shicheng Xu Philip J R Roche Michel Lortie Stephane Gagnon David V Plant Real-Time Cardiac Beat Detection and Heart Rate Monitoring from Combined Seismocardiography and Gyrocardiography Sensors (Basel) 2019 Aug; 19(16): 3472

[4] Balamurali R Maheswaran Prashanthi A R Roshan Fathima S Varshini S 2020 Enhanced Automation Using Controller and Conscripts International Journal of Advanced Science and Technology 29(3) 8620 – 8627

[5] Peng Wang Matthew B Green QiangJi Automatic Eye Detection and Its Validation San Jose State University California United States of America 2005

Contigency Analysis in Multiline Distributed System using Dynamic Voltage Restorer

A Antony Charles[1], D Joshua Jeyasekar[2], G Ramya[3], P Suresh[4], B Manimaran[5]

[1]Assistant Professor, EEE, Jeppiaar Institute of Technology, Chennai.
[2]Assistant Professor, ECE, Jeppiaar Institute of Technology, Chennai.
[3]Assistant Professor, EEE, SRMIST, Ramapuram, Chennai.
[4]Assistant Professor, EEE, St. Joseph College of Engineering, Chennai.
[5]Assistant Professor, EEE, Rajalakshmi Institute of Technology, Chennai.

Corresponding mail id: antonycharlesa@jeppiaarinstitute.org

Abstract: Power system security analysis involves a vital role in improving the system security and helps preventing the system collapse condition. This paper shows the conception and simulation of nine bus system with generator outage contingency. The objective of this work is to improve the real and reactive powers by employing DVR. Nine bus systems with generator outage contingency condition healthy, unhealthy without IDVR and unhealthy with DVR are simulated. The comparative study is presented to demonstrate the improvement in the power quality by adding DVR. The results indicate that real and reactive powers are improved by adding DVR.
Keywords: *Dynamic Voltage Restorer, Contigency, Generator Outage, Power Quality.*

1. Introduction

The two basic priorities for the evolution of modern power systems are reliable delivery of power and operating the system economically. The power system build with less redundancy and more failures makes the system highly uneconomical [1]. The system can be designed rather with small possibility of dropping load to avoid uneconomical conditions. Contingency analysis is done to analyze stability of power system which belongs to the power system security category. Failure of generation units and transmission line outages are the two different security threats for the failure of the system. Removal of generator sources from the system makes the system unhealthy [2]. The variation laws explained in the flux controlled memductance may not be same always as and when there is a change in excitation voltage [3,6]. Contingency analysis is generally employed to predict the outages effect and single failure events and operating personnel must be aware of which generator outages cause failure and fall outside the limits [4]. This paper focused on the performance of the system during addition and removal of the generator sources and how well the system compensates during out of service of generators [5]. The performances are analyzed for the IEEE 9 bus scheme during healthy, unhealthy and unhealthy with DVR system using MATLAB Simulink.

2. System Description

In this paper, Contingency analysis is conducted for the IEEE 9 bus scheme at different operating conditions. The single line diagram of the IEEE 9 bus scheme with three generator sources under healthy condition is depicted in Figure 1. Suppose in a condition, where Generator source 2 is unavailable to the IEEE 9 bus scheme makes the arrangement unhealthy. Figure 2 displays the single line illustration of unhealthy IEEE 9 bus scheme.

IOP Conf. Series: Materials Science and Engineering **994** (2020) 012018 doi:10.1088/1757-899X/994/1/012018

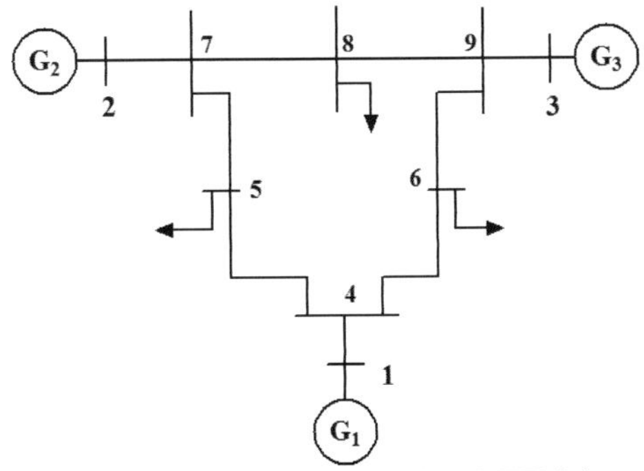

Fig.1 One line illustration of healthy IEEE 9 bus system

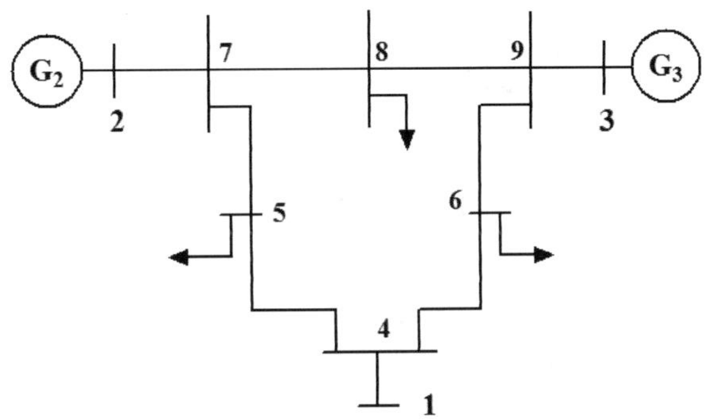

Fig.2 One line diagram of unhealthy IEEE 9 bus system

Figure 3 shows the one line diagram of DVR connected in unhealthy IEEE 9 bus scheme. The technique and procedure for analyzing contingencies in multibus system is shown as flowchart in Figure 4.

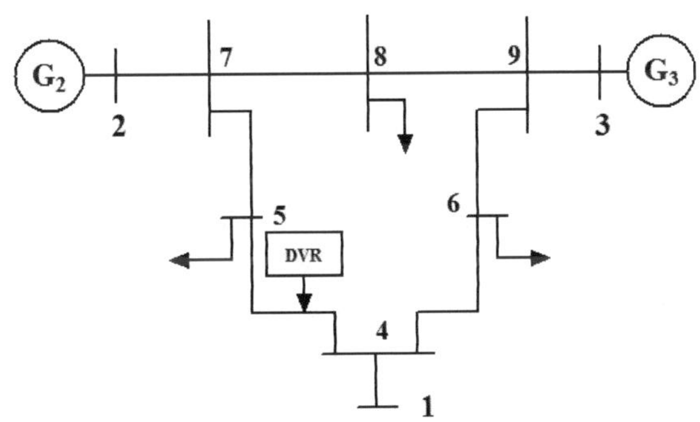

Fig.3 One line diagram of unhealthy IEEE 9 bus system with DVR

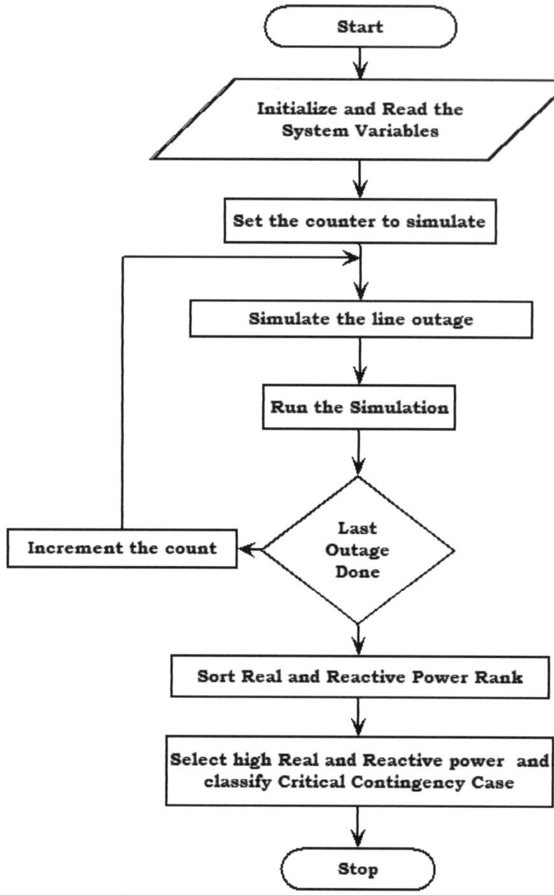

Fig.4 Flowchart of contingency analysis

3. Simulation Results and Discussion

The simulation results are discussed in this section. Figure 5 displays the Simulink diagram of 9 bus system. The real and reactive powers are measured at bus 5,6 and 8 of IEEE 9 bus system under healthy condition are shown in Figures 6 and 7. Generator 2 open in the IEEE 9 bus scheme denotes the unhealthy condition. Figures8 and 9 depict the real and reactive power comparisons at bus 5,6 and 8 of unhealthy 9 bus system. The figures reveal that lag of real and reactive powers during unhealthy conditions. DVR is employed to improve the real and reactive powers. The real and reactive power comparisons at bus 5,6 and 8 of unhealthy 9 bus with DVR system are depicted in Figures 10 and 11 respectively. Figures 10 and 11 show that DVR improve the real and reactive powers due to unhealthy conditions.

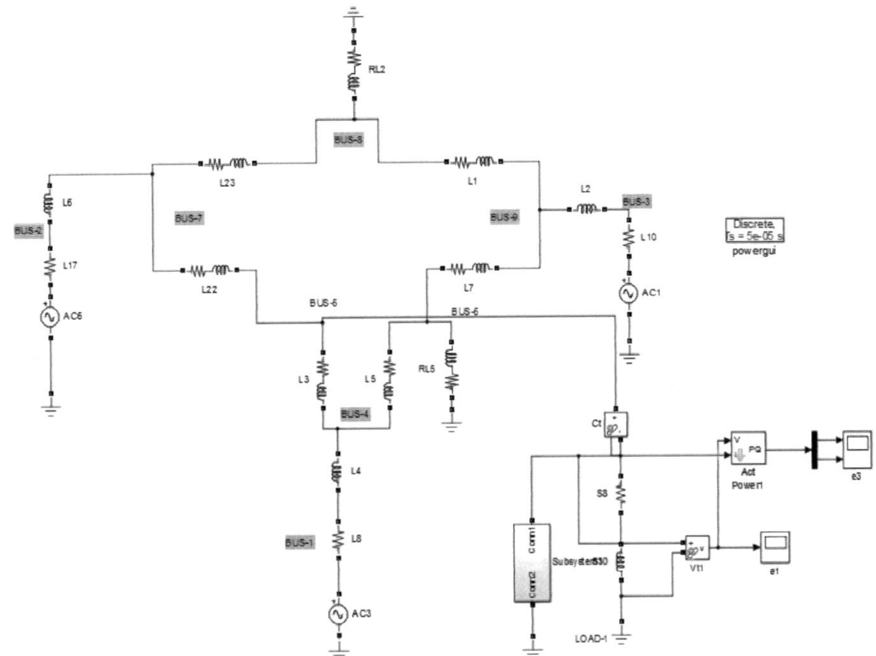

Fig.5 Simulink diagram of 9 bus system

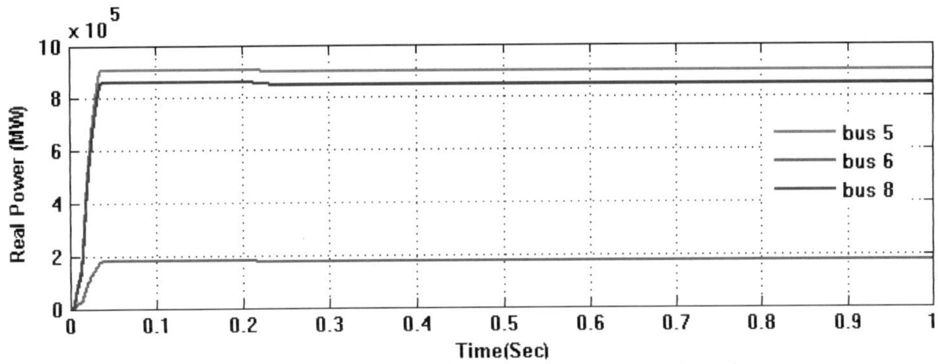

Fig.6 Real power comparison at bus 5,6 and 8 of healthy 9 bus system.

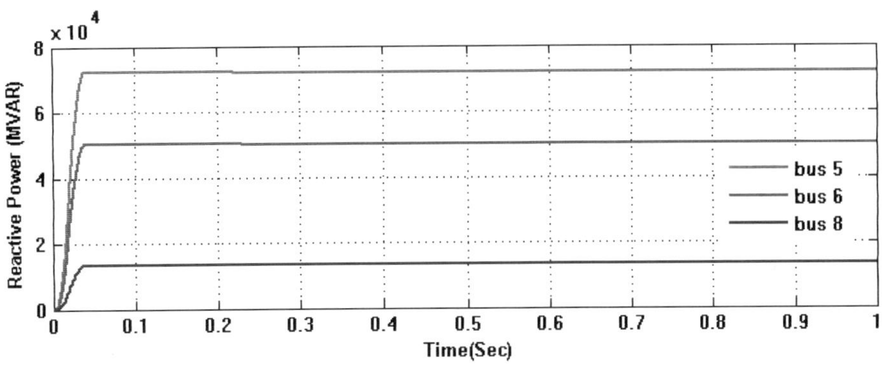

Fig.7 Reactive power comparison at bus 5,6 and 8 of healthy 9 bus system.

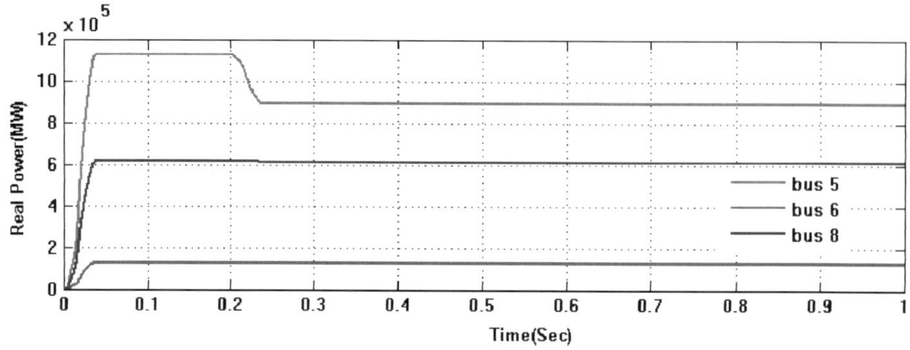

Fig.8 Real power comparison at bus 5,6 and 8 of unhealthy 9 bus system.

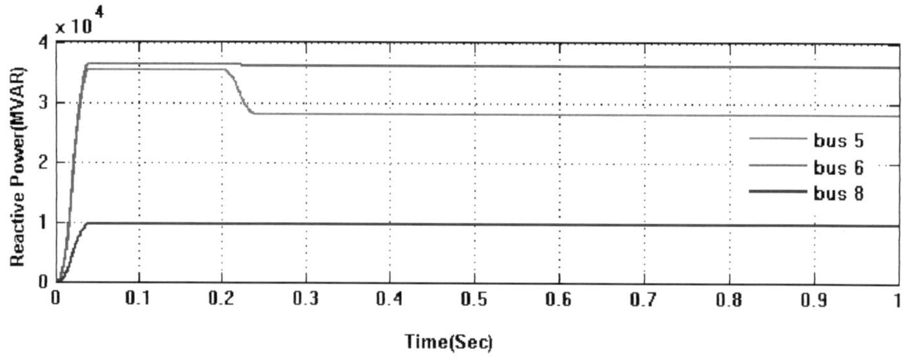

Fig.9 Reactive power comparison at bus 5,6 and 8 of unhealthy 9 bus system.

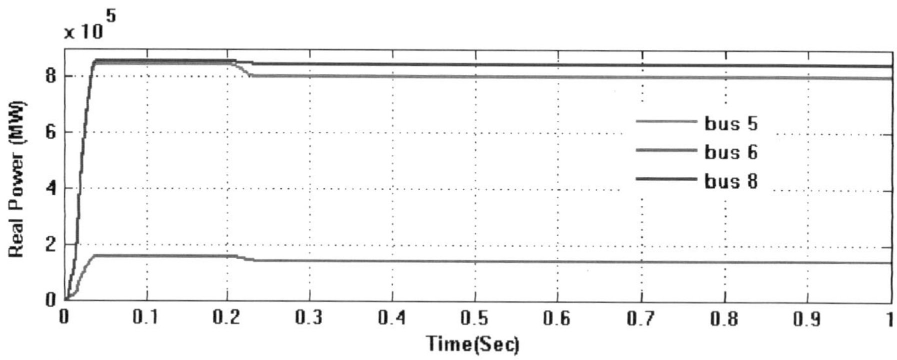

Fig.10 Real power comparison at bus 5,6 and 8 of unhealthy 9 bus with DVR system

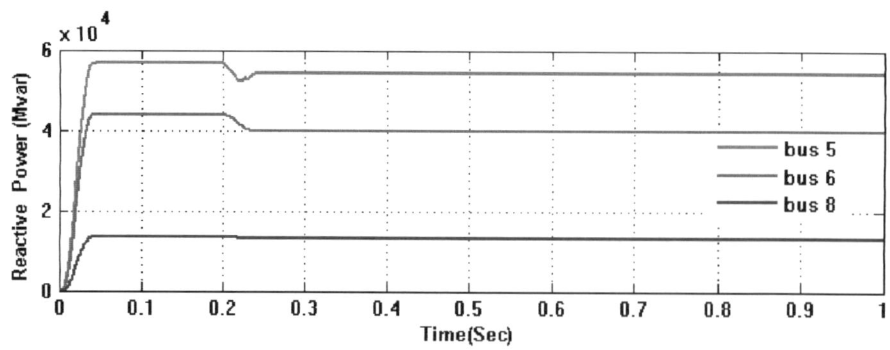

Fig.11 Reactive power comparison at bus 5,6 and 8 of unhealthy 9 bus with DVR system

The RMS voltage comparison of 9 bus system at bus 5,6 and 8 under different conditions are shown in Table 1. It reveals that IDVR in unhealthy condition can improve the RMS voltages at bus 5, 6 and 8. The real and reactive powers of the 9 bus arrangement at bus 5,6 and 8 under different conditions are shown in Table 2 and 3. All the tables reveal that DVR in unhealthy condition can improve the real and reactive powers at bus 5, 6 and 8.

Table 1. RMS voltage comparison of the Multibus system

CONDITIONS	RMS Voltage (Volts)		
	Bus 5	Bus 6	Bus 8
Healthy	4371	4382	4123
Unhealthy	2988	4149	3913
Unhealthy with DVR	3892	3912	4110

Table 2. Real power comparison of the Multibus system

CONDITIONS	Real power (MW)		
	Bus 5	Bus 6	Bus 8
Healthy	0.08981	0.01769	0.08497
Unhealthy	0.08992	0.01275	0.06123
Unhealthy with DVR	0.07971	0.01402	0.08446

Table 3. Reactive power comparison of the Multibus system

CONDITIONS	Reactive power (MVAR)		
	Bus 5	Bus 6	Bus 8
Healthy	0.7173	0.5002	0.1335
Unhealthy	0.2803	0.3604	0.9619
Unhealthy with DVR	0.5410	0.4050	0.1327

4. Conclusion

In this paper, the performances of IEEE9 bus system during healthy, unhealthy ad unhealthy with DVR conditions have been studied and the roles of DVR on the compensation of the powers have been found out. The RMS voltage of unhealthy condition with DVR has improved voltages of 3892V, 3912V and 4110V at bus 5, 6 and 8 respectively. 0.07971 MW, 0.01402 MW and 0.08446 MW are the real powers of unhealthy 9 bus system with DVR at bus 5, 6 and 8 correspondingly. The reactive powers at bus 5, 6 and 8 of unhealthy system with DVR are 0.5410 MVAR, 0.4050 MVAR and 0.1327 MVAR respectively. The results prove that DVR in unhealthy system can improve RMS voltages, real and reactive powers.

References

[1] Hedman KW O'Neill RP Fisher EB and Oren SS 2009 Optimal transmission switching with contingency analysis *IEEE Transactions on Power Systems 24*(3) pp1577-1586

[2] Ramya G Ganapathy V and Suresh P 2017 Power quality improvement using multi-level inverter based DVR and DSTATCOM using neuro-fuzzy controller *International Journal of Power Electronics and Drive Systems* 8(1) p316

[3] Ramakrishnan B Durdu A Rajagopal K and Akgul A 2020 Infinite attractors in a chaotic circuit with exponential memristor and Josephson junction resonator AEU-International Journal of Electronics and Communications 123 p153319

[4] Magadum RB Patil BK Dodamani SN and Shetty VJ 2019 August Contingency based UPFC Placement for Steady State Analysis In 2019 2nd International Conference on Power and Embedded Drive Control (ICPEDC) (pp 450-453) IEEE

[5] Suresh P Baskaran B and Ramya G 2017 Fuzzy Logic Controller Based IDVR in IEEE 30 Bus System for Voltage Sag Compensation *Indian Journal of Science and Technology* 10 p29

[6] Janarthanan R Doss S and Balamurali R 2020 Robotic-based nonlinear device fault detection with sensor fault and limited capacity for communication Journal of Ambient Intelligence And Humanized Computing

HARDWARE DESIGN AND IMPLEMENTATION OF MICROCONTROLLER BASED REDUCED SWITCH MULTILEVEL INVERTER TOPOLOGY FOR HARMONIC STUDIES USING MRPWM

[1]Uma D, Senior Assistant Professor, umavijay@eee.sastra.edu,
[1]Augusti Lindiya S, lindiya, Senior Assistant Professor, lindiya@eee.sastra.edu,
[1]subashini N, Associate Professor, nmnsi@eee.sastra.edu,
[2]A.Vijayakumar, Professor, viji1me@gmail.com
[1]SASTRA Deemed University,Thanjavur
[2]Anjalai Ammal Mahalingam Engineering College,Tiruvarur

ABSTRACT: Multilevel inverter (MLI) is the recent advancement in power electronics which employed to synthesize a high-quality output voltage waveform. This paper focuses on developing a single phase modified nine-level inverter with the reduced number of switches for improving the quality of output voltage waveform. Single carrier and multireference Pulse Width Modulation (PWM) technique is utilized for generating the switching pulses. The proposed topology consists of a single DC source, level generation part (balancing capacitor and bidirectional switches) and polarity generation part (H-Bridge). The operation of the proposed topology is examined for generating 9-level output voltage using MATLAB/SIMULINK. The result of the proposed topology is verified through laboratory-based prototype model. For experimental setup, the switching pulses are generated by PIC16F877A microcontroller. This proposed lab experimental and simulation can be used to study the value of %THD at output side of the inverter.

Keywords: Multilevel Inverter Cascaded H-Bridge Multilevel Inverter, Total Harmonic Distortion, Multi Reference Pulse Width Modulation, Microcontroller

I. INTRODUCTION

MLIs has many attractive features as compared with two-level inverter such as low switching losses, low electromagnetic interference and capable of generating the quality output voltage waveform (Najafi and Yatim, 2012; Hinago and Koizumi, 2010). The conventional MLI topologies are Cascaded H-Bridge MLI (CHBMLI), Diode Clamped Multilevel Inverter (DCMLI) and Flying Capacitor Multilevel Inverter (FCMLI) (Jami et al, 2014) FCMLI and DCMLI require a high number of clamping capacitors and clamping diodes, which are heavy and bulky as the level increases. CHBMLI does not require clamping capacitors and clamping diodes, but it requires separate DC sources for generating the required output voltage level. The main drawbacks of conventional MLIs are utilizing the higher number of components such as power semiconductor switches, clamping diodes and balancing capacitors. Therefore, it reduces the reliability of the inverter (Jami et al, 2014; Panda and Mohapatra, 2019; Naggar and Abdelhamid, 2008) CHBMLI involves a series connection of single-phase H-Bridge inverter which generates the medium output voltage levels by utilizing the low voltage components. Depending upon the output voltage levels, the number of H-Bridge can be connected in series, which leads to increase in the utilization of power semiconductor switches (Najafi and Yatim, 2012; Hinago and Koizumi, 2010). To overcome the drawbacks mentioned above, a new topology is developed with a reduced number of switches for generating the required output voltage level. Compared with conventional MLIs, the proposed topology requires a minimum number of semiconductor devices to achieve the same number of output levels. As a result, the %THD of output voltage waveform is reduced, and thus, the quality of the waveform could be improved.

Several modulation and control techniques have been adopted for MLI, which includes sinusoidal PWM, Space Vector Modulation (SVM), and Selective Harmonic Elimination (SHE) (Tolbert and Habetler, 1999). SPWM is generally employed for its simplicity and less control complexity. The different carrier arrangement in SPWM techniques are Phase Disposition PWM (PDPWM), Phase Opposition Disposition PWM (PODPWM), Alternate Phase Opposition Disposition PWM (APODPWM) and Carrier Overlapping PWM (COPWM) (Gupta and Jain, 2014]. To obtain lower %THD, the overmodulation technique has adopted and thereby reducing the size of passive filters (Mohammed et al, 2018) A new modulation method known as Variable Amplitude Phase Shifted PWM (VA-PSC-PWM) has employed in micro grid connected CHBMLI, which provides superior THD profile (Biswas and Khan, 2017). some investigators (Podder et al, 2016), describe the carrier signal is modified to reduce the duration of the volt-sec output of the CHBMLI. By reducing the volt-sec output, the output voltage becomes near sinusoidal and reduces harmonic content drastically (Harbi et al, 2018). A low frequency transformer based reduced switch cascaded H bridge MLI has been presented (Khounjahan et al, 2015). SHEPWM technique has been employed in the developed topology to mitigate lower order harmonic component (Khounjahan et al, 2015). In a discussion (Suroso et al 2017), SPWM method has been employed with low pass filter. This method has utilized two triangular waves and two sinusoidal waveforms with opposite phase to generate the triggering pulses for five-level reduced switch inverter with a voltage balancing circuit. SHE method has used to mitigate the third and fifth harmonic components. Newton Raphson method has employed to solve the nonlinear equations (Gobinath et al, 2013) Phase Shift PWM (PSPWM) method requires '(n-1)/2' carriers for 'n' level inverter. Triggering pulses have been generated by comparing the sinusoidal reference wave with the carriers, which are phase shifted each other by an angle of 360°/[(n-1)/2]. In modified PSPWM method, two sets of carrier waveforms have been used where one set has the magnitude from 0 to 1 and another set has the magnitude from 0 to -1 and phase shifted by 180° with respect to first set. These carriers have been compared with a bipolar sinusoidal waveform. It is discussed in one article [Kirthika Devi and Srivani, 2017), the modified PSPWM method provides lower THD when compared to conventional PSPWM. In order to obtain the desired 5 level output, two reference sine wave and one carrier wave have been utilized (Panchal et al, 2014). Level shifted Phase Disposition technique has been chosen for the developed reduced switch MLI in (Shuvo et al, 2019).

Multiple Reference PWM (MRPWM) technique is adopted for generating the switching pulses in proposed MLI. In this technique, multiple unipolar sinusoidal reference signals with a single carrier are utilized. This PWM technique requires '(n-1)/2' reference signals for generating 'n' number of levels. The reference signals have the same frequency, same

IOP Conf. Series: Materials Science and Engineering **994** (2020) 012019 doi:10.1088/1757-899X/994/1/012019

amplitude, and in phase with each other, but it have offset by a value equal to the magnitude of carrier signal (Rajalakshmi and Rangarajan, 2016). A single-phase reduced switch MLI is proposed to reduce the harmonic content in the output voltage. It operates in both high and low frequency switching, which cause improves efficiency, reduce the capital and maintenance cost. Also, it has some additional benefits such as the minimum number of switches, utilizing a single DC source for generating the required output voltage level, high efficiency, low cost and simple control scheme as compared with conventional PWM method. The proposed topology could be employed in different high and medium power applications such as HVDC, FACTS and UPS.

II. PROPOSED MLI TOPOLOGY

MLIs have some important features such as improved staircase waveform, reduction in common mode voltage, low distortion and there exists no EMI problem. The number of levels is indirectly proportional to %THD, and the number of power semiconductor device is directly proportional to the number of levels. A block diagram for the proposed topology is shown in Fig. 1 which comprises of DC source, reduced switch MLI with balancing capacitor, single carrier multi reference PWM and load. MRPWM (Multi Reference Pulse Width Modulation) technique is used to generate the switching pulses for the required output voltage level with reduced harmonic content.

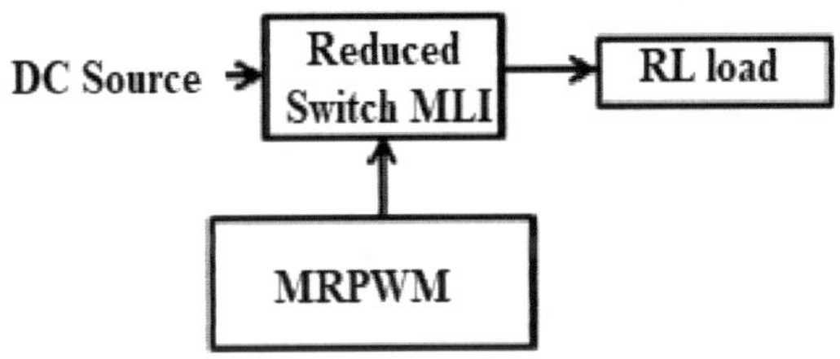

Fig. 1. General block diagram for the proposed work

The proposed topology is shown in Fig. 2. It consists of single-phase cascaded H-bridge inverter (S_1, S_2, S_3 and S_4), three bidirectional power semiconductor switches (S_5, S_6, S_7) and four capacitors (C_1, C_2, $C_3$and C_4). The bidirectional power semiconductor switch consists of the combination of a single switch and four diodes. The phase output voltage level is given by,

$$N = 2n+1$$

Where 'n' represents the number of capacitors, and 'N' represents the number of levels. The switching combinations for generating the nine-level output voltage, is shown in Table 1. Comparison of the number of components utilized in the proposed inverter is compared with the components used in the conventional MLIs for generating nine-level output voltage is given in Table 2. In the proposed MLI, each bidirectional switch consists of one transistor and four diodes. For example, generating a 9-level output voltage requires three bidirectional switches. Here, bidirectional switch consists of one switch with four diodes.

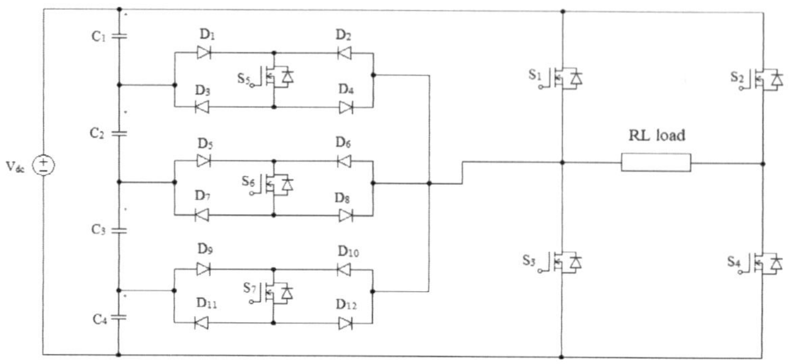

Fig. 2 Proposed topology of MLI

Table 1 Switching combinations to create a nine-level output voltage

V_0	S_7	S_6	S_5	S_4	S_3	S_2	S_1
V_{dc}	0	0	0	1	0	0	1
$3V_{dc}/4$	0	0	1	1	0	0	0
$2V_{dc}/4$	0	1	0	1	0	0	0
$V_{dc}/4$	1	0	0	1	0	0	0
0	0	0	0	1	1	0	0
$-V_{dc}/4$	0	0	1	0	0	1	0
$-2V_{dc}/4$	0	1	0	0	0	1	0
$-3V_{dc}/4$	1	0	0	0	0	1	0
$-V_{dc}$	0	0	0	0	1	1	0

Table 2 Comparison of Components

Type of component	**Conventional MLIs**			**Proposed MLI**
	Diode Clamped MLI	**Flying Capacitor MLI**	**Cascaded H bridge MLI**	
Main switches	16	16	16	7
Main Diodes	16	16	16	12
Clamping diodes	56	0	0	0
DC bus capacitors	8	8	0	4
Balancing capacitors	0	28	0	0
Isolated DC source	1	1	4	1

2.1 PRINCIPLE OF OPERATION

The different modes of operation for generating the nine-level output voltage are shown in Fig. 3. Fig. 3a to Fig.3i show that the current flow path in different mode of operation using a solid line. The voltage across each capacitor is ($V_{dc}/4$). In mode 1, the power semiconductor devices S_1 and S_4 are conducting and the remaining devices are in OFF state. The voltage available across the capacitors (C_1, C_2, C_3 and C_4) is V_{dc} which appears across the load. In mode 2, the conducting devices are S_5 and S_4. The capacitors C_2, C_3 and C_4 are connected across the load. Therefore, the voltage available at the load is ($3V_{dc}/4$). In mode 3, the power semiconductor devices S_6 and S_4 are turned on. The capacitors C_3 and C_4 are connected across

IOP Conf. Series: Materials Science and Engineering **994** (2020) 012019 doi:10.1088/1757-899X/994/1/012019

the load. The voltage available at the load is ($V_{dc}/2$). In mode 4, the power semiconductor devices S_7 and S_4 are turned on. Only the capacitor C_4 is connected across the load and the current flow path is shown by the solid line. The voltage available across the load is ($V_{dc}/4$). In mode 5, S_3 and S_4 are conducting so that the load is short circuited. Therefore, the volatge available across the load is zero. In mode 6, the negative polarity output voltage can be obtained when the devices S_5 and S_2 are turned on. The voltage available at the output is ($-V_{dc}/4$). In mode 7, S_6 and S_2 are turned on so that the load voltage is equal to ($-V_{dc}/2$). Similarly, the S_7 and S_2 power conductor switches are turned on in the 8th mode whereas the S_2 and S_3 are conducted in the 9th mode. The voltage across the load in 8th and 9th mode is ($-3V_{dc}/4$) and $-V_{dc}$ respectively.

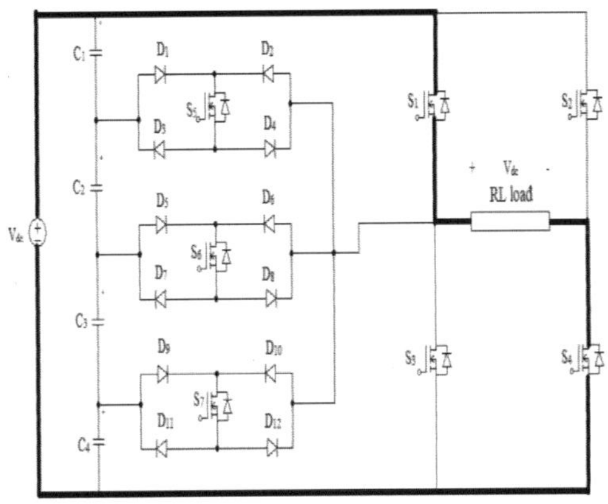

(a)

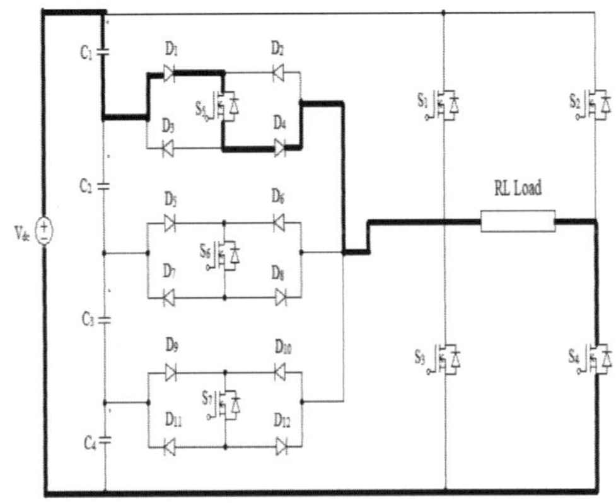

(b)

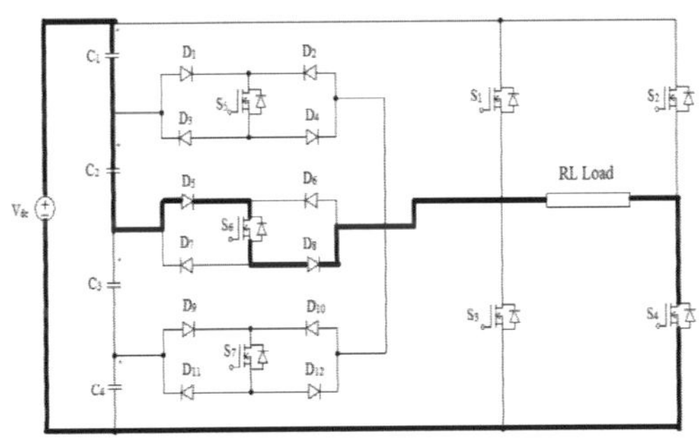

(c)

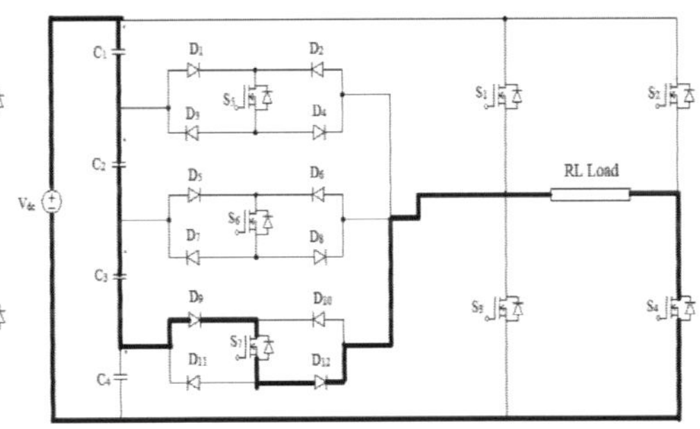

(d)

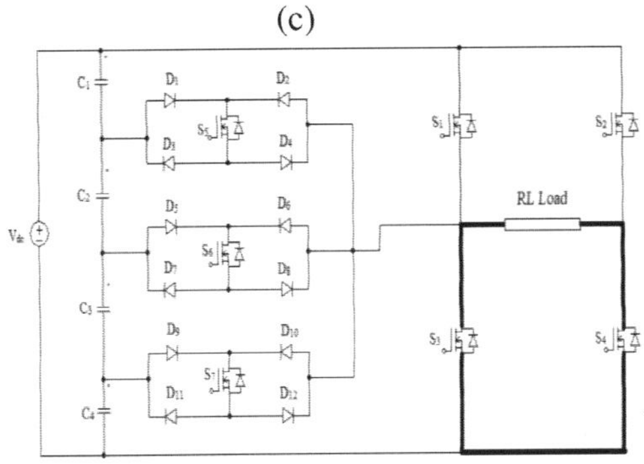

(e)

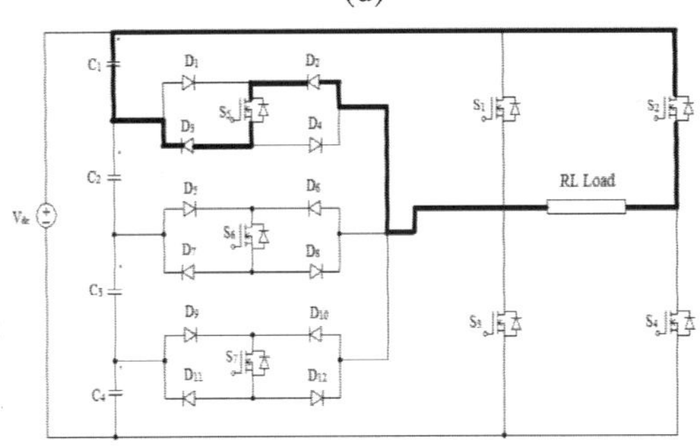

(f)

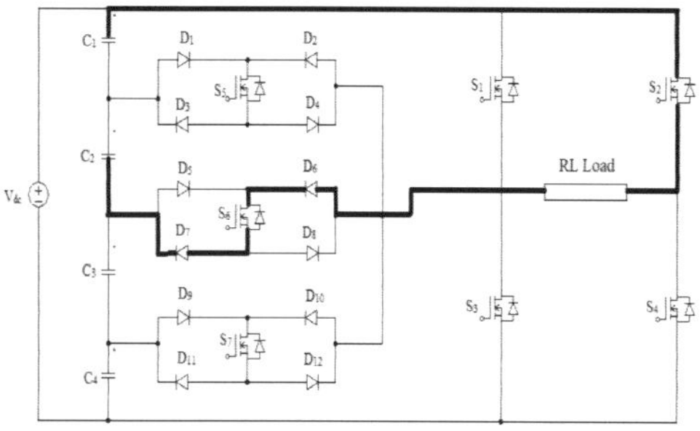

(g)

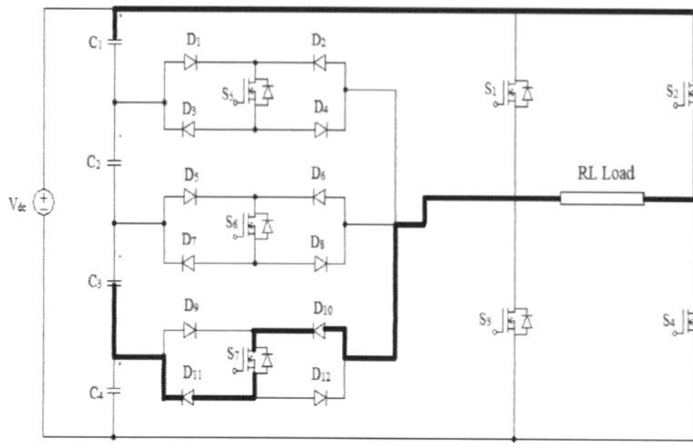

(h)

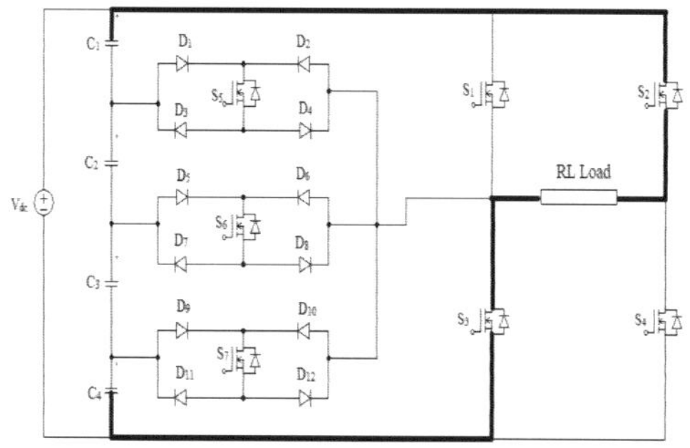

(i)

Fig. 3. Modes of operation: (a) Mode 1:V_L= V_{dc} (b) Mode 2:V_L= 3V_{dc}/4 (c) Mode 3:V_L= V_{dc}/2 (d) Mode 4:V_L= V_{dc}/4 (e) Mode 5:V_L= 0 (f) Mode 6:V_L= -V_{dc}/4 (g) Mode 7:V_L= -V_{dc}/2 (h) Mode 8:V_L= -3V_{dc}/4 (i) Mode 9:V_L= -V_{dc}

III. MULTI REFERENCE PULSE WIDTH MODULATION TECHNIQUE

Multiple Reference PWM technique (Jayapalan and Edward, 2017) is used for generating the switching pulses of the proposed MLI topology. Here, single triangle carrier and multiple unipolar sine reference are adopted in the pulse generation. Here, an 'N' level inverter requires '(N-1)/2' reference signals with the same amplitude and frequency which are in phase. But these signals have an offset whose value is equal to the magnitude of carrier signal. Therefore, a nine-level inverter requires four reference signals and a single carrier signal. The pulses are generated by comparing the reference signals V_{ref1}, V_{ref2}, V_{ref3} and V_{ref4} with a carrier signal are shown in Fig. 4.

The reference voltage V_{ref1} is compared with a carrier signal and a pulse is generated when its amplitude exceeds the carrier signal. V_{ref1} is compared with the carrier signal till V_{ref2} exceeds the carrier signal. Similarly, V_{ref2} exceeds the carrier signal which generates pulses till V_{ref3} exceeds the carrier signal. This procedure is repeated for generating the switching pulses to all the power semiconductor switches. Fig. 5 shows the switching pulses for the proposed topology switches. The modulation index for multi reference PWM is given by,

$$M = \frac{A_r}{A_c} \qquad (1)$$

where 'A_r' is the peak to peak value of the reference signal and 'A_c' is the peak value of the carrier signal. The phase displacement for modulation index more than 0.75 is determined by referring the article (Jayapalan and Edward, 2017).

$$\theta_1 = \sin^{-1}\left(\frac{A_c}{A_r}\right) \tag{2}$$

For a reference wave whose amplitude is 5 times the triangular carrier, the angle θ_1 is calculated as follows

$$\theta_1 = \sin^{-1}\left(\frac{1}{5}\right) = 11.54° \tag{3}$$

Similarly, the remaining angles can be calculated using the following formula

$$\theta_2 = \sin^{-1}\left(\frac{2A_c}{A_r}\right) \tag{4}$$

$$\theta_3 = \sin^{-1}\left(\frac{3A_c}{A_r}\right) \tag{5}$$

$$\theta_4 = \sin^{-1}\left(\frac{4A_c}{A_r}\right) \tag{6}$$

$$\theta_5 = \pi - \theta_4 \tag{7}$$

$$\theta_6 = \pi - \theta_3 \tag{8}$$

$$\theta_7 = \pi - \theta_2 \tag{9}$$

$$\theta_8 = \pi - \theta_1 \tag{10}$$

The theoretical waveform of nine-level output voltage with the corresponding switching angles is shown in Fig. 6.

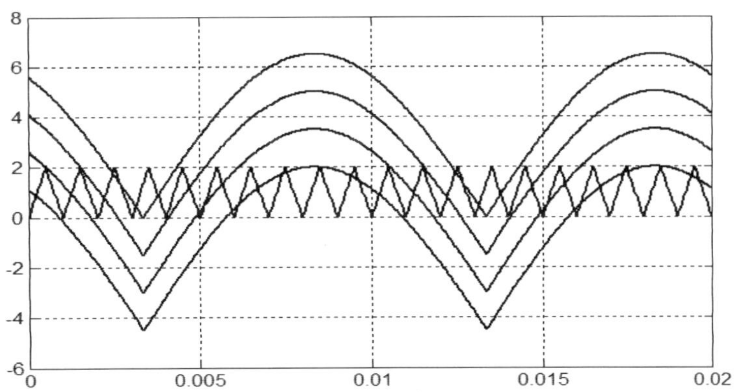

Fig.4. MRPWM Technique

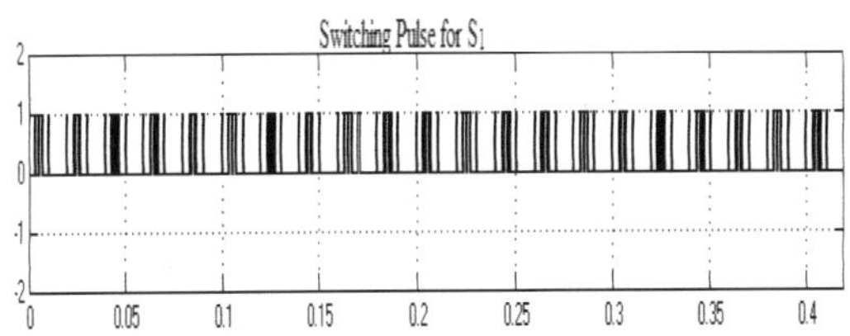

(a)

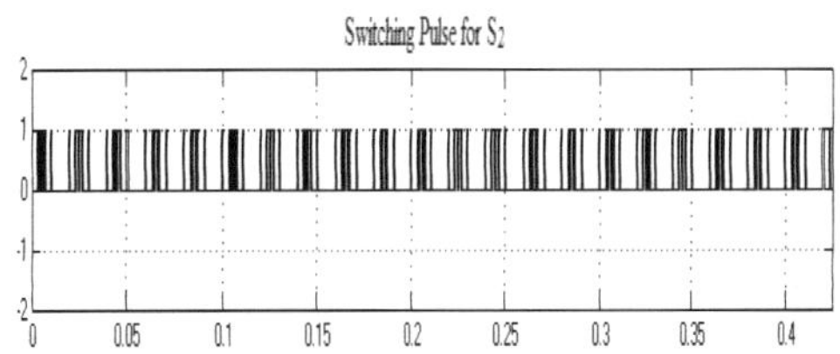

(b)

IOP Conf. Series: Materials Science and Engineering **994** (2020) 012019 doi:10.1088/1757-899X/994/1/012019

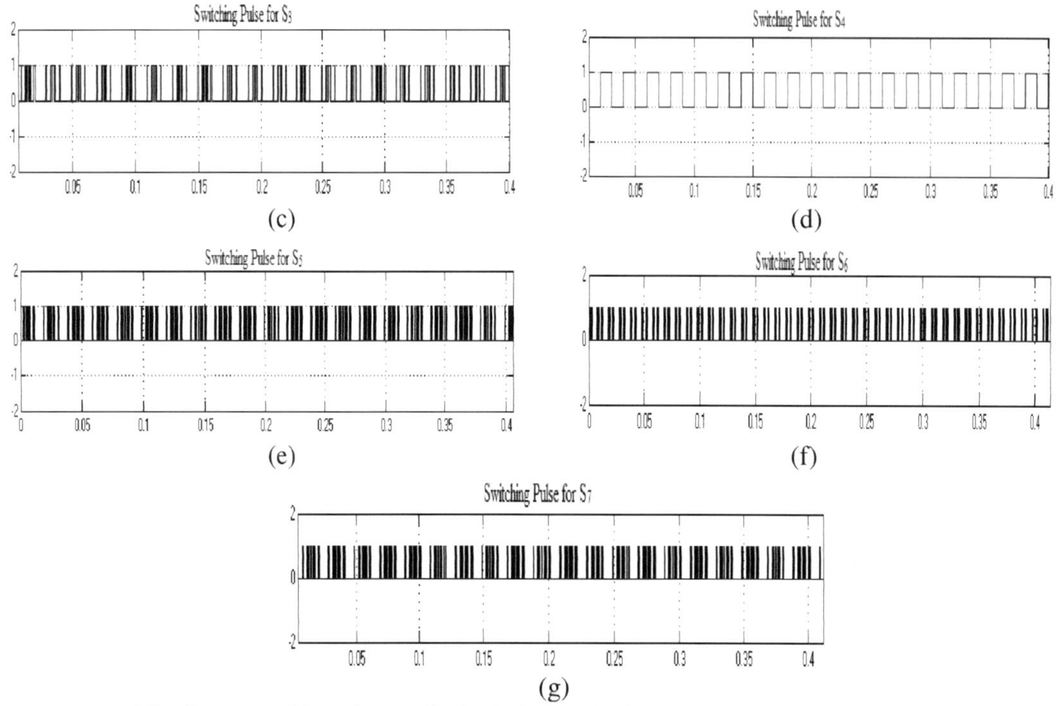

Fig.5. PWM Signals: (a) Switch S_1 (b) Switch S_2 (c) Switch S_3 (d) Switch S_4 (e) Switch S_5 (f) Switch S_6 (g) Switch S_7

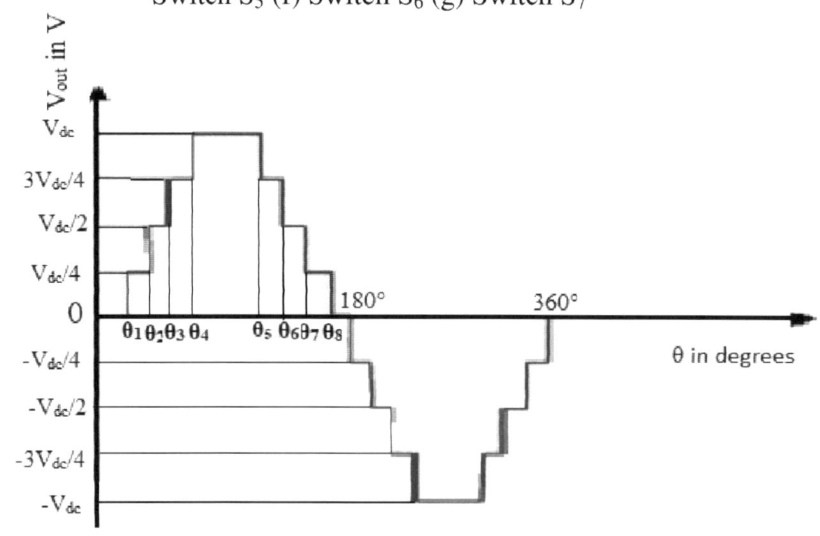

Fig. 6. Theoretical 9-level output voltage with the appropriate switching angle

IV. SIMULATION AND EXPERIMENTAL RESULTS

The simulation model for proposed MLI using MATLAB is shown in Fig 7. The switching pulses are generated using Multi Reference and single carrier PWM technique. The carrier switching frequency is 1K Hz. The simulation parameters are as follows input voltage (DC source value) =230 V, capacitor = 470 μF and load resistance = 100 Ω and L=100 mH. The 9-level output voltage waveform and its FFT spectrum for proposed MLI are shown in Fig. 8 and 9 respectively. The %THD of the output voltage waveform is 16.34%. Fig. 10 shows the load current waveform.

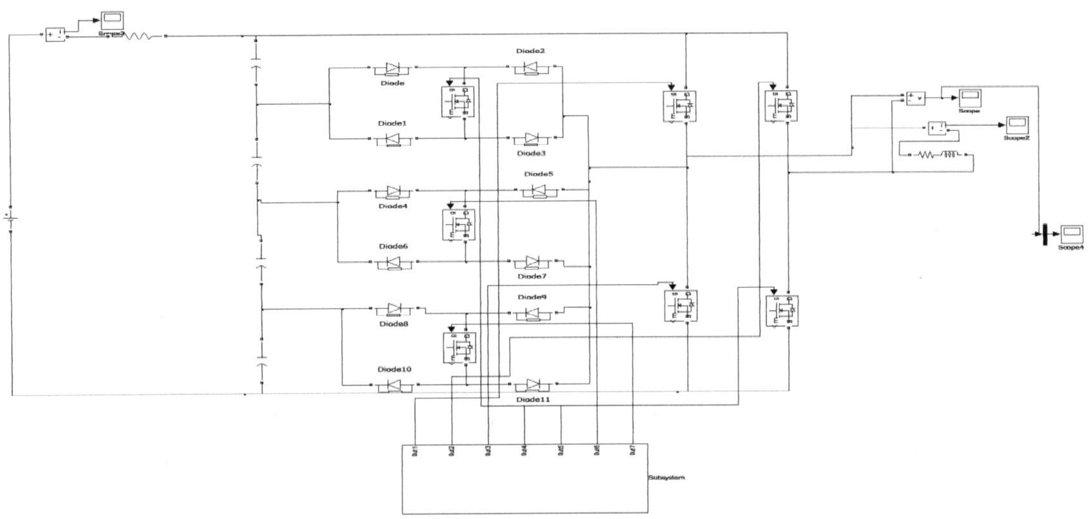

Fig. 7. Simulation Diagram for proposed MLI

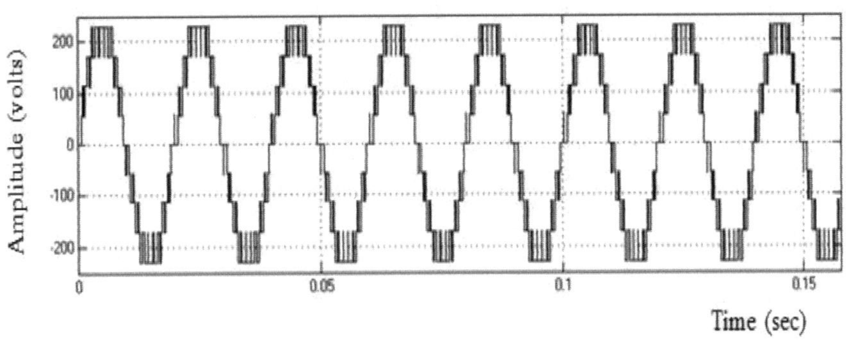

Fig. 8. 9-level output voltage waveform for proposed MLI

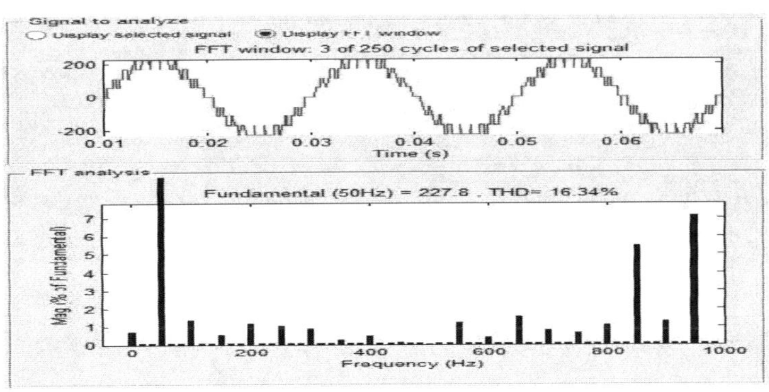

Fig. 9. FFT Analysis for proposed MLI

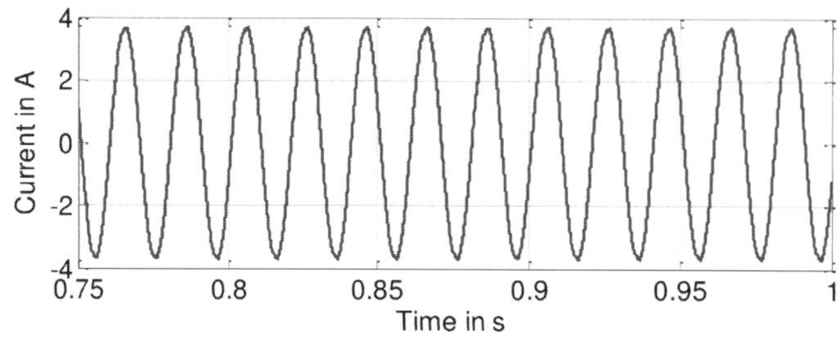

Fig. 10. Load current waveform for proposed MLI

Fig. 11 shows the laboratory based experimental setup of the proposed MLI. The supply is fed to rectifier which converts AC into DC power using a diode bridge rectifier. The obtained output is fed to a proposed MLI that converts DC into AC power. PIC16F877A microcontroller is utilized for generating the switching pulses in the experimental setup. Fig. 12(a-g) show the switching pulses are obtained from the PIC16877A which are fed to the proposed MLI switches through the driver circuit. In experimental setup, %THD is obtained using power quality analyzer. Fig. 13 and Fig. 14 show the experimental 9-level output voltage and its corresponding FFT analysis respectively. The %THD obtained from the experimental prototype using power quality analyzer is compared with the simulation results using MATLAB as shown in Table 3. The %THD attained from the hardware setup is 22.76% whereas the %THD achieved from the simulation is 16.34%.

Table 3 Comparison of %THD for 9-level output voltage

S.NO	Type	%THD
1	Simulation	16.34
2	Hardware	22.76

Fig.11. Experimental Prototype of Hardware

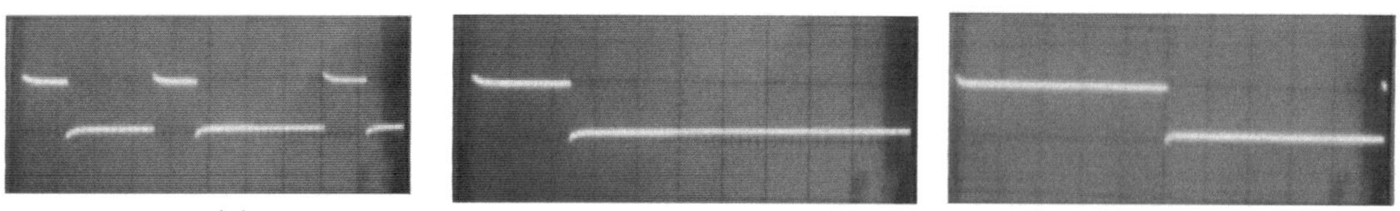

(a) **(b)** **(c)**

IOP Conf. Series: Materials Science and Engineering **994** (2020) 012019 doi:10.1088/1757-899X/994/1/012019

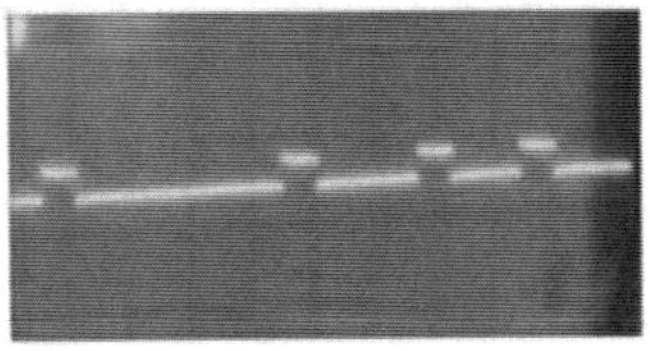

(d)

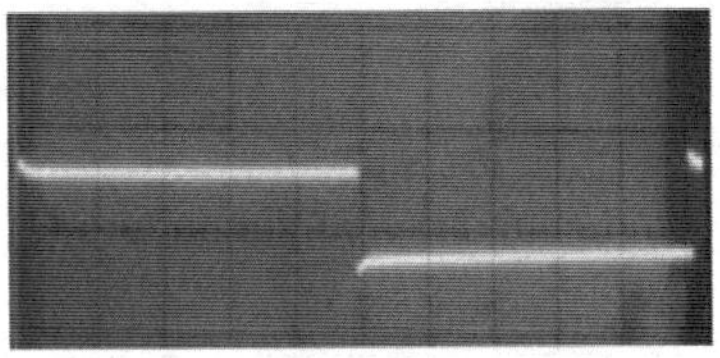

(e)

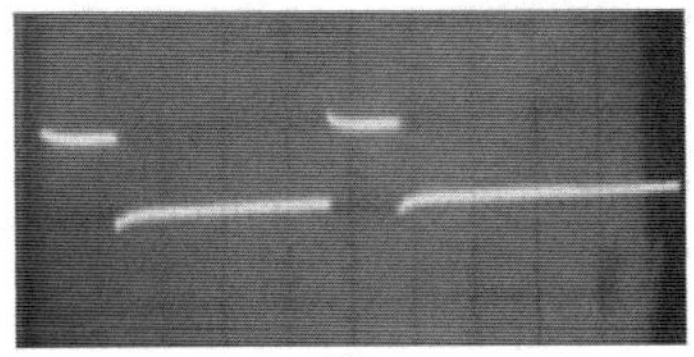

(f)

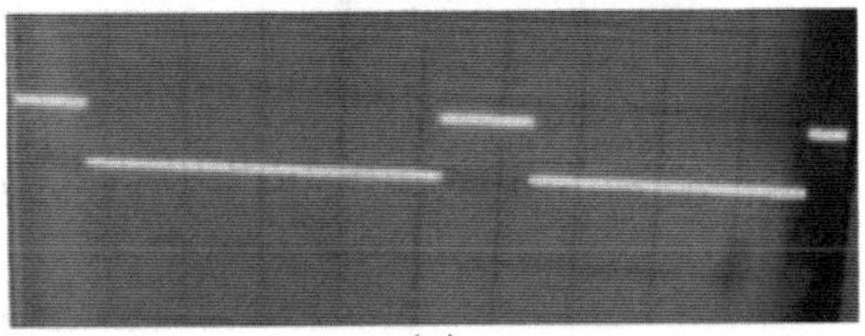

(g)

Fig.12. PWM pulses using microcontroller: (a) Switch S_1 (b) Switch S_2 (c) Switch S_3 (d) Switch S_4 (e) Switch S_5 (f) Switch S_6 (g) Switch S_7

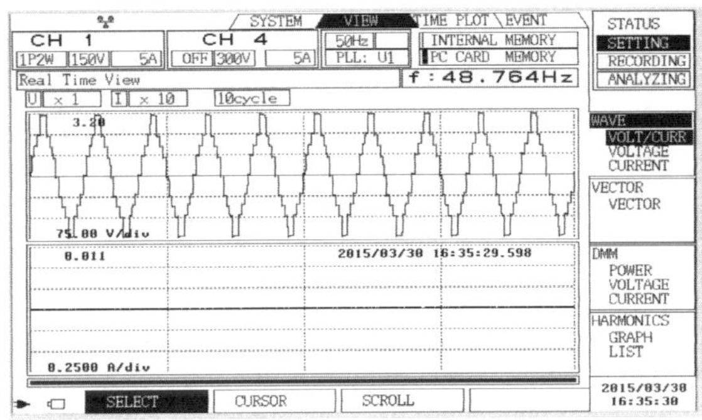

Fig.13 Experimental Result: (a) Hardware Prototype with output voltage (b) 9-level output voltage waveform for the proposed MLI

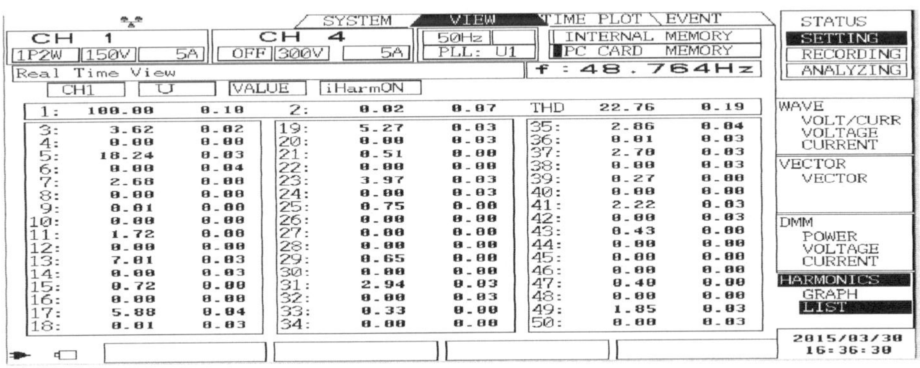

Fig.14 Output voltage FFT Analysis for proposed MLI

V. CONCLUSION

The reduced switch MLI topology has been proposed and its operation has been examined for generating the 9-level output voltage in both simulation and experimental setup. MRPWM technique has employed for generating the switching pulses for the proposed MLI. The main feature of this proposed technique is that it has only one carrier and multiple unipolar sinusoidal reference waveforms. The %THD of output voltage is 16.34% and 22.76% in simulation and experimental results respectively. As a result, the proposed topology provides better quality of output voltage waveform with reduced harmonic content. As there is a growing concern over power quality, this experimental setup may be utilized in power electronics laboratories as harmonic study purpose.

REFERENCES

[1] Bayat Z & Babaei E (2012 February) A new cascaded multilevel inverter with reduced number of switches. In: Power electronics and drive systems technology (PEDSTC), 2012 3rd (pp. 416–421) IEEE

[2] M M Biswas and M Z R Khan Amended THD with modified phaseshifted PWM for micro-grid connected multilevel inverter in Proc. IEEE Power Energy Conf Illinois (PECI) Feb 2017 pp 1-6

[3] K Gobinath S Mahendran and Dr I Gnanambal Novel cascad-ed H-Bridge multilevel inverter with harmonics elimination Pro-ceedings of 2013 International Conference on Green High Perfor-mance Computing IEEE 2013

[4] K K Gupta and S Jain A novel multilevel inverter based on switched DC sources IEEE Trans Ind Electron vol 61 no 7 pp 3269-3278 Jul 2014

[5] A Harbi Haitham Z Azazi Azza E Lashine and Awad E Elsabbe A higher levels multilevel inverter with reduced number of switches International Journal of Electronics 2018 VOL 105, NO 8 1286–1299

[6] Y Hinago H.Koizumi A single Phase Multilevel Using Switched Series/Parallel DC Voltage Sources IEEE Trans. Industrial Electronics Aug.2010 vol 57 no. 8 pp 2643-2650

[7] A Jami MRJ.Oskuee A Mokhberdoran A Van den Bossche Developed cascaded multilevel inverter topology to minimize the number of circuit devices and voltage stress of switches IET Power Electronics Feb 2014 pp 459-466

[8] Gowrishankar Jayapalan and Belwin Edward, "Implementation of 9-Level Hybrid Inverter using Multi Reference Pulse Width Modulation Technique with Flyback Converter for Photovoltaic System International Journal of Control Theory and Applications 2017

[9] V S.Kirthika Devi S G Srivani Modified phase shifted PWM for cascaded H Bridge Multilevel Inverter IEEE 2017

[10] H Khounjahan M R Banaei Amir Farakhor A new low cost cascaded transformer multilevel inverter topology using minimum number of components with modified selective harmonic elimination modulation Ain Shams Engineering Journal (2015) 6, 67–73

[11] Malinowski M Gopakumar K Rodriguez J Marcelo AP A survey on cascaded multi level inverters IEEE Trans. Ind. Electron. 2010 57 2197–2206

[12] Mustafa Fawzi Mohammed Ali Husain Ahmad AbdulRahim Thiab Humod Harmonics reduction of a five-level inverter by unbalanced car-riers and over-modulation techniques International Journal of Engineering & Technology 2018

[13] Naggar E N Abdelhamid T.H. Selective harmonic elimination of new family of multilevel inverters using genetic algorithms. Energy Convers. Manag 2008 49 89–95

[14] E.Najafi A.H.M.Yatim Design and Implementation of a New Multilevel Topology IEEE Trans Industrial Electronics Nov 2012 vol 59 no 11 pp 4148-4154

[15] Panda KP Mohapatra SK Anti-predatory PSO technique-based solutions for selected harmonic elimination in cascaded multilevel inverters Int J Ind Electron. Drives 2016 3 78–88

[16] Leon M Tolbert and Thomas G. Habetler Novel multilevel inverter carrier based PWM method IEEE Transaction On Industry Application Vol 35 No 5 Sep 1999 pp 1098-1107

[17] S Podder M Biswas and Z R. Khan A modi_ed PWM technique to improve total harmonic distortion of multilevel inverter in Proc. 9th Int Conf Elect Comput Eng (ICECE) 2016 pp 515-518

[18] Tejas M Panchal Rakesh A Patel Hiren S Darji Simulation of Modified Cascaded H-Bridge Multilevel Inverter for 3-Phase Asynchronous Motor IEEE 2014.

[19] J.Rodriguez Jih-Sheng Lai,Fang Zheng Peng Multilevel inverters: A survey of topologies controls and applications IEEE Trans. Industrial Electronics Aug.2002 vol.49 pp 724-738

[20] Sambasivam Rajalakshmi and Parthasarathy Rangarajan Analysis of Reduced Switch Topology Multilevel Inverter with Different Pulse Width Modulation Technique and Its Application with DSTATCOM Circuits and Systems 2016 7 2410-2424.

[21] Shuvangkar Shuvo Eklas Hossain Tanveerul Islam, Abir Akib Sanjeevikumar Padmanaban and Md Ziaur Rahman Khan Design and Hardware Implementation Considerations of Modified Multilevel Cascaded H-Bridge Inverter for Photovoltaic System IEEE Access 2019

[22] Suroso Abdullah Nur Aziz and Toshihiko Noguchi Five-level PWM inverter with a single DC power source for DC-AC power conversion International journal of power electronics and drive system IJPEDS Vol 8 No 3 2017 pp 1230-1237

[23] Manjusha S Nair and Dr. A.Rajaram, Low power receiver using envelope Detector converters International Journal of Advanced Information Science and Technology, 3(3): 50-57, March 2014

Design and methodology of algorithm for the enhancement of breast tumor images

Ms. Shwetha S V[1], Dr. Dharmanna L[2],Dr.Basavaraj S Anami[3]

[1] Research Scholar, Department of CSE, SDMIT, Ujire-574240
[2] Associate Professor & Head, Department of ISE, SDMIT, Ujire-574240
[3]Principal, KLEIT ,Hubballi, Karnataka
shweth4a@gmail.com,dharmannasdmit@gmail.com,anami_basu@hotmail.com

corresponding author email:shweth4a@gmail.com

Abstract. Breast cancer is the second largest disease in the world having high mortality rate among ladies. Breast image enhancement is required for the foundation of high accuracy segmentation. Several features are extracted from the ROI of mammogram in order to detect the cancer tumor in the early stage. Ultrasound breast images have the speckle noise during the capturing of image. In mammogram images also, the noises like Guassian, Normal and Poisson are introduced. There are no proper enhancements algorithms are designed to enhance the images for the preprocessing of the images. The image enhancement is broadly categorized into two different domains such as spatial and frequency. The spatial domain algorithms will smoothen the image quality and frequency domain algorithms will sharpen the image. Till today there is no concrete work on suitable algorithm methods using different domains of enhancement. The research work has been implemented in two stages. In the first stage we are identifying various noises in the mammogram and ultrasound images. In second stage the algorithms like mean, median, Homomorphic, Weiner, Modified Gabor, Anisoid diffusion filter, Contrast limited Adaptive Histogram equalization (CLAHE), Bilateral filter and Guided filter are worked out. Among all these filters we modified Gabor filter that clearly highlights the portion of the tumor. For all these filters Mean Square Error (MSE), RMSE, PSNR,VAR, SSIM,SD are computed. The best PSNR is obtained for the modified Gabor filter.

Keywords: MSE, PSNR, Filter, Mammogram, Homomorphic, Weiner, Malignant, Benign

1. Introduction

Breast cancer detection is a major issue in women community. Due to low energy X-ray image formation the mass in breast is not clearly visible. If we use high energy X-ray instruments the radiation will introduce cancer. If cancer does not exist or exist during screening process in mass (breast tumor).There are two kinds 1) Benign 2) Malignant. Benign tumor is a just mass with normal cells that do not spread across the body of the women 2. Malignant tumors are cancerous and they will spread across the body and causes death. In order to extract the exact portion of the mass for the analysis to detect either benign or malignant preprocessing stage is very much essential for proper visibility of the tumor for segmentation algorithm. The clear separation of the mass portion from the other parts of the breast for segmentation and analysis will assist the radiologist for the diagnosis in convenient method and till date no author has commandingly defined which method is suitable for the mammogram enhancement. In this work Weiner, Gabor, Mean, Modified Gabor, Homomorphic etc methods are implemented and the quality had been measured with PSNR and MSE as metrics. Fig. 1 shows a) the normal breast image, b) Mammogram with two benign and one malignant tumor and marked by the circle. c) Mammogram with cancer third stage with eight tumors. d) Malignant and e) Benign tumor. From the figure it is very difficult to identify the location and portion of the mammogram affected by the cancer. Figure 2 shows the different types of the cancerous tumors in the breast cancer.

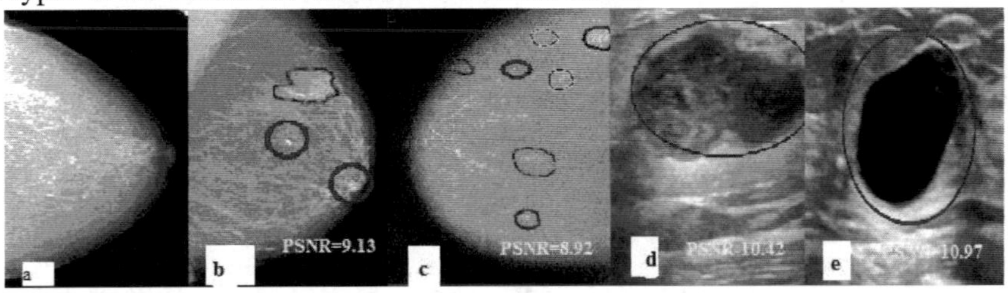

Figure 1. DDSM noisy mammogram images a) Normal b) two benign and one malignant tumor c) third stage eight tumor image, ultrasound d) Malignant e) benign tumor

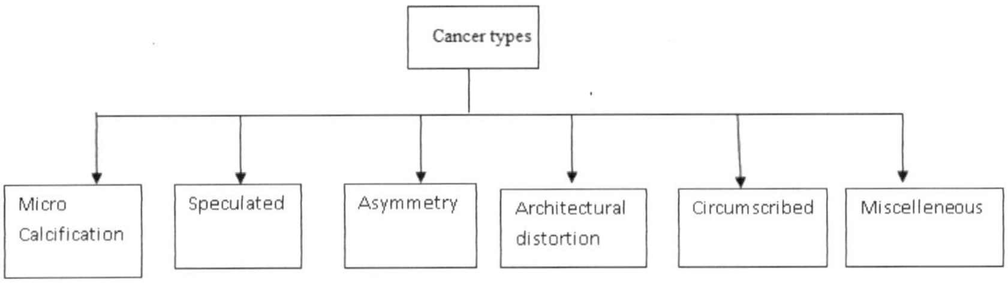

Figure 2. Different types of Breast Cancer

2. Previous Works

This section describes the work carried out by several authors for the enhancement of mammogram and ultrasound images.The section also reveals the limitations of the work carried out so far. Aziz Makhander & Bhagirathi Halalli presented work on various denoising and contrast enhancement algorithm to identify the breast mammogram enhancement methods for x-ray images. Breast x-ray images are denoised by linear or nonlinear techniques and accuracy was measured through RMSE and PSNR.

Vishnukumar et. al. emphasized on improving visual quality of mammograms. The smoothening and sharpening methods was proposed and the technique's significance was assessed as a beneficially mammogram image enhancement technique. In this technique the random noise was eliminated from x-ray images.

Yufeng Zheng demonstrated a work on breast cancer detection through Gabor features from digital mammograms. In this work, x-ray images were down sampled, denoised, quantized and enhanced. Nonlinear diffusion technique was adapted to suppress the noise. Sara Dehghani et.al. revealed work on "A method after improve preprocessing mammography" and the author divided work in four stages

1.Removing excessive parts of the mammography 2.Differentiation of the breast direction 3.Gray level threshold limit 4. Breast region segmentation from the background this was achieved by series of point operation and RGB techniques.

Deokate B H et. al. worked on image compression using Gabor filter. In this work author presented filtering technique can be used for image compression for large amount of graphical data or information that can be transferred over internet. Compression ratio of the technique is 68.65.Neha N et.al. worked on "filtering technique for preprocessing mammogram images to breast cancer detection, the author believes that enhancement of the mammogram image leads to the better accuracy of the segmentation which will benefit to the computer aided diagnosis, essential to provide doctors. The authors worked on several filters like median, Weiner, DWT and Anisotropic diffusion filter .In this work author claims DWT2Haar is the best technique for enhancement.

Tomislav Stoic & Braniar Redjin, presented on "Enhancement of microcalcification in digitized mammogram, multi-fractal and mathematical morphology approach". In the first method from original mammogram a corresponding multi-fractal images were created. Hence doctors has a freedom of segmentation in a user interactive way. The second technique enables user to high local contrast enhancement then followed significant suppression of background tissue of the breast. In this work the author mini MIAS has been used.

3. Methodology& Implementation

The section describes the various enhancement techniques for mammograms and ultrasound images. The enhancement of the image reduces the noise and increases the contrast with the tumor and also improves the visual appearance of the image

3.1 Weiner

The Weiner filter has mean square error. The optimal stationary linear enhancement filter for mammogram is degraded by additive noise and blurring. In the uncorrupted image f, the mean square error between f and f' is minimized. This error measure is given by

$$e^2 = E\left\{\left(f - \hat{f}\right)^2\right\}.........................eq(1)$$

Where E{.} is the expected value of the argument. The minimum of the error function in equation 1 is given in the frequency domain by the expression.

$$\hat{F}(u,v) = \left[H*(u,v)s_f(u,v)/s_f(u,v)|H(u,v)|^2 + s_\eta(u,v)\right]G(u,v)$$
$$= \left[H*(u,v)s_f(u,v)/|H(u,v)|^2 + (s_\eta(u,v)/s_f(u,v))\right]G(u,v)$$
$$= \left[1/H(u,v)*|H(u,v)|^2/(|H(u,v)|^2 + (s_\eta(u,v)/s_f(u,v)))\right]G(u,v)$$

This result is called least square filter or Weiner filter.

IOP Conf. Series: Materials Science and Engineering **994** (2020) 012020 doi:10.1088/1757-899X/994/1/012020

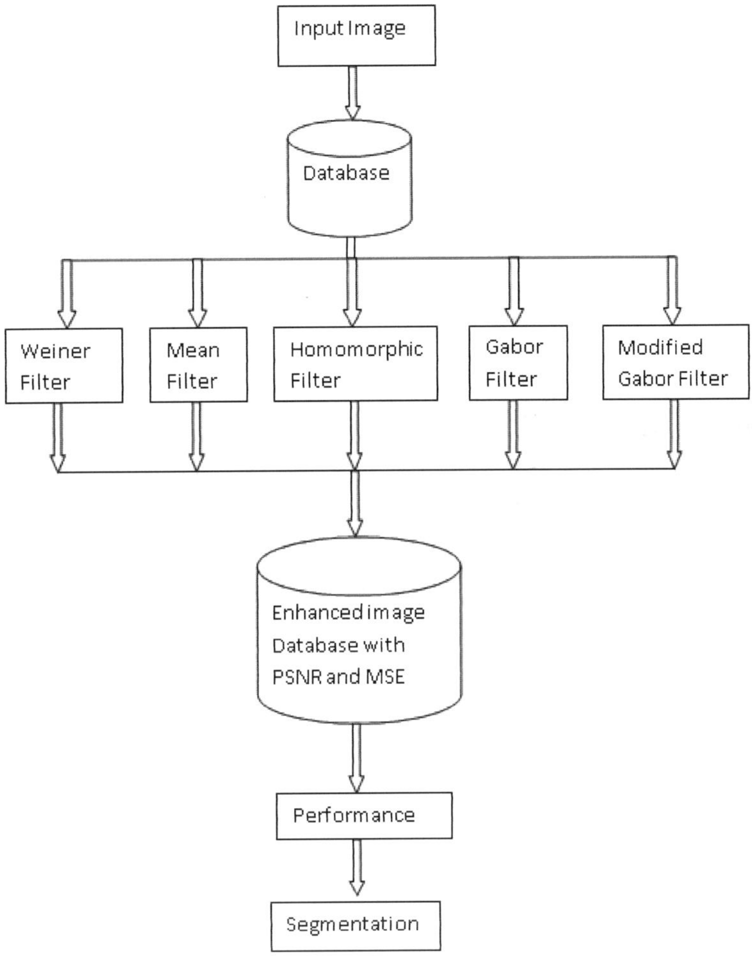

Figure 3. Dataflow diagram of the proposed work

3.2 Mean Filter

The main idea of mean image enhancement filter is to replace each pixel value in a mammogram with the mean value of its 3x3 kernel neighbors and including itself. This filter removes pixel values that are unrepresentative of their surroundings. The mean filtering is also refereed as a convolution filter. The kernel of the mean filter is as follows and this is adopted throughout the images.

1/9	1/9	1/9
1/9	1/9	1/9
1/9	1/9	1/9

The algorithm of the mean filter is as shown below

```
Algo.Mean Filter()
//The algorithm is about enhancement of image
//The input is image img
For any image,
 for i=0 to m-1
  for j=0 to n-1
   for k=i to i+3
    for l=j to j+3
       sum= sum(img(k,l))
      end for
    end for
   img(i,j)=1/9*(sum)
  end for
end for
```

Append two rows and columns at the end for the proper adjustment of last element of the image and kernel.

3.3 Homomorphic filter

Homomorphic filtering is used for image and signal processing which involves a nonlinear mapping to a different domain in which linear filter technique are adopted.
If we process z(u,v) by means of a filter function H(u,v)

$$s(u,v) = H(u,v).z(u,v)$$

$$= H(u,v)F_i(u,v) + H(u,v)F_r(u,v) \text{ ....equation 5}$$

Where s (u,v) is the Fourier transform of the result in the spatial domain

$$s(x,y) = \phi^{-1}\{s(u,v)\}$$

$$= \phi^{-1}\{H(u,v)F(u,v)\} + \phi^{-1}\{H(u,v)F_r(u,v)\} \text{ ....equation 6}$$

Assume,

$$i^{-1}(x,y) = \phi^{-1}\{H(u,v)F_i(u,v)\} \text{ ....equation 7}$$

and $r^{-1} = \phi^{-1}\{H(u,v)F_r(u,v)\}$ ....equation 8

Equation 6 can be expressed in the form

$$s(x,y) = i^{-1}(x,y) + r^{-1}(x,y) \text{ ....equation 9}$$

The inverse operation yields the desired image denoted by g(x,y)

$$g(x,y) = e^{xs(x,y)}$$

$$= e^{i^{-1}(x,y)}.e^{r^{-1}(x,y)}$$

$$= i_0(x,y)r_0(x,y)$$

Where,

$i_0(x,y) = e^{i^{-1}(x,y)}$ **and** $r_0(x,y) = e^{r^{-1}(x,y)}$

3.4 Gabor Filter

Gabor filter is a linear filter used for texture analysis that analyzes whether there is any separate frequency involvement in the mammogram in specific direction in a localized are around the region or analysis. It's a complex exponential modulated by a Gaussian function. Its impulse response in the 2d-plane has the following general form
Where u=radial frequency of the Gabor function

σx, σy = Spread of Gabor envelop along x and y axes

Self similar filter bank can be obtained by scaling and motion of the following function

3.5 Anistropic diffusion filter combined with Modified Gabor

This method is used to remove the noise without losing the content of mammography typically the edge, line or other part of the objects. The ADT Filter is linear filter which is used for highly blurred images for the enhancement.

4. Result and Discussion

This work is carried out on the mammogram samples from MIAS,DDSM databases in the Mat lab software in laptop. There are required different metric qualities to estimate the effectiveness of the proposed enhancement techniques. The rigorous experimental result analysis carried out to take decision which enhancement technique is able to generate significant enhanced results for breast cancer assessment.

In the experimental work DDSM dataset of mammogram images are involved. In the field of medical image analysis application for breast mass, lung cancer, Brain tumor analysis is required. The input image quality estimation is an open research and more challenging task for several applications like image denoise, image enhancement, image restoration adjusting image quality, image registration etc. that the enhancement accuracy is observed in terms of PSNR, Standard deviation, SNR (self simulating index measure (SSIM), IMAE, VAR, Root mean square error, AMBE, mathematical formulation of above parameter are as follows,

$$SD = \frac{\sqrt{\sum |x-u|}}{N}$$

Where n is value in data point and u is mean

$$SSIM = \frac{(2\mu_X\mu_Y + C_1)*(2\delta_{X,Y} + C_2)}{(\mu_x^2 + \mu_y^2 + C_1)*(\delta_x^2 + \delta_y^2 + C_2)}$$

$$PSNR = 10\log_{10}\left(\frac{MAX^2}{MSE}\right)$$

$$MSE = \frac{1}{MN\sum_{i=0}^{m-1}\sum_{j=0}^{n-1}[(i,j)-k(i.j)]^L}$$

$$RMSE = \sqrt{\frac{\sum_{j=1}^{n}(p_i - o_i)^2}{n}}$$

The quantitative results of the enhancement breast image is observed applying several parameters and provided in fig. 4 to fig.8. The figure 4 depicts,(a) the original noisy image with PSNR=10.19 , (b) Result of the mean filter with PSNR =61.10 (c) Result of modified Gabor with PSNR=73.79 d) Result of Homorphic filter with PSNR=64.58 (e) Result of Weiner filter with PSNR=34.96 (f) Result of Gabor with PSNR=28.61 (g) Result of Guassian filter with PSNR= 15.13h) Result of ADT filter with PSNR=17.25

Similarly the PSNR values for the malignant mammograms, ultrasound benign and malignant images.

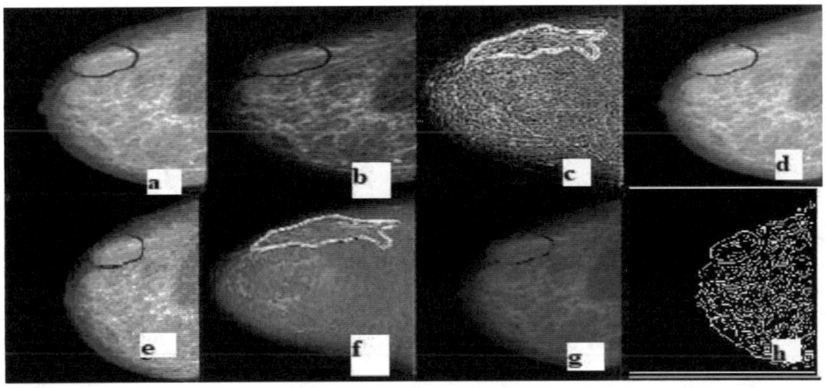

Figure 4. Benign Mammogram-a) Original b) Result of Mean filter c) Result of Modified Gabor d) Result of Homomorphic filter e) Result of Weiner filter f) Result of Gabor filter g)Result of Guassian filter h) Canny edge detector

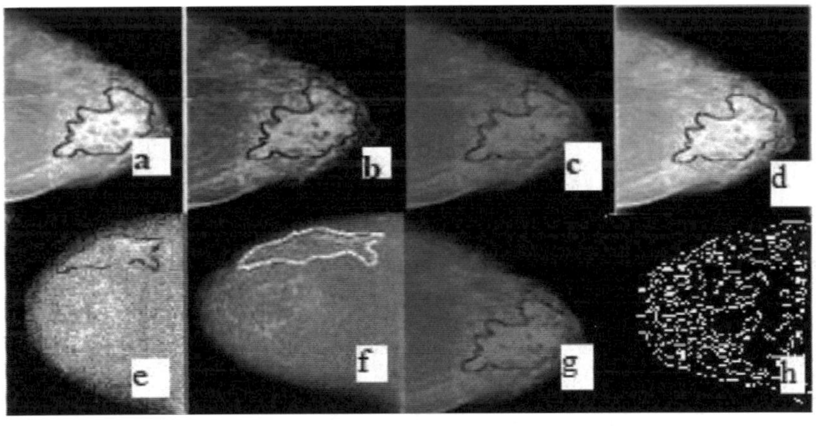

Figure 5. Malignant Mammogram-a) Original b) Result of Mean filter c) Result of Modified Gabor d) Result of Homomorphic filter e) Result of Weiner filter f) Result of Gabor filter g)Result of Guassian *filter h) Canny edge detector

From the above figures, it is observed that the result of the modified Gabor filter (Anistropic diffusion with Gabor filter) is better in performance on comparison with the other filtering methods in most of the parameters. Here the important advantage of Gabor filter is that it assists to extract multi scale feature related to edge present in breast image. The qualitative comparison of several enhancement filtering techniques on mammogram from DDSM dataset is provided in the table 1. From the qualitative and quantitative results, It is examined that the filtering method based upon Anisotropic diffusion filter with Gabor filter is able to generate significant result and has provided edge contrast information to separate edges of the breast and tumor. The parameters value like SNR,VAR, MSE, PSNR, RMSE etc. of the proposed modified Gabor filter is better than other filtering techniques and the improvements of the results (quantitatively & qualitatively) are the proof depicting the ADT with Gabor filter is the excellent among the other filtering techniques.

The table 1 describes the Poisson signal to Noise Ratio are calculated for the benign and malignant samples in mammogram and ultrasound images. The mean and Standard deviation of PSNR is calculated. The average of MSE is calculated.

Table 1. Performance of the filters

Sl.No.	Image	PSNR of Original Image	Weiner Filter	Homomorphic Filter	Mean filter	Gabor Filter	Modified Gabor
1.	Benign 1	10.69	35.26	64.58	63.44	27.63	70.36
2.	Benign 2	11.56	34.80	65.59	64.23	28.18	70.65
3.	Benign 3	9.45	33.83	64.82	63.42	29.76	82.17
4.	Benign 4	10.52	35.29	64.18	62.73	29.07	73.30
5.	Benign 5	9.69	34.88	64.38	63.31	29.61	81.85
6.	Malignant 1	11.22	35.99	62.41	61.28	27.12	70.39
7.	Malignant 2	11.65	34.86	64.57	62.67	27.02	72.19
8.	Malignant 3	9.85	35.70	61.90	30.41	26.36	81.72
9.	Malignant 4	9.36	35.75	63.18	61.69	27.37	70.37
10	Malignant 5	10.27	36.09	63.51	62.38	27.43	71.38
11	Benign 6	10.21	34.21	65.33	62.24	29.64	70.55
12	Benign 7	9.19	35.37	66.43	63.18	30.31	71.78
13	Benign 8	11.22	36.72	64.55	64.22	28.34	79.89
14	Benign 9	10.32	35.29	63.83	63.27	27.44	70.76
15	Benign 10	11.56	34.88	64.82	62.77	29.17	69.67
16	Malignant 6	9.45	35.26	64.18	63.23	31.29	70.39
17	Malignant 7	10.52	34.80	64.38	61.15	30.19	72.19
18	Malignant 8	9.69	34.21	64.58	62.33	29.18	71.72
19	Malignant 9	10.45	35.37	65.59	61.28	28.19	70.11
20	Malignant 10	7.82	33.69	64.89	62.67	29.67	79.51
21	Benign 11	8.66	33.29	66.37	59.41	28.19	70.39
22	Benign 12	9.59	32.99	66.64	61.44	29.23	79.79

23	Malignant 11	10.63	35.78	65.11	62.25	28.38	70.69
24	Malignant 12	11.09	34.90	64.27	61.47	28.04	79.16
	Mean(Avg)	**10.19**	**34.96**	**64.58**	**61.10**	**28.61**	**73.79**
	SD	**0.96**	**0.90**	**1.14**	**6.62**	**1.22**	**1.04**
	MSE	**30**	**20.16**	**0.02**	**0.03**	**0.90**	**0.01**

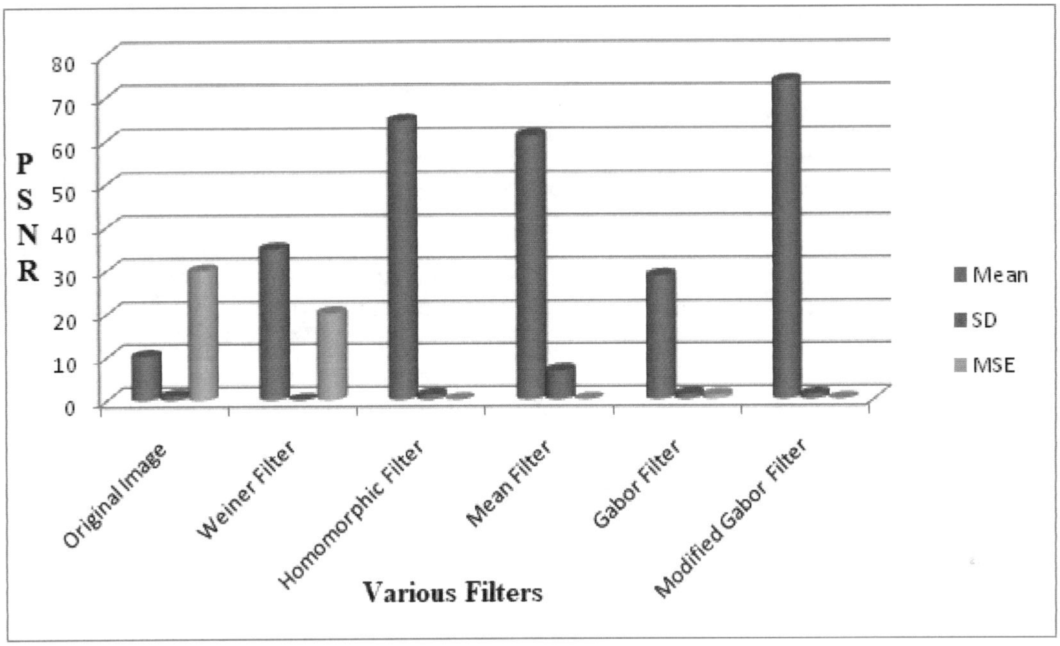

Figure 6. Graph of various metrics to measure quality of the enhanced image using different filters

5. Conclusion

Original x-ray images have very noisy, blur, low contrast and fuzzy nature. The conventional image characterization techniques fail to identify tiny lumps in the breast. It is very much essential to highlight the tiny lumps of the breast, so that it helps the segmentation algorithm for accurate extraction of all the regions of interest. The better feature is extracted and assists the radiologist for the analysis in order to diagnose cancer in earlier stage. The modified linear and nonlinear filtering algorithms performed well and have shown good result with respect to parameters PSNR and MSE.

6 . References

[1] Stojić, Tomislav, and BranimirReljin. "Enhancement of microcalcifications in digitized mammograms: Multifractal and mathematical morphology approach." FME Transactions 38.1 (2010): 1-9.

[2] Bhelke, Gayatri P., M. V. Sarode, and H. B. Nadiyana. "Detection of Noise in Degraded Images by Efficient Noise Detection Algorithm: A Survey." International joural of Application Or Innovations in Engineering & Management 2.4 (2013).

[3] Makandar, Aziz, and Bhagirathi Halalli. "Breast cancer image enhancement using median filter and CLAHE."International Journal of Scientific & Engineering Research 6.4 (2015): 462-465.
[4] Patel, Vishnukumar K., Syed Uvaid, and A. C. Suthar. "Mammogram of breast cancer detection based using image enhancement algorithm." Int. J. Emerg. Technol. Adv. Eng 2.2012 (2012): 143-147.
[5] Zheng, Yufeng. "Breast cancer detection with Gabor features from digital mammograms." algorithms 3.1 (2010): 44-62.
[6] Dehghani, Sara, and MashallahAbbasiDezfooli. "A method for improve preprocessing images mammography." International Journal of Information and Education Technology 1.1 (2011): 90.
[7] Wani, InamUl Islam, M. C. Hanumantharaju, and M. T. Gopalkrishna. "Contrast enhancement of mammograms images based on hybrid processing." Proceedings of the 3rd International Conference on Frontiers of Intelligent Computing: Theory and Applications (FICTA) 2014. Springer, Cham, 2015.
[8] Ramani, R., N. SuthanthiraVanitha, and S. Valarmathy. "The pre-processing techniques for breast cancer detection in mammography images."International Journal of Image, Graphics and Signal Processing 5.5 (2013): 47.
[9] Erwin, Saparudin, Adam Nevriyanto, and DiahPurnamasari. "Performance Analysis of Comparison between Region Growing, Adaptive Threshold and Watershed Methods for Image Segmentation."Proceedings of the International MultiConference of Engineers and Computer Scientists 2018.Vol. 1. 2018.
[10] Mr. B H Deokate,Dr. P M Patil, Mr. S SMajgaonkar,Image compression using Gabor Filter,International journal of Emerging Trends in Electrical and Electronics (IJETEE),Vol. 2,Issue 3, April -2013
[11] AshwiniGadge,Dr.S S Agarwal,"Guided Filter for Color Image",International journal of innovative research in electrical,Electronics,Instrumentation and control Engineering,Vol 4,issue 6,june 2016
[12] Rasha abdulRazak A P,HarishBinu K P, "A survey on various image filtering techniques for image sharpening",International journal of Sciences and research (iJSR)
[13] Neha N Ganvir,D. M. Yadav,"Filtering Method for preprocessing mammogram images for breast cancer Detection", International journal of engineering and advanced technology,(IJEAT),Volume 9,issue 1 ,October 2019.

Radio Frequency Identification (RFID) based ubiquitous health care data handling

Tamilvizhi T[1], Surendran R[2], Bommi R M[3]

[1]Department of Information Technology, Vel Tech Multi Tech Dr.Rangarajan Dr.Sakunthala Engineering College, Chennai, India

[2]Department of Computer Science & Engineering, Chennai Institute of Technology, Chennai, India

[3]Department of Mechatronic Engineering, Chennai Institute of Technology, Chennai, India

Email: tamilvizhi.mtech.it@gmail.com

Abstract. The Health care data is a most important and essential data compare with other types of data in the world. So the Health care Industry need to increase the data handling ability with a help of modern techniques. The Health care data handling must be improve with modern techniques like Cloud Computing and Radio Frequency Identification (RFID). The proposed framework merge the features of Cloud computing and RFID. The proposed framework provide the unique features to achieve the efficient data handling. RFID establish the data collection process from varies medical devices or centers. The cloud computing provide storage facility for Healthcare. The security feature is provided by the RFID for Healthcare data. Here Ubiquitous health care data accessibility represents the health care data can be access at anywhere, anytime, and from anything in Cloud computing. The proposed framework provide security in the form of the RFID tag to the patients and medical devices /centers. The RFID provides the security for the following data's like patient records, doctors and nurse details caring the patient and hospital bed details. The proposed data handling produces the higher data accuracy, richer security and less energy consumption with help of Quality of Service factors.

Keywords: Radio Frequency Identification (RFID), Ubiquitous, Health care, data handling, Cloud computing, Quality of service factors.

1. Introduction

Today world data is an assets of every organization. So every organizations are keep trying to collect, store and access more numbers of data for their development purpose. The Healthcare data must be secure with Radio Frequency Identification (RFID) technique. An Android application to allow doctors to scan the chips from the bed and review all details about patient. The RFID chips can act as a calling card for patient's data, which are stored on secure servers and updated regularly and in Real time via the patients' health care providers. At present most of the nations are proposed and implement Regional Collaborative Healthcare (RCH). The RCH permits hospitals to synchronous the patients' data between them, alongside pharmacies and medicinal services units to reduce the pressure. The patients experiencing when purchasing physician recommended prescription, moving starting with one clinic then onto the next or regular medical check-up at the nearest health care unit [11].

The ideas of time and refreshed information are extremely significant with regards to medical care, just as nearness and importance; with the framework proposing a doctor or allotted nurses would have the option to get to an information base containing the entirety of the patient's basic and delicate information essentially through the sweep of a RFID tag by means of Near-field communication (NFC) innovation. This will effectively enable time efficiency provided that a patient can be treated instantly with no more paperwork but simply through the reliable world of technology. The proposed work is implemented by RFID tag and receiver to provide the secure data collecting, cloud data storing, data access, Monitoring, and Controlling.

The RFID host system manages the flood of data between the reader and tags [4]. A RFID have system can similarly have a muddled structure, where distinctive reader are arranged across over different areas and data streams to the host PC through LANs or the web. The RFID reader initiates the tag and recuperates the information put away in its inner circuit. A RFID reader imparts and get signs with the help of the receiving wire. A RFID reader, in any case called an examiner, is essentially a scaffold between the strikes have system and the reader reception apparatus. Near Field Communication (NFC) is a procedure plan of short range remote programming.

2. Review of Literatures

In existing health care data handling works facing more challenges to achieve the data access. As indicated by Healthcare Industry Standard, a gathering representative should average 4 to 6 minutes while checking in a patient; which means 12 patients for every hour per worker. Meeting rooms, particularly in huge clinical buildings, are understaffed contrasted with the everyday convergence of patients, which must be that the business standard isn't even distantly accomplished [5]. This issue emerges because of patients having to physically round out structures, and the staff having to physically check for ID and Insurance status (on the off chance that relevant) and afterward physically book the facility for the patient and add them to the Queue. Gathering staff are additionally entrusted with noting calls and booking arrangements for the patients that bring in or check the status of accessible medical services suppliers and answer the guest's request [1].

The Patients can book an arrangement on the web, yet that scarcely spares time; a patient is needed to introduce the important records and the gathering staff need to check the subtleties before adding the patient to their center's Queue. This problem is even present in the Emergency Rooms (ER), where a relative or a friend or family member needs to remain behind and check in the patient who is in basic condition and must be dealt with right away [12]. In the ER, a progression of tests and revelations must be made on an unregistered patient before overseeing any prescription or playing out any tasks to maintain a strategic distance from any unsavoury astonishments during therapy, while the patient's clinical records are as of now accessible, complete, and refreshed in another office [6].

The remote innovation has been continuously progressing in the realm of medication administrations applications where the clinical gadget, information innovation and correspondences have started to join and can change. The innovation of Radio Frequency Identification (RFID) has shown overpowering assurance in helping clinical administrations upgrades patient's prosperity and achieving gifted activity. Due to the increase of human missteps and patients similarly the decrement of restorative staff, investigators found that the RFID development is been a sensible plan.

In any case, the current RFID tag clearly has a couple of challenges in getting a profitable and definite data trade in like manner to help steady data transmission. So as to direct the most ideal consideration, medical care experts ought to have the option to have simple admittance to the patient's data in a hurry instead of heft folds of desk work around, additionally they ought to have the option to follow their in-patients around the office and ensure they are accepting the correct therapies at their booked occasions, just as discover who are their going to doctors, attendants, and screens. Gear task, feast plans, meds, details, and so on are altogether basic during the time spent treating a patient, and representing everything is a significant factor. The Hospitals, Clinics, Health Care Units, and Major Pharmacies need the capacity to convey, and share significant patient information essential for encouraging the patient's admittance to quality consideration, right doctor prescribed prescription, and their entitlement to their own data [2].

RFID framework shows critical potential for changing therapeutic administrations, yet barely any examinations assess this potential. Our examination assessed the viability of utilizing RFID as a bed tag: an instrument to stimulate conspicuous proof of void beds [3]. We affected a little change in the releasing to technique to relate RFID tag with patients and made a RFID-based system that subsequently chosen discharge time [13]. For each patient, we assess the distinction in the releasing occasions recorded physically by the current methodology and the RFID-based system. The investigation was led on 43 patients over 2 months in 2 genuinely separate multi-guarantee to units. Contrasted and the earlier technique, the RFID-based system recognized release beds >15 minutes sooner 61% of the time with a typical of 27 minutes and middle of 10 minutes sooner. Medical clinics activity portrayed a difference in a short time as critical. With insignificant speculation, our little scope considers lead clinic position to begin masterminding RFID organization [8].

3. Proposed RFID based Ubiquitous Health Care Data Handling

The primary target of the proposed work to help Healthcare framework into a more complex and appropriately created substance that gives most extreme consideration and quality support of its patients. The feasibility of a proposed work is measure with following Quality of Service (QoS) factors, Cost refers to the value of each individual component like Cloud Storage Cost and Internet cost [10]. RFID techniques are very notable to be cost productive, so despite the fact that it is to be executed on a major scale it would at present be considered as a spending well-disposed venture. Availability refers to a complete time span for the gadget to arrive at the defender area. Being a RFID based framework the data is consistently accessible and prepared to recover just by examining a patient's tag. Reliability refers to the disappointment rate in which segments working under determined conditions will perform inadmissibly for a given timeframe. The proposed framework doesn't have a ton of provisos, likewise the foundation wherein the proposed framework will be executed should guarantee the framework's security by taking the proper measures.

The making of Tags and additionally Labels that can store a remarkable Retrieval ID that will be allotted to every patient; every Retrieval ID is utilized to get patient information from the cloud server [9]. RFID label readers to be utilized at the front counter for recovery of patient's information from the information cloud storage space. RFID readers at centers, division gatherings and attendant stations just as diagnostic rooms for confirmation of patient data, recovery of documents, and constant updates or progress, drugs, activity booking, and remedies. The software will be installed on all the computers used by the medical staff, as well as mobile phones/assigned tablets for attending Physicians and Nurses for faster data input/retrieval and configuration [7]. A secure server that stores all the data related to this system. The server should be synced with a main server that connects all the hospitals and healthcare facilities for the globalization of patient data, which will support the Regional Collaborative Healthcare (RCH) theory.

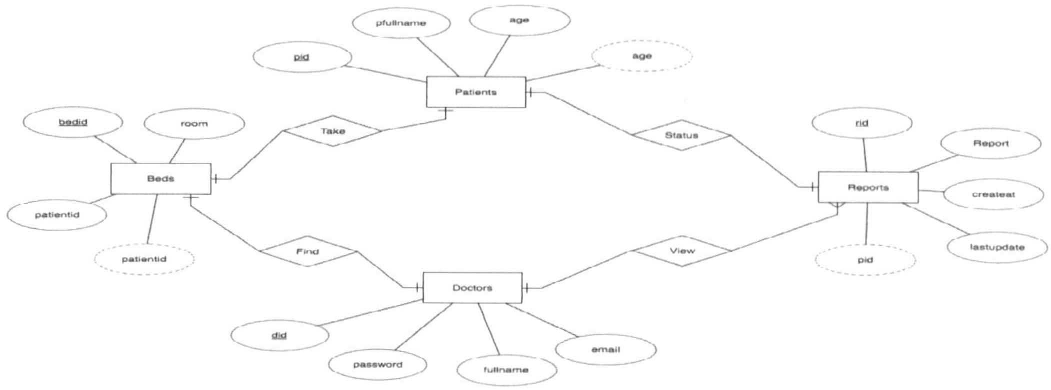

Figure 1. Entity-Relationship Diagram of proposed framework

The figure 1 shows an Entity-Relationship diagram to represent the connection between entities with relations. The figure 2 shows the Data Flow Diagram the proposed framework, where the information will originate from and go to, and where the information will be put away.

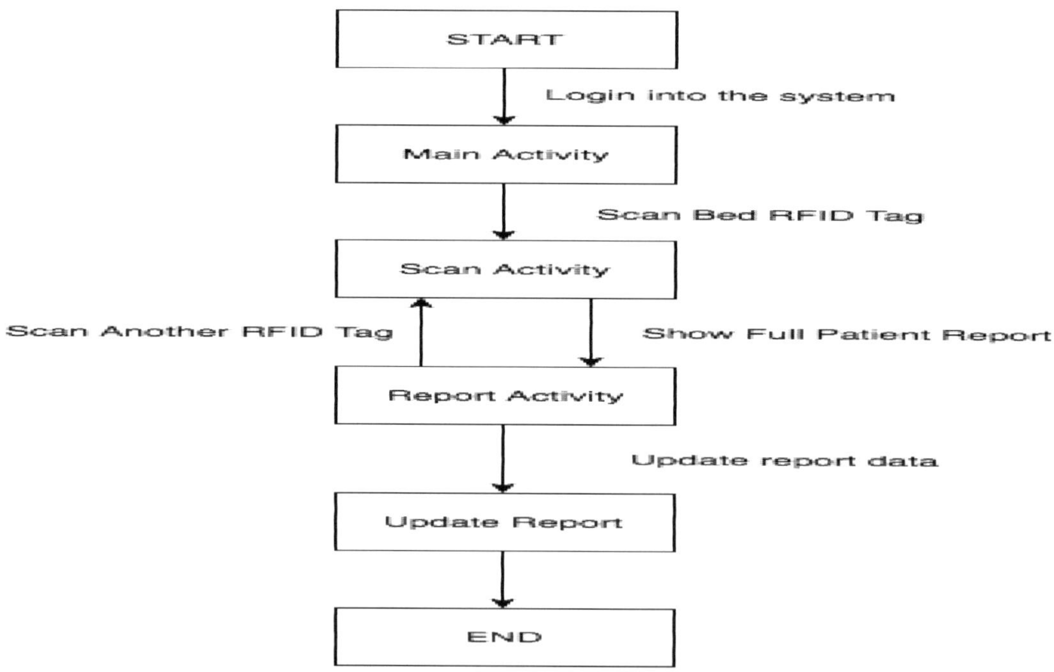

Figure 2. Data Flow Diagram of the proposed framework

The proposed framework is implemented as an Android application Android Studio (Java), PHP and MySQL database to store all the scanned information from the patient, doctors and bed. All the registered users their data will be stored in the database and the password will be decrypted. The Patients records can be viewed through RFID. The proposed system Performance produce a quick response time. It will be used by the health section of the hospitals. Since this application is subject to process patient documentation matters, this should be reliable to the users of this application. A substance relationship Entity-Relationship (ER) chart is utilized to imagine the framework and speak to the client's prerequisites. The ER outline is utilized to speak to elements and how they identify with

IOP Conf. Series: Materials Science and Engineering **994** (2020) 012021 doi:10.1088/1757-899X/994/1/012021

each other. The ER chart additionally demonstrates the connections between the elements, their event (multiplicities) and characteristics. The figure 3 and figure 4 display the Table description of entire healthcare database and Patient Bed.

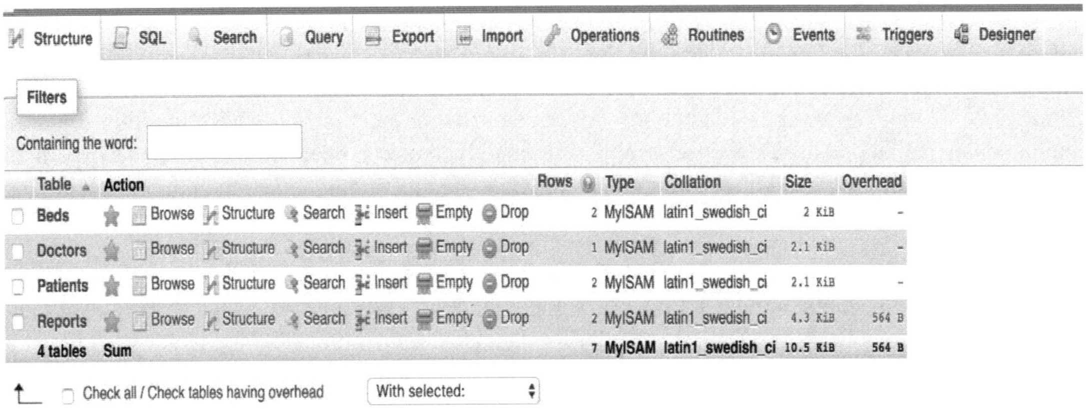

Table	Action	Rows	Type	Collation	Size	Overhead
Beds	Browse Structure Search Insert Empty Drop	2	MyISAM	latin1_swedish_ci	2 KiB	-
Doctors	Browse Structure Search Insert Empty Drop	1	MyISAM	latin1_swedish_ci	2.1 KiB	-
Patients	Browse Structure Search Insert Empty Drop	2	MyISAM	latin1_swedish_ci	2.1 KiB	-
Reports	Browse Structure Search Insert Empty Drop	2	MyISAM	latin1_swedish_ci	4.3 KiB	564 B
4 tables	Sum	7	MyISAM	latin1_swedish_ci	10.5 KiB	564 B

Figure 3. Table description of entire healthcare database

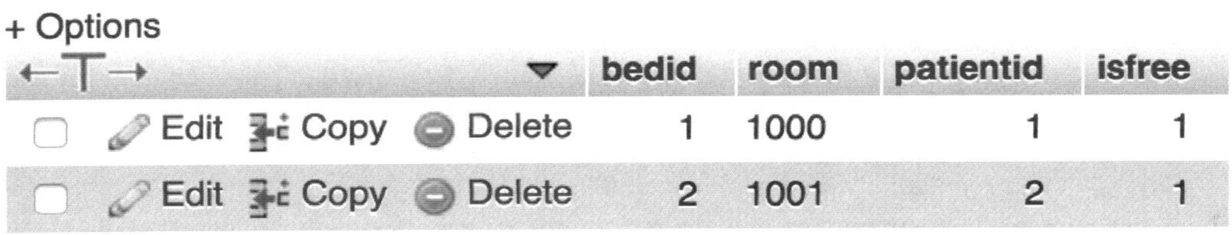

	bedid	room	patientid	isfree
Edit Copy Delete	1	1000	1	1
Edit Copy Delete	2	1001	2	1

Figure 4. Patient Bed Table description of healthcare database

The result of the proposed work provides the following activities

Action 1: Shows the main page for the user to sign in by entering the email address and password to the account to gain access to the patient's information.

Action 2: The user will scan the tag that is placed next to the patient bed using the mobile to get the patient information.

Action 3: After scanning the tag the information about the patient will appear in the application, the doctor/nurse can update the report and details saved in the tag.

Action 4: the doctor or nurse will update and save the patient information.

The proposed Health care data handling work consists of data collection, data storage and data access with high security.

4. Conclusion and Future direction

The proposed Health care data handling work efficiently covered the data collection, data storage and data access with high security. The innovation creates every day, this has made a gigantic impact, analysts can discover huge amounts of data's in regards to a particular illness, and patients can discover heaps of arrangements or prescriptions with respect to a particular malady. Shockingly, for patients to go starting with one medical clinic then onto the next the Doctor. Needs to conclusion them again to affirm the Patients record could get lost. The structured task bring extraordinary achievement on ubiquitous Health Care Data Handling. Since it can support numerous emergency clinics, hospitals and so forth. Aiding in putting away records, overseeing them, seeing them and refreshing them all simultaneously. In future implement as a real time system for any healthcare centers with help of Cloud of Medical Things [CoMT].

5. References

[1] Abugabah Ahed Nishara Nizamuddin and Alaa Abuqabbeh 2020 A review of challenges and barriers implementing RFID technology in the Healthcare sector *Procedia Computer Science Bol.*170 pp 1003-1010

[2] Al Nahas, Huzaifa and Jitender S. Deogun 2007 Radio frequency identification applications in smart hospitals *Twentieth IEEE International Symposium on Computer-Based Medical Systems (CBMS'07).*

[3] Buettner, Ricardo, et al. 2020 RFID Tracking and Localization Technologies in Healthcare *2020 IEEE Symposium on Industrial Electronics & Applications (ISIEA).*

[4] Coustasse Alberto, Shane Tomblin and Chelsea Slack 2013 Impact of radio-frequency Identification (RFID) technologies on the hospital supply chain: a literature review *Perspectives in Health Information Management* **10**.

[5] Gu, Vicky Ching, and Ken Black 2020 Integration of TTF and network externalities for RFID adoption in healthcare industry *International Journal of Productivity and Performance Management.*

[6] Haddara, Moutaz, and Anne Staaby 2020 Enhancing Patient Safety: A Focus on RFID Applications in Healthcare *International Journal of Reliable and Quality E-Healthcare (IJRQEH)* **9** 1-17.

[7] Naresh, Vankamamidi S., Sivaranjani Reddi, and Nistala VES Murthy 2020 Secure Lightweight IoT Integrated RFID Mobile Healthcare System *Wireless Communications and Mobile Computing.*

[8] Noori, Davood, Hassan Shakeri, and Masood Niazi Torshiz 2020 Scalable, efficient, and secure RFID with elliptic curve cryptosystem for Internet of Things in healthcare environment *EURASIP Journal on Information Security* **1** 1-11.

[9] Surendran R. and Tamilvizhi 2020 Cloud of Medical Things [CoMT] based Smart Healthcare Framework for Resource Allocation *3rd Smart Cities Symposium*, University Of Bahrain, IET Inspec.

[10] Tamilvizhi T Parvatha arthini.B 2016 Online vaccines and immunizations service based on resource management techniques in cloud computing *BIOMEDICAL RESEARCH-INDIA; Special Issue, Special Section:Health Science and Bio Convergence Technology: Edition-I,* pp. S392-S399.

[11] Wicks Angela M. John K. Visich, and Suhong Li 2006 Radio frequency identification applications in hospital environments *Hospital topics* **84** 3-9.

[12] Yao, Wen, Chao-Hsien Chu, and Zang Li 2012 The adoption and implementation of RFID technologies in healthcare: a literature review *Journal of medical systems* **36** 3507-3525.

[13] Muthukumar B, Dhanagopal R and Ramesh R 2019 KYP modeling architecture for cardiovascular diseases and treatments in healthcare institutions *J Ambient Intell Human Comput* https://doi.org/10.1007/s12652-019-01653-z

[14] Zhu, Feng, et al. 2020 A Novel Lightweight Authentication Scheme for RFID-Based Healthcare Systems *Sensors* **20** 4846.

Mathematical planning of a multi-factor experiment and optimization of the feed extrusion process

Marzhan Y. Kizatova[1], Madina Zh. Sultanova1, Alexander Y. Borovsky1, Nurzhan Zh.Muslimov[1], Zhibek A. Nokusheva[2], Yevgeny S. Aitzhanov[2]

[1]Astana branch of the LLP “Kazakh Research Institute of Processing and Food Industry”, Kazakhstan, 010000, Nur-Sultan, Akzhol Street, 26
[2]“North Kazakhstan research Institute of agriculture” ltd Rebublik of Kazakhstan, 150700, nort Kazakhstan region, kyzylzhar district, beskol, Institutskaya street 1
Corresponding author mail id: ackcoe@gmail.com

Abstract. To increase the level of protein and fat in milk, it is necessary to increase the total amount of feed fed, the diet must be balanced in energy, protein, fiber, minerals and vitamins; the feed must be of high quality. Oilseed waste produced during post-harvest processing and industrial processing contains a sufficient amount of protein and raw fat that can be used in animal feeding diets to increase milk productivity, and when using advanced processing methods, such as extrusion, it becomes a source of environmentally friendly vegetable protein that can increase the nutritional value of the feed several times. In order to study the efficiency of the technological process, the dependencies of quality parameters were determined. Using mathematical methods of planning a multi-factor experiment, a mathematical model of the extrusion process of feed raw materials was developed. The obtained model is used to identify recommended modes of extrusion of feed raw materials from post-harvest waste.
Keywords: extruded feed additive, modes, raw materials, extrudate, flax cleaning waste.

1. Introduction

One of the problems of modern animal husbandry is to increase the productivity of animals due to higher efficiency of using feed nutrients. This can be achieved by increasing the metabolism of the animal's body and the metabolizable energy of feed, increasing the transformation of feed nutrients into products through the use of new technologies for preparing feed for feeding (Shvetsov *et al.*, 2011. Shevchenko, 2012. Shvetsov and Ievlev, 2011. Yacko *et al.*, 2013. Lyapchenko *et al.*, 2014).

Unfortunately, the modern feed base does not allow to organize a full and balanced feeding of animals, which entails the use of the genetic potential of livestock productivity by only 40-60%. To increase the level of protein and fat in milk, the following main recommendations should be followed: increase the total number of feeds fed; provide the necessary variety of feeds; the diet should be balanced in energy, protein, fiber, minerals and vitamins; the feed should be of high quality. The production of mixed feeds is based on the use of raw materials, mainly crop products (Gaivoronsky and Ramazanov, 1998. Lanetsky, 2008).

To increase the nutritional value of feed in general and its structural components, special processing methods are used, the most common of which is thermal (Mishuro, 2006).

Analysis of technologies for heat processing of mixed feeds has shown that the most promising at present are granulation, extrusion, which contribute to increasing the nutritional value of the original components of mixed feeds, their disinfection, which ultimately significantly improves the quality of mixed feeds (Mishuro *et al.*, 2019).

Compared to granular feed, extruded feed is exposed to higher temperature, pressure, and shear forces during extrusion, which contributes to the breakdown of covalent crosslinking disulfide bonds in the protein (Aslaksen *et al.,* 2006). This contributes to the hydrolysis of protein and starch, and reduces the concentration of food anti-nutritional factors (Venou *et al.,* 2003. Venou *et al.,* 2009. Francis *et al.,* 2001. Drew *et al.,* 2007. Adamidou *et al.,* 2009). Studies have shown that protein, starch, and non-starch polysaccharides are improved during the extrusion process (Glencross *et al.,* 2008. Gaylord *et al.,* 2008. Venou *et al.,* 2009. Glencross *et al.,* 2012. Sorensen *et al.,* 2020, Balamurali *et al.*, 2020). After extrusion, the product improves the digestibility of its nutrients, increases storage stability, fat digestibility, and increases the volume mass (Shevtsov *et al.,* 2014). The variable parameters of the extrusion process are the composition of the raw material, moisture, pre-processing methods, as well as changes in temperature, pressure, and duration of the intensity of mechanical action on the product during extrusion (Topuria *et al.,* 1990).

The quantity of pressure created in the extruder depends on a set of interrelated parameters - the nature of the initial product, the extrusion temperature, the product hydration, the rotation speed of the working body of machine, the size and shape of the die. Improving the theory and methods of designing extrusion machines is an important task that allows optimizing the design of its units, the purpose of which is to produce products of the required quality (Ostrikov *et al.,* 2018).

It was also found that there is directly proportional dependence between viscosity of the final product, and its moisture content, lipids and simple sugars, and inversely proportional to the dependence of viscosity on temperature extrusion process and screw rotation speed. In addition, the viscosity is affected by the design of the extruder and its working bodies. Special attention should be paid to the higher digestibility of the resulting product (25-30% higher than usual), which allows to increase the weight gain during feeding (compared to traditional feed) and dramatically reduce the amount of waste (Zhigankov, 1981),(Kiran George,2020).

Thus, extrusion allows to purposefully change the properties of the product and its constituent components and is increasingly widely used in many sectors of the national economy, including the feed industry. The solution to this problem was obtained by planning multi-factor experiments, statistical processing of experimental data, and search engine optimization (Kizatova *et al.,* 2017).

2. Materials and methods

2.1 Methods of research

The objects of research are waste from post-harvest processing of oilseed flax produced in Northern Kazakhstan, extruded feed based on waste from post-harvest processing of oilseed flax.

2.2 Methods of conducting research on the extrusion of post-harvest processing waste of oilseeds

To study the quality indicators, well-known methods were used, including: GOST 11549-76 “Linen flax seeds. Industrial raw material. Specifications”; GOST 10582-76 “Oilseed flax seeds. Industrial raw material. Specifications”; GOST 10583-76 “Rapeseed for industrial processing. Specifications”; GOST 22391-89 “Sunflower. Industrial raw material. Specifications”; GOST 10854-2015 “Oilseeds. Methods of determining the weed, oilseed and specially considered impurities”; GOST 27988-88 “Oilseeds. Methods of determining the color and smell”; GOST 10856-96 “Oilseeds. Methods of determining the moisture content”; GOST 10857-64 “Oilseeds. Methods of determining the oil content”; GOST R 51410-99 “Oilseeds. Determination of the acidity of oils”; GOST 10852-86 “Oilseeds. Acceptance rules and sampling methods”; GOST 29142-91 “Oilseeds. Sampling”; GOST 13496.15-97 “Crude fat content”; GOST 13496.18-85 “Acid number of fat”; GOST 13496.2-91 “Crude fiber content”; GOST 26226-95 “Crude ash content”; GOST 26657-97 “Phosphorus content”; GOST 26593-85 “Peroxide number of fat”; NFES (nitrogen-free extractive substances) were determined by calculation; digestible protein was determined using the coefficients of digestibility; flowability – according to V.E. Pestov; hygroscopicity – by establishing the equilibrium moisture by statistical method; granulometric - determined according to GOST 13496.8-72.

Working out the extrusion modes of oilseed flax waste was carried out at the production site of the scientific and innovative complex “North Kazakhstan scientific research institute of agriculture” on the feed extrusion line, which consists of a crusher, mixer, hoppers for dewatering and an extruder “PE-150”. The line capacity is 150 kg/hour.

Feed analysis was performed on the device InfraXact FOSS in accordance with GOST 32040-2012 “Feed, mixed feed, feed raw materials. Method of determining the content of crude protein, crude fiber, crude fat, and moisture using near infrared reflectance spectroscopy”.

In order to determine the optimal mode of obtaining feed based on quality and economic indicators, experimental studies on the extrusion of flax cleaning waste were conducted. The results obtained were recorded in tables in text processor Microsoft Excel, then the data obtained was studied the influence dependence of die diameter (D, mm), moisture content of the feedstock (W %), the feed rate of raw materials (v, s^{-1}) specific energy costs per unit of production (W_P, W/kg) and for a period of time (W_T, kW/h) and metabolizable energy (q, mJ/kg).

To obtain the mathematical model of technological process for receiving extruded feed material from flax cleaning waste, which represents the regression equation used rotatable plan of the second order (plan Box) when the number of factors K=3, the number of experiments plan more than 20, the number of experiments at the zero point was 6 and the number of coefficients is 10. The coded values for obtaining a mathematical model are shown in Table 1.

Table 1 Coded values for obtaining a mathematical model

Coded values		
x_1	x_2	x_3
-	-	-
-	-	+
-	+	-
-	+	+
+	-	-
+	-	+
+	+	-
+	+	+
-1,68	0	0
+1,68	0	0
0	-1,68	0
0	+1,68	0
0	0	-1,68
0	0	+1,68
0	0	0

3. Results

A number of key components for the production of extruded feed additives from oilseeds were considered.

To investigate the interaction of various factors that affect the extrusion process, mathematical methods of experiment planning were used.

Coding of intervals and levels of variation of input parameters is shown in Table 2.

Table 2 Coding of intervals and levels of variation of input factors

Factors		Variation levels					Variation intervals
natural	coded	-1,68	-1	0	+1	+1,68	
D, mm	x_1	5.3	6	7	8	8.6	1

The upper and lower levels of factors are shown in Table 3.

Table 3 Upper and lower levels of factors

Indicators	Coded value	Planning			
		X_1 die diameter, mm	X_2 moisture, %	X_3 frequency of rotation of the feed drive, s^{-1}	X_4 size, mm
Upper level	+	8	10	8,2	3
Zero level	0	7	9	5,8	2,25
Lower level	-	6	8	3,4	1,5

The rotatable planning matrix is presented in Table 1. Results of the experiments of two variants (with the particle size of the feed mixture - 3 mm and the particle size of the feed mixture – 1.5 mm) obtained on the basis of the planning matrix are presented in Table 4.

Table 4 Results of experimental studies of extrudate production from flax cleaning waste 2 variants

Natural values			Experimental values	
D, mm	W, %	*v*, s^{-1}	1-variant q, mJ/kg	2-variant q, mJ/kg
6	8	3,4	8,5	8,86
6	8	8,2	8,96	8,8
6	10	3,4	8,97	8,76
6	10	8,2	8,9	8,63
8	8	3,4	8,94	8,84
8	8	8,2	8,8	8,71
8	10	3,4	9,03	8,76
8	10	8,2	8,8	8,72
5,3	9	5,8	9,09	8,87
8,6	9	5,8	8,59	8,68
7	7,3	5,8	8,7	8,87
7	10,6	5,8	8,78	8,66
7	9	1,7	8,64	8,49
7	9	9,8	9,35	9,2
7	9	5,8	8,97	8,94

The total number of experiments for one size was 27: 8 experiments in the full-factor zone of the matrix, 2 variants of 6 experiments in the zone of star points and 7 experiments in the center of the plan.

Table 5 shows the values of the confidence intervals of the process optimization criteria for obtaining extrudate from the flax cleaning waste of 2 variants.

Table 5 Values of confidence intervals of the optimization criterion

Variants	The process of obtaining extruded feed		Input parameter	Confidence intervals			
				Δb_0	Δb_i	Δb_{ii}	Δb_{ij}
1	Exchange energy	*q*, mJ/kg	y	±0,17	±0,11	±0,11	±0,15
2	Exchange energy	*q*, mJ/kg	y	±0,12	±0,08	±0,08	±0,11

The coefficient of the regression equation is significant if its absolute value is greater than the confidence interval (b_i>Δb_i). Otherwise, it is considered insignificant and can be excluded from further consideration of the mathematical model. Comparing the values of confidence intervals from Table 5 with the corresponding regression coefficients in Table 6, it can be concluded that the interaction effects of input factors are insignificant, and they could be ignored.

Table 6 Coefficients of regression equations for output parameters

Optimization criterion	Coefficients	The process of obtaining extruded feed from flax cleaning waste	
		variant 1	variant 2
	For coded values of factors		
Obtaining extruded feedstock from flax cleaning waste of 2 variants	b_0	9,010637974	8,888907786
	b_1	-0,0593652	-0,021929256
	b_2	-0,006986208	-0,000967704
	b_3	0,105716904	0,113768904
	b_{12}	0,01375	0,025
	b_{13}	-0,04125	0,0025
	b_{23}	-0,02375	0,0025
	b_{11}	-0,04780719	-0,047027785
	b_{22}	-0,08317124	-0,05056419
	b_{33}	0,007007088	-0,02227295
	For natural values of factors		
	B_0	-0,1530	3,9177
	B_1	0,585872958	0,405418074
	B_2	1,451241943	0,728146058
	B_3	0,239312214	0,075592291
	B_{12}	0,01375	0,025
	B_{13}	-0,0171875	0,001041667
	B_{23}	-0,00990	0,00104
	B_{11}	-0,04780719	-0,047027785
	B_{22}	-0,08317124	-0,05056419
	B_{33}	0,001216508	-0,003866832
F_p*		1,004116599	1,305305175

* - no more than the table value of the Fisher criterion F_{table}=5,05

Next, searched for the optimal response functions with the highest possible accuracy (solving a compromise problem), while taking into account insignificant coefficients for building a mathematical model, which will represent the regression equations (1):

$$y_1 = b_0 + b_1x_1 + b_2x_2 + b_3x_3 + b_{12}x_1x_2 + b_{13}x_1x_3 + b_{23}x_2x_3 + b_{11}x_1^2 + \\ + b_{22}x_2^2 + b_{33}x_3^2; \quad (1)$$

Thus, the regression equation for modes of obtaining extrudate from flax cleaning waste of variant 1, based on the metabolizable energy content in the final product is represented in the formula (2) and regression equations for modes of obtaining extrudate from flax cleaning waste of variant 2 based on the metabolizable energy content in the final product is represented in the formula (3):

$$y = 9{,}010637974 - 0{,}0593652x_1 - 0{,}006986208x_2 + 0{,}105716904x_3 + 0{,}01375x_1x_2 - \\ - 0{,}04125x_1x_3 - 0{,}02375x_2x_3 - 0{,}04780719x_1^2 - 0{,}08317124x_2^2 + 0{,}007007088x_3^2; \quad (2)$$

$$y = 8{,}888907786 - 0{,}021929256x_1 - 0{,}000967704x_2 + 0{,}11368904x_3 + 0{,}025x_1x_2 + \\ + 0{,}0025x_1x_3 + 0{,}0025x_2x_3 - 0{,}047027785x_1^2 - 0{,}05056419x_2^2 - 0{,}02227295x_3^2; \quad (3)$$

After decoding the independent variables in equations (1-2 and 1-3), we get the regression equations for the natural values of the factors (4):

$$y = -0{,}1530 + 0{,}585872958D + 1.451241943W + 0{,}239312214v + 0{,}01375DW - \\ - 0.0171875Dv - 0.00990Wv - 0{,}04780719D^2 - 0{,}08317124W^2 + 0{,}001216508v^2; \quad (4)$$

The adequacy of the obtained mathematical regression models was evaluated using the Fischer F_p criterion. The obtained calculated values of F_p are shown in Table 6.

Thus, given that the $F_p < F_{table}$ model of the nutritional value of extrudate from the flax cleaning waste of variant 1 can be considered adequate with a 95% confidence probability.

After the canonical transformation of the second-order models, regression equations were obtained in canonical form, the values of optimization parameters were calculated on an electronic computer of the word processor Microsoft Excel, on the basis of which a model was built in three-dimensional space, representing a plane that characterizes the dependence of the influence of the die diameter (D, mm), the moisture of the feedstock (W,%), the rotation speed of the feed drive (v, s^{-1}) on the specific energy costs per unit of production (WP, W/kg) and for a period of time (WT, kW/h) and metabolizable energy (q, mJ/kg). Figures 9-14 show graphical representations of the dependency graphs $y_n=f(D,W)$, $y_n=f(D,v)$ and $y_n=f(W,v)$ of the 2 variants.

Analysis of three-dimensional spatial models shows that the necessary values of the optimization criterion y are achieved in the search area under consideration. This means that the levels of variation of input factors in the planning of experiments are taken quite correctly.

Analysis of the presented graphs showed that on the three-dimensional model in space there are optimal regions of variable values of the magnitudes D (mm), W (%), v (s^{-1}) at which the technological process of obtaining extrudate from the flax cleaning waste of 2 variants is carried out with the maximum values of the metabolizable energy content in the final product (q, mJ/kg).

In variant 1, analysis of the behavior of resulting three-dimensional surface showed the presence of three distinct zones: the 1st with metabolizable energy of 9-9.5 mJ/kg; the 2nd with metabolizable energy of 8.5-9 mJ/kg; the 3rd is characterized by values of 8-8.5 mJ/kg (Figure 1).

Figure 1 shows a three-dimensional model that characterizes the dependence of die diameter and the moisture content of feedstock on the metabolizable energy of the extrudate of variant 1. Monitoring of obtained results showed that increasing the diameter of the die D from 5 to 9 mm leads to an increase in moisture W up to 9%, after which this indicator begins to decrease.

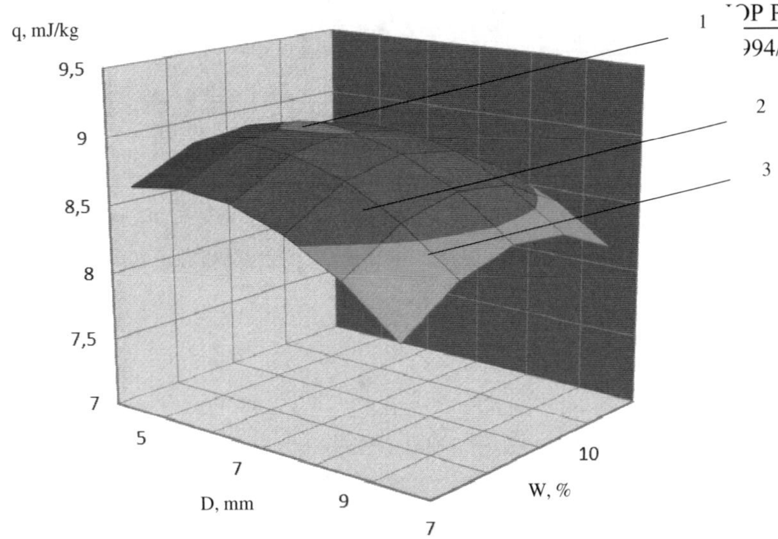

1 - optimum zone and zone with metabolizable energy of 9-9.5 mJ/kg;
2 - reduction zone and zone with metabolizable energy of 8.5-9 mJ/kg;
3 - minimum zone and zone with metabolizable energy of 8-8.5 mJ/kg

Figure 1 Three-dimensional model in space that characterizes the dependence $y_n=f(D,\varphi)$ of die diameter (D, mm) and the moisture content of feedstock (W,%) on the exchange energy (q, mJ/kg) of variant 1

In this case, the metabolizable energy *q* decreases from 9.35 to 8.59 mJ/kg with an increase in the die diameter, and an increase in moisture to 9% leads to an increase in *q* and with a further increase in φ, the metabolizable energy begins to decrease.

For example, at D=5.3 mm and W=9%, the metabolizable energy is 9.09 mJ/kg, while the increase in the die diameter values led to a harmonic oscillation of the metabolizable energy values with the maximum value of studied indicator at the peak point: at D=6 mm, q=8.85-8.97 mJ/kg; D=7 mm, q=8.64-8.74 mJ/kg; D=8 mm, q=8.8-9.03 mJ/kg; D=8.6 mm, q=8.59 mJ/kg.

However, increasing the values of die diameter (D, mm) to 8.6 mm and increasing the moisture (W, %) of feedstock to 10.6 %, led to a decrease in the exchange energy (q, mJ/kg) to 8.59 mJ/kg.

The optimal technological mode, based on the consideration of the dependence of die diameter and raw material moisture on the metabolizable energy of extrudate, is D=6-7 mm and W=8-9%. The results of experimental studies indicate that the main indicator that affects the reduction of metabolizable energy is the diameter of die.

In variant 2, analysis of the obtained results showed the presence of two distinct zones: the 1st with a metabolizable energy content of 8.5-9 mJ/kg, in the range of values D=5-8 mm and W=7-10%; the 2nd is characterized by values of 8-8.5 mJ/kg in the range of values D=9 mm and W=7-10% (Figure 2).

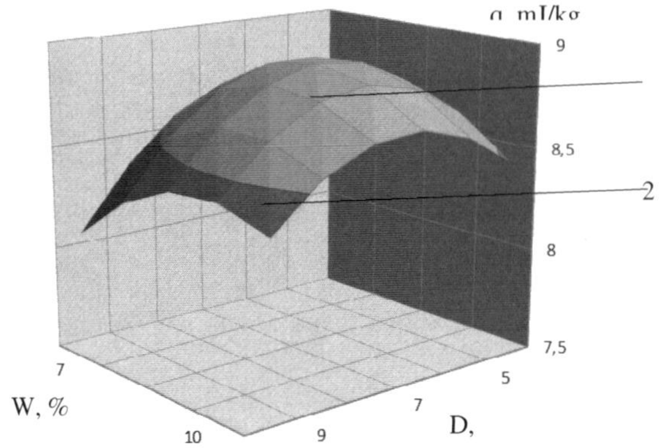

1 - zone with a metabolizable energy content of 8.5-9 mJ/kg
2 - zone with a metabolizable energy content of 8-8.5 mJ/kg

Figure 2. Three-dimensional model in space that characterizes the dependence $y_n=f(D,\varphi)$ of die diameter (D, mm) and the moisture content of feedstock (W,%) on the metabolizable energy (q, mJ/kg) of 2 variants

The dynamics of changes in the metabolizable energy content (q, mJ/kg) is observed in a sinusoidal form at a fixed moisture with an increase in the die diameter, as well as at a fixed value of the die with an increase in moisture. The indicator at which the metabolizable energy content is the maximum q=8.88 mJ/kg and the rise of sinusoid goes into decline is the values of D=7 mm and W=9%.

Further, in order to determine the optimal content of metabolizable energy, conducted experimental studies to study the degree of influence of the die diameter and the rotation speed of the feed drive on the metabolizable energy (q, mJ/kg) in the obtained extrudate.

Figure 3 shows a three-dimensional model that characterizes the dependence of die diameter and rotation frequency of the feed drive on the metabolizable energy of the resulting extrudate of variant 1.

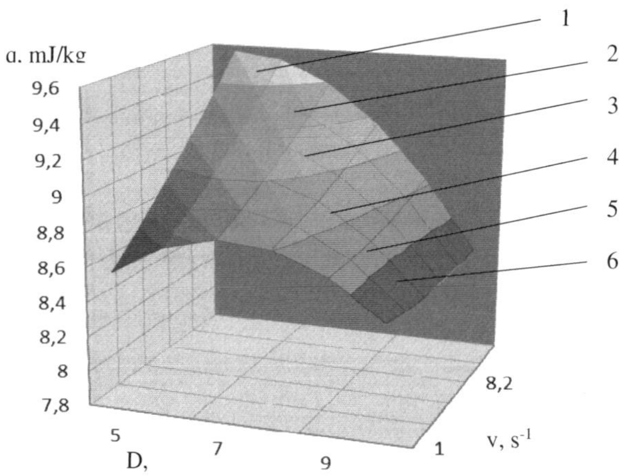

1-2 - optimum zone; 3-4 - decrease zone; 5-6 - minimum zone
1 - zone with metabolizable energy of 8.4-8.6 mJ/kg;
2 - zone with metabolizable energy of 8.6-8.8 mJ/kg;
3 - zone with metabolizable energy of 8.8-9 mJ/kg;
4 - zone with metabolizable energy of 9-9, 2 mJ/kg;
5 - zone with metabolizable energy of 9.2-9.4 mJ/kg;
6 - zone with metabolizable energy of 9.4-9.6 mJ/kg

Figure 3. Three-dimensional model in space that characterizes the dependence $y_n=f(D,\varphi)$ of die diameter (D, mm) and the rotation speed of the feed drive (W,%) on the metabolizable energy (q, mJ/kg) of variant 1

Analysis of the behavior of the resulting three-dimensional surface (figure 3) also showed the presence of six distinct six zones: the 1st with metabolizable energy of 8.4-8.6 mJ/kg; the 2nd with metabolizable energy of 8.6-8.8 mJ/kg; the 3rd corresponded to 8.8-9 mJ/kg; the 4th zone, characterized by values of 9-9, 2 mJ/kg; the 5th zone with metabolizable energy of 9.2-9.4 mJ/kg, and the 6th zone, respectively, 9.4-9, 6 mJ/kg.

Analysis of the obtained results showed that the increase in die diameter D from 5 to 8.5 mm and rotation speed of the feed drive v from 1.7 s^{-1} to 9.8 s^{-1} leads to lower values of the metabolizable energy content in the extrudate, while q varies from 9.35 mJ/kg to 8.59 mJ/kg. For example, when v=9,8 s-1 and D=7 mm metabolizable energy obtained extrudate amounted to q=9.35 mJ/kg, with a further increase in the values of the die diameter D (mm) led to the decrease of the values of metabolizable energy at D=6 mm, q=of 8.96 mJ/kg; D=8 mm, q=8,8 mJ/kg; at D=8.6 mm, q=8,59 mJ/kg. A decrease in the rotation frequency of the feed drive (v, s^{-1}) in a constant value of the die diameter (D, mm) led to a decrease in the values of metabolizable energy (q, mJ/kg).

The optimal technological mode, based on the consideration of the dependence of die diameter and the feed rate of feedstock on the metabolizable energy of extrudate, is D=5-6 mm and v=9.8 s^{-1}. The obtained results of experimental studies indicate that the factors that affect the reduction of metabolizable energy equally are the die diameter and the feed rate of the raw material.

In variant 2, analysis of the behavior of resulting three-dimensional surface (Figure 4) also showed the presence of six distinct zones:

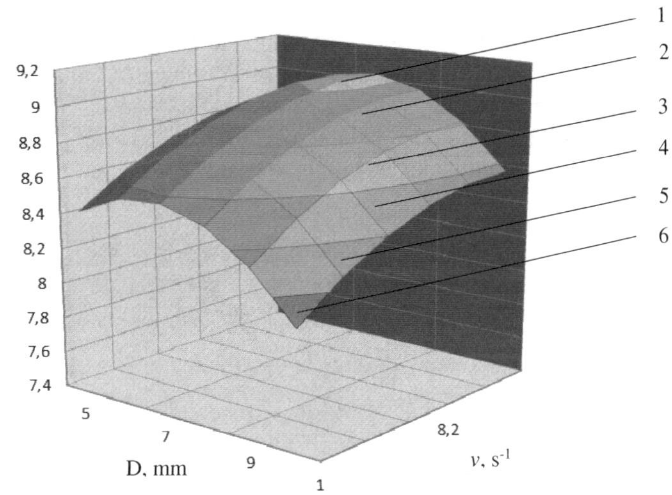

1 - zone with a metabolizable energy content of 9-9. 2 mJ/kg
2 - zone with a metabolizable energy content of 8.8-9 mJ/kg
3 - zone with a metabolizable energy content of 8.6-8.8 mJ/kg
4 - zone with a metabolizable energy content of 8.4-8.6 mJ/kg
5 - zone with a metabolizable energy content of 8.2-8.4 mJ/kg
6 -zone with a metabolizable energy content of 8-8. 2 mJ/kg

Figure 4. Three-dimensional model in space that characterizes the dependence $y_n=f(D,v)$ of die diameter (D, mm) and the rotation speed of the feed drive (v, s^{-1}) on the metabolizable energy (q, mJ/kg) of variant 2

-1st with a metabolizable energy content of 9-9. 2 mJ/kg, in the range of values D=6-7 mm and v=8.2-9.8 s^{-1};

- 2nd is characterized by values of 8.8-9 mJ/kg in the range located on the periphery of the 1st zone, values of D=5-8 mm and v=3.4-9.8 s^{-1};

- 3rd, respectively, 8.6-8.8 mJ/kg in the range located on the periphery of the 2nd zone, values of D=5-9 mm and v=5.8-9.8 s^{-1};

- 4th is determined by the values of 8.4-8.6 mJ/kg in the range located on the periphery of the 3rd zone, values of D=5-9 mm and v=1-9 s^{-1};

- 5th zone with metabolizable energy of 8.2-8.4 mJ/kg, in the range of values D=9 mm and v=1-3, 4 s^{-1};
IOP Conf. Series: Materials Science and Engineering **994** (2020) 012022 doi:10.1088/1757-899X/994/1/012022
- 6th zone with metabolizable energy of 8-8. 2 mJ/kg, in the range of values D=9 mm and v=1 s^{-1}.

The dynamics of changes in the metabolizable energy content (q, mJ/kg) is ascending at a fixed value of the die diameter and an increase in the rotation frequency of the feed drive. The indicator for which the metabolizable energy content is the maximum q=8.98 mJ/kg is D=7 mm and v=8.2 s^{-1}.

The next stage of the study was to determine the optimal content of metabolizable energy (q, mJ/kg) based on experimental data on the degree of influence of the rotation frequency of the feed drive and the moisture of feedstock.

In variant 1, analysis of the behavior of the resulting three-dimensional surface showed the presence of five distinct zones: the 1st with metabolizable energy of 8.4-8.6 mJ/kg; the 2nd with metabolizable energy of 8.6-8.8 mJ/kg; the 3rd is characterized by values of 8.8-9 mJ/kg; the 4th, respectively, 9-9.2; the 5th has indicators of 9.2-9.4 mJ/kg.

Monitoring of the obtained results showed that increasing the rotation speed of the feed drive v from 1.7 to 9.8 s^{-1} leads to an increase in moisture W in the final product up to 9%. At the same time, the metabolizable energy q increases from 8.59 to 9.35 mJ/kg with an increase in the rotation frequency of the feed drive with a constant moisture index.

For example, at v=9.8 s^{-1} and W=9%, the metabolizable energy is 9.35 mJ/kg, while a decrease in the rotation frequency of the feed drive at a constant moisture value led to a sinusoidal decrease in the studied indicator q: at v=8.2 s^{-1}, q=8.96 mJ/kg; v=5.8 s^{-1}, q=9.09 mJ/kg, v=3.4 s^{-1}, q=8.85 mJ/kg; v=1.7 s^{-1}, q=8.64 mJ/kg. However, changing the moisture values W from 7 to 10% led to a harmonic oscillation of the metabolizable energy values with the maximum value of the studied indicator q at the peak point W=9.

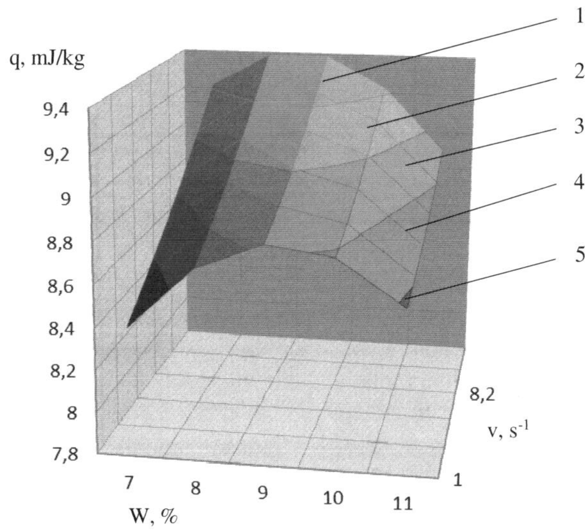

1 - optimum zone; 2-3 - decrease zone; 4-5 – minimum zone.
1 - zone with metabolizable energy of 8.4-8.6 mJ/kg;
2 - zone with metabolizable energy of 8.6-8.8 mJ/kg;
3 - zone with metabolizable energy of 8.8-9 mJ/kg;
4 - zone with metabolizable energy of 9-9, 2 mJ/kg;
5 - zone with metabolizable energy of 9.2-9.4 mJ/kg

Figure 5. Three-dimensional model in space that characterizes the dependence $y_n=f(W, v)$ of the raw material moisture (W,%) and the feed drive (v, Hz) on the metabolizable energy (q, mJ/kg) of variant 1

The optimal technological mode, based on the consideration of the dependence of moisture and rotation frequency of the feed drive on the metabolizable energy of extrudate, is v=8.2-9.8 s^{-1} and W=8-9%. The results of experimental studies indicate that the main indicator that affects the reduction of metabolizable energy is the rotation frequency of the feed drive.

In variant 2, the analysis of the behavior of the resulting three-dimensional surface showed the presence of five distinct zones:

-1st with a metabolizable energy content of 9-9.2 mJ/kg, in the range of values W=9% and v=8.2-9.8 s^{-1};

-2nd is characterized by values of 8.8-9 mJ/kg in the range located on the periphery of the 1st zone, values W=7-11% and v=5.8-9.8 s^{-1};

-3rd, respectively, 8,6-8,8 mJ/kg in the range located on the periphery of the 2nd zone, values W=7-11% and v=3,4-8,2 s^{-1};

-4th is determined by the values of 8.4-8.6 mJ/kg in the range located on the periphery of the 3rd zone, values of W=7-11% and v=1-3, 4 s^{-1};

-5th zone with metabolizable energy of 8.2-8.4 mJ/kg, in the range of values W=11% and v=1 s^{-1}.

Figure 6 shows a three-dimensional model that characterizes the dependence of the feed drive rotation frequency and raw material moisture on the metabolizable energy of the extrudate of variant 2.

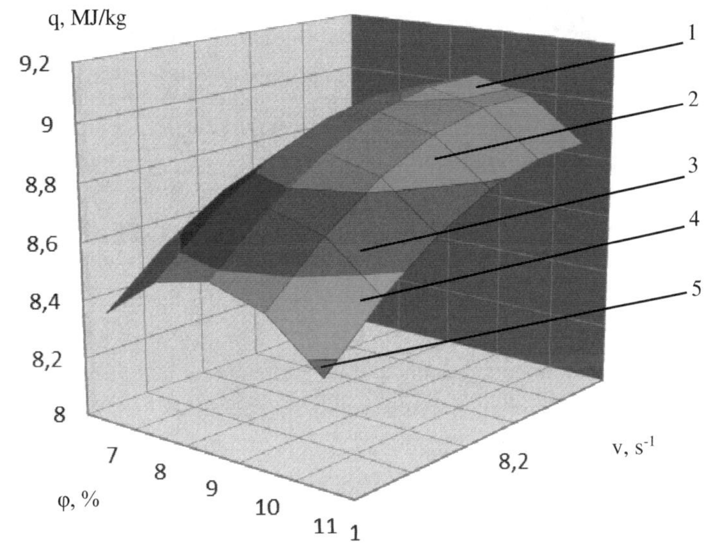

1 - zone with a metabolizable energy content of 9-9.2 mJ/kg
2 - zone with a metabolizable energy content of 8.8-9 mJ/kg
3 - zone with a metabolizable energy content of 8.6-8.8 mJ/kg
4 - zone with a metabolizable energy content of 8.4-8.6 mJ/kg
5 - zone with a metabolizable energy content of 8.2-8.4 mJ/kg

Figure 6. Three-dimensional model in space that characterizes the dependence $y_n=f(W,v)$ of of the moisture content of feedstock (W,%) and the rotation speed of the feed drive (v, s^{-1}) for metabolizable energy (q, mJ/kg) of variant 2

The dynamics of changes in the metabolizable energy content (q, mJ/kg) is ascending at a fixed value of moisture and an increase in the rotation frequency of the feed drive. The indicator for which the metabolizable energy content is the maximum q=9.02 mJ/kg is W=9% and v=10.6 s^{-1}.

The given dependencies are $y_n=f(D,\varphi)$, $y_n=f(D,v)$, $y_n=f(W,v)$ from the variable parameter of the technological process for obtaining extrudate from flax cleaning waste of 2 variants - D (mm), W (%), v (s^{-1}), allow to predict with sufficient accuracy the change in the values of the optimization criteria y in the studied range of values of factors - metabolizable energy q (J/kg).

It is possible to establish a dominant effect of each investigated factor on the quality of the final product, that allows with sufficient approximation to describe the process of obtaining an extrudate from the flax cleaning waste of 2 variants (particle size – 3 mm and 1.5 mm). The results obtained will allow to optimize the process under study by applying the developed mathematical model.

Next, the identification of factors affecting the change in the values of optimization criteria y in the investigated range of values of extrudate factors from raw materials of 2 options is carried out.

As a result of analysis of the dependence $y_n=f(D,\varphi)$, $y_n=f(D,v)$, $y_n=f(W,v)$ of a variable parameter of technological process of production of extrudate from flax cleaning waste of variant 1, the optimal mode, providing moderate specific energy costs per unit of production W_P=60-80 W/kg, and moderate specific energy costs for a period of time WB=6-8 kW/h at optimum metabolizable energy content q=of 8.5-9 mJ/kg is the die diameter D=7 mm, moisture of the raw materials W=9 %, rotation frequency of the drive feed v=5,8 s^{-1}.

Due to the fact that the grinding of raw materials increases costs: specific energy costs per unit of production W_P=25 W/kg and specific energy costs for a period of time W_T=4 kW/h; to obtain a pilot batch, the optimal mode for obtaining extrudate from flax cleaning waste of variant 1 (with a size of 3 mm), with parameters: the die diameter D=7 mm, moisture of the raw material W=9%, rotation speed of the feed drive V=5.8 s^{-1} was selected. These parameters correspond to the experimental batch number 14, which is in the results of the experimental data of variant 1.

Thus, the recommended modes of extrusion of feed raw materials from the waste of flax post-harvest processing are presented in Table 7.

Table 7. Modes of extrusion of feed raw materials from waste of flax post-harvest processing

Size of raw material, mm	The moisture content of feedstock, %	Die diameter, mm	Rotation frequency of the feed drive, Hz	Raw material heating temperature, °C
3	9	7	5,8	90-95

4. Discussion

The studies were conducted building models in three-dimensional space representing a plane, which characterizes the dependence of the effect of die diameter (D, mm), the moisture content of feedstock (W %), rotation frequency of the feed drive (v, s^{-1}) specific energy costs per unit of production (W_p, W/kg) specific energy costs for period of time (WB, kW/h), the metabolizable energy (q, mJ/kg).

5. Conclusion

The study on optimization of waste from the flax cleaning in which the optimum mode of extrusion are the die diameter D=7 mm, moisture of raw materials W=9%, rotation frequency of the feed drive v=5,8 s-1, with a size of 3 mm. Production testing of the research results showed that the mathematical regression model is adequate with 95% confidence probability, and the results can be used to optimize the extrusion process of waste post-harvest processing of oilseeds with the aim of obtaining highly nourishing combined feed.

References

[1] Adamidou, S., Nengas, I., Henry, M., Grigorakis, K., Rigos, G., Nikolopoulou, D., Kotzamanis, Y., Bell, G.J. and Jauncey, K., (2009). Growth, feed utilization, health and organoleptic characteristics of European seabass (Dicentrarchus labrax) fed extruded diets including low and high levels of three different legumes. *Aquaculture* 293, 263-271. http://dx.doi.org/10.1016/j.aquaculture.2009.04.045

[2] Aslaksen, M.A., Romarheim, O.H., Storebakken, T. and Skrede, A., (2006). Evaluation of content and digestibility of disulfide bonds and free thiols in unextruded and extruded diets containing fish meal and soybean protein sources. *Anim. Feed Sci. Technol.* 128, 320-330. http://dx.doi.org/10.1016/j.anifeedsci.2005.11.008

[3] Drew, M.D., Borgeson, T.D. and Thiessen, D.L., (2007). A review of processing of feed ingredients to enhance diet digestibility in finfish. *Anim. Feed Sci. Technol.* 138, 118-136. https://doi.org/10.1016/j.anifeedsci.2007.06.019

[4] Francis, G., Makkar, H.P.S. and Becker, K., (2001). Antinutritional factors present in plantderived alternate fish feed ingredients and their effects in fish. *Aquaculture 199*, 197-227. https://doi.org/10.1016/S0044-8486(01)00526-9

[5] Gaylord, T.G., Barrows, F.T. and Rawles, S.D., (2008). Apparent digestibility of gross nutrients from feedstuffs in extruded feeds for rainbow trout, Oncorhynchus mykiss. *J. World Aquac. Soc.* 39, 827-834. https://doi.org/10.1111/j.1749-7345.2008.00220.x

[6] Gaivoronsky B. A. and Ramazanov A. U. (1998). Ways to increase the usefulness of feeding dairy cattle in Northern Kazakhstan. Petropavlovsk, 3-4. https://doi.org/636.2.084(574.2)

[7] Glencross, B., Blyth, D., Tabrett, S., Bourne, N., Irvin, S. and Anderson, M., (2012). An assessment of cereal grains and other starch sources in diets for barramundi, lates calcarifer-implications for nutritional and functional qualities of extruded feeds. Aquaculture Nutrition, *18*(4). https://doi.org/10.1111/j.1365-2095.2011.00903.x

[8] Glencross, B., Robert, M., Kate, A. and Rhys, H., (2008). Productivity, carcass composition, waste output and sensory characteristics of large barramundi, Lates calcarifer fed high-nutrient density diets. *Aquaculture* 284, 167-173. https://doi.org/10.1016/j.aquaculture.2008.07.031

[9] Kizatova M.E., Medvedkov Y.B., Shevtsov A.A., Drannikov A.V. and Tlevlessova D.A., (2017). Experimental-Statistical Analysis and Multifactorial Process Optimization of the Crust from Melon Pulp Separation Process. *Journal of Engineering and Applied Sciences*, 12 (7),1762-1771. https://doi.org/10.36478/jeasci.2017.1762.1771

[10] Lanetsky, V. A. (2008). Use of waste oil and fat industry. *Food industry* (pp. 14-16). Moscow.

[11] Lyapchenko, V. A., Artyukhov A. I. and Sorokin A. E. (2014) Effective feeding of highly productive dairy herd. *Husbandry*, *6*, 8-9. https://doi.org/636.2.034

[12] Mishurov, N. P., (2006). Perspective technologies of heat treatment of compound feeds. p. 82. https://doi.org/636.085.55

[13] Mishurov, N. P., Davydova, S. A. and Davydov A.A., (2019). Promising technologies for improving the quality of feed. *Bulletin of all-russian scientific research institute of mechanization of animal husbandry*, 3(35), 4-11. https://doi.org/636.085.55:001.891

[14] Ostrikov A.N., Shakhov S.V., Ospanov A.A., Muslimov N.Zh., Timurbekova A.K., Jumabekova G.B. and Matevey Y.Z., (2018). Mathematical modeling of product melt flow in the molding channel of an extruding machine with meat filling feeding. *Food process Engineering*, 12 (7), 1-11. https://doi.org/10.1111/jfpe.12874

[15] Sorensen, M., Storebakken, T., Penn, M., Krogdahl, A. and Skrede, A., (2009). Pea protein concentrate substituting fish meal or soybean meal in diets for Atlantic salmon, Salmo salar-effect on growth performance, nutrient digestibility, carcass composition, gut health, and physical feed quality. *Aquaculture 288*, 305-311. https://doi.org/10.1016/ j.aquaculture.2008.12.012

[16] Balamurali R, Maheswaran, Prashanthi A R, Roshan Fathima S, Varshini S. 2020. Enhanced Automation Using Controller and Conscripts. International Journal of Advanced Science and Technology, 29(3), 8620 - 8627.

[17] Shevchenko, N. I. (2012). The use of nutrients in diets by black-and-white cows. *News Orenburg state agrarian University,* 35-1 (3), 105-108. https://doi.org/636.2.084.41:636.087.7

[18] Shevtsov, N. N., Zuyev, N.P., Naumov, M.M., Shvetsova, M.R., Salamakhin, S.P., Zuyeva, Ye.N., Zuyev, S.N., and Shumskiy, V.A. (2014). Milk performance of cows when feeding concentrated formula feed with extruded grain. *Bulletin of the Altai state agrarian University*, 12 (122), 100-102. https://doi.org/636.2.085.55

[19] Shvetsov, N. N. and Ievlev, M. Yu. (2011). Dairy productivity of cows when feeding with forage mixtures with sprouted extruded grain components. *Proceedings of the Kuban state agrarian University*, 31, 208-211. https://doi.org/636.2.085.55

[20] Shvetsov, N. N., Ievlev, M. Yu. (2011). *Kursk state agricultural Academy: Use of sprouted extruded grain feed in feed mixes for dairy cows.*Kursk, RF.

[21] Topuria O., Katsitadze D., Parlagashvili Yu. and Katsitadze V. (1990). Extrusion of grain components. *Feed industry* (pp. 18-19). Voronezh.

[22] Venou, B., Alexis, M.N., Fountoulaki, E., Nengas, I., Apostolopoulou, M. and Castritsi-Cathariou, I., (2003). Effect of extrusion of wheat and corn on gilthead sea bream, Sparus aurata growth, nutrient utilization efficiency, rates of gastric evacuation and digestive enzyme activities. *Aquaculture* 225, 207-223. https://doi.org/10.1016/S0044-8486(03)00290-4

[23] Venou, B., Alexis, M.N., Fountoulaki, E. and Haralabous, J., (2009). Performance factors, body composition and digestion characteristics of gilthead sea bream, Sparus aurata fed pelleted or extruded diets. *Aquaculture*. Nutr. 15, 390-401. https://doi.org/10.1111/j.1365-2095.2008.00603.x

[24] Yacko, N. A., Suchkova, I. V. and Letunovich, E. V. (2013). Quality characteristics of the" protected " protein of rapeseed feeds and their influence on the dairy productivity of cows. *Scientific notes of the educational institution "Vitebsk order" Badge of honor "state Academy of veterinary medicine"*, 49, 206-210. https://doi.org/636.2.03.086.72

[25] Zhigankov, B. V., (1981). Granulation of husks at grain plants. *Storage and processing of grain* (pp. 61-66). Moscow.

[26] Kiran George,Immaculate Joy S.,Nisha Susan Thomas, Balamurali R. &Baskaran K.., 2020. Gender-Based Vegetarian and Nonvegetarian Dietary Impact on Cardiac Autonomic Function of Heart Rate Variability. Journal of the American College of Nutrition, pp.1-5.

Mathematical model of dependence of factors for chromatographic separation of fructose from glucose-fructose syrup

Alibek O. Baikenov1, NurzhanZh. Muslimov1,Kadyrbek A. Baigenzhinov1,Zhazira A. Yessimova1,Yuliya V. Kim1

[1]Astana branch of the LLP "Kazakh Research Instituteof Processing and Food Industry",Kazakhstan, 010000, Nur-Sultan, Akzhol Street, 26

Abstract. The aim of this study is to determine the optimal parameters for the separation of glucose-fructose syrup sugars into glucose, fructose, and oligosaccharides. A mathematical model of chromatographic separation of glucose-fructose syrup (GFS) was constructed. During the optimization of the chromatographic separation process, it was found that the most significant parameter affecting the purity of the fructose yield is the dry matter content. In the course of research, it was found that the dry matter content of 30% is optimal for separating glucose-fructose syrup, at separating the use of this parameter made it possible to obtain a carbohydrate composition with a content of fructose 93%, glucose 5% and oligosaccharides 2%. This is the highest yield of fructose when separated by ion exchange resins of company "Purolite" brand PCR641Ca.
Keywords: glucose-fructose syrup, fructose, ion chromatography.

1. Introduction

Using chromatographic systems to separate sugars, such as fructose and glucose, is a fairly effective and feasible method (Mostafazadeha*et al.*, 2011). Chromatographic separation of glucose and fructose in the production of high-fructose corn syrup (HFCS) is one of the most commonly used in chromatography (Ganetsos and Barker, 1993).

Classically, the separation of sugars is carried out using resins in the form of calcium, while the separation of sugar/non-sugar is achieved using resins in the form of potassium. Resins in the form of calcium simultaneously produce two types of separation. First, the resin acts as a molecular sieve. Large molecules that cannot penetrate the resin granules are more or less excluded because of their size. Second, the separation is due to the difference in the stability of the sugar complex with calcium: only polyols and some sugars (fructose, galactose) are able to form such a complex. Other components (sucrose, glucose) do not form a complex with the resin, which explains the separation (Paillat*et al.*, 2000).

Among the main and socially important food products, sugar and sugary substances are one of the first places, and, at the current level of purchase power of the population, they are the most affordable products in the diet of people (Uglov*et al.*, 2017). Alternative natural sources of edulcorating components are sugary starch products: glucose, maltose, glucose-fructose syrups, maltodextrins and others that are obtained from corn, wheat, tapioca, triticale, sorghum, etc. (Polumbrik*et al.*, 2016).

Glucose-fructose syrup is a plant ingredient used in the food industry, mainly derived from wheat and corn. Glucose-fructose syrup has a composition similar to table sugar (also called "sucrose"), which is obtained from sugar cane or beet. Sucrose and glucose-fructose syrups consist of glucose and fructose, but in different proportions (https://starchinfood.eu/ingredient/glucose-fructose-syrup/). GFS

production is highly dependent on the local sugar industry and on agricultural raw materials (Vuilleumier, 1993), (kiran ,2020).
GFS is obtained from corn starch by first hydrolyzing the starch into glucose in an enzymatic manner, which is then partially converted to a fructose solution containing 42% fructose (HFCS-42) (Palazzi et al., 1999). GFS is cheaper than beet and cane sugar. Large-scale implementation of GFS in the food industry of developed countries began in the late 1960s (Ventura et al., 2011). Sucrose, GFS, and crystalline fructose give the body comparable energy: calories are 3.9 kcal/g, 3.7 kcal/g, and 3.6 kcal/g, respectively. GFS contain both glucose and fructose, most glucose and fructose are in the free form of monosaccharides. GFS contains a small amount of residual glucose oligosaccharides (White, 2014).
GFS production was made possible by parallel developments in refining, isomerization, and separation technologies in the 1960s. Fructose contributes many useful physical and functional properties to foods and beverages, including sweetness, improved taste, moisture, color and flavor development, reduced freezing point, and osmotic stability. GFS is widely used in carbonated beverages, baked goods, canned fruits, jams and jellies, as well as in dairy products. The use of crystalline fructose and crystalline fructose syrup has recently expanded from pharmaceutical and specialty foods to basic foods and beverages (Hanover and White, 1993).
Glucose-fructose syrups need special storage conditions, which exclude the possibility of glucose crystallization, increasing the product color or its infection. Glucose-fructose syrup can be stored for a long time at a temperature of 25-30°C. At temperatures below 25°C, the process of glucose crystallization begins (when the content of dry matter (DM)=70% saturation occurs at 27°C). At temperatures above 30°C, color increases over time due to the thermal decomposition of monosugars, as well as their reaction with amino acids. In factories, for syrup storage, stainless steel tanks with autonomous heating are used. Sometimes storage containers are placed in an isolated room where the temperature is maintained at 35-34°C. This eliminates the need for heating elements and isolation of individual containers. It should be noted that in mixtures of glucose-fructose syrup (42% fructose) with sucrose, there is no probability of crystallization due to the greater solubility of the mixture of components compared to the solubility of each of them separately. At a temperature from 22.8 to 50°C, the maximum solubility is a syrup with a sucrose content of 33-47%, so the storage of such syrups is simplified (Guljuka, 1985).
Since 1978, the United States, Japan, and other countries have started producing syrups with a higher fructose content and, consequently, a lower glucose content.
These are so-called second-generation syrups. High-fructose syrups are obtained mainly by chromatographic separation of ordinary glucose-fructose syrup into fructose and glucose in a special separation column, where ion-exchange resins in Ca2+form are used as an adsorbent. Having an affinity for the adsorbent, fructose is retained on the resin. Glucose and polysaccharides pass at high speed through the adsorbent layer. Subsequent elution with 230 deposited desorbent produces an enriched fructose fraction containing up to 90% fructose. Enriched glucose syrup contains up to 89% glucose. The glucose fraction is returned to isomerization, and the fructose fraction is mixed with the original glucose-fructose syrup after purification with special ion-exchange resins. In this case, a syrup with a fructose content of 55% or higher is obtained (Guljuka, 1985). Due to its advantages, high-fructose syrup (HFS) is currently the leader among sweeteners in the food industry, especially in the beverage industry (White, 2014). First, these are relatively stable beverages with a low pH and require less processing costs. In addition, it has a high tendency to change color, which is useful in the manufacture of baked products. In addition, the sugars are already pre-dissolved and resistant to crystallization and growth of microorganisms as a traditional method of preservation. It should be noted that fructose is the sweetest dietary sugar, which is almost 1.2 times sweeter than sucrose (Hanover and White, 1993).
Now the most popular and effective way to separate glucose-fructose syrup is chromatography. In Finland, "SuomenSokeri" started producing pure crystalline fructose for the first time in the world. This technology is based on a method for distributing invert syrup into fructose and glucose fractions using a chromatographic method. Using this technology, the world's largest "Ksjurofin" plant was built in the United States [https://nomnoms.info/poluchenie-glyukozo-fruktoznyh-siropov-i-fruktozy/].

The use of the high-performance liquid chromatography method allows monitoring changes in the composition of carbohydrates and organic acids of the grain wort, and promptly identifying and eliminating the causes that lead to a decrease in the quality of the finished product. The developed new methods for determining the mass concentration of carbohydrates (maltose, glucose, fructose) and organic acids (acetic, lactic) are based on the processes of ligand exchange and ionic excretion (Medrish*et al.,* 2016).

Usually, cation exchange resin (Ca2) with pure water as an eluent are used at separation (Verhaar and Kuster, 1981).

In the "Ca-pillar" system, the main function of ion exchangers is to immobilize Ca^2, whereas separation is the result of different complexing abilities of polyols with Ca^2. However, the degree of crosslinking of divinylbenzene is also important; polymer sugars are better separated at low degrees, and monomeric and dimeric sugars at high crosslinking degrees (Salah, 2006).

Chromatographic processes for separating sugars use layers of cation exchange, which are in the form of calcium (Ca2), which is known to form a weak complex with fructose, which leads to a predominant slowdown of fructose, while glucose is carried away with the mobile phase (Barker and Joshi, 1991). The advantages of these processes are low-cost eluent, high performance and stability of column, and complete elution of all added sugars (Angyal et al., 1979).

Fructose distribution coefficients depend on temperature; fructose retention is lower in the chromatographic column at a higher temperature (Viard and Lameloise, 1992). In practical use, calcium forms of ion-exchange adsorbents are preferred because of their low cost and non-toxic nature (Lloyd and Nelson, 1984).

Significant research has been conducted on the separation of GFS on ion-exchange resins in the world, but there is insufficient data on the separation of GFS 42 into glucose and fructose by "Purolite" resins of PCR641 brand. The main purpose of these studies is mathematical model of the process of separating GFS 42 used to produce a syrup with a high fructose content.

2. Mathematical model

To construct a mathematical model of the chromatographic separation of GFS, which is a regression equation used rotatable plan of the second order (Box plan) when the number of factors to equal 3, the number of experiments is not more than 20, the number of experiments at the zero point was 6 and the number of coefficients is 10.

A mathematical and statistical method was used as a mathematical apparatus and a system of regression equations was obtained that models the relationship of the most preferred optimality criterion with the rest.

Specific factors that determine specific production conditions were used as optimality criteria that affect the process of chromatographic separation of GFS. In this regard, it is advisable to adjust the system of regression equations in accordance with the factors taken.

The regression equation of the mathematical model of separation in the chromatographic column will have the form:

$$y=b_0+b_1x_1+b_2x_2+b_3x_3+b_{12}x_1x_2+b_{13}x_3+b_{23}x_2x_3+b_{11}x_1^2+b_{22}x_2^2+b_{33}x^2_3 \quad (1)$$

where b (i = *0, 1, .., 33*) represent the regression coefficients of the model.

3. Experimental part

3.1 Materials

Glucose-fructose syrup containing 52% glucose, 42% fructose and 6% oligosaccharides was used as the starting product. For the separation of glucose-fructose syrup, the method of separation with a strong-acid cation-exchange resin PCR641Ca of the company "Purolite International Ltd" (Great Britain) was used. In chromatographic separation, the following parameters were monitored: DE concentration in glucose-fructose syrup, separation temperature, and mobile phase speed. Dependent variables are the volume of elution (ml) and time (min) for the fructose-rich fraction, the return fraction, and the glucose-rich fraction, as well as the concentration of glucose and fructose in each of the fractions.

IOP Conf. Series: Materials Science and Engineering **994** (2020) 012023 doi:10.1088/1757-899X/994/1/012023

Description of resin

PCR641Ca is a resin consisting of polystyrene cross-linked with divinylbenzene, the functional group is sulfonic acid. The resin is presented in the form of spherical grains, the diameter was 310 mm with a total exchange capacity of 1.5 g-eq/l (32.8 Kilogran/cubic foot) (H+ form) and a maximum operating temperature of 120 °C. The functional group of the resin was sulfonic acid, and the ionic form in Ca^{2+}.

3.2 Experiments in chromatographic packed column

Process of wetting the resin

A glass column with a length of 20 cm and a diameter of 3 cm was used as a device for chromatography. The resin slightly retains glucose with its layer in the calcium-fructose complex. In this way, fructose will remain bound to the resin until the flow of eluting water weakens their bond by eluting fructose with water.

A weighted amount of resin (1 g/ml of column content) was placed in a separating chromatographic column. PCR641Ca resin separates fructose from glucose by molecular elimination. A peristaltic eluent pump (Buchi Chromatography B-688 Pump) was connected and deionized water was pumped through a column to wet the resin. In a resin-filled container, water was filled to activate the resin, swell it, and form uniform pores between the grains, after which the water was drained.

The separation steps

Glucose-fructose syrup (eluate) was diluted in deionized water to a dry matter content of 30-50% and passed through a 2-mm cellulose filter, then the solution was heated to a temperature of 40 to 48 °C. Volume of the syrup solution added into the column was 5% of BV. The eluent was deionized water and heated to a temperature higher than the column temperature to remove any dissolved gases that were released.

The resulting solution of glucose-fructose syrup was fed into a column with flow rates from 4 to 8 milliliters per minute. After the batch of syrup solution, the calculated amount of deionized water (elution) was fed to the column with the same flow rate equal to the flow rate of glucose-fructose syrup. The elution time was calculated from the time when the water pump was turned on. The solution flowing out of the column was sent to the fraction collector (Buchi B-684 Fraction Collector) at a speed of 60 s per tube. When the separation was complete, the column was washed with deionized water at a flow rate of 10-15 milliliters per minute.

The collected effluents from the column (in 60-second portions) were analyzed using high-performance liquid chromatography (HPLC) to determine the carbohydrate content.

3.3 Sample analysis

In the obtained samples, the carbohydrate composition was analyzed for the content of glucose in the syrup. The fructose concentration was determined by high-performance liquid chromatography (Waters HPLC "Waters Corporation", USA) with a differential refractometric detector Waters 2414 using a column (Aminex HPX-87H (300*7.8 mm), Bio-Rad Labs, USA). The studies were performed under the following parameters: flow rate: 0.6 ml/min; detector: refractive index (RI), 40°C; column temperature: 60°C; mobile phase: 5 mmol/l H2SO4; injection: 10 µl; preprocessing of the sample: dilution.

4. Results and discussions

4.1 Mathematical model and optimization of the chromatographic separation process

During experimental studies of chromatographic separation, samples of glucose-fructose solution were selected and analyzed. To build a mathematical model, the following factors were established: the heating temperature of GFS (T, °C), the GFS feed rate to the chromatographic column (V, gal/min), and DM content (DM,%). All these factors influence the optimization criteria – the purity of fructose yield in solution (n,%).

At the beginning, the intervals and levels of variation of the parameters taken were coded, which are presented in Table 1. The experiment planning matrix is shown in Table 2.

Table 1. Coding of intervals and levels of the variation of input factors

Factors		Variation levels					Variation intervals
natural	coding	-1,68	-1	0	+1	+1,68	
T, ^{0}C	x_1	40	42	44	46	48	2
V, gal/min	x_2	4	5	6	7	8	0,1
DM,%	x_3	30	35	40	45	50	5

Table 2. Matrix of rotatable planning of experimental studies of the chromatographic separation

No.	Coded values			Natural values			Experimental values
	x_1	x_2	x_3	T,^{0}C	V, gal/min	DM,%	n, %
1	2	3	4	5	6	7	8
1	-	-	-	42	5	35	88
2	-	-	+	42	5	45	89
3	-	+	-	42	7	35	84
4	-	+	+	42	7	45	84
5	+	-	-	46	5	35	85
6	+	-	+	46	5	45	80
7	+	+	-	46	7	35	85
8	+	+	+	46	7	45	83
9	-1,68	0	0	40	6	40	85
10	+1,68	0	0	48	6	40	88
11	0	-1,68	0	44	4	40	84
12	0	+1,68	0	44	8	40	90
13	0	0	-1,68	44	6	30	85
14	0	0	+1,68	44	6	50	93
15	0	0	0	44	6	40	88
16	0	0	0	44	6	40	86
17	0	0	0	44	6	40	87
18	0	0	0	44	6	40	87
19	0	0	0	44	6	40	85
20	0	0	0	44	6	40	87

Table 3. shows the values of confidence intervals of the chromatographic separation optimization criteria.

Table 3. Values of confidence intervals of the optimization criterion for y (n, %)

Chromatographic separation		Parameter	Confidence intervals			
			Δb_0	Δb_i	Δb_{ii}	Δb_{ij}
Fructose yield purity	n, %	y	±0,84	±0,56	±0,54	±0,73

The coefficient of the regression equation is significant in cases where its absolute value is greater than the confidence interval. Otherwise, it is considered insignificant and can be excluded from further consideration of the mathematical model.

When comparing confidence intervals from Table 3 with regression coefficients from Table 4, it can be said that the effects of the interaction of input factors are negligible, and they can be neglected.

Substituting the resulting encoded values in the regression equation (1), then the equation of the fructose yield purity in the solution will have the form:

$$y=86{,}80348128-0{,}28987x_1+0{,}002928x_2+0{,}447984x_3+1{,}625x_1x_2-1{,}375x_3+0{,}125x_2x_3-0{,}48695x_1^2-0{,}83975x_2^2+0{,}218646x_3^2$$

After decoding the equation, we get the regression equation for natural values of the factors:

$$n=189{,}3959+11{,}48293T-26{,}6759V+5{,}110737CB+0{,}8125TV-0.1375CB+0.02500VCB-0.12174T^2-0.83975V^2+0{,}008746CB^2$$

Table 4. Coefficients of regression equations of the yield parameters for y (n, %)

Optimization criterion	Coefficients	Yield of finished product
	For coded values of factors	
	b_0	86,80348128
	b_1	-0,28987
	b_2	0,002928
	b_3	0,447984
	b_{12}	1,625
	b_{13}	-1,375
	b_{23}	0,125
	b_{11}	-0,48695
	b_{22}	-0,83975
Fructose yield purity, %	b_{33}	0,218646
	For natural values of factors	
	B_0	-189,3959
	B_1	11,48293
	B_2	-26,6759
	B_3	5,110737
	B_{12}	0,8125
	B_{13}	-0,1375
	B_{23}	0,02500
	B_{11}	-0,12174
	B_{22}	-0,83975
	B_{33}	0,008746

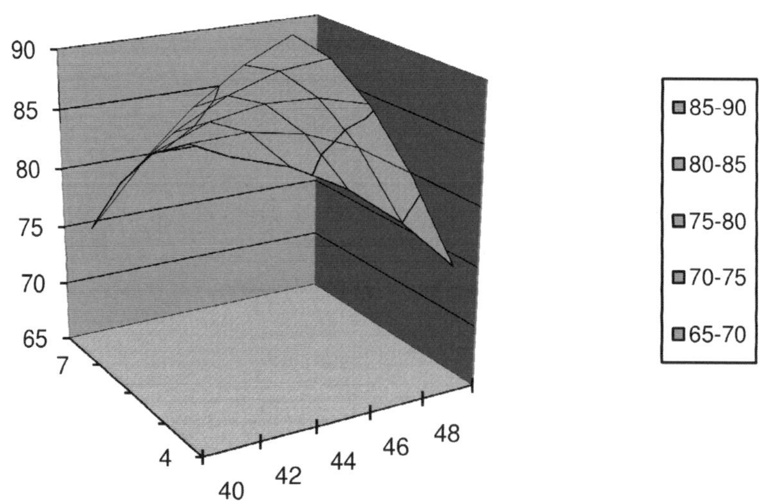

Figure 1. Three-dimensional model in space that characterizes the dependence of $y_n=f(T, V)$ temperature and syrup feed rate

IOP Conf. Series: Materials Science and Engineering **994** (2020) 012023 doi:10.1088/1757-899X/994/1/012023

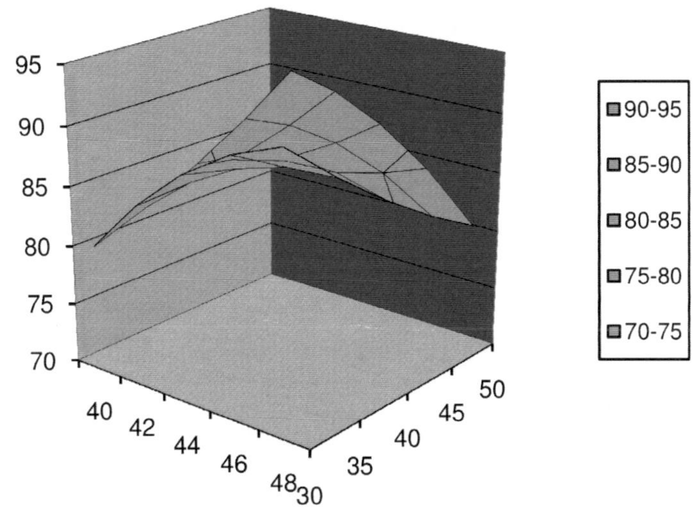

Figure 2. Three-dimensional model in space that characterizes the dependence of y_n=f(T, DM) temperature and dry matter content

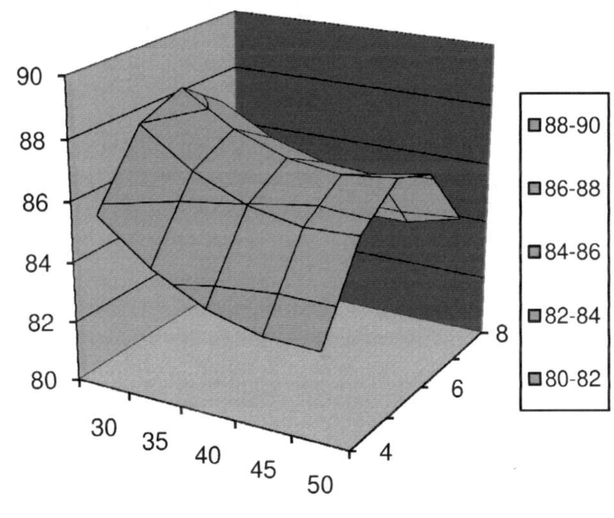

Figure 3. Three-dimensional model in space that characterizes the dependence of y_n=f(V, DM) of dry matter content and feed rate

At the stage of optimization of chromatographic separation, the concentration of dry matter (DM) of the supplied glucose-fructose syrup solution in the separation column is of particular interest, as a characteristic that determines one of the main parameters that affect the fructose content purity in the solution at the outlet of column. The dependences (Figures 1-3) of influence of the main parameters (temperature, solution feed rate and dry matter content) of the separation are obtained. Figure 1 shows that the fructose yield purity decreases with increasing temperature and increasing the solution feed rate to the separation column, the value of the solution feed rate close to 4 ml/min is observed the highest fructose yield purity.

Figure 2 shows that at the lowest DM of 30°C, the yield of pure fructose in the solution increases.

The three-dimensional model of Figure 3 shows how the solution feed rate and the dry matter content affect the fructose yield purity in the solution. In this model, it is clearly seen that when the dry matter content exceeds 50%, the separation process is worse, as shown in Figure 2. When the minimum and maximum value of the solution feed rate is equal to 4 and 8 milliliters per minute, the yield of fructose decreases. From the model of Figure 3, it can be concluded that the most optimal value is the speed equal to 6 ml/min and the DM content equal to 30%. The maximum fructose yield purity was 93% of the fructose content in the solution.

4.2 High-performance liquid chromatography

At the end of optimization of the chromatographic separation process, control separation experiments were performed using the obtained parameters. After that, the resulting separated solution was analyzed. The obtained data on the carbohydrate composition of the separated solution were compared with the data on the carbohydrate composition of GFS.

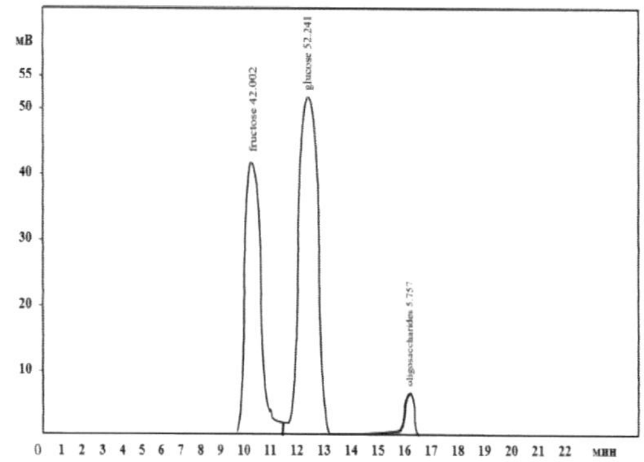

Figure 4. Chromatogram of carbohydrate composition of GFS before chromatographic separation

The chromatogram of GFS carbohydrate composition (Figure 4) shows that the glucose content was 52%, fructose 42%, and other sugars (oligosaccharides) 6%.

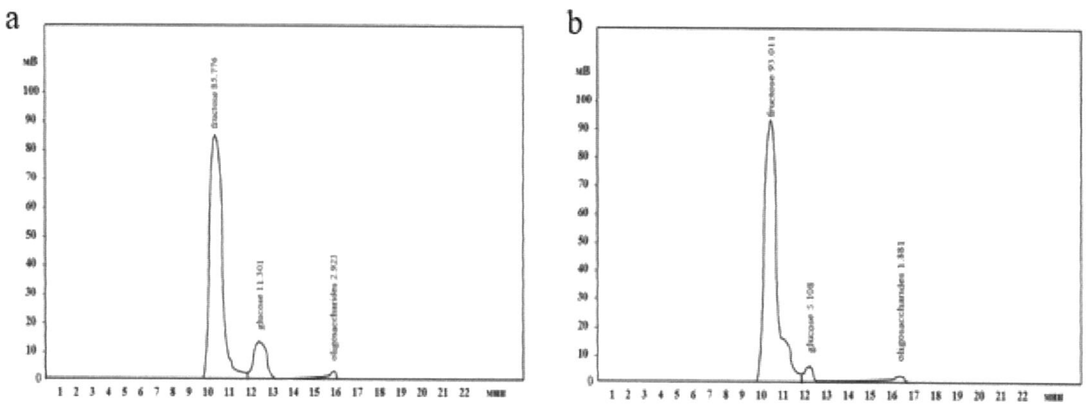

Figure 5. Chromatogram of the carbohydrate composition of solution after chromatographic separation of GFS at the DM content of 50% (a) and 30% (b)

During the optimization of the chromatographic separation process, it was found that the most significant parameter affecting the purity of the fructose yield is the dry matter content. In this regard, control experiments were performed in several repetitions with different dry matter content. The chromatograms (Figure 5 (a) and (b)) show the carbohydrate composition of high-fructose syrup after GFS separation. Figure 5 (a) shows a chromatogram of the carbohydrate composition of the experiment at a 50% DM content, which shows that the yield of fructose was 85%, glucose 12% and oligosaccharides 3%. In order to increase the yield of pure fructose in the syrup, the DM content was lowered to 30% and separated. The chromatogram of Figure 5 (b) shows the carbohydrate composition of the syrup divided at a 30% DM content. The carbohydrate composition showed a fructose content of 93%, glucose of 5% and oligosaccharides of 2%. When the DM content decreases below 30%, the glucose contained in the solution does not have time to react with the resin. This in turn increases the concentration of fructose in the solution. This is the highest yield of fructose when separated by "Purolite" cationic resins of PCR641Ca brand.

5. Conclusion

Fructose is supplied as a solution, and in this case its degree of purity must be at least 80%, so the fructose solution obtained from the column by chromatographic method is a ready-made solution that meets the requirements of commercial quality.

During the chromatographic separation of GFS to obtain high-fructose syrup, a number of experiments were conducted using ion-exchange resins of PCR641Ca brand of company "Purolite". From the results, it became clear that fructose reacts more strongly than glucose, which means that it leaves the column a little later. It follows that this method can be used to obtain a fraction saturated with fructose from glucose-fructose syrup for the food industry. One of the problems, as it turned out, is the concentration of syrup, that is, the content of dry matter in the syrup. For the purpose of separation, DM content was lowered to 30%. It is also important to apply the solution to the resin surface.

Optimization of chromatographic separation parameters was carried out by constructing a mathematical model. The model represented the dynamics of the column behavior with varying input optimization parameters from the zero point.

References

[1] Angyal S J Bethell G S Beveridge R J (1979). The separation of sugars and polyols on cation-exchange resins in the calcium form. Carbohydrates Research, 73, 9-18. https://doi.org/10.1016/s0008-6215(00)85471-3.

[2] Barker, P.E, Joshi, K. (1991). The recovery of fructose from inverted sugar beet molasses using continuous chromatography. Journal of Chemical Technology and Biotechnology, 52, 93-108.https://doi.org/10.1002/jctb.280520107

[3] Glucose-Fructose Syrup and Fructose-Glucose Syrup 2020 Retrieved from https://starchinfood.eu/ingredient/glucose-fructose-syrup

[4] Ganetsos G and Barker P E 2019 Semicontinuous Countercurrent Chromatographic Refiners. Preparative and Production Scale Chromatography pp 233–256 DOI: 10.1201/9780367811150-15

[5] Guljuka N G 1985 Starch and starch products. Moscow RF Agropromizdat

[6] Hanover L M White J S 1993 Manufacturing composition and applications of fructose The American Journal of Clinical Nutrition Vol 58 pp 724S–732S https://doi.org/10.1093/ajcn/58.5.724S

[7] Lloyd N E Nelson W J 1984 Glucose and fructose containing sweeteners from starch In Whistler R L Bemiller J N Paschall E F Starch chemistry and technology 2nd ed New York: Academic Press

[8] Medrish M Je Abramova I M Poljakov V A Gavrilova D A and Pavlenko SV 2016 Exposure to the content of carbohydrates and organic acids in grain wort and brew with the use of PEX Innovative technologies 12 Retrieved from https://cyberleninka.ru/article/n/opredelenie-soderzhaniya-uglevodov-i-organicheskih-kislot-v-zernovom-susle-i-brazhke-s-primeneniem-vezhh/viewer

[9] Mostafazadeh A K Sarshar M Javadian S Zarefard M and Haghighi Z A 2011 Separation of fructose and glucose from date syrup using resin chromatographic method: Experimental data and mathematical modeling Separation and Purification Technology vol 79 pp72–78 doi: 10.1016/j.seppur.2011.03.014

[10] Paillat D Cotillon M and Theoleyre M 2000 Technology of chromatographic separation in glucose syrup processes In Association Avh 7th Symposium Reims France: ValentinHauyAssocation p 62-67

[11] Palazzi E and Converti A 1999 Generalized linearization of kinetics of glucose isomerization to fructose by immobilized glucose isomerase Biotechnology and Bioengineering vol 63 pp 273-284 https://doi.org/10.1002/(SICI)1097-0290(19990505)63:33.0.CO;2-P

[12] Obtaining a syrup of glucose and fructose 2018 Retrieved November 27 2018 from https://nomnoms.info/poluchenie-glyukozo-fruktoznyh-siropov-i-fruktozy/

[13] Polumbrik M O Litvyak V V Lovkis Z V and Kovbasa V N 2016 Carbohydrates in food Minsk

[14] Salah M and Al Eid Chromatographic separation of fructose from date syrup International Journal of Food Sciences and Nutrition 57 1/2 pp 83-96 https://doi.org/10.1080/09637480600658286

[15] Deepa N R Nagarajan N PalanivelRajan D and Balamurali R 2019 Image Processing System for Detecting and Recovering from Cotton Leaf Disease using The Android Application. International Journal of Innovative Technology and Exploring Engineering IJITEE vol 8 pp3366 – 3381

[16] Uglov B A Boroday E V and Aksenov V V 2017 Production of sugary starch products Modern trends in technical sciences: Materials of 5th International Science Conference pp 57-60 Kazan Tatarstan

[17] Ventura E E Davis J N Goran M I Sugar content of popular sweetened beverages based on objective laboratory analysis: focus on fructose content Obesity Silver Spring 19 Retrieved from https://www.ncbi.nlm.nih.gov/pubmed/20948525

[18] Verhaar L A and Kuster B F M 1981 Improved column efficiency in chromatographic analysis of sugars oncation-exchange resin by use of water-triethylamine eluents Journal of Chromatography A vol 210 pp 279-290 https://doi.org/10.1016/S0021-9673(00)97838-5

[19] Viard V and Lameloise M L 1992 Modeling glucose-fructose separation by adsorption chromatography on ion exchange resins Journal of Food Engineering 17(1) 29-48. https://doi.org/10.1016/0260-8774(92)90063-c

[20] Vuilleumier S 1993 Worldwide production of high-fructose syrup and crystalline fructose The American Journal of Clinical Nutrition vol 5 pp 733-736 https://doi.org/10.1093/ajcn/58.5.733S

[21] White J S 2014 Sucrose HFCS and fructose: history manufacture composition applications and production In:Rippe J M Ed Fructose High Fructose Corn Syrup Sucrose and Health Springer pp 13-33 https//doi.org/10.1007/978-1-4899-8077-9

[22] Kiran George Immaculate Joy S Nisha Susan Thomas Balamurali R and Baskaran K 2020 Gender-Based Vegetarian and Nonvegetarian Dietary Impact on Cardiac Autonomic Function of Heart Rate Variability Journal of the American College of Nutrition pp1-5

Solution of Tasks of Energy Efficiency of the Process of Drying Raw Plant Materials by Methods of Mathematical Modeling

Alexander A. Shevtsov1, Aleksey V. Drannikov2, Larisa I. Lytkina3, Anna A. Derkanosova4, Aleksey V. Pribytkov5

[1]Doctor of Engineering, professor, Department of Technologyof Fat, Processes and Apparatuses of Chemical and Food Production, Voronezh State University of Engineering Technologies, Revolution Av., 19, Voronezh, 394036, Russia, shevalol@rambler.ru
[2]Doctor of Engineering, professor, Department of Machines and Apparatuses of Food Production, Dean of the Faculty of Food Machines and Automata, Voronezh State University of Engineering Technologies, Revolution Av., 19, Voronezh, 394036, Russia, drannikov@list.ru
[3]Doctor of Engineering, professor, Department of Technology of Bakery, Confectionery, Pasta and Grain Processing Industries, Voronezh State University of Engineering Technologies, Revolution Av., 19, Voronezh, 394036, Russia, larissaig2410@rambler.ru
[4]Doctor of Engineering, Department of Service and Restaurant Business, Voronezh State University of Engineering Technologies, Revolution Av., 19, Voronezh, 394036, Russia, aa-derk@yandex.ru
[5]Doctor of Engineering, Department of Machines and Apparatuses of Food Production, Dean of the Faculty of Food Machines and Automata, State University of Engineering Technologies, Revolution Av., 19, Voronezh, 394036, Russia, Pribytkov_aleks@mail.ru
FSBEE HE "Voronezh State University of Engineering Technologies", Russia, 394036, Voronezh, Revolution Avenue, 19.

Abstract. Beet pulp has a high feed value, but after pressing it has a sufficiently high humidity, which leads to rapid spoilage. One of the ways to stabilize the quality of beet pulp is drying with superheated steam of reduced pressure. Mathematical modeling of heat transfer during convective drying of beet pulp is proposed. A feature of the model is the description of the drying process, which ensures its high intensity due to high heat and mass transfer ratios and preserves the quality of the finished product as a result of the process at a reduced pressure of transfer medium. In the period of a constant drying speed, dehydration of macroscopic pores topologically related to the sample surface, which are characterized by a relatively low binding energy of water molecules with the liquid phase, is taken into account. According to experimental data, the proportion of water content in these pores is 60...70% of the total moisture content. In the final stage of this period, an increase in the binding energy occurs due to the removal of water from microcapillaries, which is explained by the formation of structured clusters in thin wetting films of water molecules. The change in steam temperature as a result of passing through the product layer was 8 ... 10 K, and the relative temperature change was 5 ... 10%. The parameters of the models were identified in the following range of parameters: superheated steam temperature 393 ... 453 K; speed of superheated steam 3 ... 5

m/s; the pressure inside the drying chamber is 40 ... 100 kPa; the specific load of the product on the grate is 8 ... 24 kg/m2. The interpretation of the simulation results is presented by the dependences of the temperature difference between beet pulp and heat carrier ΔT on the temperature of superheated steam Ts at various pressures P: 1 - 100 kPa; 2 - 80 kPa; 3 - 60 kPa; 4 - 40 kPa; υ = 4 m/s; at various specific loads of beet pulp on the gas distribution grid qsp: 1 - 8 kg/m2; 2 - 16 kg/m2; 3 - 24 kg/m2; υ = 4 m/s; P = 60 kPa and various speeds of superheated steam υ: 1 - 5 m/s; 2 - 4 m/s; 3 - 3 m/s. A sufficient convergence of the results was ensured, in which the deviation of the calculated data from the experimental data did not exceed 12,5% in absolute value. The obtained results of the simulation can be used in the operational control of the technological parameters of the process of drying beet pulp with superheated steam of reduced pressure with a restriction on the temperature mode due to the receipt of a high quality product.
Key words: drying, beet pulp, superheated steam of reduced pressure, kinetics.

1. Introduction

The problem of the production and use of non-traditional types of raw materials, such as waste from the sugar industry, in particular beet pulp, which, having high feed value and taking into account the scale of production, can greatly contribute to solving one of the most important issues in the development of the country's agriculture - creating a solid fodder base for livestock breeding. However, the pulp after pressing has a sufficiently high humidity (72 ... 75%), resulting in its rapid deterioration and souring during storage and, therefore, reduction efficiency of its use as a valuable fodder raw material or as filler in the production of biologically active additives used in the organization of full feeding of different species and age groups of agricultural animals and poultry [1, 2, 3].

It has been established that due to the application of thermal effect, it is possible to ensure high quality of beet pulp, increase shelf life and be effectively used in animal feed technology. One of the ways to stabilize the quality of beet pulp is drying with superheated steam of reduced pressure, which largely provides a high-quality product without reducing the intensity the process [4, 5].

The temperature modes during drying of beet pulp determine the intensity and duration of the process, and the temperature gradients in the product, as a rule, determine its quality. Due to the complexity of the experimental determination of temperature fields in single particles, it is crucial to develop modeling methods to determine the temperature fields in the product, which as a rule, they allow the analysis of many alternative options for hardware and technological design of drying processes and are optimal in terms of heat and power costs and indicators of quality of the product to be dried.

2. Materials and methods

In this regard, mathematical modeling of thermal exchange in convective drying of beet pulp due to the transfer of thermal energy of superheated steam of reduced pressure through the surface of the product particle and the endothermic phase transition of water molecules from the bound state to the gaseous state is proposed [6],[7].

3. Results and discussion

When formulating the simulation problem, the following simplifying assumptions were made:

1. The particle shape of the product is represented by a thin plate, which is blown with superheated steam at a speed of υ. The plate is located in the plane (x, y) of the coordinate system, and the heat carrier flow is directed along the z axis.
2. The characteristic time of heating of the samples is determined by the ratio, s:

$$\tau_f = \frac{l^2}{a},$$

where$l$- thickness of the plate material, m; a - coefficient of thermal diffusivity of the material, m^2/sec.

Since the drying process time exceeds the sample heating time by almost two orders, the heat flux quantity entering the product particle from the heat carrier is controlled by the thermal conductivity of superheated steam, and the temperature of the product particle is assumed to be the same throughout its volume.

3. The change in temperature of superheated steam in the working volume of the drying unit is taken into account only from one spatial variable z. The z axis is directed along the steamspeed vector, and the coordinate origin coincides with the location of the gas distribution grid. The steam temperature depends on two variables: the z coordinate and time t: $T_s = T_s(z, t)$.
4. The studies on the kinetic laws of the beet pulp drying process in an environment of superheated steam of reduced pressure show the presence of two drying periods: a period of constant and decreasing drying speed [11].

In the period of a constant drying speed, dehydration of macroscopic pores topologically related to the sample surface, which are characterized by a relatively low binding energy of water molecules with the liquid phase, is taken into account. According to experimental data, the proportion of water content in these pores is 60...70% of the total moisture content. In the final stage of this period, an increase in the binding energy occurs due to the removal of water from microcapillaries, which is explained by the formation of structured clusters in thin wetting films of water molecules [8]. In the period of decreasing drying speed, water is removed from the pores due to the diffusion of water molecules through the walls of the product tissue.

The general problem of heat exchange during drying of a particle located in a moving heat carrier flow is that after the drying time some thermal equilibrium is established and it is necessary to determine the resulting temperature difference $(T_s - T)$.

The kinetic equation for changing the particle temperature T is considered, based on the balance of thermal energy:

$$Vc\frac{\partial T}{\partial \tau} = aS(T_S - T) - S\varepsilon j, \quad (1)$$

where V- volume of one particle of the product, m^3; c-specific heat of the product, J/(kg·K); T_s-temperature of the heat carrier near the product particle, K; S- area of the product particle, m^2; α- heat transfer coefficient from the heat carrier to the product particle, W/(m^2·K); ε- binding energy of the water molecule with the liquid phase; j- the flux density of water molecules evaporated from the sample; $\dot{T} = \frac{\partial T}{\partial \tau}$-change speed of the sample temperature, K/s; $aS(T_S - T)$ - heat energy flow from steam to the productparticle; $S\varepsilon j$- endothermic process of water evaporation from the surface of a particle.

The left-hand side of equation (1) represents the change speed of the thermal energy of the particle, the first term on the right-hand side determines the heat transfer of the thermal energy of superheated steam to the product particle, the second term - the decrease in the thermal energy of the particle as a result of evaporation [9],[10].

The equilibrium concentration of water molecules in steam at the temperature of the sample is presented in the following form, 1/cm^3:

$$n_s(T) = n_{atm}\left(e^{u/kT_{atm} - u/kT}\right), \quad (2)$$

whereT_{atm}- boiling point of water at atmospheric pressure (T_{atm} = 373K); n_{atm}- concentration of water molecules under these conditions (n_{atm}= 2 · 1019 1/cm^3); u- parameter found from the experimental data.

The temperature dependence of the kinetic coefficient k is determined by the expression:

$$k = R\frac{v_m}{4},$$

where v_t- thermal speed of the movement of water molecules; R- condensation coefficient, which represents the probability of "sticking" of a water molecule from the heat carrier to the surface of the condensed state of the substance upon impact on the surface.

The equation for changing the steam temperature based on the balance of the thermal energy of the heat carrier was considered similarly (1):

$$3n_s kT_S = -3n_s kv\frac{\partial T_S}{\partial Z} - \alpha S n_f(T_S - T). \quad (3)$$

On the right side of expression (3), the change speed of the thermal energy density of water steam is recorded. $\dot{T}_s = \frac{\partial T_S}{\partial Z}$- the change speed of the steam temperature, K/s; n_f- the number of product particles per unit volume, 1/m^3; υ – steam speed in the drying chamber, m/s.

The first term in (3) represents the change speed of thermal energy at some point in the volume of the drying apparatus due to the movement of steam at a speed υ. The second term takes into account the decrease in thermal energy of steam due to heat transfer to heat the product.
Substituting (2) in (1) taking into account (3) allows us to formulate a model of the drying process of beet pulp in a closed system of differential equations in which the quantities to be found are T, T_s, j:

$$T = \frac{\alpha S}{Vc}(T_s - T) - \frac{Suk}{Vc}\left(n_p - n_s\right), \tag{4}$$

where n_s- the concentration of water molecules in the superheated steam in the drying chamber, 1/cm^3.

$$T_s = -\upsilon\frac{\partial T_s}{\partial Z} - \frac{\alpha S n_f}{3kn_s}(T_s - T). \tag{5}$$

Under the conditions under consideration, other characteristics of the system, such as the speed of the heat carrier and its concentration, may also depend on the z coordinate. The change n_s with height z is due to changes in steam pressure and its temperature. The change in temperature of the steam as a result of passing through the product layer was 8 ... 10 K, and the relative change in temperature was 5 ... 10 %. The same quantity is the relative seal of the heat carrier. Therefore, in the system (4), (5), the change in the steam density was neglected.
A similar situation arose in connection with a change in steam pressure before and after the product layer. This meant that a decrease in the density of water molecules in the heat carrier arising from a decrease in pressure can also be neglected. For the same reason, in the system of equations (4), (5), the steam density and its speed are considered constant over the entire thickness of the product layer. These characteristics, together with the initial temperature T_{0s} and the mass of the product, were the controlling parameters of the process and their values are set during an experimental study of the kinetics of the drying process.
Based on the above assumptions, the solution of system (4), (5) is determined for the drying process of beet pulp in a stationary mode, i.e. $\dot{T} = 0$, $T_s = 0$. Hence follows:

$$T_s - T = \frac{uk}{\alpha}\left(n_p(T) - n_s\right). \tag{6}$$

$$\frac{\partial T_s}{\partial Z} = -\frac{\alpha S n_\phi}{3kn_s \upsilon}(T_s - T). \tag{7}$$

To find a solution to system (6), (7), a function is introduced that represents the temperature difference between the heat carrier and the product:

$$\theta = T_s - T = \frac{u\chi}{\alpha}\left(n_p(T) - n_s\right), \tag{8}$$

where χ- the kinetic coefficient relating the flux density of evaporating water molecules from the product to the deviation of the concentration of water molecules in superheated steam from equilibrium.
Then the system of equations (6), (7) takes the form:

$$T_s = T + \Theta(T); \tag{9}$$

$$\frac{\partial T_s}{\partial Z} = -\frac{1}{\lambda}\Theta(T), \tag{10}$$

where λ characterizes the linear section along the height of the drying chamber, on which the steam temperature decreases. This quantity is determined by the expression:

$$\lambda = \frac{3kn_s\upsilon}{\alpha S n_f}. \tag{11}$$

Equation (9) was differentiated according to the z coordinate:

$$\frac{\partial T_s}{\partial z} = \frac{\partial T}{\partial z} + \frac{\partial \Theta}{\partial T}\frac{\partial T}{\partial z}, \tag{12}$$

and after substituting (12) into (10), we obtained:

$$\frac{\partial T}{\partial z}\left(1 + \frac{\partial \Theta}{\partial T}\right) = -\frac{1}{\lambda}\Theta(T). \tag{13}$$

The solution of equation (13) is presented in the form:

$$\int \frac{dT}{\Theta(T)} + \ln\Theta(T) = -\frac{z - z_0}{\lambda}, \tag{14}$$

where z_0- the integration constant, which is found from the condition $T(0) = T_0$, here T_0- the temperature of the product near the gas distribution grid, K.

The drying model of beet pulp (6) - (12) reflects the situation with a low value of the degree of layer expansion, in which the mixing of product particles is poorly expressed. This may occur, including due to the dispersion of product particles in size. Larger particles occupy relatively low positions in the heat carrier flow, and the temperature of the product particles becomes dependent on the height *z*.
Since the numerical estimates of the integrand in (14) take a value significantly less than unity, this integral is neglected. Asaresult, weget:

$$\Theta(T) = \Theta(0)e^{-\frac{z}{\lambda}}.$$

Hence, the dependence *T (z)* was found by the formula:

$$T = \frac{T_k}{1-\frac{kT_k}{u}\ln\left(\frac{n_S}{n_{atm}}\left(1+\frac{\alpha(T_{0S}-T_O)}{uk}e^{-\frac{z}{\lambda}}\right)\right)}, \quad (15)$$

hereT_{0s}- the temperature of the steam at the input to the drying plant, K; T_k- the final temperature of the product, K.
When drying beet pulp in a fluidized bed, the particle temperature $\overline{T}$ is averaged over the time spent in different parts of the drying chamber and takes a fixed value. Then the solution of the kinetic equation (3) for the steam temperature at each section of the drying chamber in the stationary mode $\dot{T}_s$ =0 has the form:

$$T_s = \bar{T} + (T_{0s} - \bar{T})e^{-\frac{z}{\lambda}}. \quad (16)$$

The average temperature of the product particles $\overline{T}$ is found from the balance condition for the absorbed heat and energy that goes to the water evaporation.
The density of thermal energy per unit time absorbed by the particles of the product at a height *z* is equal, W/m^3:

$$q = \alpha n_f S(T_s - \bar{T}). \quad (17)$$

As a result of substituting (16) into (17), we obtain:

$$q(z) = \alpha n_f S(T_{0s} - \bar{T})e^{-\frac{z}{\lambda}}.$$

Integrating this expression over the height of the product layer *d,* m, the amount of heat energy *Q,* W, transferred by the heat carrier to the product in the layer having a unit area is obtained:

$$Q = \int_0^d q(z)dz = \lambda\alpha n_f S(T_{0s} - \bar{T})\left(1 - e^{-\frac{d}{\lambda}}\right).$$

The number of water molecules n_{lay} evaporated in this layer:

$$n_{\text{lay}} = kSn_f\left(n_p(\bar{T}) - n_s\right)d,$$

where*d* - the height of the product layer, m
Hence we obtain the equation for the definition $\overline{T}$.

$$ukd\left(n_p(\bar{T}) - n_s\right) = \lambda\alpha(T_{0s} - \bar{T})\left(1 - e^{-\frac{d}{\lambda}}\right). \quad (18)$$

Substituting in (18) $n_p(\overline{T})$ from (4), we obtain:

$$\frac{ukd}{\alpha\lambda}\left(n_{\text{atm}}e^{\frac{u}{k}\left(\frac{1}{T_{\text{atm}}}-\frac{1}{T}\right)} - n_s = (T_{0s} - \bar{T})\left(1 - e^{-\frac{d}{\lambda}}\right). \quad (19)$$

According to the experiment, a linear approximation is established:

$$d = d_0 + d_1 \upsilon = (1 + 2\upsilon), \quad (20)$$

где d_0 – начальная высота слоя продукта, м;d_1 – высота слоя продукта при скорости υ, м, а также зависимость коэффициента теплоотдачи от скорости пара:
whered_0- the initial height of the product layer, m; d_1- the height of the product layer at a speed υ, m, and also the dependence of the heat transfer coefficient on the steam speed:

$$\alpha = (1 + 8\upsilon). \quad (21)$$

Taking into account (19) and (20), equation (18) is reduced to the form:

$$he^{\frac{u}{kTatm}-\frac{u}{kT}} - 1 = \frac{T_{0s}-\bar{T}}{T_*}, \quad (22)$$

where $n = \frac{n_{atm}}{n_s} > 1$;$T_* = \frac{\lambda u S d n_f}{3kv(1-e^{-g})}$;$g = \frac{\alpha S d n_f}{3kn_s v}$.

Thus, the average temperature of the product particles depends on three control parameters T_{0s},$d$andT_*

Simplifying the product $S\,d\,n_f$, we get:

$$s = \frac{S d n_f S_{as}}{S_{as}} = \frac{N}{S_{as}/S}, \quad (23)$$

whereS_{as}- the cross-sectional area of the drying chamber, m^2; N- the total number of product particles in the drying chamber, pcs.; $\frac{S_{as}}{S}$- the number of product particles stacked in one layer on the cross-sectional area of the drying chamber, pcs.; S - the number of product layers in the apparatus, pcs.

Substituting (11) in (19) we obtain:

$$T_* = \frac{kus}{3kv(1-e^{-g})};g = \frac{\alpha s}{3kn_s v}. \quad (24)$$

Based on the simulation results, calculation formulas were obtained for determining the temperature of a beet pulp particle during drying under reduced pressure with superheated steam, both in the filter layer along one spatial coordinate in the direction of heat carrier motion (15) and in a fluidized bed for various sections of the drying chamber (20).

The parameters of the models were identified in the following parameter variation range: superheated steam temperature 393 ... 453 K; superheated steam speed 3 ... 5 m/s; pressure inside the drying chamber 40 ... 100 kPa; specific product load on the grate 8 ... 24 kg/m^2.

The interpretation of the simulation results is presented by the dependences of the temperature difference between beet pulp and heat carrier ΔTon the temperature of superheated steam $T_s$at various pressures P: 1 - 100 kPa; 2 - 80 kPa; 3 - 60 kPa; 4 - 40 kPa; υ = 4 m/s; q_{sp} = 16 kg/m^2 (Table 1); at various specific loads of beet pulp on the gas distribution grateq_{sp}: 1 - 8 kg/m^2; 2 - 16 kg/m^2; 3 - 24 kg/m^2; υ = 4 m/s; P = 60 kPa (Table 2) and various speeds of superheated steam υ: 1 - 5 m/s; 2 - 4 m/s; 3 - 3 m/s; q_{sp} = 16 kg/m^2; P = 60 kPa (Table 3). A sufficient convergence of the results was ensured, in which the deviation of the calculated data from the experimental data did not exceed 12.5% in absolute value.

Table 1. The temperature difference between beet pulp and superheated steam at various pressures of superheated steam in the drying chamber

Temperature	*Superheated steam pressure in the drying chamber P, kPa*			
T_s, K	*100*	*80*	*60*	*40*
393	***24***	***32***	***40***	***52***
413	***35***	***42***	***53***	***64***
433	***49***	***58***	***67***	***79***
453	***67***	***77***	***86***	***96***

Table 2. The temperature difference between beet pulp and superheated steam at different specific product loads on the gas distribution grid

Temperature	The specific load of the product on the grate q_{sp}, kg/m^2	
T_s, K	8	16	24
393	***35***	***40***	***46***
413	***45***	***53***	***58***
433	***60***	***67***	***72***
453	***80***	***86***	***91***

Table 3. The temperature difference between beet pulp and superheated steam at different speeds of superheated steam in the drying chamber

Temperature	The speed of superheated steam in the drying chamber υ_s, kg/m^2		
T_s, K	3	4	5
393	***38***	***40***	***44***

413	***48***	***53***	***55***
433	***63***	***67***	***70***
453	***82***	***86***	***90***

4. Conclusion

The obtained simulation results can be used in the operational control of the technological parameters of the process of drying beet pulp with superheated steam of reduced pressure with a restriction on the temperature mode due to the receipt of a high quality product.

References

[1] Preparation and application of fodder vitamin additive choline chloride B4 on the basis of dried beet pulp in premix composition / Shevtsov AA Drannikov AA Derkanosova AA Borodovicyn AM Glebova IV Gurin AG // International Journal of Pharmaceutical Research & Allied Sciences – 2017№ 6(1 P 217-226

[2] Experimental and Analytical Study of the Beet Pulp Drying Process by Overheated Steam in Active Hydrodynamic Conditions / ShevtsovAA DrannikovAV DerkanosovaAA MuravevAS KvasovAV // JournalofEngineeringandAppliedSciences – 2017 № 12(SpecialIssue 1 P 5754-5760 DOI: 1036478/jeasci201757545760

[3] Mathematical model of the drying process of Capillaryporous materials particles in the apparatus with asuspended-swirled flow of heat - carrying medium/ AntipovST Drannikov AV Poymanov VV Pribytkov AV Yurova IS / EurAsian Journal of BioSciences – 2019 – T 13 – № 2 – C 1051-1056

[4] Ostrikov AN Shevtsov SA A formal approach to the management of extreme drying process plant material with super-heated steamProceedings of the Voronezh State University of Engineering Technologies2013;(3:7-12 (In Russhttps://doiorg/1020914/2310-1202-2013-3-7-12

[5] ChusovaAE AgafonovGV ZelenkovaAV PozhalovaJV Changeintheenzymaticactivityofgluten-freemaltduringgerminationanddrying Proceedings of the Voronezh State University of Engineering Technologies 2017;79(2:198-203 (In Russhttps://doiorg/1020914/2310-1202-2017-2-198-203

[6] Shishatskiy Y I Migration of extractantintheporousstructureoftheplanttissue / Y I Shishatskiy S V Lavrov S Y Pluhaetal // Storageandprocessingofagriculturalproducts – 2011 – № 5 – P 40-42

[7] Balamurali R Maheswaran Prashanthi A R Roshan Fathima S Varshini S 2020 Enhanced Automation Using Controller and Conscripts International Journal of Advanced Science and Technology 29(3 8620 - 8627

[8] Calin-Sanchez A et al Drying of garlic slices using convective pre-drying and vacuum-microwave finishing drying: kinetics energy consumption and quality studies //Food and Bioprocess Technology – 2014 – T 7 – № 2 – C 398-408

[9] Prakash O Kumar A Environomical analysis and mathematical modelling for tomato flakes drying in a modified greenhouse dryer under active mode //International journal of food engineering – 2014 – T 10 – № 4 – C 669-681

[10] Kiran GeorgeImmaculate Joy SNisha Susan Thomas Balamurali R &Baskaran K 2020 Gender- Based Vegetarian and Nonvegetarian Dietary Impact on Cardiac Autonomic Function of Heart Rate Variability Journal of the American College of Nutrition pp1-5

[11] Seiiedlou S et al Convective drying of apple: Mathematical modeling and determination of some quality parameters //International Journal of Agriculture and Biology – 2010 – T 12 – № 2 – C 171-178

Innovative and Smart Methodology towards Kidney Disease Detection in Earlier Stage

S P Karthi[1*], S Guganesh[2],K Kavitha K[3], J R Dinesh Kumar[4] & D Selvakumar[5]

[1,4]Assistant Professor, [5]Associate Professor, Department of Electronics and Communication Engineering, Sri Krishna College of Engineering and Technology, Coimbatore.
[2]P.G. Student, Department of Electronics and Communication, Engineering, Sri Krishna College of Engineering and Technology, Coimbatore
[3]Professor, Department of Electronics and Communication, Engineering, Kumaraguru College of Technology, Coimbatore.

Corresponding Authour mail Id : karthisp@skcet.ac.in

Abstract: In modern era of medical field, kidney diseases are the foremost common diseases found in the majority of the populations in subsequent years. The essential is diagnosing the problem and however advanced task that ought to be accomplished precisely with proficiently. In most of the cases, kidney condition leads to loss of life. Diagnosing could be a tough thus the intention of the research is to develop a neural-trained system to spot kidney condition risk of the patients. The Artificial Neural network will collect the data from the previous method and it will be helpful to identify the chance factors on the premise of provided data. This paper uses neural techniques for diagnosing of the kidney disease. MATLAB simulation was employed as a development tool.
Keywords: Artificial Neural Network, Chronic Kidney Disease.

1. Introduction

Now, on a daily basis the employment of technology is crucial in each filed associate degreed diagnosis space isn't an exception [1]. We have a tendency to all know okay that these areas, inside which the modern computer area unit employed, possess terribly higher quality, irregularity, conjointly the employment of skilled units like mathematical logics, artificial networks and genetic algorithmic rules are established [2]. Mathematical logic may be a powerful reasoning technique which will handle uncertainty of acceptable knowledge. One among the foremost vital causes of death worldwide is kidney failure. Therefore, identification of the center disease is terribly important demand in existence to millions of uncertainty, and risk issue the cardiovascular disease identification is difficult for specialists. Once a kidney failure is known, the speed of detection is very essential to avoid wasting the lifetime of kidney failure patient and to stop kidney harm. Soft computing will be well-defined as a community of "imprecise" computational techniques to model and examine terribly advanced issues that square measure intelligently. Classical and reliable research approaches have been unable to offer low-priced, systematic, and detailed solutions for these advanced problems. Among the various components of soft computing, square measurements of mathematical logic (FL), artificial neural networks (ANNs) [3] and organic process algorithms (EAs) are considered because of the simple soppy computing practices.

2. Existing System

The maximum necessary claims of fuzzy system (fuzzy logic) are towards unsure problems. Once retardants have lively behaviors, mathematical logics could be an appropriate means that agreements with this downside. Beginning of fuzzy knowledgeable system coming up with its purpose of input/output parameters. Eleven input variables and one output variable are considered. Then, we have the tendency of all variables to model membership functions (MFs). These MFs confirm that objects are part of fuzzy sets [4,5,6]. Cotton disease and its symptoms have been discussed with the detection algorithm [7]. We have an inclination at first to describe the input variables by their membership functions. At this point, fuzzy sets wouldn't overlap and sets outline in positive predictor.

3. Proposed System

Neural networks have very good generalization capabilities, if it is trained properly. In this paper, a neural grounded structure for sorting of Kidney condition level with the assistance of 24-input characteristics using neural based system. The proposed algorithm incapacitates the shortcomings in the learning process used for current approaches. In general, the question for classification is as follows. However, the conceptualization of our system is created on; complete design is completely new based on the optimum learning algorithm described in the next sub-section. The proposed architecture includes learning neural networks with typical database with enough training pairs to solve the problems faced by current approaches. Neural network has 24 inputs from the regular database, and 1 goal value from the disease group. Neural network processes information and controls cardiac production illnesses as normal or abnormal rates. The training range of the input information can be kept low enough for the neural network to take extreme benefit of the learning capabilities of the neural networks. The data sets selected for this study are occupied as database of heart diseases. This involves the sorting of person into normal and abnormal people.

Chronic Kidney Disease has also been identified as a Chronic Renal Dysfunction, and is still growing an alert area globally. Chronic Kidney Failure has become a progressive kidney injury for quite a while. Chronic kidney disease is a benign illness that spoils the kidneys and foreign compounds cannot easily wash out of our bodies. An individual acknowledged that chronic kidney disease is well figured-out when the function of the kidneys is below than 25 percent of the normal, kidney disease as one of the main health problems of the person. MATLAB is the method included in this paper to get the accurate effect based on the sum of input parameters as long as the set of information is specified. MATLAB provides simplicity for obtaining chronic kidney disease's accurate estimation. The subsequent modules are used: (1) data acquisition and collection, (2) pre-processing, (3) splitting data set as training/testing, (4) classifier training, (5) test attribute prediction.

Artificial Neural Network helps to identify kidney defects. A neural network has multiple nodes for input, hidden, and output. On some results, the respective node applies a value (could be soft max, linear, logistic) and generates an output. Each node in the next layer takes a weighted average of the prior layer outputs before it achieves an output. The cognitive would be that multiple nodes will collectively raise the perception that an individual node is unable to solve a problem (such as classification).

Artificial neural network with back propagation in order to classify Landsat data, one different way of ANN with BP is proposed in the request of neural network. The back-propagation algorithm is used for neural network processing. For multispectral image classification, a further version of ANN with BP is proposed in. The BP is trained on a classical image area, and the neural network is then used to classify the image.

Improved Back propagation algorithm addresses neural network programming using gradient delta rule with back propagation algorithm. It is particularly applicable for architecture of parallel hardware. Rather than being kept constant, the momentum factor is calculated at each step. Better BP has improved speed and stability to convergence than traditional BP.

The meta-heuristic algorithms yield approximate results and are applicable to any field. Such algorithms are used where maximum local production is achieved by conventional algorithms.

Traditional algorithms also increase the cost of computation, and use more time to produce results. Several researchers in previous studies merged ANN with these meta-heuristic algorithms to address its limitations.

4. Results and Discussions

A. Prediction of Test Attribute

Prediction is based on feedback from the user. Compare user feedback with qualified dataset. The trained dataset is nothing but the neural network, regression tree classifier used to test data.

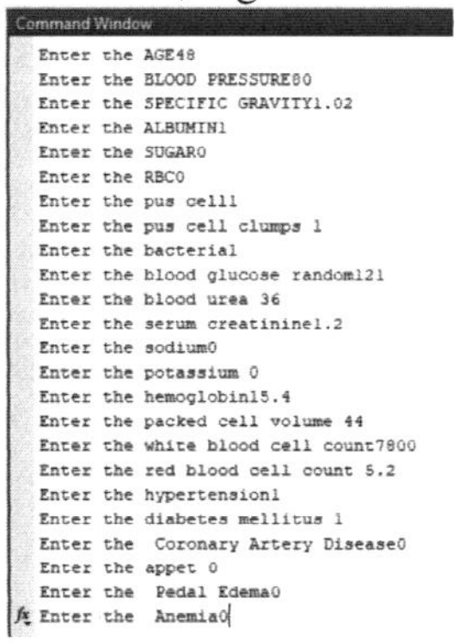

Fig.1 Values from the test report of patient (for abnormal condition)

The result will then be provided after comparison. Fig.1 shows the values from the test report of patient (for abnormal condition) and Fig.2 shows the test output for kidney abnormal condition. Fig.3 shows the values from the test report of patient (for normal condition) and fig.4 shows the test output for kidney normal condition

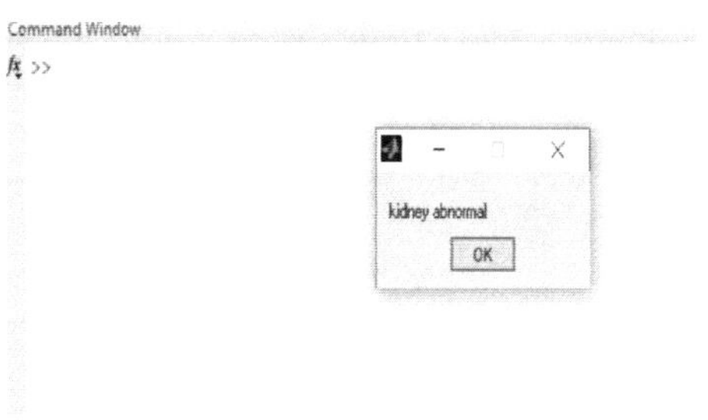

Fig.2 Test output for kidney abnormal condition

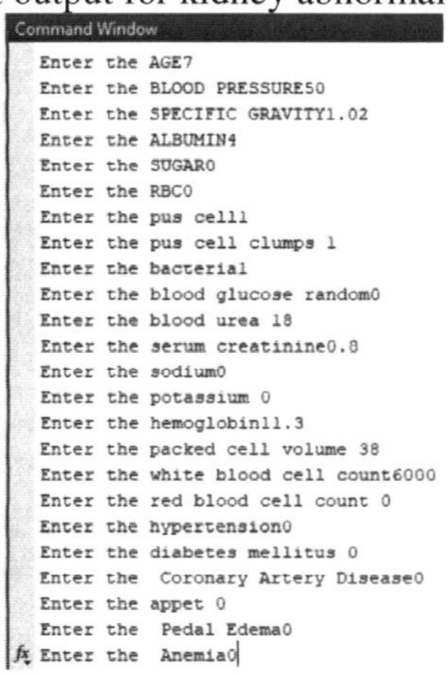

Fig.3 Values from the test report of patient (for normal condition)

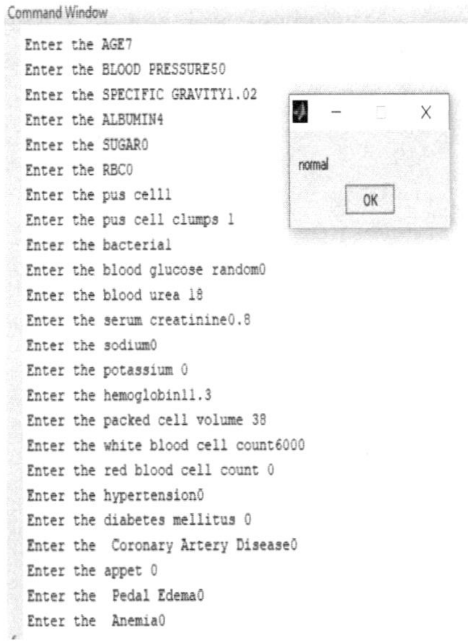

Fig.4 Test output for kidney normal condition

B. Result Analysis

Table 1 shows the Comparative analysis of results of proposed and existing system.

Table 1. Comparison of proposed and existing system

Methods	Attributes	No. of pairs	Training & Testing
Existing system	13	303	92.3% & 82.20%
Proposed system	24	400	99.7% & 100%

5. Conclusion

ANN plays a major role in detecting kidney disease. This scheme plays prominent role in medical field. In this proposed method, the user can periodically monitor the patient's health conditions which make user to easily detect the kidney conditions. This proposed scheme gives accurate results for kidney diseases.

References

[1] ASai Sabitha Abhay Bansal Khushboo Chandel Veenita Kunwar Chronic Kidney Disease Analysis Using Data Mining Classification Techniques 6th Int Conf on Cloud System and Big Data Engineering 2016

[2] AgarwalY Pandey HM Performance evaluation of different techniques in the context of data mining-A case of an eye disease In Confluence the Next Generation Inf Tech Summit (Confluence) 5th Int Conf -IEEE 2014

[3] Ahmed S Tanzir Kabir M Tanzeer Mahmood N Rahman RM Diagnosis of kidney disease using fuzzy expert system In Software Knowledge Inf Management and Applications (SKIMA) IEEE 2014

[4] Janarthanan R Balamurali R Annapoorani A and Vimala V 2020 Prediction of rainfall using fuzzy logic Materials Today: Proceedings

[5] Alfisahrin SDNN Mantoro T Data Mining Techniques for Optimization of Liver Disease Classification In Adv Computer Sci Appl and Technologies (ACSAT)

[6] Amin SU Agarwal KBeg R Genetic Neural Network based data mining in prediction of heath disease using risk factors Inf and Comm Tech(ICT) Conference 2013

[7] NR Deepa N Nagarajan D PalanivelRajan R Balamurali 2019 July Image Processing System for Detecting and Recovering from Cotton Leaf Disease using The Android Application International Journal of Innovative Technology and Exploring Engineering (IJITEE) 8(9) pp 3366 – 3381

IOP Conf. Series: Materials Science and Engineering **994** (2020) 012026 doi:10.1088/1757-899X/994/1/012026

Deep Learning Based Lung Cancer Detection and Classification

N Kalaivani[1*], N Manimaran[2], Dr. S Sophia[3]&D Devi D[4]
[1]Assistant Professor,Sri Krishna College of Engineering and Technology,
[2]Assistant Professor,SACS MAVMM Engineering College,
[3]Professor,Sri Krishna College of Engineering and Technology,
[4]Assistant Professor,Sri Krishna College of Engineering and Technology.

Corresponding author mail id: kalaivanin@skcet.ac.in

Abstract: Lung diseases are indeed the lung-affecting diseases which impair the respiratory mechanism. Lung cancer has been one of the leading causes of mortality in humans worldwide. Early detection can enhance survival chances amid humans. If the condition is diagnosed in time, the average survival rates for people with lung cancer rise from 14 to 49 percent. While computed tomography (CT) is far more effective than X-ray, a thorough diagnosis includes multiple imaging approaches to support each other. A deep neural network for detecting lung cancer from CT images is developed and evaluated. For the classification of the lung image as normal or malignant, a densely connected convolution neural network (DenseNet) and adaptive boosting algorithm wasused.A dataset of 201 lung images is used in which 85% of the images are used for training and 15% of the images are used for testing and classification. Experimental results showed that the proposed method achieved an accuracy of 90.85%.
Keywords: DenseNet, Image Processing, Deep Learning, Convolution Neural Networks (CNN).

1. Introduction

Lung cancers have been identified as one of the world's most serious causes of death [1]. It is among the most malignant tumors that can affect human wellbeing. Its death rate scores among all tumor deaths, and is also the top killer towards male and female cancer death [2-3]. There have been nearly 1.8 million fresh cases of lung cancer annually (13 percent among all cancers), 1.6 million deaths worldwide (19.4 percent among all cancers). Lung cancer is a proliferation of expanding and developing irregular cells into a tumour. Of the other forms of cancer, the death rate of lung cancer is the greatest. Cigarette smoke induces an approximate 85 percent of cases of lung cancer in males and 75 percent in females. Lung cancer is amongst the most terrible illnesses in the developing countries, with a death rate of 19.4 percent. Lung cancer is among the most dangerous cancer worldwide, with lowest success rate following diagnosis, with a steady rise in casualty count per year [4-6]. Advantages of Fuzzy logic in the earlier predictions will lead to result oriented analysis [5]. Survival of lung cancer as a result of diagnosis is directly related to its progress. Yet individuals have a greater success rate it will be found in the early stages of life. Cancer cells are distributed in blood from the lungs, the lymph fluid that covers the lung tissues. The lymph passes into lymph vessels that discharge through lymph nodes in the lungs and chest region. Examination and treatment of lung disease has become one of the biggest obstacles that humanity faces in recent years. Early tumor diagnosis will reliably promote its survival of vast numbers of life around the world. This paper introduces a method that uses a convolutional neural Network (CNN) to identify the lung tumors as malignant/benign. The results achieved by CNN are 81 percent, which are more effective than the results achieved by the traditional neural network.

2. Existing System

Sharma and Jindal have suggested a method for identifying lung cancer tumor by using fuzzy interference system and the active contour model. This device makes use of gray level transformation to improve the image contrast. Binarization of images is achieved prior to segmentation, and the resulting image is divided by using active contour model. Moreover, the system efficiency is 94.12 percent. Bhatnagar et al., devised a technique utilizing watershed segmentation. It utilizes Gabor filter in preprocessing to improve the image quality. They compared the effectiveness of the algorithm with neural fuzzy and the region growing method. The accuracy was observed as 89.1 percent which is significantly greater than the method of segmentation using neural fuzzy and region growing method. This framework being the current best solution, but it has certain drawbacks. Just a few characteristics of cancer nodules have been identified. No preprocessing features such as noise reduction and image smoothing has been applied which will potentially help improve the effective identification of nodules. There has been no proper classification method as benign or malignant. This approach varies from the current best solution, and proposed new method. Median filter and Gaussian filter is introduced in preprocessing stage rather than Gabor Filter. The best system ends after cancer nodule detection, extraction of feature and accuracy estimation. Yet, it has not enforced its classification as normal or irregular.

3.Proposed Model

Convolutional neural networks are designed for minimizing the number of parameters and adjusting the architecture of the network for image classification. Convolutional neural networks are made up of a series of layers organized according to their features and functionality. A ConvNet 's architecture is close to that of the human brain connectivity pattern of neurons inspired by the Visual Cortex structure. Data augmentation is the process by which data quantity and complexity increase. We will collect fresh data rather than converting the data already available. Data augmentation is an important phase in deep learning, since we require vast quantities of data in deep learning and in certain instances it's not really possible to capture thousands or even millions of images, so this data augmentation comes to the scene. It allows us to maximize the dataset size and add uncertainty within the datasets. In addition to this, webpage is created to store the database of the patients, here the patients can login to their page at any time for the future references. Hospital locality server details can also be inferred easily.Fig.1 depicts the proposed model.

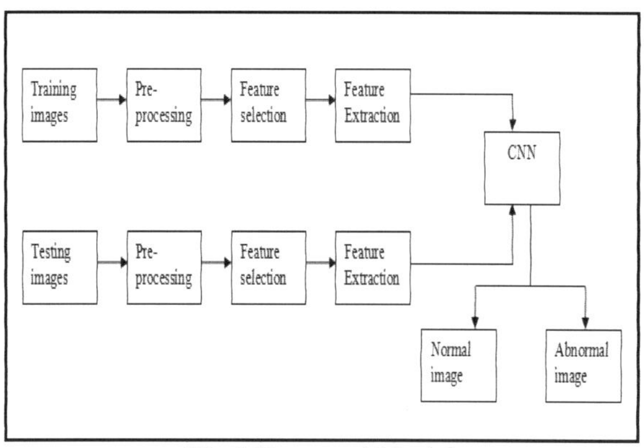

Fig.1 Proposed Model

4. Operation of the Proposed Model

The working is divided into two parts, i.e., training and testing parts. A sample of 201 images is used where 171 images are used for training and the remaining 30 images are used for testing. In testing part, images are first pre-processed, where resizing and blur removal of images are done by using histogram equalization due to which non-linear images are stretched and pixel values are redistributed. Fig.2shows the articulation of data processing.

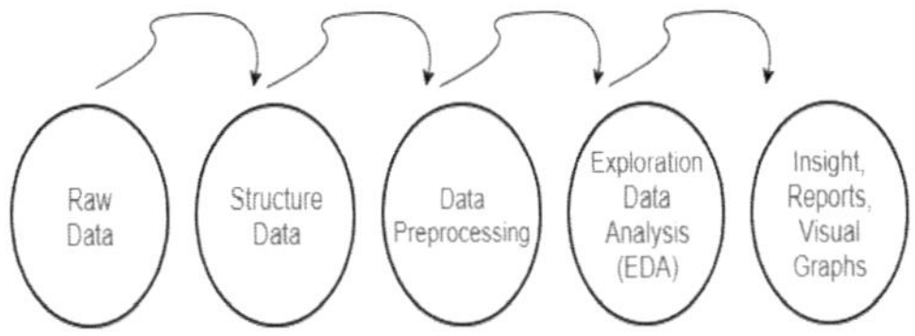

Fig.2 Data Processing

The results obtained from training and testing part are fed into in CNN layers where the images are classified and the output is obtained. The algorithm used for classification is ADABOOST (Adaptive Boosting), where accuracy calculation is of the images is done based on the sample weights of images. Fig.3shows the conventional CNN architecture.

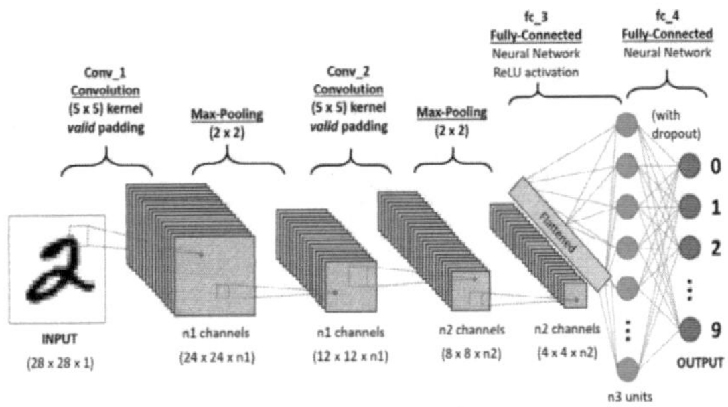

Fig.3 CNN Architecture

The convolutional layer is the first and foremost layer and is the essentiallayer of a CNN. The layer's parameters contain a set of filters that are referred as kernels. Then the output obtained from ReLu layer is passed on to the next Pooling layer.There are two common pooling approaches, they are average(optimized) pooling and max pooling that condense the average presence of a feature.Fully connected layer is the most essential layer component of Convolutional Neural Networks (CNNs). Which has been successful in recognized for classifying images for computer vision. AdaBoost algorithm is the best usage of algorithm in deep learning. XAMPP is an open-source and cross-platform web server stack package. XAMPP was mainly developed for Apache HTTP Server, scripts written in the PHP etc., XAMPP is used as web server in most cases, and thus it is designated as the local test server to a live server. Thus the web page consists of home page which included the hospital login and user login. Patients can login to the portal using their patient id at any time to have their results.

5. Results and Discussions

Total number of images taken is about 201, among which 171 images are been trained and the remaining 30 images are tested accordingly. The trained images are then moved on to the classification part where each image is classified as either normal or malignant. Output of each image is been displayed with its accuracy as either normal or malignant. The formula to calculate accuracy is as follows:

$Accuracy = \frac{TP}{TP+FP}$, *here TP – True Positive and FP – False Positive*

Output of each image has its own true positive(TP),true negative(TN), false positive(FP) and false negative(FN).

S. No.	Parameters	Accuracy
01	Trained images	171
02	Tested images	30
03	True Positive	5
04	True Negative	20
05	False Positive	4
06	False Negative	1
07	Accuracy	90.85%

Thus, by this approach the output is obtained using deep learning, which includes detection and classification of the images. Finally, the overall accuracy obtained is about 90.85%. This is obtained at an accurate ratio. The contents are then stored to the web server which contains hospital and admin users.

Dataset Image Number	Accuracy of each image
15	99.8%
30	96.41%
22	78.92%
01	99.89%
10	98.17%
24	99.90%
19	59.56%
21	99.76%
16	98.84%
03	74.93%
02	93.23%
Overall accuracy	**90.85%**

Accuracy of 11 images are calculated for predicting the overall accuracy, the average of all individual accuracy's of each image is been taken. Thus, the overall accuracy obtained is the average of accuracy of individual images is obtained as 90.85%.

6. Conclusion

The greatest benefit of deep learning above other algorithms on machine learning is its ability in executing feature engineering on its own. This examines the data to search for associated features and incorporates them to provide for quicker learning.It takes advantage of spatial coherence in the input. The training and testing of images are done where images are pre-processed and feature selection and feature extraction of images are done. Once training and testing part is done successfully, the CNN algorithm classifies the input lung image either as normal or abnormal and the output will be displayed. Hence, a Deep CNN network is used for the classification of lung images for the detection of cancer.

References

[1] RLSiegel KDMiller AJemal Cancer statistics 2017*Cancer Journal for Clinicians* 67(1): 7–30 2017

[2] RLSiegel KDMiller SAFedewa Colorectal cancer statistics *Cancer Journal for Clinicians* 68(1): 7130 2018

[3] JLortet-Tieulent ISoerjomataram JFerlay MRutherford EWeiderpass FBray "International trends in lung cancer incidence by histological subtype: Adenocarcinoma stabilizing in men but still increasing in women Lung Cancer 84(1): 13-22 2014

[4] DSharma GJindal ''Computer aided diagnosis system for detection of lung cancer in CT scan images *International Journal of Computer and Electrical Engineering* 3(5): 714-718 2011
[5] Janarthanan R Balamurali R Annapoorani A and Vimala V 2020 Prediction of rainfall using fuzzy logic Materials Today: Proceedings
[6] DBhatnagar AKTiwari VVijayarajan AKrishnamoorthy ''Classification of normal and abnormal images of lung cancer *IOP Conference Series: Materials Science and Engineering* Vol 263 2017

A CONTRAST ADJUSTED WAVELET AND CURVELET TRANSFORM BASED FUSION OF BREAST IMAGES

Ms.G.Vijayalakshmi 1 , Dr.M.Shanthakumar 2 ,Mr.K.Muthukumar 3
Research Scholar, Department of Computer Science, Kamban College of Arts and Science,Coimbatore-641669
Assistant Professor, Department of Computer Science. Kamban College of Arts and Science, Coimbatore-641669
Research Scholar , Department of Computer Science , Kamban College of Arts and Science, Coimbatore -641669

vijayalakshmi3377@yahoo.com , Professorshanth@gmail.com , muthukutty42@gmail.com

ABSTRACT: Restoration is one of the image processing techniques to restore the distorted image. The distorted image can be any type like blurred image, noise added either while storing or capturing of the images. the traditional restoration techniques are geometric correction, different types of filters to remove the noises. Generally, the restoration is applied to the single image to retrives its information. In this, the restoration process is applied in the application of image fusion of medical images. the reason to chose the image fusion process is to improve its entropy. The medical images used in this paper is the mammogram and MRI-breast image. In this, the image fusion is performed with the help of wavelet and curvelet transform. The curvelet transform is applied to the restored mammogram image with the help of contrast adjusted top-hat filter process. The wavelet transform is applied to the MRI image because it has more volume of data to acquire important details. The fused image is evaluated with the help of entropy, mean standard deviation and variance. The fused image has efficient value as compared to its original image but it is higher than the DTCWT-fused image. To verify the quality of the proposed method, the psnr is evaluated to determine the noise present in the image. From the evaluation, the noise is less and entropy is higher in the proposed method. Hence, the contrast adjusted wavelet curvelet fusion is the best restoration and fusion process in the breast images.

Keywords: restoration, x-ray, mri , breast image, wavelet, curvelet, top hat filtering, evaluation metrics.

Introduction:

In recent days, the analysis of images are performed with the help of machines prior to the human observation. To support such an observation, the imaging system should be highly processed one to capture the image noise and blurr free. But, it is not that much possible due to the motion of the human especially in the medical imaging systems.

In such a medical imaging method , the noise is obtained from the human interventions or the initial error in the machine manufacturing. To overcome this problem the image has to be enhanced which is performed by different process namely filtering, enhancement of the contours and contrast, and fusion process. All these process are to improve the low resolution images into high resolution images [12].

The two dimensional images such as x-ray the image restoration is performed with the help of filters like mean, median and weiner filters. [16]. The three dimensional images like MRI the image enhancement is performed with the help of multi level resolution and adaptive smoothing [17].

IOP Conf. Series: Materials Science and Engineering **994** (2020) 012027 doi:10.1088/1757-899X/994/1/012027

The contours are enhanced with the help of transformations like contourlet, curvelet and active contour transform to detect the edges. The contrast is enhanced with the help of the histogram equalization process. The histogram is the count of the pixel values. By adjusting the histogram bins the contrast is enhanced. The portable devices using converters and filters [20].

The other approach for the contrast enhancement is the multiscale resolution and transformation with the help of transforms like wavelet, cosine and KL transforms.

The image fusion is done by combining the two or multiple images into a single image. the single output image has more details than the input images. the fusion is done either directly on the images or the transformed images. the fusion process can be performed in the window based or pixel based manner. Mostly, the pixel based fusion is preferred due to its easily applicable and increase the entropy of the fused image.

The major pixel level fusion methods are multi scale decomposition, sparse representation, different transforms and non- transforms method.

The restoration process is evaluated with the help of stastical parameters like entropy, mean and so on. The visual quality of the image is evaluated using peak signal to noise ratio, mean square error and structural similarity index.

This paper proposed a new restoration approach to enhance the image along with the higher details of the abnormalites of the breast cancer in the image with the help of fusion. The paper is organised by explaining the related works of the paper in section 2, the detailed explanation of the proposed method in the section 3, the implementation and evaluation is explained in the section 4 and it is concluded in the section 5.

Literature survey:

[2] paper described about the two types of image restoration techniques namely model based optimization and discriminative based restoration. Both the methods are able to restore the image but it requires larger time for operation in model based optimization and designed for specific range of applications in the discriminaive based method. to overcome this, they introduced a CNN based method to design the denoiser prior to the model based optimization approach to remove the noises and restore the images. This method able to remove the noise faster but it require large set of data for the prior noise information.

[3] surveyed about the pixel level image fusion and its advantages in various applications like medical imaging, survelliance, remote sensing etc., the main advantage of this type of fusion is it able to combine multiple images into a single image. The fused image will have more information as compared to the individual input images. this information is helpful for the interpretation or machine. The major pixel level fusion methods are multi scale decomposition, sparse representation, different transforms and non- transforms method.

[4] explained about the neccessity of the registration and fusion process in the medical images. both the process can be used individually or in the combined form in the medical images. these process helps the practioner to diagonse the disease and analyze the images easily in various domain. In registration, the image in one domain is registered to another target image to improve the visibility of the image. In fusion, the multiple images are combined to improve the information of the image.

[5] defined an approach of image fusion using cross bilateral filter. In this , the one image is used for the kernel finding and the other is for the filtering process. The fused image is formed with the help of weighted averaging method. Its advantage is to improve the smoothness and edge enhancement. But, it suffered when there is no gray scale differences.

[6] implemented the deep learning network to improve the pixel level image fusion process. The deep learning methods such as convolution neural networks, auto encoders and so on. The deep learning helps to increase the number of images and multi scale and multi focus transformation. [7] explained about the pixel level image fusion methods and its evaluation metrices.

[8] implemented a new approach called plug and play which perform fusion using the denoising. In this, the denoising is applied in the form of black box method. the fusion of images in one of the following methods by sharpening or blurred noisy images. its drawback the fused image quality little bit lower as compared to other techniques due to the noisy fusion.

[9] explained about the image restoration of MRI brain images using the histogram based weiner filter. The coefficients of weiner filter is designed with the help of the histogram bins of the image. It is mainly designed to remove the impulsive noise in the image.

[10] explained about the image restoration techniques on natural images, satellite images and medical images. the filters considered for the analysis are the mean, median, weiner and gaussian mixture model with wavelet. Each filter able to noiseless image by filtering the noises like salt and pepper, poisson and gaussian noise. For the speckle , weiner is best. Salt and pepper and gaussian noise is removed with the help of mean and median filter in medical images.

IOP Conf. Series: Materials Science and Engineering **994** (2020) 012027 doi:10.1088/1757-899X/994/1/012027

[11] explained about the image enhancement using optimized wavelet masking. The optimization used in this process is the enhanced cuckko search with the help of genetic algorithm. The image is enhanced by combining the wavelet coefficients and enhanced cuckkoo search algorithm coefficients.

Proposed methodology:
In this, the restoration of the image is performed in the fusion application . The fusion process to improve the information in the fused image. To perform the restoration, the images are processed with the two types of filters namely top-hat filtering and MSE- filtering. Then, the fusion process is performed to produce the new fused X-MRI image. The detailed explanation of fusion process is shown in the block diagram.

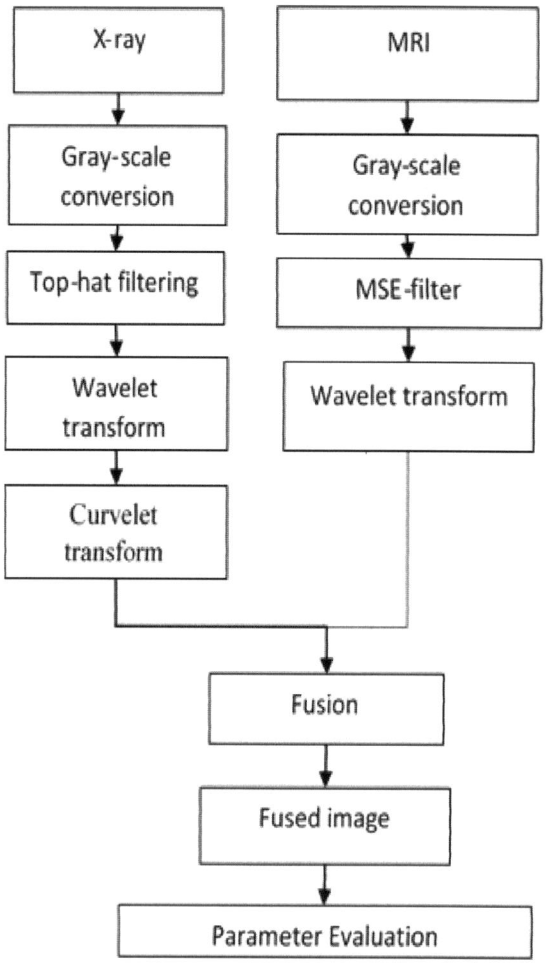

Fig. 1. Block diagram of the proposed method

Input image:
The input images are the x-ray and magnetic resonance image of breast part. The x-ray of the breast image is called mammogram which is used for earlier testing method to detect the early signs of breast cancer. The Magnetic resonance image of the breast is to support the mammography and to detect the level of abnormalities in the breast. In this, both the images are fused to improve the analysis of breast cancer and help the physician.

Pre-processing:
Pre-processing is the basic step in all the image processing applications to improve the qulaity of the image. In this, the pre-processing steps are resizing the size of the images into m*n pixels of size m and n equal to 256. The resized images pixel values are converted between 0 and 1 using gray scale conversion. The converted images are processed separately with two separate filtering process namely top-hat filtering and MSE-filtering.

$$Gray\_image = 0.2989 * R + 0.5870 * G + 0.1140 * B$$

Where R,G and B are the corresponding pixel values in the image of (m*n).

Top-hat filtering:
In this, the contrast of the image is adjusted with the help of morphological operations called opening and the subtraction of opening of the image from the the image. The opening process is done with the help of structuring elements called diamond of width 10. It main purpose to brighten the foreground brighter pixels as compared to the background pixels. Due to this reason, the tophat filtering is applied to the x-ray image to brighter its pixels. [12]

$$THF_{image} = Gray_{image} - imopen(Gray_{image})$$

MSE filter:
In this, the mean square error of the original image and noiseless image is reduced in the process of inverse filtering. It preserves the texture properties of the image [13]. Hence, the MRI breast image is subjected to this filtering to preserve its depth of the abnormalities in the image. The noiseless image is obtained by the linear estimation and its orthogonality principle is given below:

$$M(I_1, I_2) = \frac{H^*(I_1, I_2)S_{xx}(I_1, I_2)}{|H(I_1, I_2)|^2 S_{xx}(I_1, I_2) + S_{\eta\eta}(I_1, I_2)'}$$

Where S indicates the power spectra of the original and noise. H denotes the value of the blurring filter.

Wavelet transform:
Wavelet transform is the shorter form of waves like fourier transform which starts and ends rapidly. Due to this shorter period, it able to differentiate in the time and location. [19]
The shifting factor of the wavelet helps to make the waves as finite which helps to capture the inner details of the image. The scaling factor helps to stretch or decrease the waves frequencies.
The scaling of different types of wavelet transform is shown in the below figure.

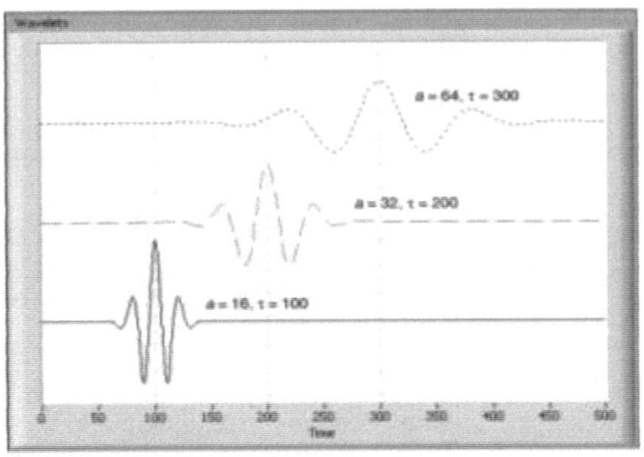

Fig.2. Scaling of wavelet transform

The wavelet transform exhibits a property called isotropic which helps to analyze in the directions like vertical , horizontal or diagonal in a given level. But it is not able to capture the edges of the image. based on the frequency of the wavelet, the classification of wavelts is shown in the below figure.

IOP Conf. Series: Materials Science and Engineering **994** (2020) 012027 doi:10.1088/1757-899X/994/1/012027

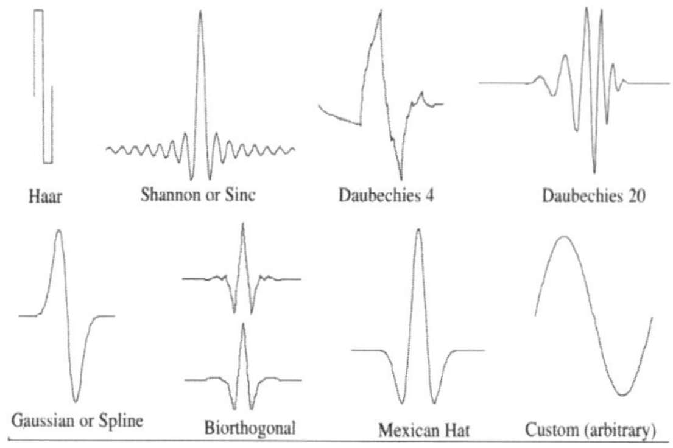

Fig.3. Types of wavelet transforms

These wavelets are able to capture the finer details that is the depth of the pixels in the image. Due to this, the MRI and X-ray is subjected to wavelet transform.

In the above wavelet families, the daubechies wavelet produced the best result due to its scaling and translation property which helps to increase the entropy as compared to the other wavelet families.

By wavelet domain theory, the larger magnitude wavelet coefficients are important image data and smaller magnitude coefficients are considered to be insignificant image data. Also, small image details are more observable at finer scales. The fine details may be preserved in an image by exploiting the dependencies between two adjacent scales called inter-scale dependency and the correlation among the wavelet coefficients in the same scale is called intra-scale dependency. If the sub-bands at various scales are of the same size it is possible to find inter-scale dependencies also.

A suitable DWT (Discrete Wavelet Transform) is used to decompose the noisy image into the required number of levels. Each level consists of one approximation level LL and the other three details of the image LH, HL, and HH. Always the number of coefficients in all the levels is equal.

Curvelet transform:

Curvelet transform is the extension of the wavelet transform to represent the edges in the images which is not able to describe by the wavelet transform. The first level of the curvelet transform is not able to observe the edges as effectively. To overcome this the second level of the curvelet transform is introduced. [18]

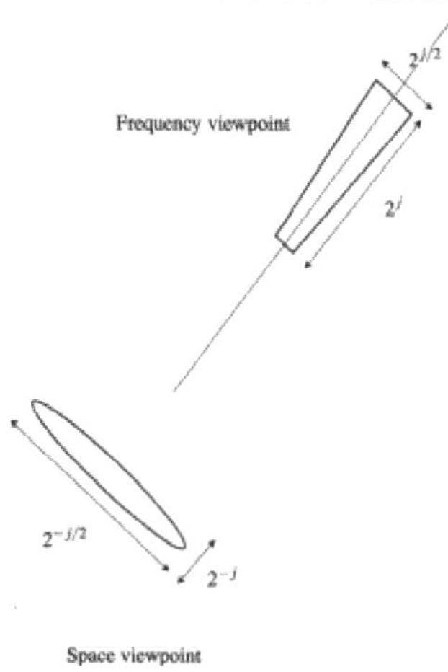

Fig.4. Curvelet frequency and space point view

In this, the second generation curvelet is used which performs partitioning in the fourier space. This different partitioning helps to smooth the edges of the image in the differen directions. The partioning of the second generation curvelet is shown in the below figure.

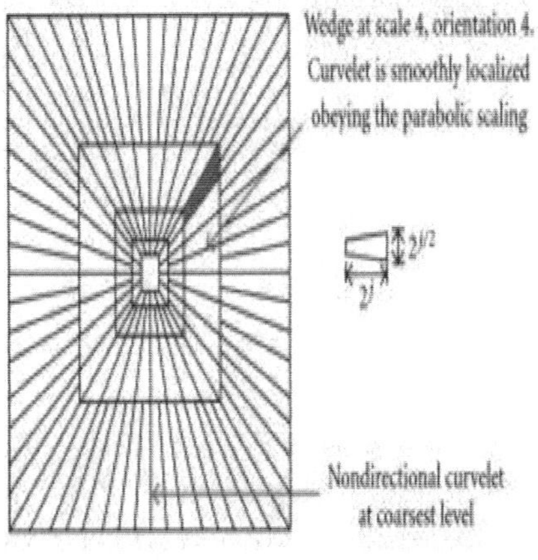

Fig.5. Second generation curvelet.

The special property of the curvelet is able to differnetiate the directions in each partioning which helps to exhibit the anisotropic nature. This property is absent in the wavelet transform. The another special property of this transform is the width of the curve is directly proportional to the square of the length of the curve. Due to this property, the two dimensional X-ray image is subjected to curvelet transform to capture the abnormalities in the image. Whereas the Magnetic resonance breast image is not subjected because its volume and its nature of exhibiting the level of abnormalities in the wavelet transform itself.

Fusion process:
The fusion of the transformed X-ray and MRI images are performed in the pixel level domain. The method used for the pixel level is the mean value of the two images. The mean value is chosen to improve the information that is the level of abnormalities of the breast image. this method improves the entropy of the image.

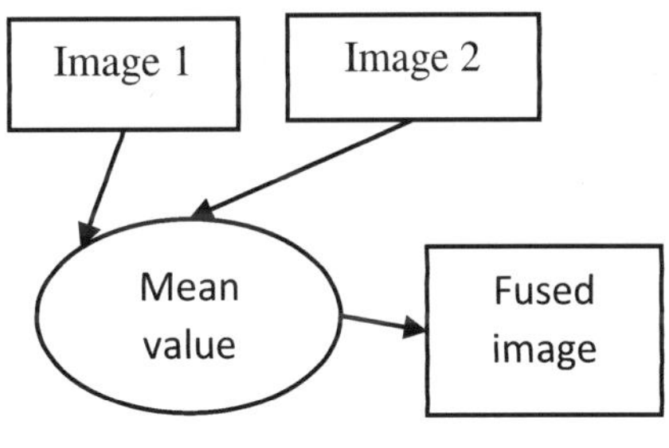

Fig.6. Pixel level fusion

Experimental Results and discussion:
The proposed method is tested on the openly available x-ray and mri breast images of the same person in the google. The proposed methodology is implemented with the help of Matrix laboratory software using version R 2018a. It is run in the windows 10 environment. The following figures are the output of the processing of x-ray images.

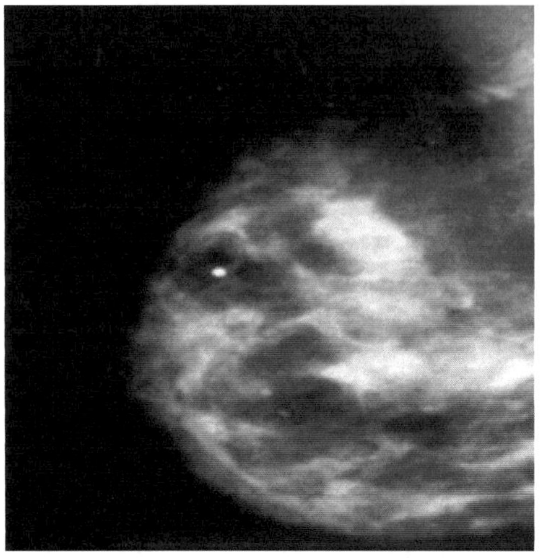

Fig .7. 2 X-ray image

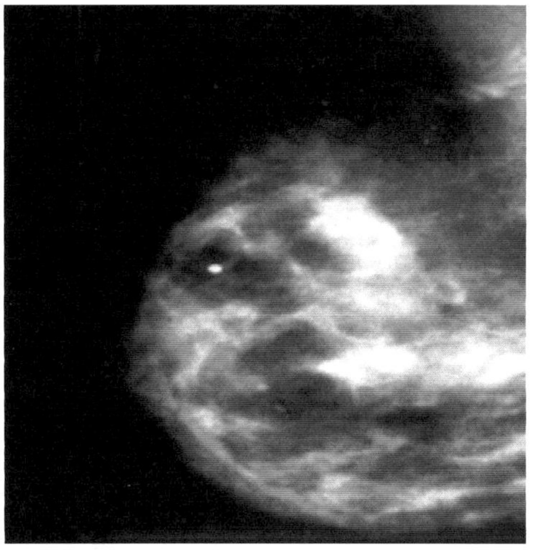

Fig 8. Gray scale and top-hat filtered image

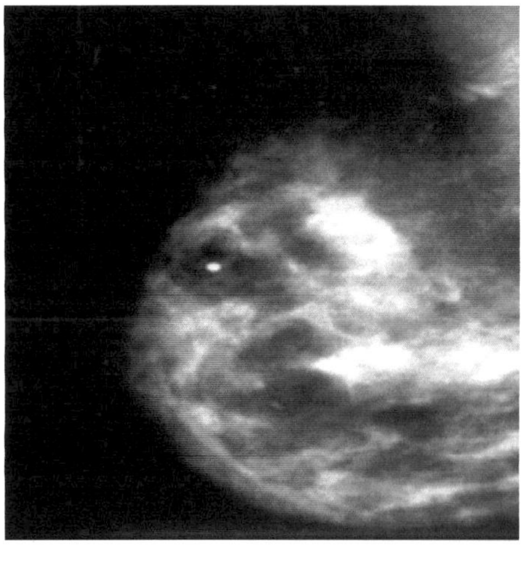

IOP Conf. Series: Materials Science and Engineering **994** (2020) 012027 doi:10.1088/1757-899X/994/1/012027

Fig 9. Wavelet transform of x-ray image

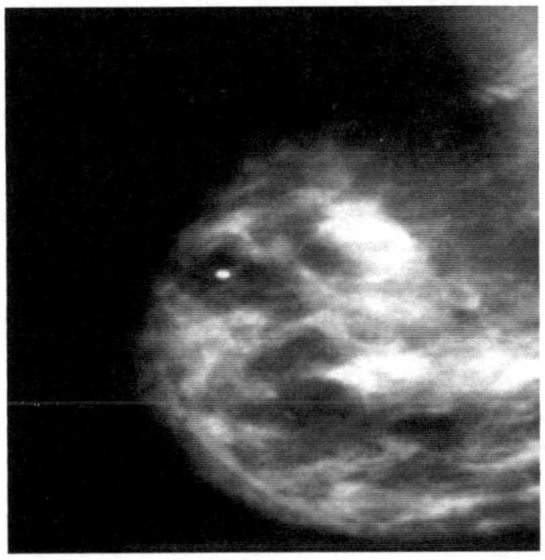

Fig.10. Curvelet transform of x-ray

The above figures are the output of the restoration of x-ray images for the fusion. The following figures are the restoration of MRI breast images.

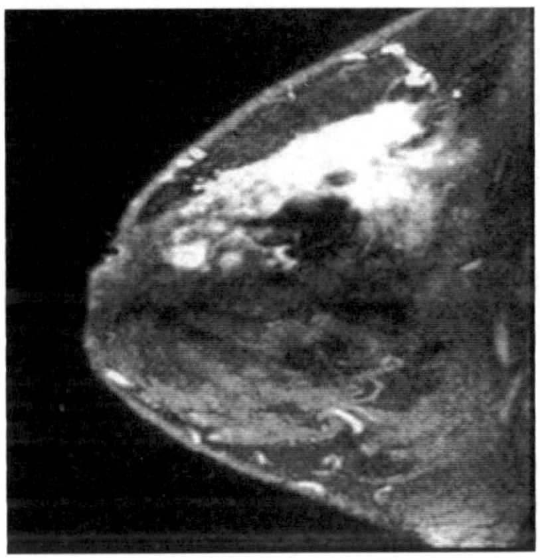

Fig.11. MRI image

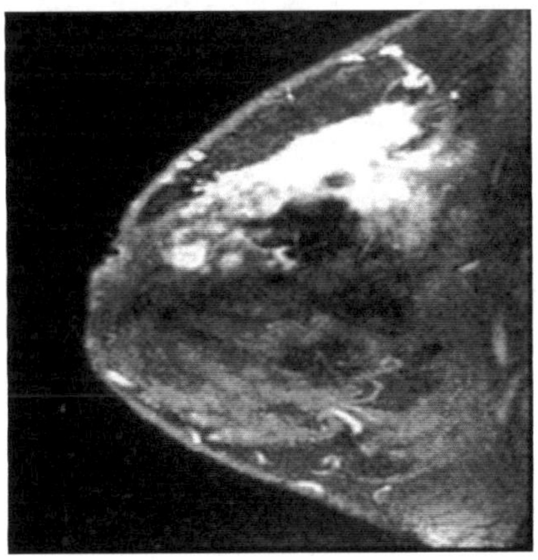

Fig.12. MRI noiseless and gray scale image

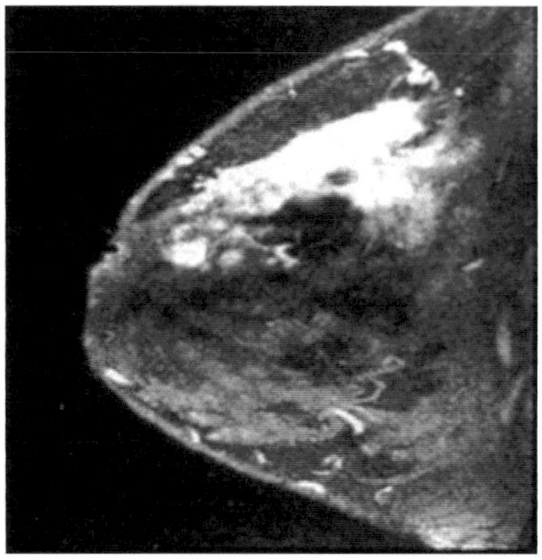

Fig.13. Wavelet transform of mri

In MRI, the wavelet transform is only applied to the image to preserve the level of abnormalities in the breast which is helpful in the fusion to hold more information of the breast. The below figure is the mean based pixel fusion of the transformed x-ray and MRI images.

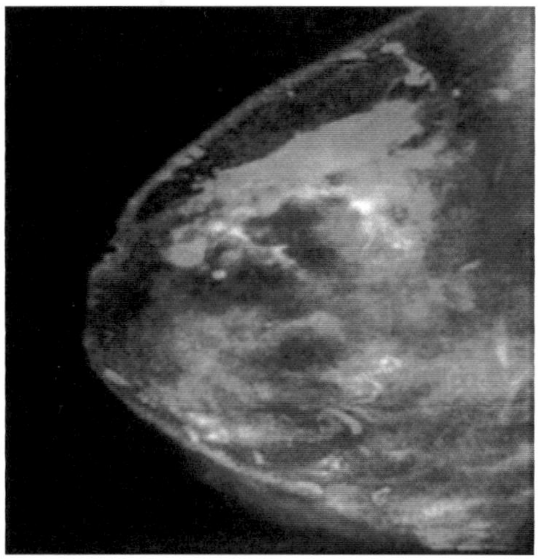

Fig.14. Fused image

The fused image is evaluated with the help of some statstical calcualations based on the pixel values as follows:

- **Entropy**
  The entropy is used to define the information present in the image using the pixel values of the histogram bins. For a good image, the entropy value should be higher.

$$E = -\sum(p.\log_2 p)$$

The p indicates the normalized histogram bin values.

- **Mean**
  The mean of the image is to define the quality of the image based on the mean difference between the pixels. The mean value will be low for the good quality image.

$$M = \frac{1}{n}\sum_{i=1}^{n-1}(x_i - y_i)^2$$

Where xi is the original image and yi is the denoised image. n is the number of pixels.

- **Standard deviation and Variance**

  The standard deviation and variance is used to differentiate the pixels of different class. For the image fusion application, these values should be low.

$$S = \sqrt{\frac{1}{n-1}\sum_{i=1}^{n} y_i - M}$$

Where yi is the denoised image and M is the mean of the image.

The variance is the square of the standard deviation.

$$V = S^2$$

These four values are tabulated in the table for the proposed method and it is compared with the existing method and original images to evaluate the proposed method. The mean of the proposed method is high because the fusion is based on the mean value of the pixels of the images.

	Entropy	Mean	STANDARD DEVIATION	VARIANCE (*10^3)
X-RAY	6.8925	63.3174	67.4471	4.5491
MRI	7.2348	71.4178	68.1574	4.6454
RES-DTCWTFUSED IMAGE	7.5127	39.9991	40.3455	1.6278
Contrast-adjusted WTCT	7.6843	69.5942	56.3199	3.1719

Even though, the mean and standard deviation is high the noise present in the image is less as compared to the traditional method. The mean value is higher because the fusion is based on the mean value of both the images. Due to this, the mean value is high which helps to improve the entropy. The reason for high entropy and less noise is due to the top-hat filtered curvelet transform. The daubechies wavelet family transform produces the best result with the average entropy of 7.5. The proposed method is better than the existing method is proved by the evaluation of the noise in the image. To evaluate the noise, the peak signal to noise ratio is used and it is calculated as follows:

- **Peak signal to noise ratio:**

  The PSNR value is to estimate the quality of the image after the removal of noise. For a noiseless image, the psnr value should be high.

$$PSNR = 10\log_{10}\frac{M^2}{mean\ square\ error}$$

The M indicates the maximum intensity of the image and the mean square error is calculated from the mean of the image.

Method	psnr
RES-DTCWTFUSED IMAGE	14.19
Contrast-adjusted WTCT	16.2651

From both the tables, the proposed method achieved the best result irrespective of its higher mean in both the fusion and image restoration. For the fusion process, the proposed method produced the higher entropy and in the restoration the psnr value is improved. Hence, the proposed Contrast-adjusted WTCT is suitable for the image restoration and fusion process.

Conclusion

The image restoration is an important process in the field of image processing. This process is to improve the quality of the image which helps the analyst to analyse the image effectively and efficiently. The paper described about the various restoration process earlier implemented in the medical field. Based on the analysis, a new restoration approach is proposed using filtering and transformation techniques. The filtering process helps to remove the noise in the image. The transformation helps to extracts the level of abnormalities in the breast. The mean-based pixel fusion helps to improve the entropy of the image. The proposed method is having the higher mean value but the level of the noise is reduced as compared to the existing method. Hence, the proposed Contrast-adjusted WTCT is suitable for the restoration and fusion process in the breast imaging. In future, the restoration is improved with the help of unsupervised learning network for the online fusion of images.

References:

[1] Katsaggelos A K (2012) Digital image restoration. Springer Publishing Company, Incorporated
[2] Zhang K Zuo W Gu S & Zhang L (2017) Learning deep CNN denoiser prior for image restoration In Proceedings of the IEEE conference on computer vision and pattern recognition (pp. 3929-3938)
[3] Li S Kang X Fang L, Hu J & Yin H (2017) Pixel-level image fusion A survey of the state of the art. Information Fusion 33 100-112
[4] El-Gamal F E Z A Elmogy M & Atwan A (2016)Current trends in medical image registration and fusion Egyptian Informatics Journal 17(1), 99-124
[5] Kumar B. S (2015) Image fusion based on pixel significance using cross bilateral filter Signal image and video processing 9(5) 1193-1204
[6] Liu Y Chen X Wang Z Wang ZJ Ward R K & Wang X (2018) Deep learning for pixel-level image fusion: Recent advances and future prospects Information Fusion 42 158-173
[7] Liu Z Blasch E & John V (2017). Statistical comparison of image fusion algorithms Recommendations. Information Fusion 36 251-260.
[8] Teodoro AM Bioucas-Dias J M & Figueiredo MA (2018) A convergent image fusion algorithm using scene-adapted gaussian-mixture-based denoising IEEE Transactions on Image Processing 28(1) 451-463.
[9] Kalavathi P & Priya T (2016 October) Removal of impulse noise using Histogram-based Localized Wiener Filter for MR brain image restoration. In 2016 IEEE International Conference on Advances in Computer Applications (ICACA) (pp. 4-8). IEEE
[10] Hatwar S & Wanare A (2015) Gmm based image segmentation and analysis of image restoration tecniques International Journal of Computer Applications, 109(16)
[11] Daniel E & Anitha J (2016) Optimum wavelet based masking for the contrast enhancement of medical images using enhanced cuckoo search algorithm Computers in biology and medicine 71 149-155
[12] Rajendran R Rao SP Agaian SS & Panetta K (2016 October) A versatile edge preserving image enhancement approach for medical images using guided filter In 2016 IEEE International Conference on Systems Man and Cybernetics (SMC) (pp. 002341-002346) IEEE
[13] Song Q Ma L Cao J & Han X (2015 August) Image denoising based on mean filter and wavelet transform. In 2015 4th International Conference on Advanced Information Technology and Sensor Application (AITS) (pp 39-42). IEEE
[14] Jangale TA & Paul RK (2019) Elimination of Impulse Noise using Mean Median filter for Image Enhancement
[15] Mukherjee PS (2017) A multi-resolution and adaptive 3-D image denoising framework with applications in medical imaging. Signal, Image and Video Processing 11(7) 1379-1387
[16] Tiwari S Bansiya A & Paul R K (2015) Elimination of Impulse Noise using Enhanced Mean Median Filter for Image Enhancement. International Journal of Engineering and Management Research (IJEMR) 5(2) 811-818
[17] Mukherjee PS (2017) A multi-resolution and adaptive 3-D image denoising framework with applications in medical imaging. Signal, Image and Video Processing 11(7), 1379-1387
[18] Ma J & Plonka G (2010) The curvelet transform IEEE signal processing magazine 27(2) 118-133
[19] Daubechies I (1990) The wavelet transform time-frequency localization and signal analysis IEEE transactions on information theory 36(5) 961-1005
[20] Manjusha S Nair and A Rajaram Low power receiver using envelope detector converters," International Journal of Advanced Info Sci and Tech 3(3)50-57 March 2014

IOP Conf. Series: Materials Science and Engineering **994** (2020) 012028 doi:10.1088/1757-899X/994/1/012028

Implementation of high performance traffic management system using novel blockade mechanism

K Ram Prasanna[1], R Menaka[2], P Pavan Ram[3], M Rithul[4]
[1]Student, Department of Electronics and Communication Engineering, Chennai Institute of Technology, India.
[2]Professor,Department of Electronics and Communication Engineering, Chennai Institute of Technology, India.
[3]Student, Department of Electronics and Communication Engineering, Chennai Institute of Technology, India.
[4]Student, Department of Electronics and Communication Engineering, Chennai Institute of Technology, India.

Abstract:Traffic in cities is one-sided during peak hours in a day, these traffics are uncontrollable with traffic management alone because it might cost fuels for opting long distance. Mainstream roads which connect important areas come under non extendable category, government spends a lot of money to buy the lands from the public and make alterations in electric cables, drainage system, platforms etc. Heart areas of cities need investments which take a big part in development budget. So in the proposed work, a novel method is introduced where movable pole blockades are controlled using IoT & intelligent traffic monitoring. As per survey, the two way roads having three lanes on each side can be used in 2:4 and 4:2 ratios, this method is cost efficient as it consumes only one third the amounts of usual practices. The traffic congestion is detected using video cameras incorporated in ATS (Area Traffic System). When the density of the vehicles cross the threshold for more than a period of time then an adaptive method is implemented to overcome one-sided traffic congestion. It can pave the way to delegates (VIP's) and even national emergency teams/ ambulances/ fire extinguishers to cross an area without disturbing the public. These poles are a one-time investment and digital signboards on it will indicate the subject change like converging/ diverging of roads, VIP usage etc. the other two lanes can be used by the public for LMV (Light Motor Vehicle) and HMV (Heavy Motor Vehicle) including public transportation. This method can be altered from signal to signal and also in between the signals in case of roads meeting perpendicular to the lanes. Mono directional traffic is common in and around the world where population densities are highly influential in time and fuel wastage.

1.Introduction

1.1 General introduction

Road traffic is a daily problem faced by all age groups of people daily irrespective of city. Many number of vehicles due to inadequate transit options or other reasons where non-extendable roads play a star factor. Causes of obstacles on the road which leads to blockage like lane closure, road narrowing down, road work, accident. Problems like inadequate green time could be solved by smart trafficking (IoT).

Overdeveloped areas where overcrowded and road systems are inadequate. Banning parking along the roadside is practiced in many cities to increase the traffic flow.
The pedestrian crossing also causes congestion during peak hours which are solved by building a pedestrian overpass or mid-block crossing. Scheduling roadwork in the middle of the night which causes less traffic fever, although there are some road works which have to be scheduled during the day time also in an emergency or rework purposes. Many cities have split lanes for motorcycles to avoid traffic clogging. Several roads are interconnected in bottleneck pattern where gridlock is practiced to relieve areas with traffic congestion and it eventually increases fuel consumption and distance travelled by the vehicles. These trafficking systems can be deployed to save time and fuel factors. A traffic system gets congested if the demand exceeds the capacity section. Congestion is a travel time delay which exceeds the normal incurred time.
A major source of anxiety, stress and anger is traffic on roads; Traffic delays are caused by clogging in highways on an average make rush-hour. The number doubles to 84 hours on the nation's 10 most clogged roads. Traffic congestion can be identified using software like MATLAB and python with the help of OpenCV. Congestion in both image and Video can be detected. The vehicles in the Image are detected based on parameters like boundary description, the motion vector, edge detection and so on, whereas in Computer vision on python, various data sets of vehicles are given as pre-input and various techniques like cascade method are used to identify the traffic. In general, western countries use traffic cameras for image or video input for identification of traffic congestion. Separate systems and Cameras are also installed to detect vehicle owners with number plates and control the violation based on speed, pollution regulation, and signboard rules etc.

1.2 Existing problems
In the construction sector, transport management can be more of a headache than in other industries. The construction problem is that each site is different, different layout of the site, different access points, different requirements, and different tasks. Traffic management designed with the concrete blocks which is cost-effective.Once after Implementation of concrete blocks further change can't be made..

Figure 1.1.a Accidents caused by concrete blocks

Traffic affects many aspects of modern society which includes economic upliftment, traffic collision, greenhouse emission increases, time and damage to immunity. Modern societies can rely on the traffic management system in this context to reduce traffic jam and its ill effects. Traffic incidents happen for several reasons. While problems with roads or safety facilities lead to some accidents.
Increased travel time caused by traffic congestion causes costs on public, in terms of both economic loss and drops quality of life and immovability as well. Due to increased traffic volume, this peaks traffic congestion leads to more accidents; however, these accidents may be less serious. Transport operations are among the main contributors to Green House Gas (GHG) emissions and the outlook is still favourable given other sectors of the economy. Transportation is directly due to an additional 30 per cent of regional

GHGs. Many transport infrastructure environmental effects include traffic jams and automotive-oriented urban sprawl, which can destroy natural farming. By reducing transport emissions globally, it is predicted that the air quality, acid rain, smog and climate change will have significant positive effects on Earth.

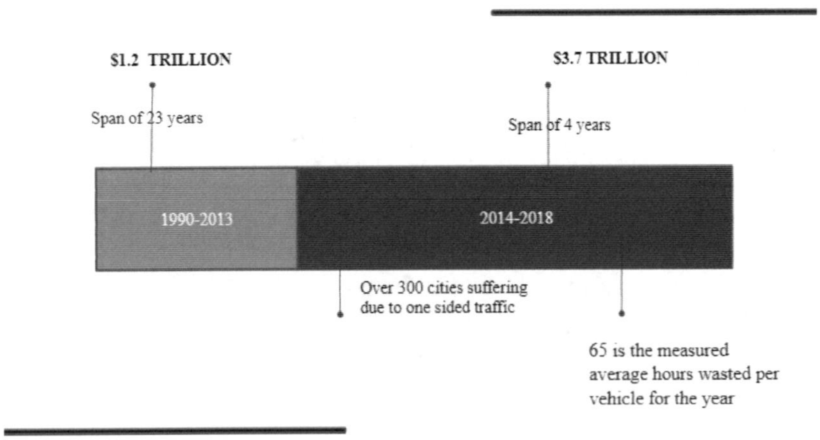

Figure 1.1.b Statistical representation on expenditure for the consumption of fuels

2. Literature Survey

2.1. Literature Survey on Image Processing in Traffic management

Al HussainAkoum [1] recommended the usage of a brilliant traffic framework utilizing picture preparing. The camera succession is analyzed utilizing different edge discovery calculations and techniques to tally objects. Beforehand they utilized a coordinating technique which implies that the camera and traffic signal will be introduced together. It catches a succession of pictures. To set a picture of an unfilled street as a source of perspective picture, the caught pictures are arrangement analyzed utilizing picture coordinating method; however in this article sifting approach that channels the picture and delivery all the waste things and just indicated the vehicles, and afterward shows the quantity of vehicles in the picture. This work is a product program that snaps a photo or a video. It is tweaked for later use to control the traffic signal signs by offering adequate opportunity to each hint, which relies upon the include of vehicles in each roadside.

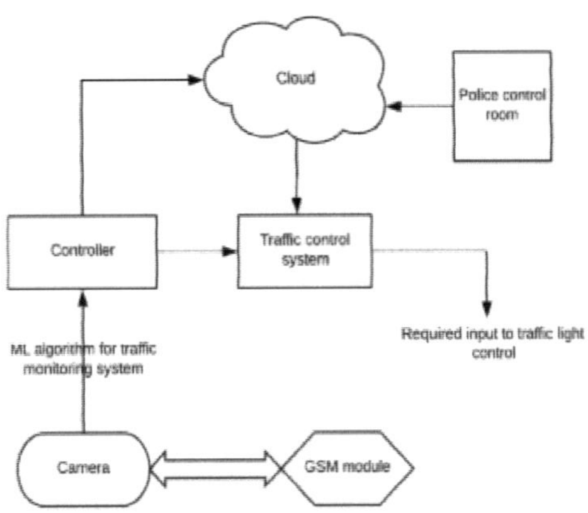

Figure 2.1.a Traffic control system

R. Krishnamoorthy [2] proposed to robotize traffic lights with the guide of a few web associated CCTV cameras to follow different intersections of streets. The cycle has two essential stages: a framework for vehicle distinguishing proof and calculation for traffic plot. Vehicle Detection in Digital Image Processing (DIP) will be done by executing a basic part based Edge Detection in Spatial Domain followed by a calculation to recognize the border of shut items in the interim actualizing the standards of Machine Learning so as to distinguish the vehicle type in the accompanying classifications of bikes; light engine vehicles and weighty engine vehicles. Ensuing techniques are done in a novel Traffic Scheduling Algorithm with the guide of a crossover Round Robin with an unpredictable time-segment acquired by utilizing Longest Remaining Job First to consistently refresh traffic light lights to ease traffic. Or maybe turning on green light for a set range of time, the range would be controlled progressively relying upon the segment of traffic on each course. This program was wanted to lessen street sticking to a noteworthy degree.

2.2. Smart Traffic Management System

Abida Sharif et al.[3] recommended that Smart Traffic System (STS) is one of the key parts of a future brilliant city. STS is all the more expensive and exceptionally configurable to give improved nature of administration to metropolitan traffic the executives. This paper proposes a minimal effort future STS to offer better assistance by promptly conveying traffic cautions. Ease vehicle sensors are fixed each 500 meters in the street. The Internet of Things (IoT) is actualized quickly gather and submit public traffic information for formulating data. Ongoing streaming information is submitted to Big Data Analytics. There are numerous experimental sacred texts to distinguish traffic power and give an answer by prescient investigation.

VidyaBilawade et al.[4] has recommended that there has been an uncommon increment in number of vehicles and there are around 1 million authorized vehicles in the most recent year since traffic issues have crested over the previous years and the current traffic signal frameworks have disadvantages because of the utilization of default equipment.Traffic the executives frameworks don't have the versatility of continuous alterations. The time ranges of the green, orange and red signs are set, so the postpone period is longer. Insightful Traffic Control System is rising; this uses sensors and implicit innovation. The traffic signals timings will be shrewdly chosen dependent on street traffic. This new framework is more dependable and versatile contrasted and past fixed mode traffic signal regulator. It likewise has the additional component to pass crisis vehicles and can likewise to recognize and follow the vehicles that are taken. This plan has space for additional development, as well.

Rizwan P [5] has talked about how the current traffic the executives and security frameworks are not skilled to address the issues of STMS. They are costlier and profoundly configurable to give better traffic the board administration. Creators likewise examined the ease shrewd traffic the executives framework Real-Time for offering better support by immediately demonstrating traffic signs to refresh. Cost productive vehicle recognition sensors are implanted in the street each 500 meters or each 1000 meters. The Internet of Things (IoT) is utilized to rapidly gain and send traffic information for preparing. Continuous streaming information is submitted to Big Data Analytics. There are numerous exact sacred texts to recognize traffic force and give an answer by prescient examination. A portable application is created as a UI for investigating traffic thickness at various areas and giving a substitutive method to deal with the traffic.

Dr.B.Prakash et al [6] recommended that every information is essentially easily accessible. However, the development signals checking are yet done genuinely. The action signals are watched truly from the control room by the heads or an anticipated time is agreed to signals rising. Instead of this, an orderly regulator based strategy checking structure will be helpful for influencing the development. This arrangement of development establishment will be valuable in diminishing the action obstruct issue in metropolitan networks. Creators proposition portrays a structure where IR sensors are consolidated with an Arduino to work the ways which measure the development thickness. This joined course of action of

development is the Internet of Things (IoT) based which similarly enables to clear the action for crisis vehicles by giving a catch in a salvage vehicle so the movement gets cleared on that side. It moreover enables the vehicles count that moves over the sensors. Consequently, development controlling gets redesigned adequately, which in the end prompts a huge change in busy time gridlock system.
TejTharangDanda [7] said that Internet of Things (IoT) is an overall organization which associates every savvy hub. The strategy expresses how everything is made to have the option to converse with one another. At whatever point those keen things associated with the Internet are restricted to vehicles just, they are called Internet Vehicles (VI). With an ever-expanding metropolitan populace and developing urban areas, vehicle proprietorship developed at an exponential rate [9]. Traffic the executives has in this way become a significant issue in the present life today. The creators work gives a traffic the board arrangement dependent on Internet of Vehicles to beat the difficult that wins in our day by day lives.

2.3. Case Study
Gridlock and mishaps are serious issues in Phnom Penh, the capital of the Kingdom of Cambodia. Gotten on 27 April 2018, an inconsistent framework and an absence of traffic signal prompted negative traffic mayhem. A region traffic light framework, which controls traffic lights in the city dependent on traffic conditions, has been actualized and different changes in rush hour gridlock the executives, for example, asphalt markings, street signs and medians, have been presented Phnom Penh Capital City it's a complete traffic the board venture comprising of a territory traffic light organization and other traffic the board works. Traffic in Phnom Penh is tumultuous, with different kinds of vehicles running on a similar street, from extravagance SUVs to bikes and cruiser drawn. These kinds of vehicles, most of which are cruisers, represent 70–90% of the volume of traffic. Different vehicle types have conflicting productivity and capacity to explore, leads wasteful development and hazardous gridlocks.
Phnom Penh has been growing more and the traffic conditions have changed significantly in past years, with the check of vehicles expanding. Built up traffic signal frameworks and offices are not in a situation to adapt to the evolving situations. Be that as it may, the circumstance has as of late improved through the execution of the task and the exchange of innovation to neighbourhood partners as a major aspect of the venture. At the point when the venture was controlled there were diverse traffic the executives' offices in the task zone, for example, traffic lights, asphalt markings, traffic signs, and middle dividers. As there were no characterized rules or techniques for signal plan, the traffic light and configuration were not discerning and sometimes fell short for the interest traffic at convergences, bringing about wasteful sign help. That downside of the styles is that the detecting territory is set once the locator is mounted and cannot be altered. The progression of traffic in Phnom Penh is recognized by variegated traffic of different verities of vehicles. Vehicles are disregarding path markings and taking any position and bikes are weaving between traveller vehicles. A vehicle identifier that has a fixed detecting region doesn't identify and precisely tally vehicles under these conditions.

3.Need for Study on Traffic Management systems
The safety is very important as is the smooth flow of traffic. It not only makes people feel safe but it also providing them security. So traffic management is required that will help coordinate traffic flow on the roads.
Traffic management centres main task is to programme,receive and transmit the required details about a certain traffic situation. It also helps direct and control the traffic on various highways. Every day of the year these traffic control centers are all staffed around the clock. Traffic management is essential because it keeps pedestrians and vehicles away from each other and it is necessary to use the right signs and equipment to reduce the risk of accidents, especially if major construction is taking place. It may include equipment such as speed bumps, flow plates, cones, traffic mirrors and separator bollards.

Ensuring the traffic for a span of time in at the momenthelps to reduce the incidence and extent of those kinds of incidents. A minute delay in traffic in any high traffic volume road creates pilling up of vehicles. A successful strategy for traffic management will reduce the delay in any part of the road network.

4. Proposed method

- Identification of traffic is feasible to work on Computer vision with python to detect traffic congestion.
- To implement high performance traffic management system with adaptable blockades that can be used to utilize lanes productively and thereby overcome the one sided traffic congestion using internet of things
- The stimulus digital output is taken if the congestion detected is above threshold (which is set on various parameters like Number of vehicles, slow movement or stagnant traffic, time period). Controller receives digital input to trigger the adaptable blockade system to pop up and pop down the movement of blockades with prior information is digital sign boards and proper trafficking of signals.

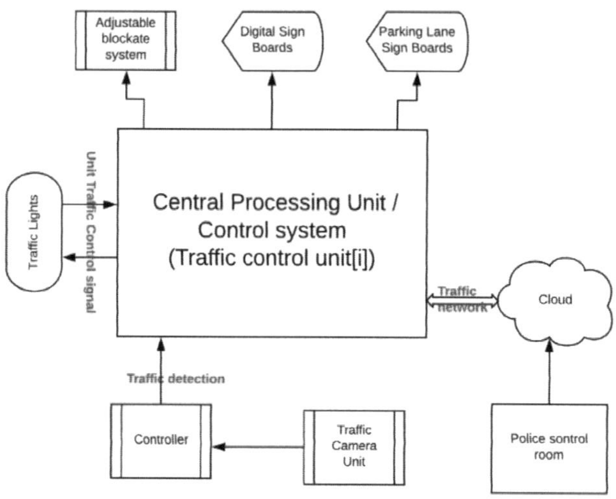

Figure 4.a Block diagram of proposed system

4.1 Adjustable blockade system

The adjustable blockade systems are cost efficient system when compared to concrete blockades. The concrete barricades laid along the roads are practically immovable and it costs from 1500 INR to 4500 INR per running feet based on the size. This system is one time investment and manufactured using heavy recycled plastics which are eco-friendly. Implementation of Adaptable blockades in every junction in cities can free mobility of vehicles giving a lot of space. Replacing the damaged blockades can be done with ease. The bases of the blockades are shunted with motor to pop up and down for lane shifting action.

IOP Conf. Series: Materials Science and Engineering **994** (2020) 012028 doi:10.1088/1757-899X/994/1/012028

Figure4.1.a side view of blockade

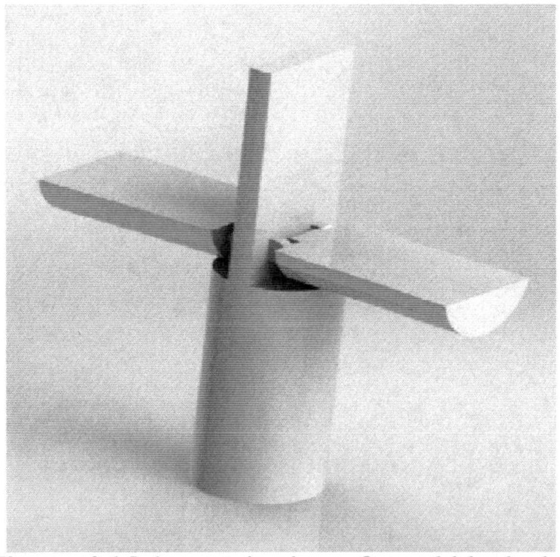

Figure 4.1.b isometric view of novel blockade

The motions of the motors are controlled by the traffic control system for lane shifting purposes. The following images show the isometric and side view motion of novel design of blockade.

Isometric view

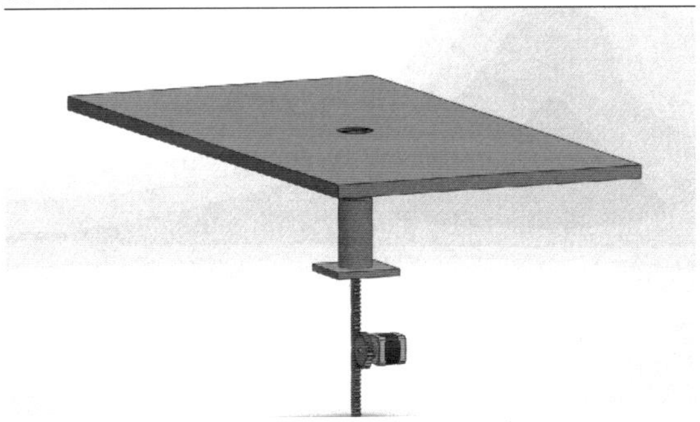

Figure 4.1.1 Pole below the road level

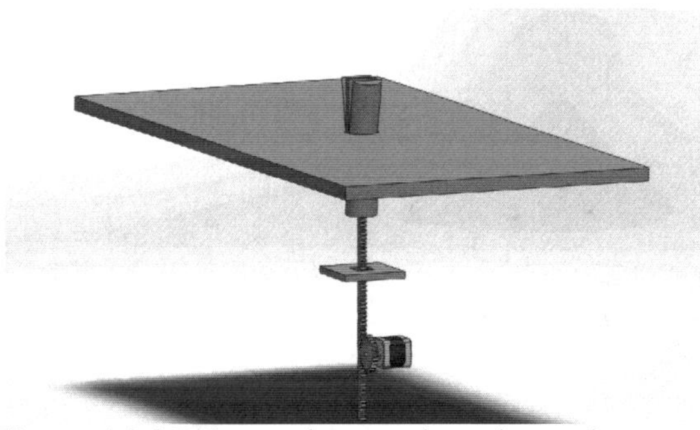

Figure 4.1.2 Pole popping up when triggered

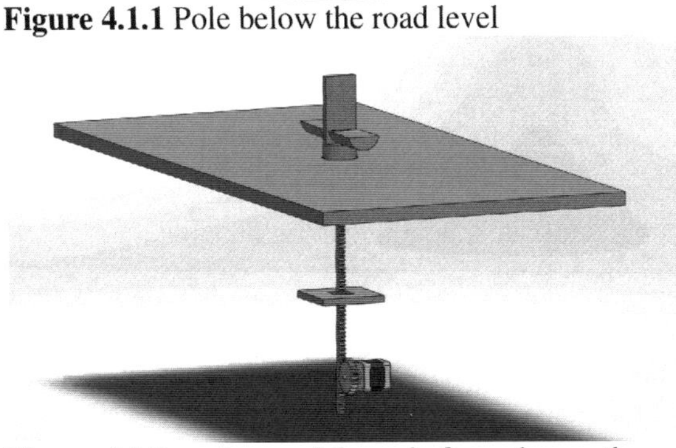

Figure 4.1.3 Arms of the pole fapped once the hieght reaches beyond half

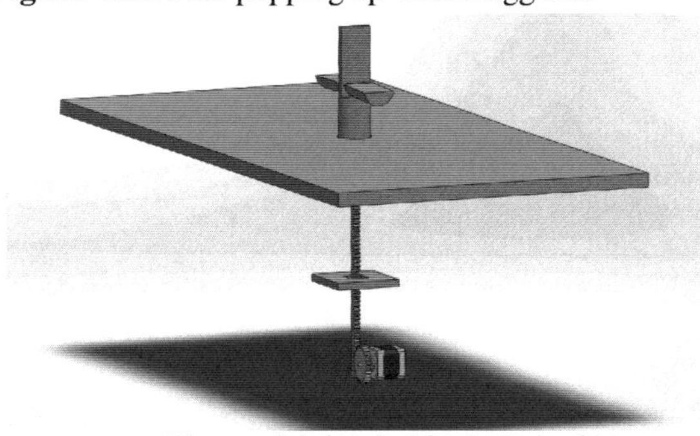

Figure 4.1.4 Pole blockade in ON state

Side view

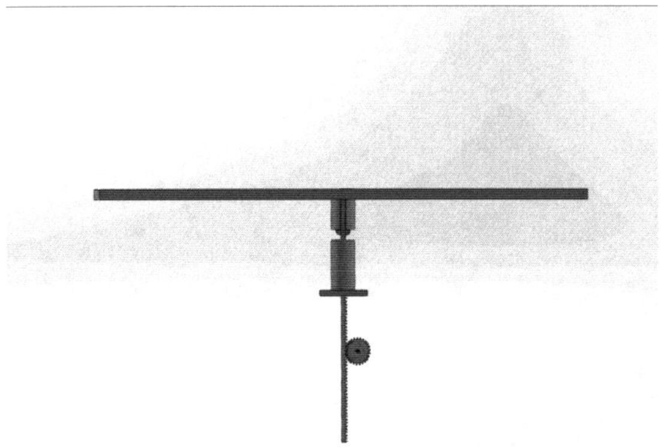

Figure 4.1.5 Pole below the road level

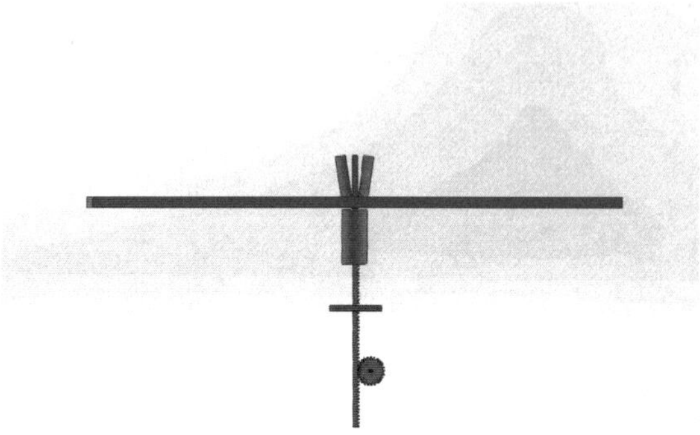

Figure 4.1.6 Pole popping up and arms are yet to flap

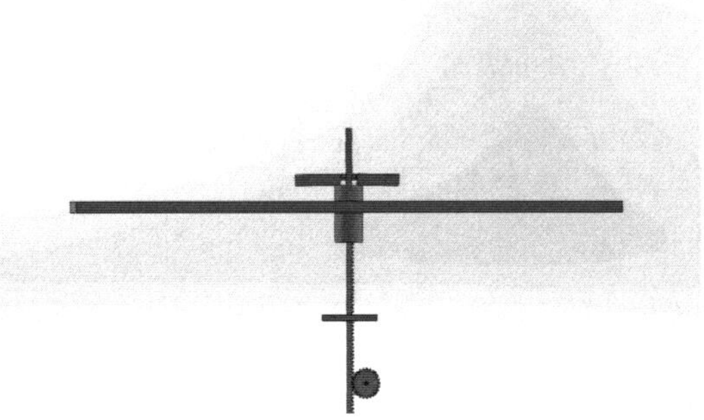

Figure 4.1.7 Pole Arms are flapped

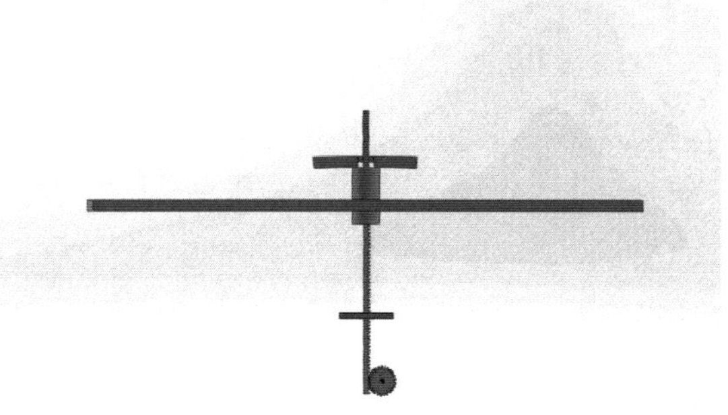

Figure 4.1.8 Pole blockade in ON state

4.2 Traffic camera unit

The traffic camera unit's main role is to play surveillance and many cities have adopted to use the cameras for detection of traffic congestion. For simulation purposes, program in python can be programmed to identify the amount of traffic congestion based on the following parameters:

- Number of vehicles
- Slow movement or stagnant traffic
- Time period

With these parameters the traffic scenario is monitored by the controller and stimulus is generated for lane shifting.

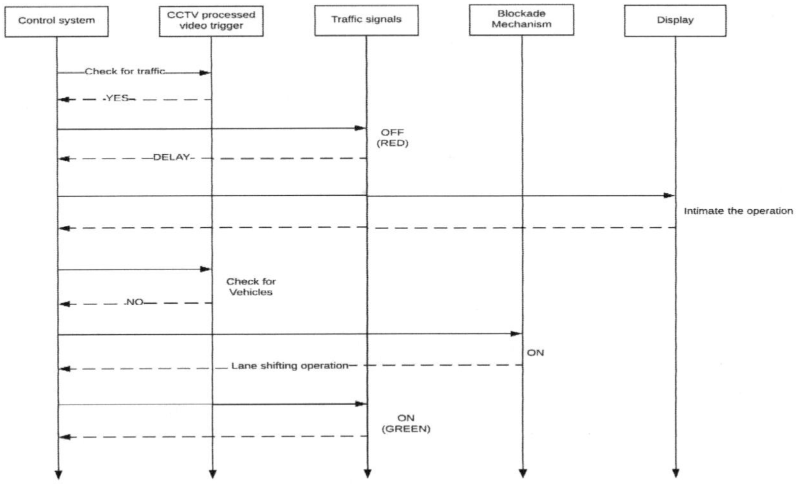

Figure 4.2.a Energy flow diagram of the system

Steps

1. The CCTV processed trigger is checked by the control system for the traffic congestion and if the congestion is identified by the factors Number of vehicles, slow movement or stagnant traffic, Time period (span of time is set by the programmer and it also depends local factors).
2. Once the controller receives the intimation, the necessary traffic lights are turned OFF(RED) for predefined delay time. This delay time is utilized to shunt lanes for wide usage.
3. The process of shifting lanes are displayed on the digital sign boards and Arch towers in order to maintain safety of the trespassers and vehicles nearby/
4. For the secondary safety purposes the lanes that are to be shifted on the other side of the road are ensured for the absence of vehicles so that poles can pop up without any obstacles.
5. After ensuring above measures the adaptable lanes are popped up, they are specially designed with flap wings to avoid trespassing of the two-wheelers.
6. Once after popping of lanes the road is added with additional lane from the other side as an advantage and the signals are turned ON (GREEN).

IOP Conf. Series: Materials Science and Engineering **994** (2020) 012028 doi:10.1088/1757-899X/994/1/012028

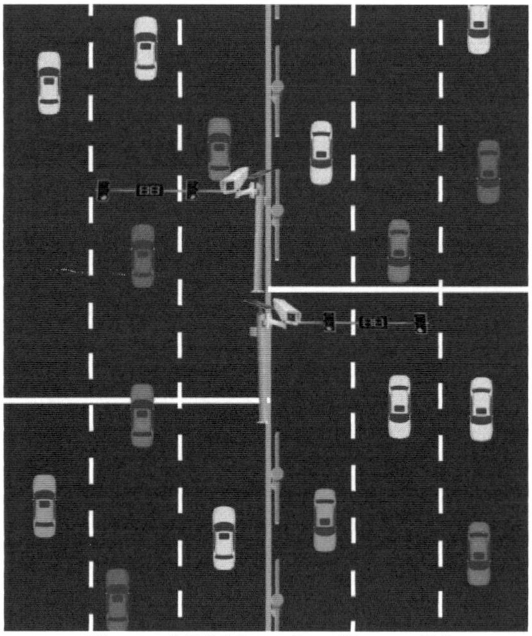

Figure 4.2.1 Normal traffic condition

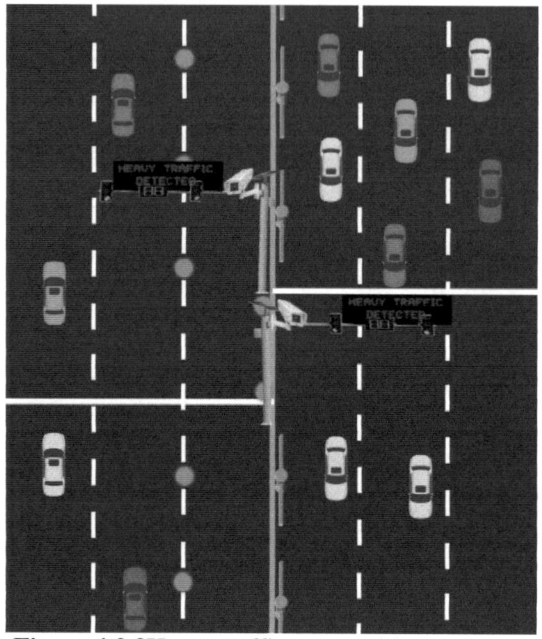

Figure 4.2.2 Heavy traffic on one side

In Figure 4.2.1 portrays the normal traffic condition, where there is no traffic on the either sides of the road lasting more the threshold time. Whereas in Figure 4.2.2 Heavy traffic is detected in the cameras incorporated, the cloud synced with the controller commands the digital signboards and displays 'Heavy traffic detected' on both sides and the signals are turned red. The lane next to the middle blockade is cross verified for the absence of vehicles. Once after evacuating the lane the trigger is set to switch the operation.

The pop up and down of poles are operated in the respective side of the lane, in addition to the system, displays can be added to the sides of the road to describe the operation clearly. Figure 4.2.3 shows the shifting of lanes. Once a lane is added to the less traffic side then the signals are turned green. Pervasive of vehicles can be operated on four lanes so that the traffic width widens and the density decreases over a period of time. This lanes can be expanded as per it is deployed. The blockade has flap wings as additional advantage and it avoids the crossing of two wheelers. By implementing this distance between two blockades can be increased and the cost is reduced eventually. The digital sign boards display the lane allotted shown in Figure 4.2.4.

IOP Conf. Series: Materials Science and Engineering **994** (2020) 012028 doi:10.1088/1757-899X/994/1/012028

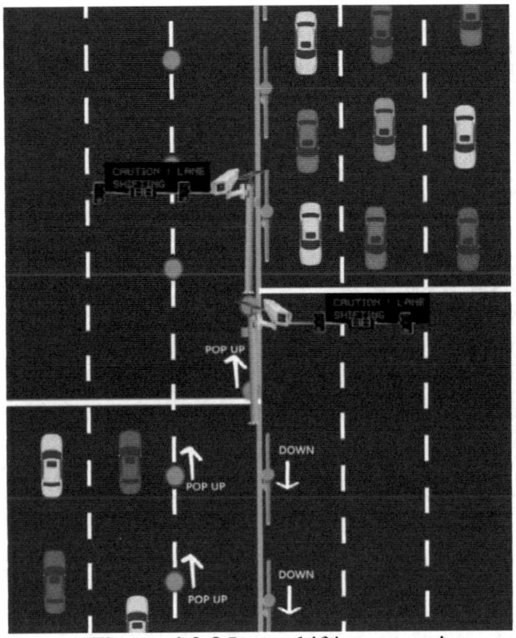

Figure 4.2.3 Lane shifting operation

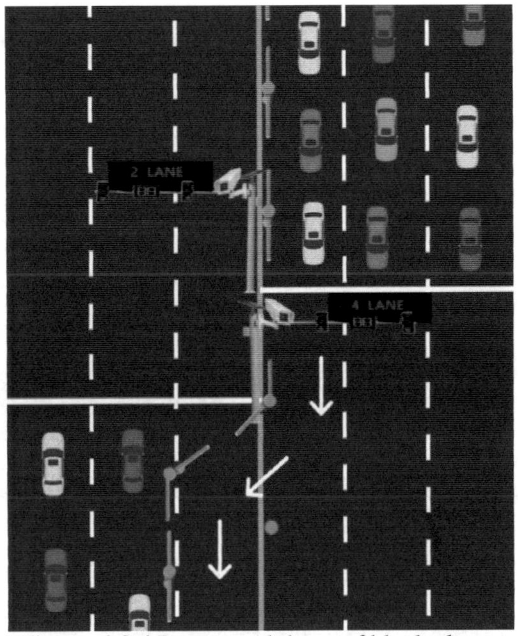

Figure 4.2.4 Pop up and down of blockades

The pop up and down of poles are operated in the respective side of the lane, in addition to the system, displays can be added to the sides of the road to describe the operation clearly. Figure 4.2.3 shows the shifting of lanes. Once a lane is added to the less traffic side then the signals are turned green. Pervasive of vehicles can be operated on four lanes so that the traffic width widens and the density decreases over a period of time. This lanes can be expanded as per it is deployed. The blockade has flap wings as additional advantage and it avoids the crossing of two wheelers. By implementing this distance between two blockades can be increased and the cost is reduced eventually. The digital sign boards display the lane allotted shown in Figure 4.2.4.

This method is cost efficient as it consumes only one third the amounts of usual practices. The traffic congestion is detected using video cameras incorporated in ATS (Area traffic system). When the density of the vehicles cross the threshold for more than a span of time then adaptive method is implemented to overcome one sided traffic congestion. It can pave way to delegates (VIP's) and even national emergency teams/ ambulances/ fire extinguishers to cross an area without disturbing the public. These poles are one time investment and digital sign boards on it will indicate the subject change like converging/ diverging of roads, VIP usage etc. the other two lanes can be used by the public for LMV (Light Motor Vehicle) and HMV (Heavy Motor Vehicle) including public transportation. This method can be altered from signal to signal and also in between the signals in case of roads meeting perpendicular to the lanes.

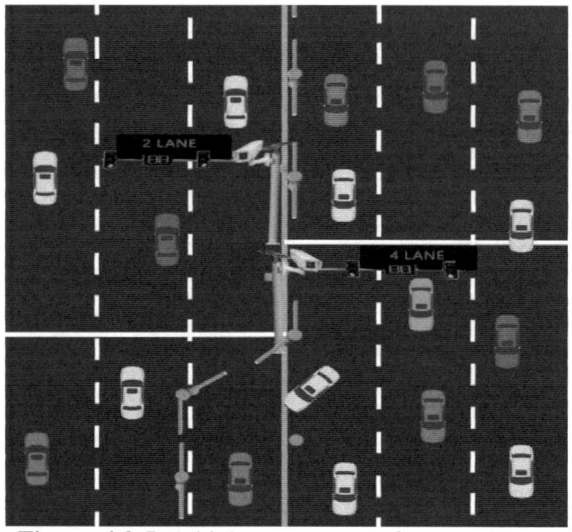

Figure 4.2.5 Vehicle movement after operations

5. Conclusion

Various Image processing techniques have been identified to detect traffic congestion. Redirecting vehicles to various other routes will consume more fuel than the threshold of actual consumption. Pre-processing images usually involves eliminating low-frequency background noise, normalizing the size of the images of the individual objects, eliminating shadows, and masking image portions. Pre-processing of images is the technique used to enhance data images before computational processing. The initial task of detecting vehicles is to transform the true colour input image into a grey image form.Traffic causes thermal & air pollution, mental stress and illness. This also limits the leverage of emergency vehicles like Ambulance, Fire service, Police patrol etc. The US spends around 416 billion dollars for maintenance of roads per year out of which 181 billion dollars are invested just to extend the road. Abolishing parking lots along the roadside, constructing alternate paths to overcome traffic add on to the extending budget.

To summarize the ill effects of congestion, lack of productive energy, increased fuel usage, increased wear and tear of car engines, increased fuel burning, high traffic accident risk, and adverse impact on the psychological condition of people, sluggish and inefficient emergency response and distribution status. The need to research those things is essential to overcome depleting petroleum resources, which is mandatory in conserving nature for the future. Though renewable energy is implemented in vast areas, traffic congestion tarnishes fuels with no valid reason. Adaptable blockades are the only solution to overcome these problems.

References

[1] AlHussainAkoum Automatic traffic using Image processing Journal of Software Engineering and Applications SRP Lebanon 14th Aug 2017

[2] RKrishnamoorthy and Sethumaanickam 2018 Automatic traffic monitoring image vision IEEE International Conference on Inventive Communication and Computational Technologies

[3] Abidasharif Jianping Li Rajesh Kumar and Atiqa Sharif 2018 Internet of things- smart traffic management system for smart cities by using Big data analytics ICCWAMTIP Conference – China November 2018

[4] Mrs. VidyaBhilawade Dr. L.K. Radha Mrs.VidyaBhilawade and Dr. L.K. Radha 2018 International Journal of Scientific and Research Publications IJSRP

[5] PatanRizwan K Suresh and Dr. M Rajasekhara Babu 2016 Real time smart traffic management system for smart cities by using IOT and Big data IEEE-International Conference on Emerging Technological Trends ICETT- Vellore
[6] Dr.B.Prakash M.NagaSaiRoopa B. Sowjanya and A. Pradyumna Kumar 2018 An Iot Based Traffic Signal Monitoring and Controlling System Using Density Measure of Vehicles International Journal of Research vol 05
[7] Dhanagopal R Krishanamurthi V and Rajaram A 2017 Energy-efficient devices using a design of reconfigurable robust integrated power converters International Journal of Control Theory and Applications vol 10 pp 67-81
[8] TejTharangDandala Vallidevi Krishnamoorthy and Rajan Alwan 2017 Internet of vehicles (IOV) for traffic management IEEE-International Conference on Computer Communication and Signal Processing Chennai
[9] Seiyamatsuoka 2018 Traffic Management Project in Phnom Penh International Association of Traffic and Safety Sciences Production and hosting by Elsevier Ltd

NAME	*PREFIX*	*RESEARCH FIELD*	*EMAIL*	*PERSONAL WEBSITE*
Ram Prasanna K	*Student*	*IoT & Embedded systems*	*ramprasannaprofessional@gmail.com*	-
Menaka R	*Professor*	*Wireless network, Signal and image processing.*	*menaka.govindaraj@gmail.com*	-
Pavan Ram P	*Student*	*Networking*	*pavanramofficial@gmail.com*	-
Rithul M	*Student*	*Embedded systems*	*rithul0206@gmail.com*	-

IOP Conf. Series: Materials Science and Engineering **994** (2020) 012029 doi:10.1088/1757-899X/994/1/012029

Multi Path Routing Technique Based RRP for P2P Network

Mary Subaja Christo1, R.M.Bommi2, K.Mariyappan3
1Associate Professor, Department of Information Science and Engineering, Faculty of Engineering Technology, Jain University, Bangalore, India.
2Associate Professor, Mechatronics Engineering Department, Chennai Institute of Technology, Kundratur, India.
3Assistant Professor, Department of Computer Science and Engineering, Jain University, Bangalore, India.

sc.mary@jainuniversity.ac.in

Abstract. All articles *must* contain an abstract. Peer to Peer (P2P) network is useful for developing applications and downloading files from the internet. In the existing P2P networks, communication is initiated and supported by peers. Security issues constitute the main challenge in the design of reliable communication techniques for P2P networks. In the past, Rumor Riding Protocol was used in distributed systems for performing coordination and communication. However, the existing Rumor Riding Protocol is used flooding technique to forward the response to the initiator it may leads high network traffic to avid this I have Enhanced Multi path forwarding technique to Rumor Riding Protocol and also enhancement with security features to make them suitable for downloading applications in P2P networks. In this method, it enhances the Rumor Riding Protocol with security features by including a Cryptographic Puzzle, a Challenge Question scheme using trust modeling and an intelligent beta reputation model for secured communication in P2P networks. For this purpose, five new techniques are proposed in this research work, all of which aim at enhancing the security of communication using Rumor Riding Protocol in the P2P networks.

1. Introduction

In distributed systems, it is necessary to perform client server communication for coordinating distributed processing. In such a scenario, a single coordinator monitors a number of participant nodes. There are situations in which a node has to act as both the client and the server (Takeda et al 2008). Such networks are called Peer to Peer (P2P) networks in which the P2P computing feature provided by the nodes are useful in resource sharing among the computers. Here, the resources may include processing power, knowledge of distribution of information, distributed disk storage and committed protocols for coordinated transaction processing in distributed databases. In a P2P networking system, the computers which participate in the distributed processing can act not only as clients but also as servers and hence the transactions can be started by any node while the participating nodes coordinate with the coordinator node for concurrent storage and retrieval of data. In a database scenario, concurrency control and recovery protocols are already available for both the centralized database and the distributed database systems. The same concept is used in P2P networks with a slight variation. This is useful for file transfer, transaction processing and distributed query processing.

In distributed systems, many facilities are provided for effective storage, maintenance, communication and retrieval of data. Among them, file sharing is one of the most important features that are highlighted by many users who are in the process of developing their applications using the P2P networking systems. In this scenario, the P2P network performs effective communication by the introduction of new protocols and also by the use of the existing protocols for network communication (Takeda et al 2008). In P2P networks, the resources, namely, CPU, memory and registers of nodes can be used effectively for speedy processing of distributed queries. In addition, it helps to perform file sharing to enable sharing of the media files such as e-books, e-learning materials, music playing systems, movies, computer games to enable cost reduction for all users. The nodes are also called peers. In P2P, networks are interconnected through local area networks, wide area networks and also the internet. Moreover, the existing Peer-to-Peer file sharing technologies have evolved through many different design stages based on the earlier networks such as Napster and also the other recent models such as Bit Torrent protocol. According to Schlosser and Kamvar (2003), many factors have contributed to the development of new protocols for effective file sharing in P2P networks. In addition, other factors, namely, digital contents, network bandwidth, storage and access techniques, text processing capabilities, image, audio and video processing techniques have contributed to the popularity of P2P networks by allowing effective file sharing.

1.1. Unstructured Peer to Peer Network

In a structured P2P network, the network has a centralized topology. On the other hand, the unstructured P2P network has a distributed topology in which the tasks to be carried out are divided among the peers in the network. The peers in a P2P network have equal privileges and hence all the nodes are responsible for the increase in the overall performance of the network. In such a scenario, the peers must get linked with various heterogeneous networks including internet and hence they require a highly efficient cooperation other networks in the network configuration.

In a Peer-to-Peer (P2P) network, the functioning of the network is dependent entirely on the client's computation power instead of its own computation power. They need not have a fixed network topology and the links can be either wired or wireless. Therefore, the clients in this network are considered as peers in the network with responsibility for coordination in communication. Hence such nodes perform a set of operations in order to maintain their link with other nodes instead of using a central server (Baptiste Pretre, 2005).

The unstructured P2P network has gained great importance when compared to other types of structured networks due to their potential for effective carrying out downloading applications. Therefore, many researchers in the area of computer networks have focused their attention on the area of P2P networks and developed new protocols which suit the requirements of the unstructured P2P networks. One of the major problems in the P2P networks is the presence of requirements which need application of the decentralized search techniques. In such systems, the existing search algorithm are highly time consuming and the search is not always successful. Therefore, it is necessary to propose a better representation and search algorithms which can perform parallel search and retrieval from distributed systems in an efficient way.

2. Problem Statement

In RR protocol follows 3 Phases they are Key generation, Query Issues, Query Forwarding in key generation phase we have used ASE algorithm to encrypt the message and the initiator forward the query to the responder basically RR protocol follows non path based techniques which is used to increase the anonymity of the message . The initiator follows query generation method to follow the query to the network. The intermediate node receive the two keys and act as the intermediate agent which is used flooding technique to forward the data the responder also respond use the flooding methods so the initiator may receive multiple response for a single request which is raised heavy network traffic so the performance of the network may get slow down . To avoid this we can use Multi path routing technique to forward the data.

3. Proposed Method

Multi Path routing protocol is important technique in Routing method. The routing method has 2 protocols they are Single Path Routing protocol and Multipath Routing protocol. In single path we have used RIP its known as Routing Information Protocol here the nodes information is passed through the Routing table which has the name of the node and IP address of the node based on this table we can understand the status of all the nodes in the network. In case of Multi cast Routing method follows Protocol Independent Multicasting (PIM) basically it follows two mode operations they are dense mode and another one is sparse mode. Both the modes are used to forward the data to group of destination. Basically these nodes are used shared tree function here the data is send to the router of shared tree and passed to the next node of the tree which nodes are not interested to belong this group they can send the prone message to the source router now that particular node may remove from the multi cast forwarding group.

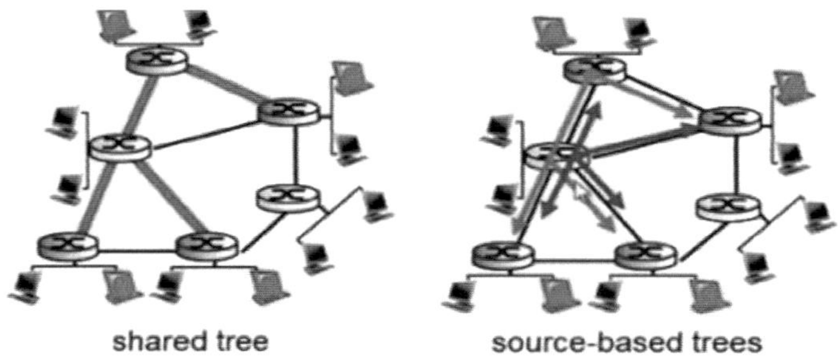

Figure 1. *Shared tree Versus Source- based tree*

Figure 1 described the features of shared tree and Source based tree. In shared tree all the routers are connected to the source routers in case of source based tree all the routers are connected to the individual parent routers.

Algorithm1:

Step 1: In Rumor riding protocol the initiator sends the query to the peer to peer network with two key values they are 1. Cipher key and Rumor Key.

Step 2: The Sender who is received the both the key values and decrypt the Rumor values with help of cipher key.

Step 3: After becoming intermediate node its forward the Rmor key the Reaming nodes in network using Multicast forwarding method.

Step 4: PIM protocol is used to prune the nodes which nodes are not interested to receive the information from the network.

Step 5: Finally all we can reduced the number of nodes from the network.

4. Performance and Results

In our experimental result we have used NS2 simulator to show the performance. Here Our proposed system shows High performance on Multi cost based RR protocol than the Flooding with RR protocol. We have taken 3 parameters they are packet delivery ratio, through put and delay. Our proposed method shows high packet delivery ratio and high through put values in case of delay has decreased.

4.1. Initiator as Repulsive Node

In our experiment we are varying the time as 5, 10,15,20,25 and 30sec. Figure 2 shows the results of delivery ratio by varying the time from 5 to 30 in Initiator attack with RR and RR protocols. When comparing the performance of the two protocols, Attack with RR protocols shows poor performance of peer to peer network.

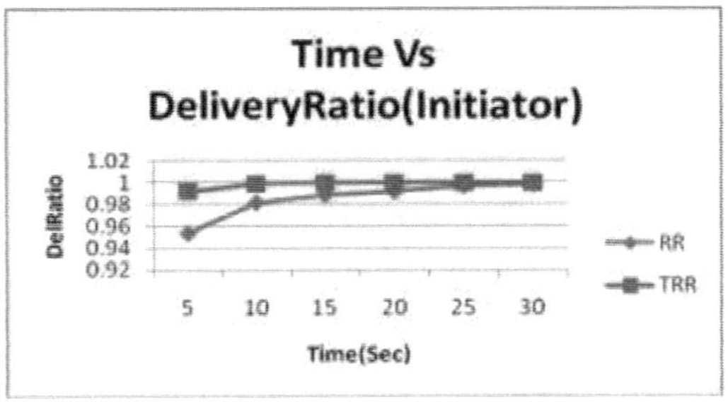

Figure 2. Delivery Ratio of RR Versus TRR

4.2. Sower as Repulsive Node

Varying the time as 5, 10,15,20,25 and 30sec Figure 3 it show the results of delay ratio by varying the time from 5 to 30 in case of packet delivery ration the X access show time in seconds and Y axis's shows the ration of delivering packets in case of throughput and in the graph X axis shows time in seconds and Y axis shows the successful received packet at a particular time (MBPS)

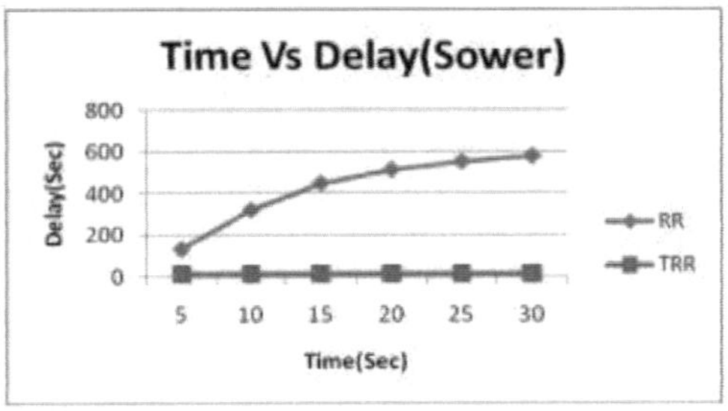

Figure 3. Delay Ratio of RR Versus TRR

4.3. Sower as Repulsive Node

Varying the time as 5, 10,15,20,25 and 30sec Figure 4 it show the results of through put ratio by varying the time from 5 to 30 in case of packet delivery ration the X access show time in seconds and Y axis shows the ration of delivering packets in case of throughput and in the graph X axis shows time in seconds and Y axis shows the successful received packet at a particular time (MBPS).

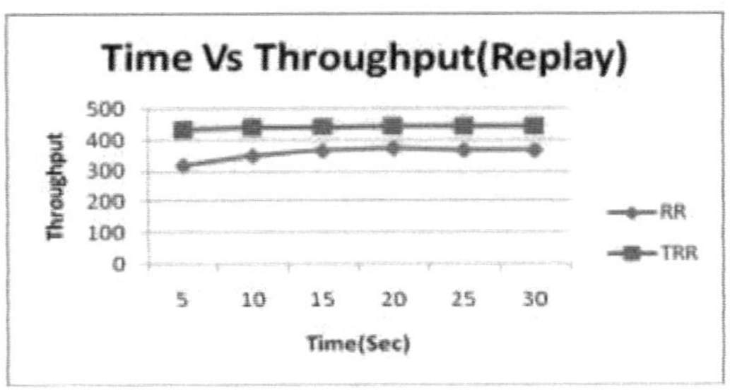

Figure 4. Throughput Ratio of RR versus TRR

5. Conclusion

In P2P network without any server all the nodes can easily send or request the file from one node to another node. RR method is consider as good protocol .for send or receive the file but here many attackers will act because every node hide their authentication to overcome we have proposed trust and cryptographic based authentication. First we should authenticate the sender and next we should authenticate the receiver node they can mutually authenticated by each with help of mutual authentication.

Reference

[1] Takeda A Chakraborty D Kitagata G Hashimoto K Shiratori N 2008 *A new scalable distributed authentication for P2P network and its performance evaluation* WSEAS Transactions on Computers. 7(10):1628-37

[2] Cheng W, Tan Z. 2013 *Correlation trust authentication model for peer-to-peer networks. Advanced Materials Research* 2013 Sep: 2237-42

[3] Arulkumar CV Jeyakumar K Malarmath M Shanmugapriya T.2012 *Secure communication in unstructured P2P networks based on reputation management and self- certification.* International Journal of Computer Applications ,44(15):1-3.)

[4] Szytula Mangi FA Memon I Jamro DA Memon MH Basit MA 2013,*The rumor riding anonymity approach for decentralized Peer To Peer systems* IJCSI. 10(3):180-7

[5] Devika P Ponmaga RS 2012 *Bruit bait: Peer To Peer systems* IJCER

[6] Gupta R, Singh YN.2015, *Reputation aggregation in Peer-toPeer network using differential gossip algorithm.* IEEE Transaction on Knowledge and Data Engineering. 27(10):2812-23.

[7] Liu Y Han J Wang J 2011 *Rumor riding: Anonymizing unstructured Peer-to-Peer systems*. IEEE Transactions on Parallel and Distributed Systems. 22(3):464-75

[8] Lu L Han J Hu L Huai J Liu Y Ni LM 2007 *Pseudo trust: Zeroknowledge authentication in anonymous Peer-to-Peer.* IEEE Transactions on Parallel and Distributed Systems.1325-37.

[9] Wang L.2006 *Attacks against Peer-to-Peer networks and counter measures*. Seminar on Network Security; 2006 Dec. p. 1-6

[10] Valarmathi R Kadhar Nawaz GM.2015 S*ecuredata transfer through audio signal with LSA*. Indian Journal of Science and Technology. 8(1):17-22

[11] Dhanagopal R and Muthukumar B 2019 A model for low power, high speed and energy efficient early landslide detection system using IoT *Wireless Pers Commun* https: //doi.org/10.1007/s11277-019-06933-7

[12] Delafrooz N, Farzanfar E. 2016, *Determining the Customer lifetime value based on the benefit clustering in the insurance industry*. Indian Journal of Science and Technology. 9(1):1-8

Gamut sensing in cognitive radio networks: A brief survey by using the concept of design thinking

K Ranjani[1], **A Rasheedha**[2] **and R Suresh Kumar**[3]
[1]Assistant professor, Department of EIE, SNS College of Technology, Coimbatore, TamilNadu, India
[2]Assistant professor, Department of BME, SNS College of Technology, Coimbatore, TamilNadu, India
[3]Associate Professor, Department of ECE, Chennai College of Technology, Chennai, TamilNadu, India

ranjani.ece.snsce@gmail.com

Abstract. With an unprecedented intensification in mobile customer and mobile multimedia services radio telecommunications in general the need for more bandwidth is escalating every day. To accommodate such a large amount of wireless technologies huge bandwidth is needed. This necessitated the need to address spectrum scarcity as the spectrum is a scarce resource. But it is evident from many gamut measurement studies that much of the gamut is used in an inefficient way and its fullest potential is not exploited. This gives rise to the technology of cognitive radio networks. Having said this let us try to brief the operation of intelligent radio networks (CRN). The overview of this paper presents a detailed account of the concept of gamut sensing which a crucial step in CRN is. We address the different performance metrics related to spectrum sensing and present a brief idea of some of the research activities that are related to spectrum sensing by different researchers.

1. Introduction

In the USA, the Federal Communications Commission (FCC) is the regulatory body that assigns spectrum to licensed users. The drastic growth in the count of wireless communication technologies and the increase in the count of devices that make use of the wireless communications concept mandate more gamut resources. Most of the gamut is licensed and used for communications by companies who have paid for the spectrum. But many of the spectrum measurement statics pronounce that the spectrum that is allocated for a particular user is not fully utilized. Most parts of the gamut remain idle generally over long period and are underutilized. Particularly those assigned for television telecast remain misspent which results in gamut wastage. This is overall scenario in all parts of the world.

The above discussion paved the way for developing technologies that could make use of the licensed gamut bands by other users when the authorization customers are not using the band. This is the concept of design thinking in which the gamut sensing is the iterative process that teams use to understand users, challenge assumptions, redefine problems and create innovative solutions to Cognitive radio Networks with an acronym CRN and it involves three phases emphasize, ideate, define and it is most useful to tackle problems that are ill-defined or unknown. In user experience (UX) design, it's crucial to develop and refine skills to understand and address rapid changes in users' environments and behaviours. The authorization customers are known as the Primary users and the denial customers are termed secondary users or cognitive users. It is to be understood that the secondary users that have not paid for the license can gain access to the authorized spectrum only on a temporary basis and the moment the primary user returns to transmit in its licensed band the secondary user will have to leave that particular gamut band and look for another gamut band to continue its transmission. Design teams use design thinking to tackle ill-defined/unknown problems because they can reframe these in human-centric ways and focus on what's most important for users. The greatest challenge in the cognitive radio networks is the sharing of the most licensed spectrum in a way of not interfering with the other primary users' transmission. The temporarily unused gamut that is licensed for an authorized user is known as the gamut hole or white space [1].

The paper is planned as follows. In section 2 a short discussion of cognitive radio network components and its working is presented. In section 3 the different phases of cognitive radio technology operation that are termed as cognition cycle are addressed. In section 4 the definition and various types of gamut holes are presented. In section 5 the gamut sensing aspect which is crucial to the performance of the cognitive radio networks is detailed. In section 6 a general model for gamut sensing used by most of the spectrum sensing approaches are presented. In section 7 sensing schedule paradigms are briefed. Section 8 presents the various performance evaluation metrics of gamut sensing. In section 9 some of the research activities carried out in the area of gamut sensing are briefed.

2. Cognitive radio network components and working

The intelligent radio network consists of two major elements namely the primary network and the secondary network or the cognitive components. The Primary elements are licensed customers and they have the top priority in gaining access to the network. The secondary components are frequency-agile radios that are equipped with the characteristic to sense the environment to detect the gamut holes and adapt its transmission parameters to make transmission in the gamut holes [GH].

Cognitive capability is the radio technology captures information about the radio environment it is working. This could be achieved by the use of highly sophisticated methodologies like autonomous learning and action decision to account for variations that occur in the radio environment in temporal and spatial domains. Dynamically programming the radio elements to meet the demands of the radio environment is termed as reconfigurability. In the Cognitive radio network paradigm, it includes programming the radio elements to perform transmission and reception in different frequencies and using different transmission access technologies.

In short, the intelligent radio network is centered on the concept of obtaining the best available spectrum to perform transmission. This is made possible by the use of cognitive capability and reconfigurability. The secondary user usually alters its transmission power level and modulation methods to avoid interference. The intelligent radio network may consist of two types of architecture namely infrastructure based CRNs and Adhoc Cognitive radio networks. The infrastructure based architecture consists of a central controller called the base station similar to the one in cellular network whereas the central entity is absent on Adhoc Cognitive Networks.

3. Cognitive cycle

The basic knowledge of the working of the intelligent radio network in the introductory section let us present a brief overview of the operations that constitute the cognitive cycle.

To adapt to the dynamic spectrum environment the intelligent radio network is in need of spectrum aware operations that are named the cognitive cycle [5]. The cognitive cycle consists of four major elements namely gamut sensing, gamut decision, gamut sharing, and gamut mobility.

The phenomenon in which the secondary user or the cognitive user continuously monitors the available spectrum band in its vicinity to detect the unused portion of the licensed gamut (ie) gamut holes is *gamut sensing*.

On identifying the spectrum holes the cognitive radio has to decide as to select which spectrum holes considering its QOS (Quality of service) requirement. This phenomenon is named spectrum decision. Apriori information of the primary users' activity is used in decision algorithms to incorporate dynamic spectrum characteristics. Spectrum decisions combine the processes of spectrum selection and route information. Gamut sharing is the phenomenon in which the spectrum is shared by multiple

Cognitive radios and coordinating the multiple CRs to prevent collisions. The process involves resource allocation so that interference to the primary users is neutralized. Gamut mobility is the phenomenon in which the secondary or the cognitive user, on the arrival of the Primary user needs to vacate the gamut and give it to the primary users and the CR user has to identify another spectrum hole that suits its QOS requirement to continue transmission.

4. The concept of spectrum holes

The basic need for the intelligent radio network operation is to identify the unused gamut bands that are not currently used by the authorised primary users. These unused portions are known as the gamut holes or white spaces.

The gamut holes may be named Frequency spectrum hole, Temporal gamut hole, and spatial gamut hole[6,7,8]. A frequency gamut hole is one that consists of a contiguous frequency band where the activities of the cognitive user cause no interference to the working of the Primary user.

A frequency band which a primary user does not occupy for a specific period of time is called a temporal gamut hole. A frequency band in a defined geographic region that is occupied by the primary user is the spatial gamut hole. If a cognitive user is in a region outside the defined geographic region it can gain access to this spectrum band.

Gamut holes are classified as black space, Gray space, and white space on the basis of spaces [2].

- Black spaces are the spectrum holes in which the dominance of high power interferers is found to dominate for a definite period of time.
- Gray spaces are the type of spectrum holes in which there is partial dominance of low power interferers.
- White spaces are the spectrum holes in which there is a complete absence of interferers. Broadband thermal noise and impulsive noise that belong to the category of natural noise is only present in the environment.

5. Gamut sensing

In cognitive radio technology, the unlicensed secondary users are allowed to access the limited licensed gamut band. Although the gamut can be shared only the licensed primary users are given priority to access the gamut. Hence the secondary users are mandated to monitor the activities of the primary user. Continuous monitoring is to be done to deal with the problem of interference and collisions which are crucial factors in determining the efficient performance of the cognitive radio networks. Gathering the results of the primary user's activity is the important task of gamut sensing. The results of gamut sensing are used by secondary users to gain access to the gamut. In a nutshell gamut sensing is the process in which the secondary users continuously monitor the activity of the primary user to detect the free bands which are called the gamut holes(GH).

Gamut sensing is a very important phenomenon for cognitive radio networks. It is very important for the cognitive radios to locate the gamut holes. Gamut sensing may be categorized as out of band sensing and in-band sensing. Out of band sensing is the sensing concept in which the secondary users look for the available unused portions of the licensed spectrum over a wide frequency range. In-band sensing is a concept in which the presence of primary users is detected by cognitive radio users in their course of monitoring the spectrum band. In-band sensing is done to mitigate the effect of interference.

Gamut sensing techniques basically aims to detect the presence or absence of the primary user. Matched filter discover [3] [4], energy discover [9], feature discover are some of the well-known gamut sensing approaches. Gamut sensing can also be carried out on the basis of the Eigen Values [10] and Momentum [11]. Autocorrelation based spectrum sensing [12] is also done in some research. Euclidian distance [13] [14] based spectrum sensing, wavelet [15] method also known as edge detection, cyclo stationary detection [16][17][18],waveform based detection, covariance-based sensing [19, 20, 21, 22] sensing based on fast sensing, learning/reasoning [23], sensing based on measurement [24], Multi taper sensing [16] [25], geolocation based sensing [26, 27] are some of the other advanced spectrum sensing methods.

The method in which the spectrum is sensed may be classified into different categories namely those that are in need of the details about the primary user in advance, those that are not dependent on the

knowledge of the primary users' signal, those that can sense signals of the primary users operating in narrowband and those that have the capability to sense the primary users' signal operating in broadband [28].
In short, the gamut sensing consists of three functions namely Primary user detection. sensing operation and cooperation.
Primary user detection involves the cognitive radio user to collect information about the local radio environment. Having gained the knowledge of the location information of itself and its neighbors the cognitive radio user arrives at a decision as to the presence of primary user transmission and hence identifies the current gamut knowledge. Sensing control involves every intelligent radio user to participate in sensing the dynamic radio environment.

6. Sensing schedule
Sensing schedule [29] may be defined as the phenomenon that combines the concept of when to sense the channel and what means need to be considered for sensing the gamut. The sensing schedule is also a very important factor that contributes to the efficient performance of a cognitive radio network.

7. Conclusion
In this paper, we have stressed the importance of the spectrum sensing process in cognitive radio networks. We have given a detailed description of the concept of gamut holes and a generic spectrum sensing model is presented. We have also presented a note on the performance evaluation metrics of spectrum sensing and a description of a few of the many research activities that are carried out by different researchers in the area of spectrum sensing. This will give other researchers to figure out novel research problems in the spectrum sensing arena. Some of the different simulation tools available for simulating cognitive radio networks are OPNET, NETSim, MATLAB, NS2, NS3, CogNs.

References
[1] Haykin S 2005 Cognitive radio Brain-empowered wireless communications *IEEE Journal on Selected Areas in Communications* vol 23 no 2 pp **201–220**
[2] Mitola J 2009 Cognitive Radio Architecture Evolution *Proc. IEEE*, vol. **97** no 4
[3] Axell E Leus G Larsson E and Poor H 2012 Spectrum Sensing for Cognitive Radio State-of the-Art and Recent Advances *IEEE Signal Processing Magazine* vol 29 no 3 pp **101 116**
[4] Stevenson C Chouinard G Lei Z D Hu W D,Shellhammer S and Caldwell W 2009 IEEE 802.22:The first cognitive radio wireless regional area network standard *IEEE Communication Mag* 47(1),**130**
[5] Tarek Mohamed Salem Sherine M, Abd El-Kader Salah Abdel-Mageid and Mohamed Zaki 2014 *Cognitive Networks: Applications and Deployments*
[6] McHency M A Tenhula P A McCloskey D Reberson D A and Hood C S 2006 Chicago spectrum occupancy measurements and analysis and a long-term studies proposal protocol *Proceedings of the 1st International Workshop on Technology and Policy for Accessing Spectrum*, Boston, MA
[7] Ma J, Li G and Juang B H 2009 Signal processing in cognitive radio *Proc. IEEE*, vol. 97, no. 5, pp. **805–823**
[8] Tandra R Sahai A and Mishra S 2009 What is a spectrum hole and what does it take to recognize one? *Proc IEEE* vol 97 no 5 pp **824–848**.
[9] Poor H V 1994 *An Introduction to Signal Detection and Estimation*, 2nd ed. Berlin: Springer Verlag
[10] Zeng Y H and Liang Y C 2009 Eigenvalue-based spectrum sensing algorithms for cognitive radio *IEEE Trans Commun* 57 1784–1793
[11] Salahdine F Kaabouch N and El Ghazi H 2017 *A Bayesian Recovery with Toeplitz Matrix for Compressive Spectrum Sensing in Cognitive Radio Networks* pp. **1–13**

[12] Reyes H Subramaniam S and Kaabouch N 2015 A Spectrum Sensing Technique Based on Autocorrelation and Euclidean Distance and its Comparison with Energy Detection for Cognitive Radio Networks *Elsevier Computers & Electrical Engineering Journal*, vol. 52, pp.**319-327**

[13] Salahdine F El Ghazi H Kaabouch N and Fihri W F 2015 Matched Filter Detection with Dynamic Threshold for Cognitive Radio Networks *International Conference on Wireless.Networks andMobile Communication*

[14] Molisch A F Greenstein L J and Shafi M 2009 Propagation issues for cognitive radio *Proc. IEEE* vol 97 no 5 pp **787–804**

[15] Yucek T and Arslam H 2009 A Survey of Spectrum Sensing Algorithms for Cognitive Radio Applications *IEEE Commun. Surv. Tutorials*, vol 11 no 1 pp **116–130**

[16] Haykin S 2009 Spectrum sensing for cognitive radio *Proceedings of the IEEE,* vol.97, pp. **849-877**

[17] Ariananda D D Lakshmanan M K and Nikookar H 2009 Spectrum sensing techniques for cognitive radio, cognitive radio and advanced spectrum management *COART2009, second International workshop* pp 74-79

[18] Tevfik Y and Arslan H 2009 Survey of spectrum sensing algorithms for cognitive radio application *IEEE Communication Surveys &Tutorials*, vol. 11, No. 1, First Quarter

[19] Kumar K.S, Saravanan R and Muthaiah R 2013 Cognitive Radio Spectrum Sensing Algorithms based on Eigenvalue and Covariance methods. *Int. J. Eng. Technol.* 5, **595–601**

[20] Zeng Y and Liang YC 2007 Covariance based signal detections for cognitive radio *Proceedings of the 2007 2nd IEEE International Symposium on New Frontiers in Dynamic Spectrum Access Networks*, Dublin, Ireland,17–20 pp. **202–207**

[21] Zeng Y and. Liang Y C 2009 Spectrum-sensing algorithms for cognitive radio based on statistical covariances *IEEE Trans. Veh.Technol.*, vol. 58, no. 4, pp. **1804–1815**

[22] Berthold U Fu F and Jondral F 2008 Detection of spectral resources in cognitive radios using reinforcement learning *Proc IEEE Dynamic Spectrum Access Networks* Chicago, USA pp. **1–5**

[23] Wang X Ho P and Wong A 2008 Towards efficient spectrum sensing for cognitive radio through knowledge-based reasoning *Proc. IEEE Dynamic Spectrum Access Networks* Chicago U.S.A., pp.**1–8**

[24] Willkomm D, MacHiraju S, Bolot J and Wolisz A 2008 Primary users in cellular networks: A large-scale measurement study *Proc IEEE Dynamic Spectrum Access Networks* Chicago USA pp.**1–11**

[25] Zhao Bing and Guo Lili 2009 Research of spectrum detection technology in cognitive Radio *International conference on network security, wireless communications and trusted computing.*

[26] Tandra R and Sahai A 2008 SNR walls for signal detection *IEEE Journal on Selected Topics in Signal Processing* vol 2 no 1 pp **4–17**

[27] Second report and order and memorandum opinion and order In the matter of unlicensed operation in the TV broadcast bands 2008 *Federal Communications Commission*

[28] Jide Julius Popoola and Rex van 2012 The performance evaluation of a spectrum sensing implementation using an automatic modulation classification detection method with a Universal Software Radio Peripheral **Elsevier**

[29] Lu Xiangwei Zhou, UzomaOnunkwo and Geoffrey Ye Li 2012 Ten years of research in spectrum sensing and sharing in cognitive radio *EURASIP Journal on Wireless Communications and Networking*

[30] Dhanagopal R and Muthukumar B 2019 A model for low power, high speed and energy efficient early landslide detection system using IoT *Wireless Pers Commun* https: //doi.org /10.1007 /s11277-019-06933-7

Effective timer count scheduling with spectator routing using stifle restriction algorithm in manet

[1]**Mr. R.P.Prem anand,** [2]**Dr.A.Rajaram,**
premanathphd1@rediffmail.com ,gct143@gmail.com
Department of Electronics and Communication Engineering,
[1]Research Scholar, SCSVMV University, Kanchipuram- 631 561
[2]Professor, Department of Electronics and Communication Engineering,

EGS Pillay Engineering College, Nagapattinam – 611002

Abstract: Mobile nodes are provides unstable communication among mobile nodes, since nodes energy levels are abnormally changed. So, difficult to manage those node activities, packets are missed to collect by target node. Multi packet is broadcasted by sender node is not easy to manage frequently, since it does not have high capacity nodes. It reduces packet integrity rate, and network lifetime. In proposed algorithm is Timer count scheduling with spectator routing (TCSSR) scheme. Spectator monitors the packet transmission use of timer count. Clustering method is used to main stability for all mobile nodes available in coverage area. Stifle restriction algorithm is used to maintaining information about communication nodes, continuous sequence used to limiting sharing of the residence state of a communicate shields. Effective path selects for communication with the support of timer count report is analyzed. It enhances packet integrity rate, and network lifetime.

Keywords: Timer count scheduling with spectator routing, Stifle restriction algorithm.

I. Introduction

Beside with the existing expansion of wireless methods, self-sufficient disseminated networks else MANET-mobile ad hoc networks that does not need sink node, should be achievement notice. Mobile networks are ordinary to worn in different sectors, such as for ranges where it is not easy to use links, throughout calamity, also to protect packet transmission in ITS-Internet technology services [1]. Though, because the topology of the mobile network alters the coverage level, those networks are updatable, so it minimizes the packet delivery ratio. Because the wireless mobile workstation node the stage functions of equally intermediate nodes with finish communication, a huge quantity of energy is used for communication. It must be a difficulty, while wireless mobile workstation relies on energy level [2].

Mobile network investigate has consequently prioritized the aim of stabilize the network and conserve energy level. Efforts to keep energy in communication by applying the increasing Ring investigate technique [3]. Increasing Ring investigate is applied to avoid the raise of request packer from overflow from the initial node while penetrating for a path in ad hoc network environment. While ad hoc routing is Time to Live use the quantity of hop as a situation, which cause traffic for each node and residual energy level as the situation ranges by the increasing Ring investigate to minimize traffic for communication also manage energy consumption.

Though, while the nodes are congested, a huge amount of packets are created, also it is not easy to contract with remaining nodes by the communication scheme unaided [4].

Consequently, to preserve energy also tries to manage energy consumption of each node that is known as the topology manage scheme. While mobile nodes are stuffed, all nodes minimize the communication energy to decrease the quantity of neighbouring nodes. Thus the direction-finding in the network is steady and all nodes' energy usage is minimized [5]. While managing message energy, event known as individually connected nodes

take place. In final unknown workstation are improved and packet drop made further regularly, disturbing packet transmission. A communication scheme is considered, which use individually connected nodes is presented [6]. While pointed for packet transmission path, Packet losing region is managed adaptively uses individually connected nodes to improve connectivity ratio. Though, because to managing packets are dropping while towards the back analyzing, the Entire network traffic occurrence of the path finding is maximum compared with previous communication technique while each node contain similar energy level with double side connection [7]. Communication technique is used to designing a path reject individually connected nodes to minimize packet drop. It also presents communication technique that estimates the individually connected nodes by deduction the nodes travelling from to any area of network, by means of the request packet and position details [8] to keep away from network. Though, it is uncertain while moreover of these technique is efficient at power management. An optional method, snooze nodes which execute separately by means of casual likelihood management in an asynchronous network is launched [9]. Individually connected node issues are not generated, because the nodes inactive. Energy saving is important. Though, since of the likelihood managing, it is a danger which an node will not be bright to accept information while node is inactive [10].
Rest of the paper is designed as given below. Part II denotes a related works. In part III, to give the details of proposed Incentive Sorted Path selection Scheme (ISPS) to obtain efficient communication in wireless sensor network also identify and remove selfish nodes in cluster. Part IV provides simulation performance results examination obtained under various parameters. Last part V concludes the paper includes future work.

II. Related Works

Liu, Jia, et al., [11] presents method merge the fixed-point with quasi-birth-and-death operation, also surrounded Markov chain to method the restrictive sharing of the tenancy state of a communicate shield. It relates the engrossing Markov chain assumption to distinguish the packet delivery procedure, for a total hypothetical structure is launched for end to end Delay investigation. Through the support of this structure, to obtain a common and accurate appearance for end to end delay depends on the modelling to packet queue latency, and receiving latency. Then reveal the request of our structure, investigate various scheme with different MAC rules to indicate minimum packet latency compared with existing method in experimental result.
Njilla, Laurent Yamen, et al., [12] proposed a packet-forwarding Issuing scheme with a negotiation game, where an authority works as group head and starts a bargain diversion subsequently, judge the option of having some cluster of nodes indicates misbehaviour and join together to undermine the mobile network. Survey the issues by result the best NE-Nash Equilibrium techniques of the arbitration diversion. Simulate the consequence of the combination of misbehaviour node in a mobile environment is easily identified by proposed distinguish with existing scheme.
Alocious, et al., [13] proposed the MAC layer misbehaving nodes on routing scheme. For those intrusions is cause by misbehaving nodes by violating their back off counters connected with MAC routing scheme. Consequently, these attacks are simulated analyse and estimate the collision on the network presentation and steadiness in mobile network. Experimental output indicates intrusion at MAC layer could considerably involve the network transmission rate, and packet loss rate. It achieves higher throughput by reject misbehaved nodes, also intrusion with the mean of whole damage that allow only minimum transmission rate for misbehaved nodes which finally led to minimize delay of the network.
Chen, Haiming, et al., [14] proposed DS-MMAC-Delay responsive Multi path MAC procedure used for Each network that obtain high packet Transmission rate and jump minimum delay for data delivered to the intermediate nodes in network. The uniqueness of the routing is that a well-organized disseminated time allocation preparation and path project scheme is used to merge with the process of path organization that take the route allocation cost into description and minimize packet latency to gather the necessary latency bound of all data transmission. Experimental output shows minimum delay, higher data reliability and packet delivery ratio.
Ravi, G., et al., [15] present OHQS-Optimized Hyper Quorum method arrangement of signal period is modified to reduce in importance the energy used by all node in the network environment. Additional OHQS is improved to create the latest encouragement organization effort successfully. It performs the nodes with various RPI-Repetition Pattern Interval transmits well with improving transmission rate. It obtains better traffic reduction among the delay and lesser energy usage. This present scheme obtains the nodes a higher probability to determine some remaining nodes that create the packet sharing perfectly. Experimental output indicates higher energy reduction, transmission rate.
Chandavale, Anjali A., et al., [16] proposed Modified Iterative scheme for identifying misbehaving node for Delay Tolerant Networks is depends on grouping of Rates as per their ratings to process contributor. The Rates are characterized as minimum probability, centre probability, and maximum probability depends on score given to process contributor. Simulation output shows the quantity of negative true rate and enhancement in process

standing accurateness. It identifies malicious node better than existing ITRM method. It also obtains accurateness of identifying misbehaving nodes.

Kaur, Satinder, Ret al., [17] present mobile network based on life span of network which additional based on several merits like, energy, speed, thickness protection of the network. Because mobile network is motorized with battery energy that is forever incomplete, constructing an energy maintenance network is forever a main mission. A power based energy efficient technique should estimate the energy of the node and the separate the process accordingly, it reduce energy usage of each node and improve network lifetime.

Sharma, Anupam Kumar, et al., [18] present the nodes are used in mobile network should stable else movable is only a network in that are acts as a host or transmitter. Communicating scheme is used in MANET is broadly separated in three group practical, Thoughtless, and combined routing scheme. Experimental output of present method is better compared with previous methods, such higher connectivity, and minimum delay. Dissimilar count of communicating nodes is reduced.

Amouri, Amar, et al., [19] present scheme provides an hierarchical arrangement which rejects using grouping scheme and, as an alternative, successively operates the capability to snuffle each message broadcasted by sender for particular coverage area of the node depends on position in the network environment. For that condition works as a PCH-pseudo cluster head organize data packets from its intermediate nodes in all Area or group. Present scheme obtains lesser overhead, and higher detection efficiency compared with existing method in experimental result.

Hayashi, et al., [20] Present latest scheme manages that Maintain the doubly connected nodes by selfish not only the Transmission energy but also the limitation of energy discovery depends on the amount of intermediate relay node. Sequence to reduce electric wave intrusion and the Time need to manage the technique, the present scheme organize the standard amount to within a area, quite than to a individual place. Experimental output shows the present scheme enhances the packet delivery ratio and reduces energy usage.

III. Overview of Proposed Scheme

In proposed method use a Timer count scheduling with spectator, reduce the time delay. Spectator closely monitors node resource utilization in different stage. The updated node resource utilization then applies probability method. In clustering some nodes cause some changes in network is an attacker. After detected attacks, all nodes go to communication and provide effective communication in MANET. To construct the Stifle restriction algorithm is applied to collect details about packet sharing nodes, uninterrupted sequence is made to restricting the sharing of data residence state of a communicate shields.

Figure 1 shows the block diagram of Timer count scheduling with spectator routing. Multi packets are broadcasted by source node using clustering scheme. Cluster head node organizes data packets from various routes. Timer count scheduling with spectator routing scheme, timer count is initiated to measure the time taken to broadcast particular data packets. Stifle restriction algorithm is designed in mobile network environment, it restrict the unwanted nodes behavior. Finally improves network lifetime and packet delivery ratio.

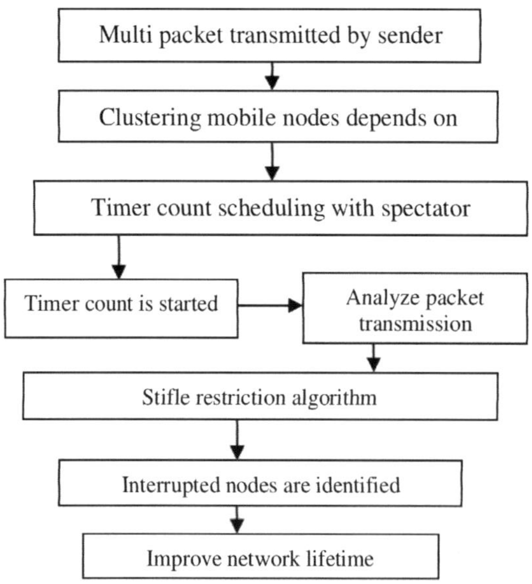

Figure 1. Block Diagram of Timer count scheduling with spectator routing

IOP Conf. Series: Materials Science and Engineering **994** (2020) 012031 doi:10.1088/1757-899X/994/1/012031

Timer count measurement is one of the best communication method, it analyze packet transmission time slot. It has stifle restriction algorithm is applied to detect the interrupted nodes available in routing path. This makes to obtain the efficient routing path. It reduce end to end packet latency, when timer count produce extract output for communication, which achieves higher transmission rate for that particular attack free route.

3.1. Multi packet transmitted by clustering nodes

In multi packets are broadcasted by clustering nodes, that analyze more areas is separated into four reasonable clusters depends on the position of mobile nodes, single cluster with non grouped areas has mobile nodes that are placed next to an inaccessible Base Station. Those mobile nodes use straight packet transmission to the Base Station nodes because the distance among that nodes and the Base station is small. Correspondingly, many ungrouped areas have mobile nodes that are placed next to the rechargeable base station. multi packet, single packet, and timer count.

– (1)

Those mobile nodes use perfect packet exchange to the base station that organize data packets, and transmit them to the Sink node. Some of the mobile nodes are placed away from out cluster, those nodes are ungrouped nodes that are identified and re-cluster those nodes efficiently. Each mobile node in a group area categorizes themselves into little clusters. In additionally all group, a scheduled node should be chooses group header node, since routing table contains entire information about communication. Furthermore, mobile nodes in these grouping areas use multi-hop connections for packet broadcasted to group header node.

$\sqrt{\ }$ – (2)

The Base Station is starting stage separates monitoring region called as clusters with a Cluster Head is arbitrarily choose all group. All nodes in a group have a similar possibility to be a cluster head. The Cluster Head is restructured in all communication tasks that are known as cycle of operation. The count of routine that neighbor node performs communication, in a particular cluster head nodes in mobile network. Related to low energy consumption cluster heads that all node in a group choose itself to be cluster head, all routing routine. Whether a mobile node is chooses to be a cluster head in routine, but does not choose a cluster head in the routine.

$\sqrt{\ }$ - (3)

$\sqrt{\ }$ - (4)

$\sqrt{\ }$ = - (5)

The possibility that the residual nodes develop into cluster head should improve in all routine, because there is a smaller amount of mobile nodes that are qualified to particular cluster to perform multi packet transmission. On the initialization of the primary routine, each cluster member nodes contains the same energy level, also contain the similar possibility to cluster head multi packet organization. Subsequent to the cluster choosing is depends on dual important metrics that are the available energy of a mobile node and the minimum remoteness among the mobile node and the rechargeable base node that is placed at the middle of the monitoring area. In initial routine, a neighbor node selects a chance quantity among minimum and maximum value. Whether the choose chance quantity for neighbor node is minimum than a before constructed limited value, then the mobile node turn into cluster head in the present routine.

3.2. Timer count scheduling with spectator routing scheme

Formerly groups are created also cluster head are choose, mobile node residue in an inactive condition and exit its two-way radio coverage analyzer. Mobile node starts the radio coverage transmitter simply while it contain packet should be transmitted also to the Base Station node that node in particular group. Control the radio coverage among start and stop conditions for mobile node have huge collision on the entire energy reduction of mobile network. For that nodes in all groups use present TCSSR method, sequence to aggregate the right of entry to the packet transmission path. It is very easy to monitor time usage that rejects broadcasting data packets based time division scheme, else overriding more energy for packet broadcasting medium as the condition in mobile network.

$$\rightarrow$$

In again grouping the out of range nodes that are depends on Timer count routing scheme, many nodes are need to distance the similar time allocation and broadcast its data packets as more distance with various conditions. By using best quantity of regularity band and packet loss is minimized that are translated to an important decrease of the packet broadcasting energy usage. Each cluster member nodes of a group broadcast its data packets to the equivalent cluster head that organize the data packets and broadcast them to the middle base station node. The middle base station node arriving message from cluster head nodes placed in that position, it organizes each data packets and broadcast them to the Base Station node. The Base Station node accepts packets from middle base station node, and the nodes that are placed in its area.

$$\rightarrow \qquad \iiint$$

Base station node collects the request message from base station node. Request packet is accepted by base station node, offers an acknowledgement packet, it retrieve the RSSI of all node to guess the connection superiority. Present routing method intend to execute data collecting with individual mobile node or base station node, there can be Dissimilar loss situation like unexpected disruption made due to blockage among the base station node and remaining inactive nodes else break down node. Such interruption may guide latency of acknowledgement else broadcast request packet from sender. To ensure undisrupted data organization, the present scheme performs communication to recognize available neighbor node to carry on packet broadcasting. The timer is planned depends on the packet generation and transmission rate for sender node.

$$\iiint \qquad \iiint \, /$$

Then execute packet aggregation, base station node is want to react node among the acknowledgement packet. The response is broadcasted by acknowledgement packet sometimes delayed, then it direct data transmission and packet loss that strength to perform communication. For this scheme, the mobile base station node organizes data packets about amount of data packets with packet size from all mobile nodes. To keep away from any intrusion, to calculate the acknowledgement time with support of timer count in all packet transmission provides constant time gap. To permit Traffic responsive packet organization and broadcasting, the acknowledgement time is chooses effectively consider current packet coming time.

Algorithm for Timer count scheduling with spectator routing
Step 1: For each nodes analyze paths
Step 2:
Step 3: sender forward multi packet to neighbor node
Step 4: Monitors time duration of particular path using timer count
Step 5: Find another path for packet transmission
Step 6: else
Step 7: *if {sender!= }*
Step 8: Single packet transmitted with minimum capacity nodes
Step 9: Group header aggregate data packets from members.
Step10: end if
Step 11: End for

3.3. Stifle restriction algorithm

Stifle restriction guarantee rearward outing scheme in network with mobile nodes, for control packets. Those packets are ensuring that the nodes available and can achieve packet exchange across mobile network environment. In universal, the broadcasting of control packets is done at convinced distinct time gap; though realize they want to hold up traffic, and

Susceptible packet organizing, in present scheme when performing route choosing should launch convinced conditions to broadcast data packets. Throughout data aggregation purpose, a single cast control packet is broadcasted from the base station node through the stable link connection. The inactive nodes request for data broadcasting reply with acknowledgement packet. To recognize the request packet also known as dual broadcast packet and acknowledgement packet, a flag taking out calculate is launched which indicates the position of all nodes in mobile network. While performing route identification, the base station node collect multi packet data from all nodes in the network, these nodes subsequently broadcast single cast acknowledgement packet to the mobile node.

– (9)

$$\iiint \quad / \qquad (10)$$

Packet aggregation method performs sender node classification. When guarantee that the base station node is under coverage area of the sender to guarantee trustworthy packet broadcasting else Packet aggregation at base station node. In packet size is measured based on byte flag
is launched to differentiate speed based data transmission and straight message packet transmitted. The flag one indicates the single packet transmission, when zero indicates multi packet broadcasting.

Stifle restriction algorithm
Step 1: for each path verify
Step 2:
Step 3: Node timer count exceed for transmission.
Step 4: Remove interrupted node
Step 5: reduce energy consumption
Step 6: else
Step 7:
Step 8: Node timer count not exceed for transmission.
Step 9: perform packet broadcasting on that path.
Step 10: Improve network Lifetime
Step11: End for
Step 12: End if.

This scheme supports to find interrupted nodes which are present in routing path, they are separated from routing. So, it easy to obtain efficient routing path and it improves network lifetime, packet delivery ratio. It also minimizes packet latency, and communication overhead.

Packet ID: Packet ID has an entire cluster based mobile node details, also cluster head placed position are mentioned.

Source ID	Destination ID	Multi packet transmitted by clustering nodes	Timer count scheduling with spectator routing scheme	Stifle restriction algorithm	Improving network lifetime
4	4	6	5	4	5

Figure 2. Proposed TCSSR Packet format

In figure 2: the present packet format is exposed. Now the source node ID field consumes 4 bytes and destination node ID field consumes 4 bytes. Third one is Multi packet transmitted by clustering nodes carries 6 bytes, clustering mobile nodes are provide multi packet transmission from sender to target node. In fourth field occupies 5 bytes. Timer count scheduling with spectator routing scheme is used to measure time taken to complete packet transmissions using timer count. In fifth have 4 bytes; Stifle restriction algorithm is launched to identify interrupted nodes in routing path. Final filed is Improve network lifetime, it carries 5 bytes and it only chooses perfect nodes to perform packet transmission.

IV. PERFORMANCE EVALUATION

a. Simulation Model and Parameters

The proposed TCSSR is simulated with Network Simulator tool (NS 2.34). In our simulation, 100 sensor nodes deployed in 1050 meter x 920 meter square region for 32 milliseconds simulation time. All sensor nodes deployed in random manner among the network. All nodes have the same transmission range of 250 meters. CBR Constant Bit Rate provides a constant speed of packet transmission in network to limit the traffic rate. DSR Dynamic source routing protocol used to provide effective communication path for packet transmission in clustering also identify selfish nodes in routing path. Indicate Table 1 network simulator initialization.

Table 1: Simulation Setup

No. of Nodes	100
Area Size	1050 X 920
Mac	802.11g
Radio Range	250m
Simulation Time	32ms
Traffic Source	CBR
Packet Size	150 bytes
Mobility Model	Random Way Point
Protocol	DSR

Simulation Output:

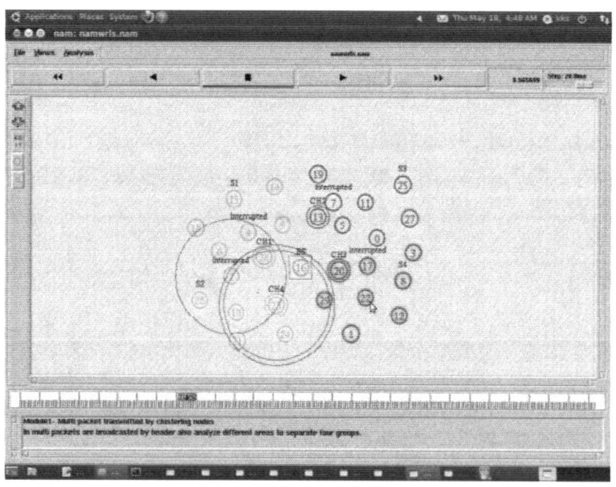

Figure 3: Proposed TCSSR Result

Simulation Result: Figure 3 show that the proposed TCSSR scheme performs packet transmission based on timer count, to analyze each packet transmission time slot, and achieve effective data aggregation compared with existing EDM [11] and GTA [12]. Present TCSSR contains Stifle restriction algorithm is used to separate the interrupted nodes from clustering to improve lifetime of network, and packet delivery ratio.

Performance Analysis

In simulation to analyzing the following performance parameters are using X graph in ns2.34.

End to End Delay: Figure 4 shows end to end delay is calculated by slot of time used for communication between sender and cluster head node, Stifle restriction algorithm to split the interrupted nodes in network environment. In proposed TCSSR method end to end delay is minimized compared to Existing method EDM, and GTA.

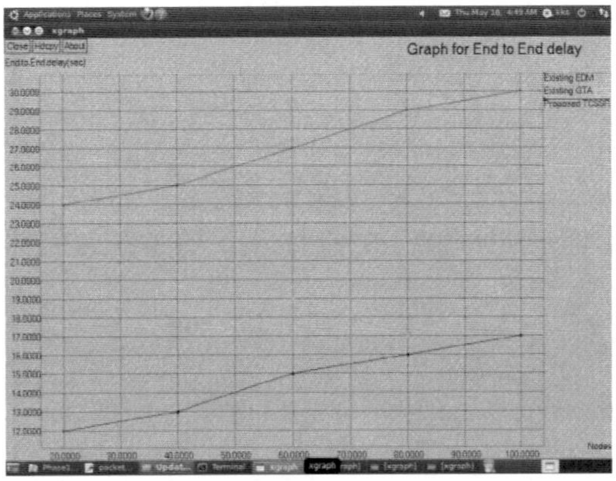

Figure 4: Graph for Nodes vs. End to End Delay

Communication overhead: Figure 5 shows communication overhead is minimized that cluster head node organize the data packet with efficient connection, the mobile node contain timer count measure every packet transmission time effectively. In proposed TCSSR scheme communication overhead is decreased compared to Existing method EDM, and GTA.

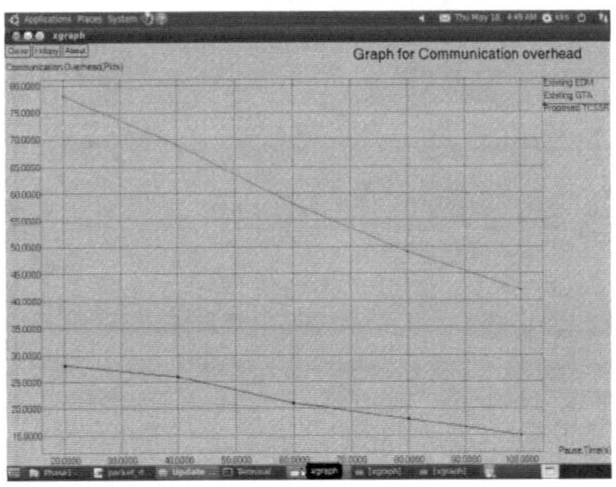

Figure 5: Graph for Pause Time vs. Communication overhead

Packet Delivery Ratio: Figure 6 shows packet delivery ratio is measured by amount of data packet received from amount of data packet sent with certain rate, Stifle restriction algorithm supports to remove interrupted nodes which are available in routing path. Generally transmission rate is set as 100, Present TCSSR method packet delivery ratio is improved compared to existing method EDM, and GTA.

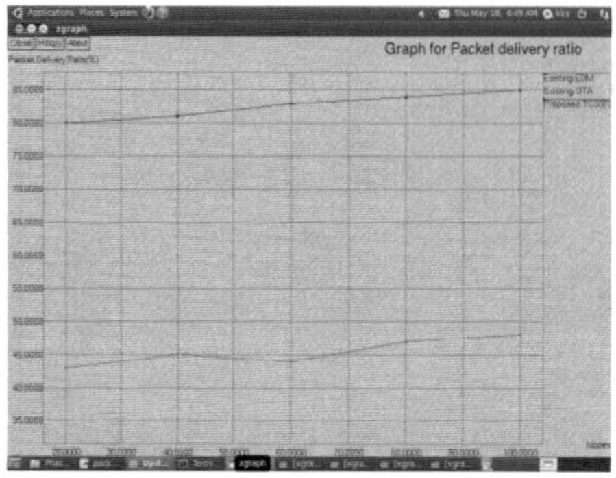

Figure 6: Graph for Nodes vs. Packet Delivery Ratio

Network Lifetime: Figure 9 show that Lifetime of the network is calculated by whole network performance, amount of energy and time taken to complete the transmission. Timer counts easy to achieve effective packet broadcasting, minimum time usage nodes are only choose to perform packet transmission. In proposed TCSSR method Network Lifetime is maximized compared to existing method EDM, and GTA.

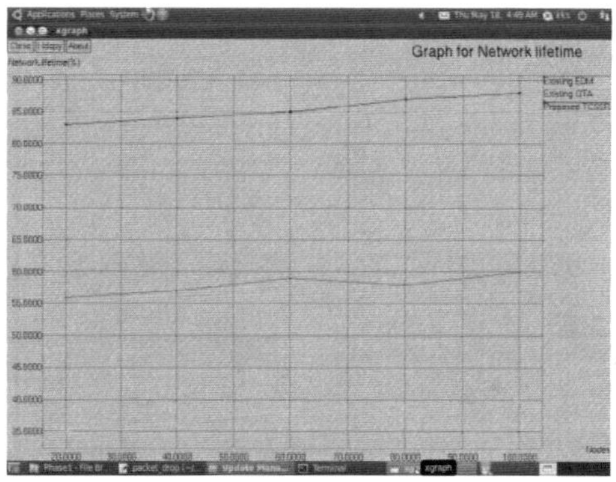

Figure 7: Graph for Nodes vs. Network Lifetime

Energy Consumption: Figure 8 shows energy consumption; amount of energy spend for starting node to ending node communication. In proposed TCSSR method analyze node frequency updates using Stifle restriction algorithm, energy consumption is reduced compared to existing method EDM, and GTA.

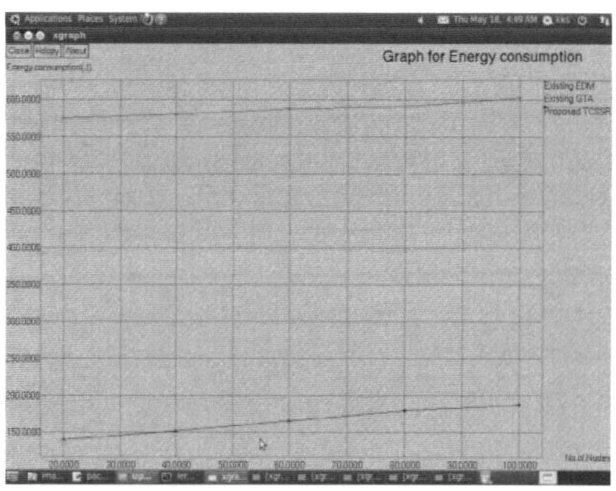

Figure 8: Graph for No. of Nodes vs. Energy Consumption

IOP Conf. Series: Materials Science and Engineering **994** (2020) 012031 doi:10.1088/1757-899X/994/1/012031

Packet drop rate: Figure 9 shows that data Packet loss is minimum it rejects the communication link among sensor nodes in cluster, also has Stifle restriction algorithm to remove interrupted nodes from routing path. In Proposed TCSSR scheme Packet drop rate is reduced compared to existing method EDM, and GTA.

(———)

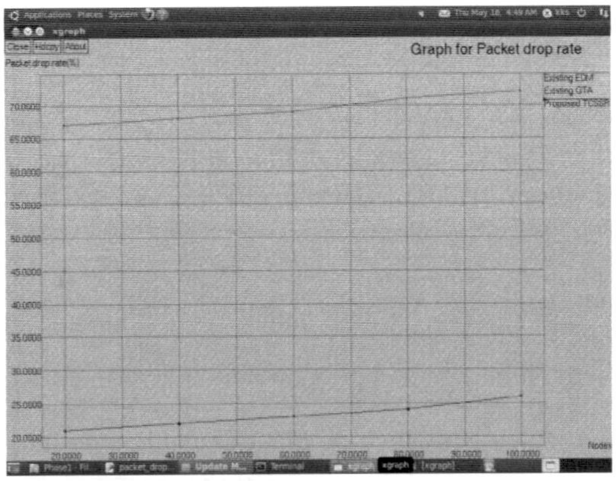

Figure 9: Graph for Node vs. Packet drop rate

V. Conclusion

Mobile nodes are unstable position node, they are updated its location from one end to another end in any place of network. For certain situation the nodes are go to out of coverage range that nodes are node joined to cluster. The unwanted nodes cause interruption, more time taken to broadcast single data packet, so minimized network lifetime. In proposed TCSSR scheme contain timer count to measure the time duration of each packet transmission from sender to neighbor node in routing path. Stifle restriction algorithm is constructed to separate the interrupted nodes from routing path, it also reduce packet latency. It improves network lifetime, and packet delivery ratio. In future work missed connection based re-clustering method to analyze several parameters.

References

[1] Yoshitugu Kawamura et.al 2007 Application of expanding ring search method on cost based routing protocol for MANET DTCOMO Symposium pp 238-244
[2] AODV RFC3561 www.ietf.org/rfc/rfc356I.txt
[3] Barbeau M Kranakis E 2007 Principles of AD HOC NETWORKING pp.I13-143 WILEY
[4] Kouji Yamada 2013 A study on topology control-aware MANET routing protocol Master Thesis, Graduate School of Engineering Gifu University Gifu Japan
[5] Fabian Fuchs Markus Volker Dorothea Wagner 2012 Simulation-based Analysis of Topology Control Algorithms for Wireless Ad Hoc Networks Design and Analysis of Algorithms pp 188-202
[6] Hiroaki MORINO Takumi MIYOSHI and Masakatsu OGAWA 2004 An efficient ad hoc routing protocol based on the relay control of route requests for handling unidirectional links IPSJSIG Technical Report 2004-M BL vol 30 pp 47-54
[7] Takumi MIYOSHI Shinsuke TERADA Koji HISAMITSU and Hiroaki MORINO Ad Hoc 2004 Routing Protocol with Area-Controlled Flooding Considering Unidirectional Links," IEICE Technical Report NS2004 vol 26 pp 5-8
[8] Yusuke FUKUI Masaki BANDAI and Takashi WATANABE 2007 Proposal and Investigation of Ad Hoc Network Routing Protocol Considering Unidirectional Links IPSJSIG Technical Report 2007-MBL vol 40 pp I13-120
[9] Yasuhiro TSUTSUI and Iwao SASASE 2005 A Stable Multicast Routing Protocol in Ad-hoc Networks with Unidirectional Links IEICE Technical Report CS2005-49 pp 73-78
[10] Yukinari HAYASHI and Michiko HARA Y AMA 2013 Electric power saving of Mobile Ad hoc Network's relay nodes by the transmission power control : Electric power saving that include reception available power IEICE Technical Report NS2013-40 pp 152-158
[11] Liu Jia et al End-to-end delay modeling in buffer-limited MANETs A general theoretical framework IEEE Transactions on Wireless Communications vol 15 pp 498-511

[12] Njilla Laurent and Yamen et al 2016 A game-theoretic approach on resource allocation with colluding nodes in MANETs Systems Conference (SysCon), 2016 Annual IEEE IEEE
[13] Alocious Chaminda Hannan Xiao and Bruce Christianson 2015 Analysis of dos attacks at mac layer in mobile adhoc networks Wireless Communications and Mobile Computing Conference (IWCMC) 2015 International IEEE
[14] Chen Haiming and Li Cui DS-MMAC 2016 A delay-sensitive multi-channel MAC protocol for Ambient Assistant Living systems China Communications vol 13 pp 38-46
[15] Ravi G and A Sakthi Saranya 2015 Optimized power management for multi hop mobile adhoc networks Communications and Signal Processing (ICCSP) 2015 International Conference on IEEE
[16] Chandavale Anjali A and Tejaswini P Chaure 2015 An approach to detect malicious node for Delay Tolerant Networks TENCON 2015-2015 IEEE Region 10 Conference IEEE
[17] Kaur Satinder RC Gangwar and Ranjit Singh 2015 A strength based energy efficient algorithmic approach in MANET Soft Computing Techniques and Implementations (ICSCTI), 2015 International Conference on IEEE
[18] Sharma, Anupam Kumar and Munesh Chandra Trivedi Performance 2016 Comparison of AODV ZRP and AODVDR Routing Protocols in MANET Computational Intelligence & Communication Technology (CICT) 2016 Second International Conference on IEEE
[19] Amouri Amar et al 2015 A simple scheme for pseudo clustering algorithm for cross layer intrusion detection in MANET Communications (LATINCOM) 2015 7th IEEE Latin-American Conference on IEEE
[20] Hayashi Yukinari and Michiko Harayama 2015 MANET topology control based on the node degree and energy detection threshold Internet Technologies and Applications (ITA) 2015 IEEE

Author Profile:

Dr.A. Rajaram received the B.E. degree in Electronics and Communication Engineering from the Government College of Technology, Coimbatore, Anna University, Chennai, India, in 2006, the M.E. degree in Electronics and Communication Engineering (Applied Electronics) from the Government College of Technology, Anna University, Chennai, India, in 2008 and he received the Full Time Ph.D. degree in Electronics and Communication Engineering from the Anna University of Technology, Coimbatore, India in March 2011. He is currently working as a professor in Department of Department of Electronics and Communication Engineering, EGS Pillay Engineering College, Nagapattinam. His research interests include Mobile Ad Hoc networks, wireless communication networks (WiFi, WiMax HighSlot GSM), novel VLSI NOC Design approaches to address issues such as low-power, cross-talk, hardware acceleration,Design issues includes OFDM MIMO and noise Suppression in MAI Systems,ASIC design, Control systems, Fuzzy logic and Networks, AI, Sensor Networks.

Article I. Design of Patch Antenna for Identification of Mal Practicing in Examination

D.S.Bhargava[#1], T.V.Padmavathy[*1], K.G.Sujanth Narayan[2], Sadagopal V[3]
[#1]Assistant Professor, *[1]Professor, , R.M.K. Engineering College [2]Research Assistant, SASTRA Deemed University,[3]Principal Engineer, Microchip Technology (India) Pvt Ltd

tvp.ece@rmkec.ac.in

Abstract. Malpractice is one of the serious issues faced in examination hall. The candidates will be examined thoroughly in the entrance itself. If they are found with possessing the electronic devices, the PGD (Paper Gadget Detector) will detect it. LCD (Liquid Crystal Display) displays the values and an alarm system will be used to give a warning to them. The proposed system is to search the candidates who are carrying the hidden written paper to the exam hall. The searching of hidden written paper can be obtained from Rectangular Microstrip Patch Antenna (RMPA). Here RMPA acts as a sensor which is used to detect the hidden material. This sensor can detect the target based on the permittivity value of each material. The sensors used in this study can be adjusted according to the values of blank paper and also to identify something written in the paper with different densities. The designed antenna has been simulated using Advanced Design System (ADS). The designed antenna (single element and 4 elements) has been compared in terms of various parameters like Gain, Efficiency, Directivity and Radiation Pattern etc.

Keywords: Paper Gadget Detector, Liquid Crystal Display, Rectangular Microstrip Patch Antenna

1. Introduction

Malpractice in examination is an act of cheating, i.e illegal action by students, violating the norms of examination to get through the exams. This affects the society in various facets like discouraging the hard work of other students, results in poor performances in jobs which affects the productivity, may involve in bribery and corruption etc., Even though there are some structural methods and arrangements been done to prevent examination malpractices, at times it is difficult to find out the people involved in it. Here we propose a RMPA design to find out the hidden papers or materials so that malpractices can be avoided to a greater extent.

2. Literature Survey

Syed An Nazmus Saqueb et al., (in 2018) proposed a rail-based Synthetic Aperture Radar (SAR) imaging system. Using a motorized linear stage, slanted object is scanned across the transmitter field of view and reflection is measured in frequency domain in two waveguide bands (220-325 GHz and 500-750 GHz). [8]. Jonathan T.Richard et al., (in 2017) used Synthetic Aperture Radar (SAR) that uses either moving platform or target to reconstruct images with higher spatial resolution. SARs operate at conventional microwave frequencies because of the high powers and sufficient bandwidth available to reconstruct large scenes over long distances more than 1km with resolution below 10 m. [4]. Joseph Landon Garry et al., (in 2015) proposed Practical Implementation of Stripmap Doppler

IOP Conf. Series: Materials Science and Engineering **994** (2020) 012032 doi:10.1088/1757-899X/994/1/012032

Imaging techniques which can generate 2D images. This method seems to be compatible in spectrally congested environments but the measurements can suffer from aliasing. [5]. A .Kannegulla et al., (in 2014) demonstrated coded-aperture imaging using the Hadamard coding. In this method the coded-aperture approach limits FOV. In spite of shortcomings, researchers for large scale applications suggest that large horizontal FOV can be achieved by combining multiple coded-apertures into a single hexagonal device. [1]. David Shrekenhamer et al., (in 2013) used Single pixel terahertz (THz) imaging technique which can work at non visible wavelengths. [2]. Gregory L. Charvat et al., (in 2008) designed a X-Band Rail SAR Imaging System which has high sensitivity in comparison with single-sideband radio receivers. [3]. R.Schneider et al., (in 2003) proposed and presented a high resolution vehicle based instrumentation radar for automobile application which performs on-line acquisition and real-time visualisation of radar images. This system has high computation efficiency, but limits data acquisition speed and also requires antenna pairs which make it suitable for stationery and slow moving targets. [7].

3. Proposed System

In the proposed system we use Microstrip patch antennas (low profile antenna mounted on flat rectangular sheet) are preferred over other antennas in today's modern wireless communication systems for their compatibility to be fit in Mobile, Aircraft, Satellites, Radars and many more applications owing to its very small size. In designing the microstrip patch antenna, selection of dielectric substrate materials and thickness are the main parameters in terms of size and compactness. Apart from compactness RMP antenna proves to be worth in terms of bandwidth, directivity and gain. Three parameters required for design are resonant frequency, dielectric constant and height of the dielectric substrate material. An inset fed of a rectangular patch antenna is designed to match the patch with a 50Ω microstrip transmission line. [6] The antenna with design dimensions is presented below in figure 1.

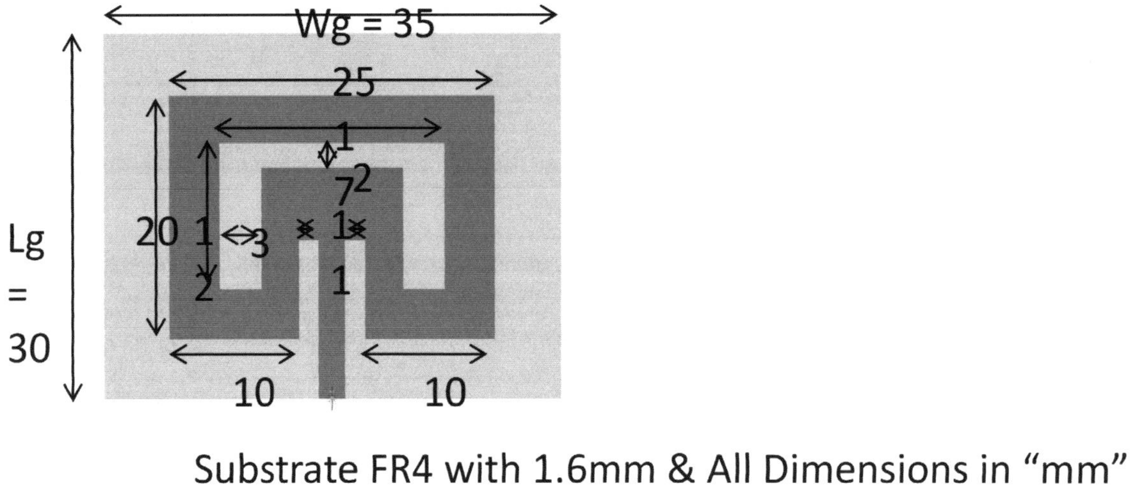

Figure 1. Antenna Design Dimensions

4. Simulation Results of RMP Antenna

The RMP Antenna Design is shown in Figure 2 and Simulation outputs are presented in this section in Figures 3 to 11 respectively.

IOP Conf. Series: Materials Science and Engineering **994** (2020) 012032 doi:10.1088/1757-899X/994/1/012032

Figure 2. RMP Antenna

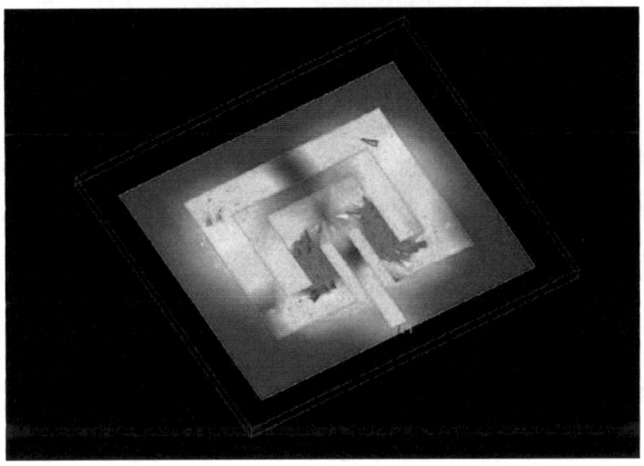

Figure 3. Current Distribution

The above figure 3 shows the current distribution of RMP antenna has maximum value at centre while current is zero at the edges

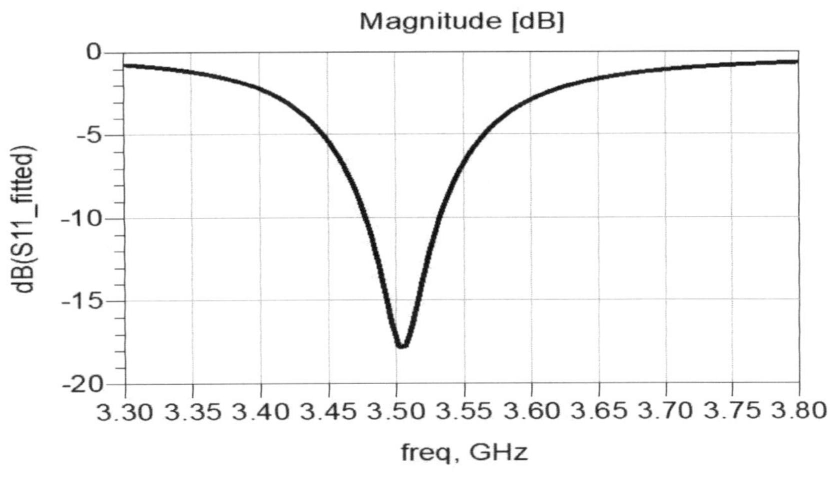

Figure 4. S-Parameters

The above figure 4 presents the S-Parameters (S11 return loss value of -18dB) of RMP antenna

IOP Conf. Series: Materials Science and Engineering **994** (2020) 012032 doi:10.1088/1757-899X/994/1/012032

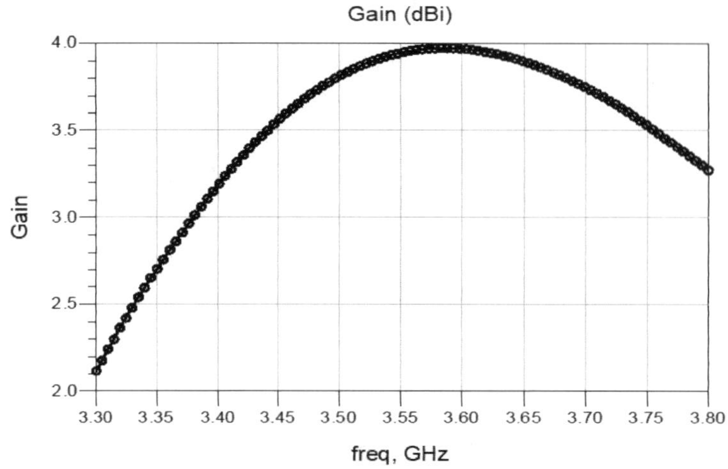

Figure 5. Gain

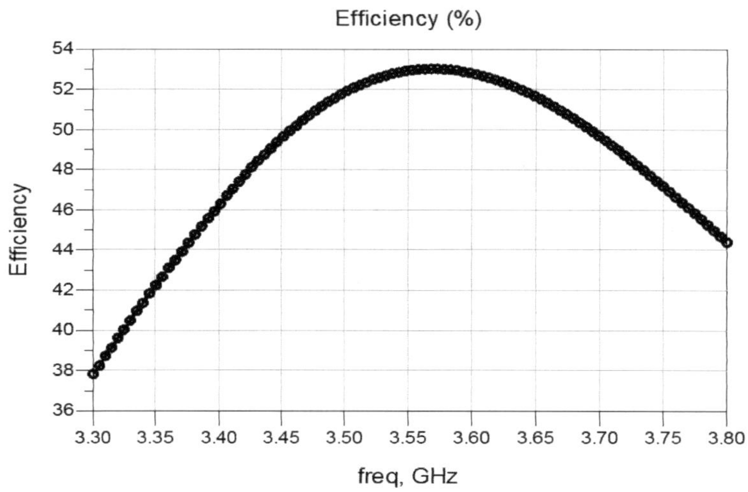

Figure 6. Efficiency

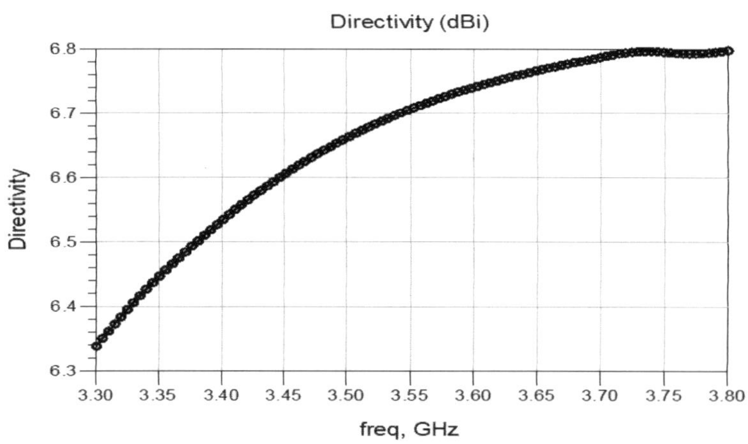

Figure 7. Directivity

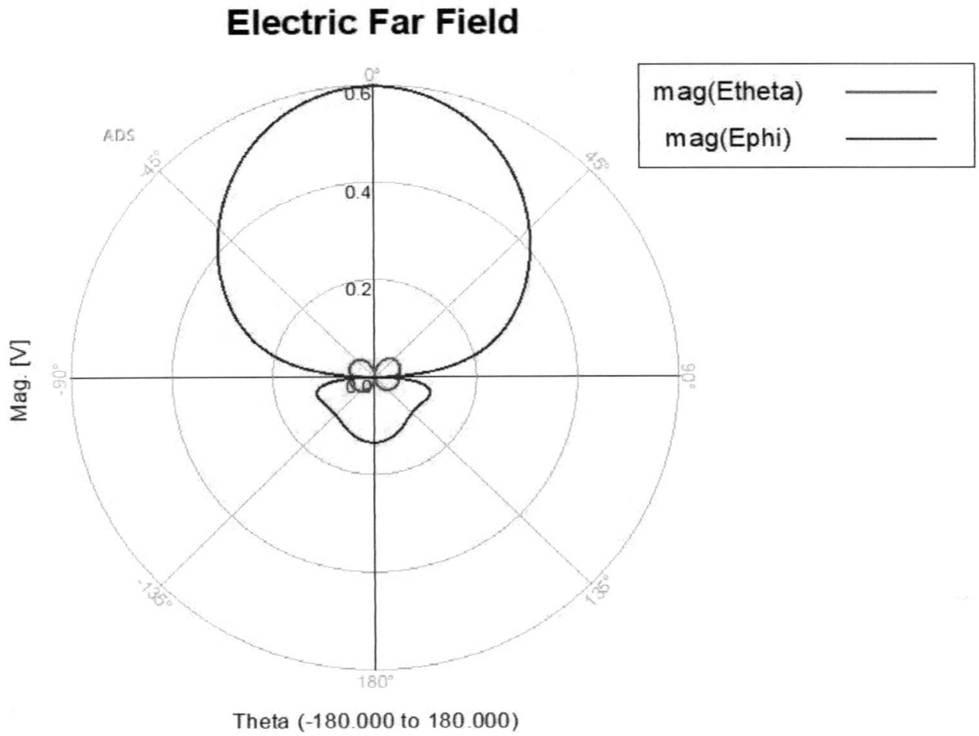

Figure 8. Electric Far Field

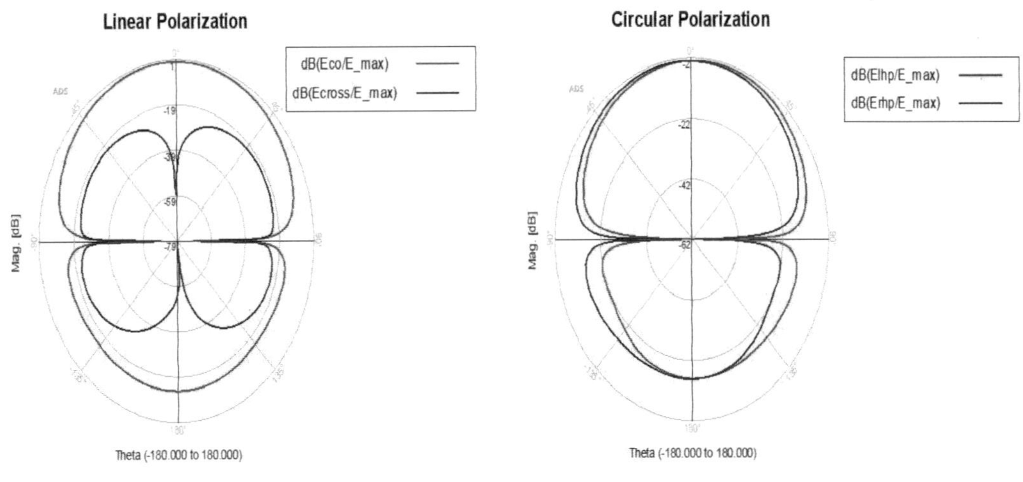

Figure 9. Polarization

Figure 8 illustrates the Electric Far Field and Figures 9 and 10 presents Polarization output and Radiation Pattern respectively

IOP Conf. Series: Materials Science and Engineering **994** (2020) 012032 doi:10.1088/1757-899X/994/1/012032

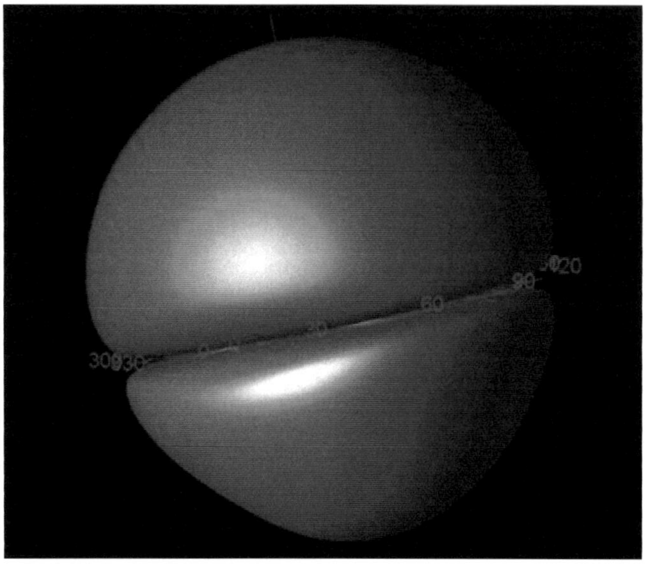

Figure 10. Radiation Pattern

Antenna Parameters

Frequency (GHz)		3.50202
Input power (Watts)		0.00245835
Radiated power (Watts)		0.00127667
Directivity(dBi)		6.66405
Gain (dBi)		3.81841
Radiation efficiency (%)		51.9321
Maximum intensity (Watts/Steradian)		0.000471276
Effective angle (Steradians)		2.70897
Angle of U Max (theta, phi)	0	270
E(theta) max (mag,phase)	0.595881	-7.45764
E(phi) max (mag,phase)	0.00373044	29.3552
E(x) max (mag,phase)	0.00373044	29.3552
E(y) max (mag,phase)	0.595881	172.542
E(z) max (mag,phase)	0	180

Figure 11. Antenna Parameters Summary

5. Simulation Results of 4 RMP Antenna Elements

The 4 element RMP Antenna Design is shown in figure 12 and Simulation outputs are presented in this section in Figures 13 to 25 respectively.

Figure 12. 4 RMP Antenna Elements

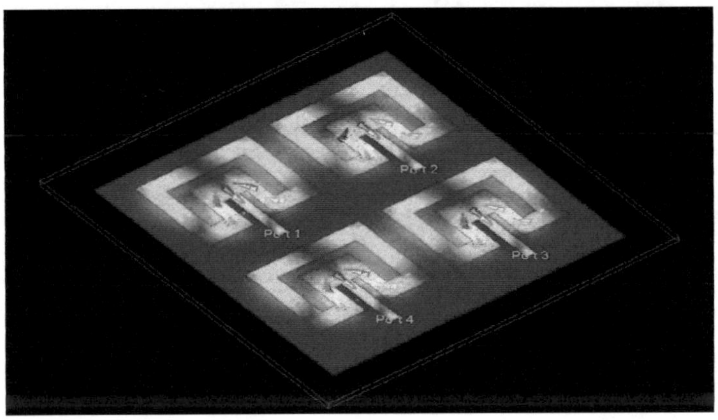

Figure 13. Current Distribution of 4 Elements

The above figure 13 shows the current distribution of 4 RMP antennas.

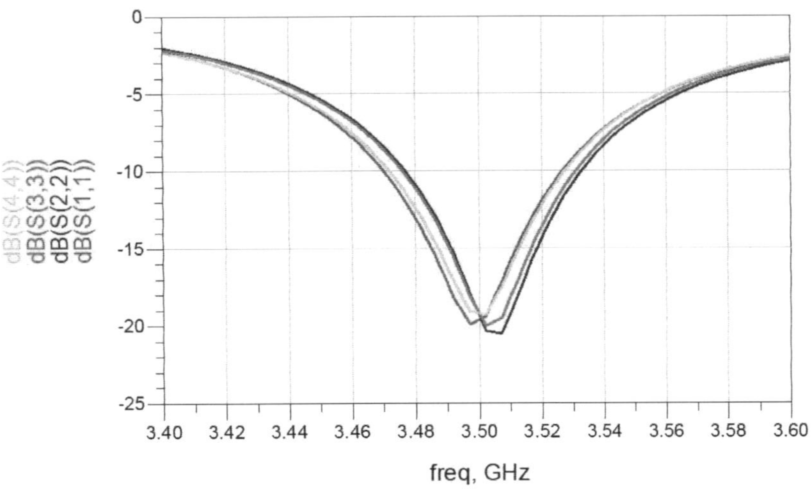

Figure 14. S-Parameters of 4 Elements

The above figure 14 presents the S-Parameters of 4 RMP antennas. (S1,1 - 19.80dB, S2,2 - 19.92dB, S3,3 - 19.57dB, S4,4 - 19.12dB)

IOP Conf. Series: Materials Science and Engineering **994** (2020) 012032 doi:10.1088/1757-899X/994/1/012032

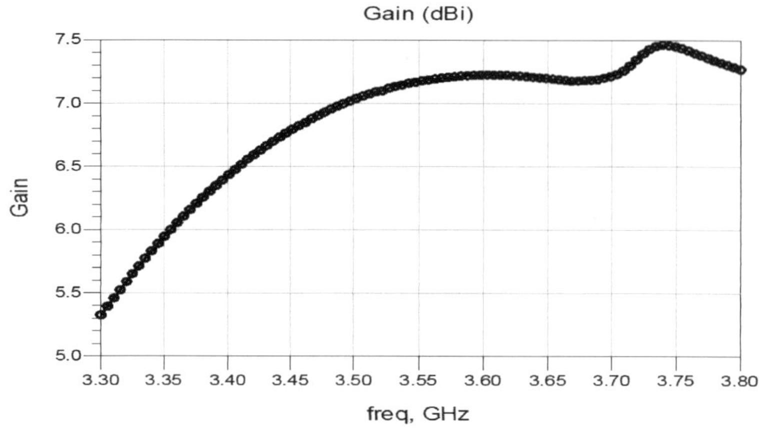

Figure 15. Gain of 4 Elements

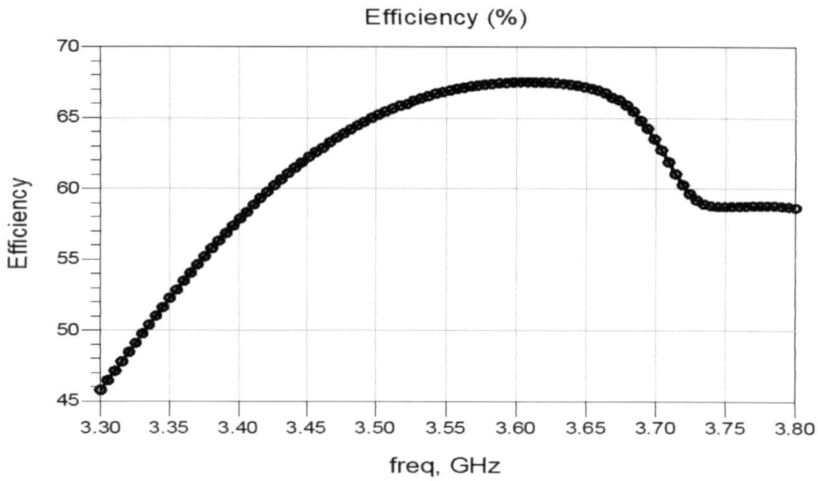

Figure 16. Efficiency of 4 Elements

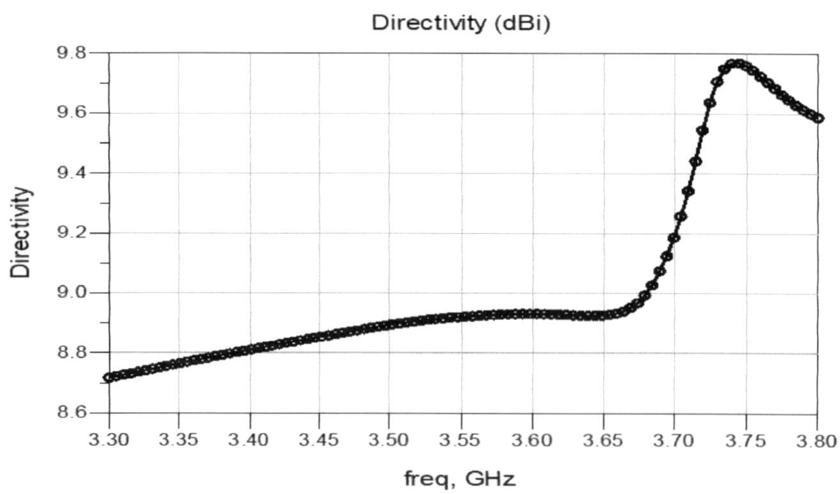

Figure 17. Directivity of 4 Elements

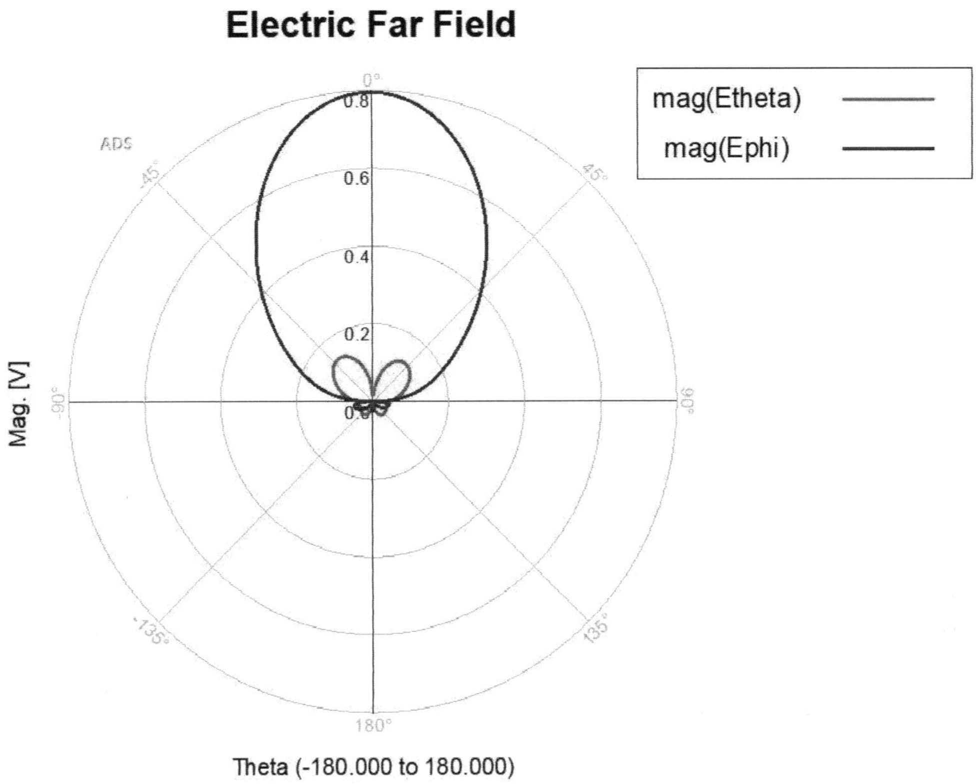

Figure 18. Electric Far Field of 4 Elements

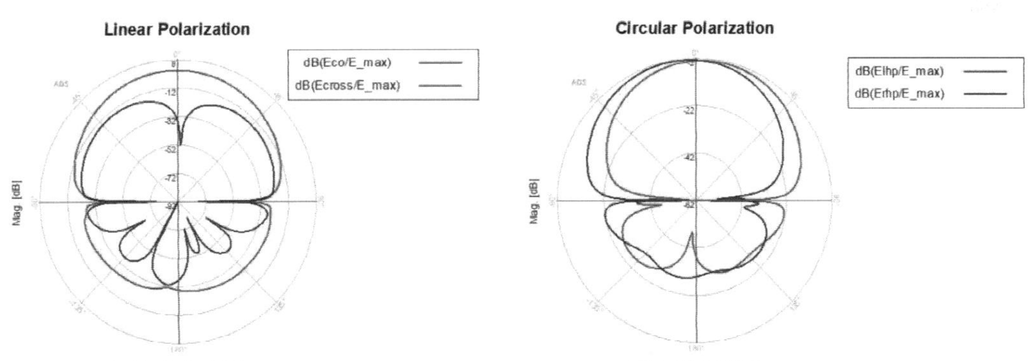

Figure 19. Polarization of 4 Elements

Figure 18 illustrates the Electric Far Field of 4 elements and figures 19 and 24 presents Polarization output and Radiation Pattern of 4 elements respectively.

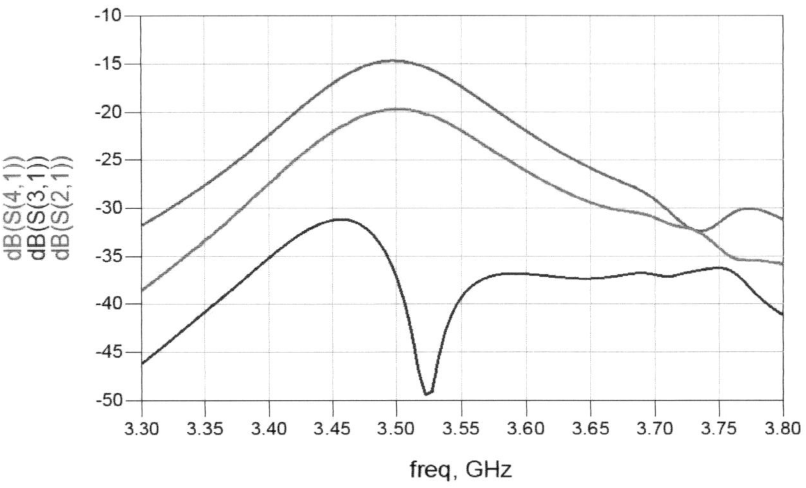

Figure 20. Mutual Coupling between Antenna 1 and other antennas

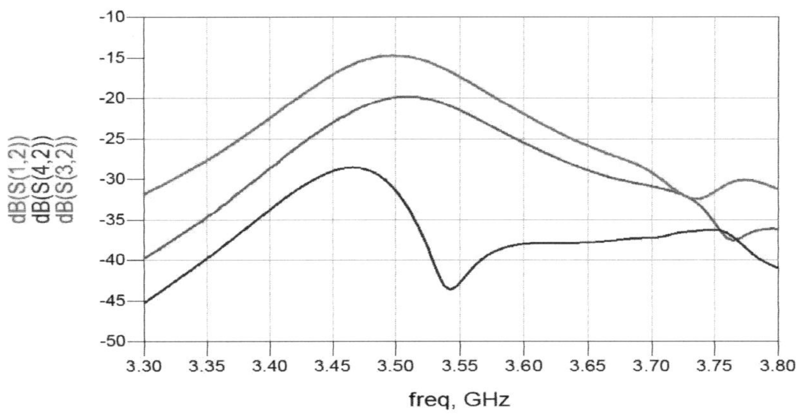

Figure 21. Mutual Coupling between Antenna 2 and other antennas

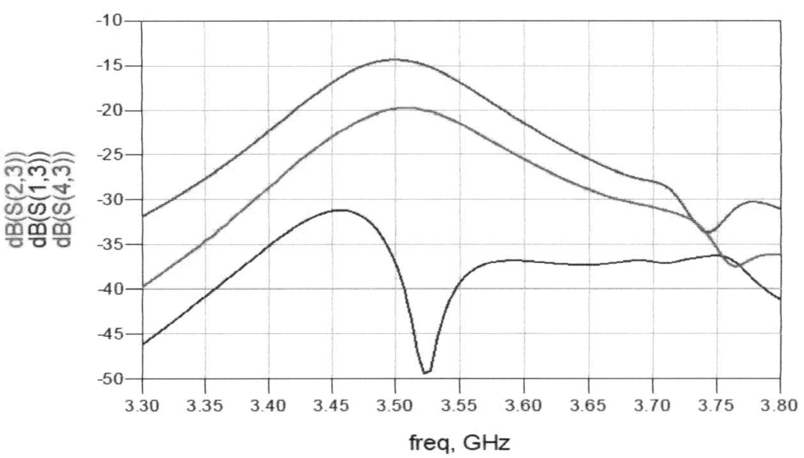

Figure 22. Mutual Coupling between Antenna 3 and other antennas

IOP Conf. Series: Materials Science and Engineering **994** (2020) 012032 doi:10.1088/1757-899X/994/1/012032

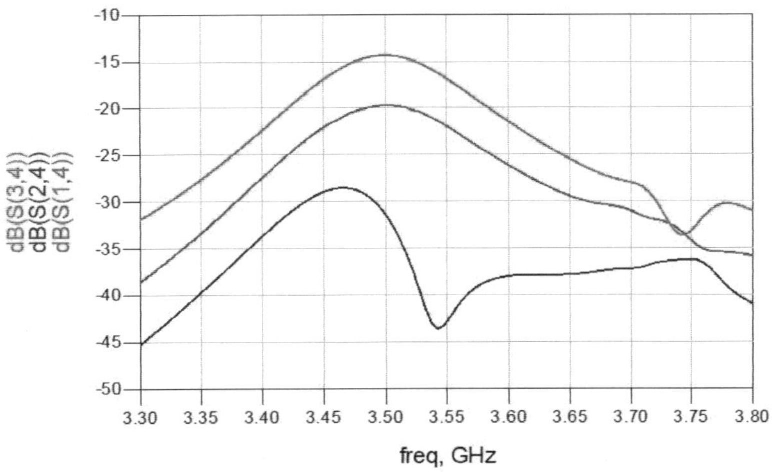

Figure 23. Mutual Coupling between Antenna 4 and other antennas

The Figures 20 through 23 illustrates the Mutual coupling between 4 antennas.

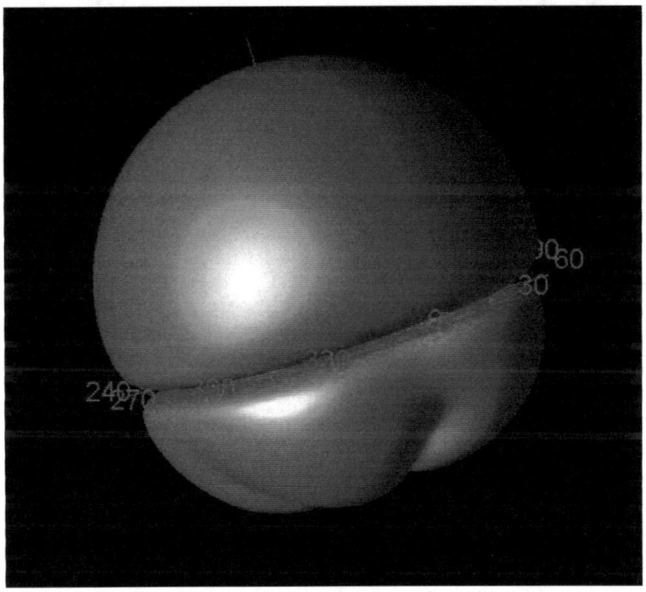

Figure 24. Radiation Pattern of 4 Elements

Antenna Parameters

Frequency (GHz)		3.50202
Input power (Watts)		0.00951106
Radiated power (Watts)		0.00620197
Directivity(dBi)		8.89592
Gain (dBi)		7.03893
Radiation efficiency (%)		65.208
Maximum intensity (Watts/Steradian)		0.00382747
Effective angle (Steradians)		1.62039
Angle of U Max (theta, phi)	1	213
E(theta) max (mag,phase)	0.939544	-21.1238
E(phi) max (mag,phase)	1.4146	-23.163
E(x) max (mag,phase)	0.0327335	-144.247
E(y) max (mag,phase)	1.69779	157.451
E(z) max (mag,phase)	0.0163973	158.876

Figure 25. Antenna Parameters Summary of 4 Elements

The simulated results show that Gain of single element RMP antenna is 3.81 dBi (presented in Figures 5 and 11 respectively) and Gain of 4 elements is 7.0 dBi (presented in Figures 15 and 25 respectively). Similarly Efficiency of single element RMP antenna is 51% (presented in Figures 6 and 11 respectively) and Efficiency of 4 elements is 65.2% (presented in Figures 16 and 25 respectively).
Further Directivity (to measure the degree to which the radiation is emitted in a single direction) of single element RMP antenna is 6.66 dBi (presented in Figures 7 and 11 respectively) and Directivity of 4 elements is 8.89 dBi (presented in Figures 17 and 25 respectively)
Mutual coupling describes the energy absorbed by one antenna's receiver when another adjacent antenna is operating. The RMPA design with 4 elements provides mutual coupling parameters as shown in the above graphs specified. The coupling values between the antennas lie well below -10dB for all its adjacent antenna elements thereby exhibiting no interference issues by the same or the nearing antenna element. This is in turn helps antennas to work with high directive beams at the finest point to be located. Further, beamforming techniques and beam switching characteristics deployed to the antenna help in achieving precise beam points to focus on the highly required spot figuring the position of the ink identification on the paper material.

6. Conclusion and Future Scope

Initially single RMP antenna has been designed for a frequency of 3.5 GHz and the results such as Gain, Efficiency and Directivity are 3.81 dBi, 51% and 6.66 dBi respectively. To further enhance the performance of antenna, the design has been extended to 4 elements and the results such as Gain, Efficiency and Directivity are 7.0 dBi, 65.2% and 8.89 dBi respectively.
This antenna can be used in the system to find hidden papers using Vector Network Analyzer. The result is determined by the rise or fall of wave found in receiver of antenna. Implementation using Antenna is Cost Efficient.

7. References

[1] Kannegulla A Jiang Z Rahman S M and Shams M I B 2014 Coded-Aperture Imaging Using Photo-Induced Reconfigurable Aperture Arrays for Mapping Terahertz Beams *IEEE transactions on the science and technology* vol 4

[2] David Shrekenhamer Claire M Watts and Willie J. Padilla 2013 Terahertz single pixel imaging with an optically controlled dynamic spatial light modulator *Optics express* vol 21 No 10

[3] Gregory L Charvat, Leo C Kempell and Chris Coleman 2008 A Low-Power High-Sensitivity X-Band Rail SAR Imaging System

[4] Jonathan T Richard and Henry O. Everitt 2017 Millimeter Wave and Terahertz Synthetic Aperture Radar for Locating Metallic Scatterers Embedded in Scattering Media *IEEE transactions on terahertz science and technology* vol 7 No 6

[5] Joseph Landon Garry Graeme Edward Smith and Christopher John Baker 2015 Practical Implementation of Strip Map Doppler Imaging *IET Radar Sonar Navigation* vol 9

[6] Osama W Ata Mohammad Salamin and Khaleel Abusabha 2020 Double U-Slot rectangular patch antenna for multiband applications *Computers & Electrical Engineering* vol 84

[7] Schneider R and Wenger J 2003 High resolution radar for automobile applications *Advances in Radio Science*

[8] Syed An Nazmus Saqueb Joseph Lgarry Graem E.Simth, Niru K Nahar and kabilay Setal 2018 THZ imaging using rail- based synthetic aperture radar for the detection of concealed objects *USNC-URSI radio science meeting (joint with AP -s symposium)*

D S. Bhargava, working as Assistant Professor in Electronics and Communication Engineering department of R.M.K. Engineering College, has 6 years of teaching experience. He received his Bachelor's degree from J.N.N Institute of Engineering in Electronics and Communication Engineering in the year 2012 and Master's degree in VLSI Design from R.M.K. Engineering College in the year 2014. His area of interests includes Cognitive radio networks and VLSI Design technology. He has published 11 research papers in International and National Journals & presented 4 research papers in National and International conferences in the area of VLSI Design, Networking and Cognitive radio networks. He is a professional member of ISSE, ACM, ISTE, and IAENG.

T. V. Padmavathy Professor, Department of ECE in R.M.K. Engineering College, has 25 years of teaching and research experience in the in the fields of Wireless sensor networks, Under Water Acoustic Sensor Networks and Antenna Design. She has graduated from Institution of Engineers (India), in Electronics and Communication Engineering. She has obtained her Master degree in Control and Instrumentation from College of Engineering, Guindy, Anna University, Chennai and Ph.D. degree from Anna University, Chennai. She has published more than 50 research papers in International and National Journals and conferences in the area of Mobile Ad hoc Networks, Wireless sensor networks, Under Water Acoustic Sensor Networks and Antenna design and she has four Patents in Wireless Sensor Networks. Her current area of research includes security and architecture issues of Mobile ad hoc networks, Wireless sensor networks and Millimeter Wave Antenna design for Wireless Communications. She is a technical paper reviewer for African Journal of Engineering Research and Journal of Engineering and Technology Management. She is recognized as Fellowship member by The Institution of Engineers (India) also she is a member of various professional bodies such as Institute of Electrical and Electronics Engineers (IEEE), Life member of Institution of Electronics and Telecommunication Engineers (IETE), International Association of Engineers (IAENG), ACM, ISSE and Life member of Indian Society for Technical Education ISTE.

K. G. Sujanth Narayan born in Tamilnadu, India. He received his B.E., degree in Electronics and Communication Engineering from R.M.K. Engineering College, Kavaraipettai, Tamilnadu in the year 2017 and M.Tech, in Communication Systems from SASTRA Deemed University, Thanjavur, Tamil Nadu in the year 2019 respectively. Currently, he is a Research Assistant, pursuing his Ph.D. in the area of Antennas and RF System from School of Electrical and Electronics Engineering, SASTRA Deemed University, Thanjavur, Tamil Nadu. He is a member of IEEE. His research areas include Antennas, Electromagnetics, RF System Design, EMI/EMC and Vehicular-Communication.

Sadagopal V working as Principal Engineer in Applications RF (RF Design and Testing) at Microchip Technology (India) Pvt Ltd. He also served as Project Lead and has 12 years of experience in RF System & Antenna Design and Testing equipments like Vector Network Analyzer, Signal Generator, Spectrum Analyzer and Noise Figure Meter. Also has good experience in wireless standards like Wi-Fi, Bluetooth, ZigBee, LoRa and GSM. He has Sound conceptual knowledge in ADS and HFSS for EM simulation and Expertise in Schematics capture and PCB Design.

Design and Simulation of MIMO Antennas for Mobile Communication

D.S.Bhargava[1], T.V.Padmavathy[2], Yadhamuri Vinitha Reddy[3], Neelapareddy Kavitha[3], Vuppalapati Hema[3]

[1]Assistant Professor, [2]Professor, [3]UG Scholars
Department of Electronics and Communication Engineering, R.M.K. Engineering College

duvurubhargava@gmail.com

Abstract. A MIMO based antenna structure for next generation mobile communication is proposed. The antenna structure consists of 4 symmetric F-shaped elements placed at the centre of the FR_4 substrate with dielectric constant of 4.4 and loss tangent value of 0.001. The thickness of the substrate is 1.6 mm and the overall dimensions of the proposed structure are 122 x 147 x 0.8 mm^3. The designed antenna works in the frequency range of 3.4-3.6 GHz (LTE band 42) with impedance bandwidth of 125MHz. The antenna is designed using CST 3D simulation software and antenna parameters namely Return Loss, Voltage Standing Wave Ratio, Gain, Directivity and Efficiency are analysed. Simulated results show better antenna performance in the desired frequency range.

Keywords: CST, MIMO, LTE, VSWR, Return Loss\

1. Introduction
Over the years, wireless systems have increased rapidly. Antenna is the most important part of wireless communication system for transmitting and receiving information in the form of Electromagnetic (EM) Waves. Advancements in communication technology have led to increased usage of wireless devices. Moreover there is always an increase in demand for miniaturization, higher data rate as well as operating speed of these wireless devices. [1] [2] [15]
In traditional SISO (Single Input Single Output) systems, there is a limitation in terms of capacity. MIMO (Multiple Input Multiple Output) antenna systems are considered to provide better performance in terms of Gain, diversity, multiplexing and to reduce multipath fading and enhance channel capacity. The MIMO technology plays a key role in Next Generation wireless/mobile communication systems by providing high data rates and capacity. [4][15]

2. Microstrip Patch Antenna
Microstrip patch antennas more commonly referred to as patch antennas; have gained popularity in today's wireless technology with its merits like low-profile, low cost, ease of integration as well as fabrication. [3] The microstrip patch antenna is a open wave guide structure has dielectric with radiating patch. The feedlines etched on one side and ground the other side. The shape of patch is not limited in particular; it could be of square, rectangular, circular, elliptical and triangular etc. The feeding methods for microstrip patch antennas can be broadly categorized into two as contacting and

non-contacting. The most commonly used feed methods are microstrip line, coaxial probe, aperture and proximity coupling. [5][9]

3. MIMO

Multiple Input – Multiple Output (MIMO) Systems are recently being used for research in next generation communication because of its significant enhancements in the channel capacity, spectral efficiency and great diversity. Independent data streams can be communicated simultaneously using MIMO systems. A major challenge in designing MIMO system for mobile phones lies in the available space within the mobile phone. [11] Various MIMO systems have been discussed in the recent literature, showing different designs of MIMO based antennas for mobile equipment. This paper proposes a new antenna structure with considerable antenna parameter values. [4][11]

4. Antenna Design

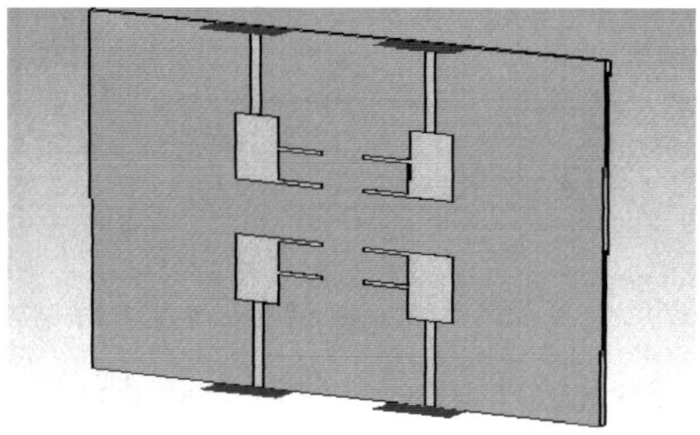

Figure 1. 4 Elements with Vertical Feed

The antenna structure consists of 4 F-shaped elements placed at the centre, fed from the top and bottom edges of the substrate as presented in above figure 1. All the elements are placed symmetric with same dimensions. The antennas are fed using edge feed technique. The feed, feed line and the patch are designed to have 50 ohms matched impedance.

5. Results and Discussion

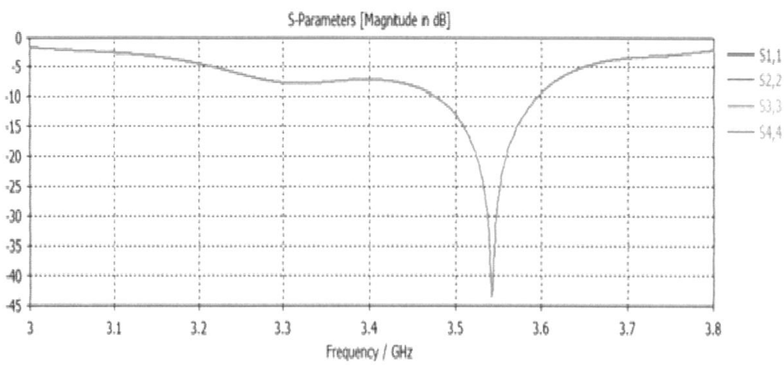

Figure 2. S-Parameters for Vertical Feed

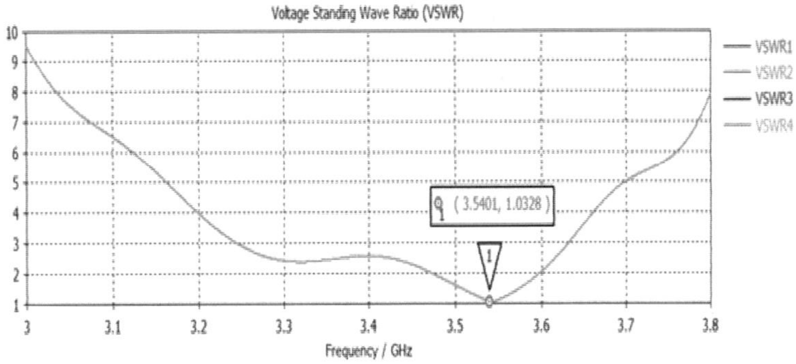

Figure 3. VSWR for Vertical Feed

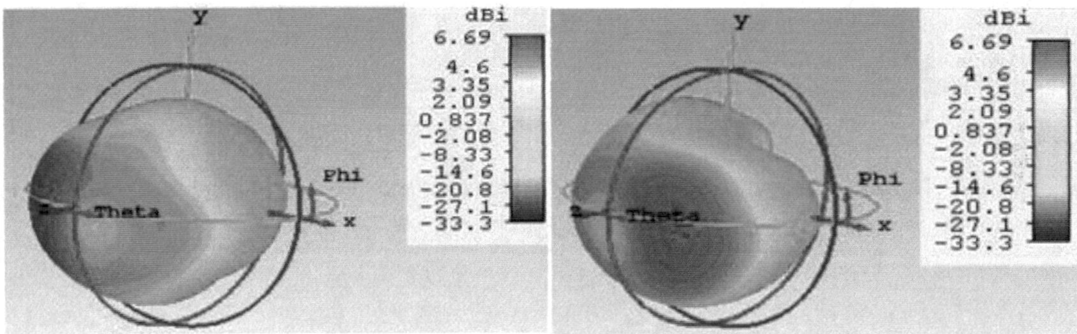

Figure 4. Radiation Pattern for Vertical Feed

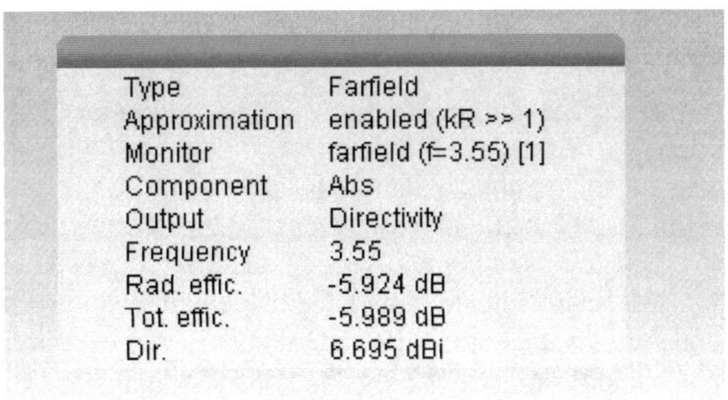

Type	Farfield
Approximation	enabled (kR >> 1)
Monitor	farfield (f=3.55) [1]
Component	Abs
Output	Directivity
Frequency	3.55
Rad. effic.	-5.924 dB
Tot. effic.	-5.989 dB
Dir.	6.695 dBi

Figure 5. Radiation and Total Efficiency

The 4 element MIMO antenna designed with vertical feed has produced gain of 0.7704 dB, directivity of 6.695 dBi, S – Parameter (return loss) value of -44 dB, VSWR of 1 as shown in figures 2 and 3 respectively, the radiation pattern is depicted in figure 4 and the values of radiation efficiency and total efficiency presented in figure 5 are -5.924 dB and -5.989 dB respectively.

6. Conclusion

The design of 2 x 2 MIMO antennas (4 vertical elements) for mobile communication has been presented in this paper. The designed antenna provided better return loss and radiation characteristics making it suitable for mobile applications. In future 2 x 2 antenna designs can be extended to 4 x 4 and 8 x 8 and the performances can be compared and further the antenna can be fabricated. By increasing the number of elements or antennas in MIMO technology, the channel capacity can be extended. In order to increase the range of the transmitted signal, MIMO antennas or array of antennas could be a better solution particularly when using battery powered devices.

7. References

[1] Aziz Dkiouak Alia Zakriti Mohssine El Ouahabi and AichaMchbal 2019 Design of a two symetrical F-shaped MIMO antenna for Wi-MAX and WLAN applications *International Conference on Wireless Technologies, Embedded and Intelligent Systems (WITS)*

[2] Aziz Dkiouak Alia Zakriti, Mohssine El Ouahabi Naima A Touhami and AichaMchbal 2019 Design of a Four-Element MIMO Antenna with Low Mutual Coupling in a Small Size for Satellite Applications *Progress In Electromagnetics Research M* vol 85 pp 95–104

[3] Banothu YVNR Swamy and Polepalli Siddaiah 2019 Design of a Compact 2×2 Multi Band MIMO Antenna for Wireless Applications *International Journal of Recent Technology and Engineering (IJRTE)* vol 7 No 6S2 pp 674–683

[4] Da Qing Liu He Jia Luo Ming Zhang Huai Lin Wen, Bin Wang and Jun Wang 2019 An Extremely Low-Profile Wideband MIMO Antenna for 5G Smartphones *IEEE Transactions on Antennas and Propagation* vol 67 No 9 pp 5772-5780

[5] Houda Werfelli Khaoula Tayari Mondher Chaoui, Mongi Lahiani and Hamadi Ghariani 2016 Design of Rectangular Microstrip Patch Antenna *2nd International Conference on Advanced Technologies for Signal and Image Processing – ATSIP*

[6] Hussein Hamed Mahmoud Ghouz 2015 Novel Compact and Dual-Broadband Microstrip MIMO Antennas for Wireless Applications *Progress in Electromagnetics Research B* vol 63 pp 107–121

[7] Kamili J Babu Rabah W Aldhaheri Mohammed Y Talha and Ibrahim S Alruhaili 2014 Design of a Compact Two Element MIMO Antenna System with Improved Isolation *Progress in Electromagnetics Research letters* vol 48 pp 27–32

[8] Kumud Ranjan Jha Bisma Bukhari Chitra Singh Ghanshyam Mishra and Satish Kumar Sharma 2018 Compact Planar Multistandard MIMO Antenna for IoT Applications *IEEE Transactions on Antennas And Propagation* vol 66 No 7 pp 3327–3336

[9] Le Huy Trinh Fabien Ferrero Leonardo Lizzi Jean-Marc Ribero and Robert Staraj 2015 4 × 4 MIMO Multiband Antenna System for Mobile Handsets *International Journal of Antennas and Propagation* vol 2015 pp 1–6

[10] Mahnoor Khalid, Syeda Iffat Naqvi, Niamat Hussain, MuhibUr Rahman, Fawad Seyed Sajad Mirjavadi Muhammad Jamil Khan and Yasar Amin 2020 4-Port MIMO Antenna with Defected Ground Structure for 5G Millimeter Wave Applications *Electronics* vol 9 No1 pp 1–13

[11] Mujeeb Abdullah, Saad Hassan Kiani, Lway Faisal Abdulrazak, Amjad Iqbal, M. A. Bashir, Shafiullah Khan and Sunghwan Kim 2019 High-Performance Multiple-Input Multiple-Output Antenna System For 5G Mobile Terminals *Electronics* vol 8 No 10 pp1–16

[12] Prashanth B U V 2014 Design and Implementation of Radar Cross-Section Models on a Virtex-6 FPGA *Journal of Engineering* vol 2014 pp 1–10

[13] Qian Wang Ning Mu LingLi Wang Safieddin Safavi-Naeini and JingPing Liu 2017 5G MIMO Conformal Microstrip Antenna Design *Wireless Communications and Mobile Computing* vol 2017 pp 1–11

[14] Satya Mitra YVSS Kalisha Sk Sindhu G Raghavendra Kumar P Satish P and Jagadeesh Babu K 2016 Design of a Novel Microstrip MIMO Antenna System with Improved Isolation *International Journal of Antennas (JANT)* vol 2 No 2 pp 19–25

[15] Vasu Babu K and Anuradha B 2018 Design of multi-band minkowski MIMO antenna to reduce the mutual coupling *Journal of King Saud University - Engineering Sciences* vol 32 No 1 pp 52–57

[16] Yanfeng Geng Runbo Ma Baoming Chen and Wenmei Zhang 2015 A Design of Double Broadband MIMO Antenna *International Journal of Antennas and Propagation* vol 2016 pp 1–6

D. S. Bhargava, working as Assistant Professor in Electronics and Communication Engineering department of R.M.K. Engineering College, has 6 years of teaching experience. He received his Bachelor's degree from J.N.N Institute of Engineering in Electronics and Communication Engineering in the year 2012 and Master's degree in VLSI Design from R.M.K. Engineering College in the year 2014. His area of interests includes Cognitive radio networks and VLSI Design technology. He has published 11 research papers in International and National Journals & presented 4 research papers in National and International conferences in the area of VLSI Design, Networking and Cognitive radio networks. He is a professional member of ISSE, ACM, ISTE, and IAENG.

T. V. Padmavathy Professor, Department of ECE in R.M.K. Engineering College, has 25 years of teaching and research experience in the fields of Wireless sensor networks, Under Water Acoustic Sensor Networks and Antenna Design. She has graduated from Institution of Engineers (India), in Electronics and Communication Engineering. She has obtained her Master degree in Control and Instrumentation from College of Engineering, Guindy, Anna University, Chennai and Ph.D. degree from Anna University, Chennai. She has published more than 50 research papers in International and National Journals and conferences in the area of Mobile Ad hoc Networks, Wireless sensor networks, Under Water Acoustic Sensor Networks and Antenna design & she has four Patents in Wireless Sensor Networks. Her current area of research includes security and architecture issues of Mobile ad hoc networks, Wireless sensor networks and Millimeter Wave Antenna design for Wireless Communications. She is a technical paper reviewer for African Journal of Engineering Research and Journal of Engineering and Technology Management. She is recognized as Fellowship member by The Institution of Engineers (India) also she is a member of various professional bodies such as Institute of Electrical and Electronics Engineers (IEEE), Life member of Institution of Electronics and Telecommunication Engineers (IETE), International Association of Engineers (IAENG), ACM, ISSE and Life member of Indian Society for Technical Education ISTE.

Yadhamuri Vinitha Reddy, currently pursuing her Bachelor's Degree in the Department of Electronics and Communication Engineering, R.M.K. Engineering College, Affiliated to Anna University Chennai. She is an active member of various technical forums, during her course of study she has organized technical event like symposiums & also attended Guest Lectures, workshops, seminars and in plant training etc. Vinitha Reddy received ISTE Chapter best student award in the year 2019. She had made a technical study on RADAR Systems during her in plant training at DRDL. Her area of interests includes Device and Circuit Simulation, Antennas and Microwave Engineering.

Neelapareddy Kavitha, currently pursuing her Bachelor's Degree in the Department of Electronics and Communication Engineering, R.M.K. Engineering College, Affiliated to Anna University Chennai. She is an active member of various technical forums, during her course of study she has attended in plant training in BSNL and she had made a technical study on RADAR Systems during her in plant training at DRDL. Her area of interests is Signals and Systems, Antennas and Microwave Engineering.

Vuppalapati Hema, currently pursuing her Bachelor's Degree in the Department of Electronics and Communication Engineering, R.M.K. Engineering College, Affiliated to Anna University Chennai. She is an active member of various technical forums, during her course of study she has organized technical event like symposiums & also attended Guest Lectures, workshops, seminars and in plant training etc. She had made a technical study on RADAR Systems during her in plant training at DRDL. Her area of interests is Antennas and Networks.

IOP Conf. Series: Materials Science and Engineering **994** (2020) 012034 doi:10.1088/1757-899X/994/1/012034

Analysis of Dynamic Frequency Reuse Techniques in LTE – A Cellular Network

S Karthika[1]*and PIndumathi[2]

[1]Associate Professor, Department of Electronics and Communication Engineering, Sri Krishna College of Engineering and Technology, Coimbatore.
[2]Professor, Department of Electronics Engineering, Madras Institute of Technology, Coimbatore.
Corresponding Author mail Id :karthikas@skcet.ac.in

Abstract: Inter-cell Interference (ICI) is a significant obstacle which is degrading the efficiency of mobile cellular networks particularly for end users. Frequency reuse is an ICI synchronization technique which helps to mitigate ICI by assigning dissimilar frequency reuse factors on the basis of its position to User equipment (UE). This paper proposes a Self-organized resource allocation (SORA) framework that chooses the reuse factors randomly. The output of this strategy is configured using MATLAB and correlated with numerous reuse factor combinations. Results from simulation show improved efficiency for Cell Edge Users (CEU).Keywords: CoMP, Frequency reuse, LTE-A, OFDMA.

1. Introduction

In 3GPP LTE-Advanced (LTE-A), Coordinated Multipoint Communication (CoMP) is considered a crucial methodology for increasing CEU in communicationsystems. Successful frequency reuse would be employed in promoting resource allocation in CoMP. OFDMA is commonly recognized as multiple access strategy towards different forms of communication schemes like mobile telecommunications networks such as Long Term Evolution (LTE) and LTE-A. In OFDMA, the frequency selective wideband channels are separated to several subcarriers of narrow bands. The orthogonality concept amongst narrow bands guarantees that the ICI is avoidable.

Fractional Frequency Reuse (FFR) is being used for efficient reduction of ICI in mobile cellular networks. The FFR definition was argued in [1]. FFR alleviated ICI by adding various reuse factors to UEs, depending on their position. FFR optimization-based process is discussed in [2-3]. Generalized Fractional Frequency Reuse (GFFR) is often used in broad and irregularly organized cellular networks [4]. A downlink LTE model metric Signal to Interference Noise Ratio (SINR), throughput and average cell energy efficiency were considered by the authors in [5]. The authors in [6] found another possibility upon integrating FFR with numerous scheduling algorithms including Round Robin Scheduling and Proportional Fairness. The non linear fault and detection algorithm has been discussed in [7]. In [8], the dynamic process is established and the optimum reuse factor is selected on the basis of user privilege.

2. Proposed Resource Allocation Scheme

Theentire band is separated into 2 parts: the inner/outer parts. Inner prepossessing frequency reuse-1 serves Cell Center Users (CCU) and Outer group serves Cell Edge Users (CEU) with frequency reuse-3. Based on SINR, the users are distinguished into edge/middle users. The β-range available is split into βF-1 and ßF-3. Frequency Reuse-1 is employed for the inner band βF-1 and Frequency Reuse-3 is used for the outer band βF-3. Fig.1 displays the flowchart of this algorithm.

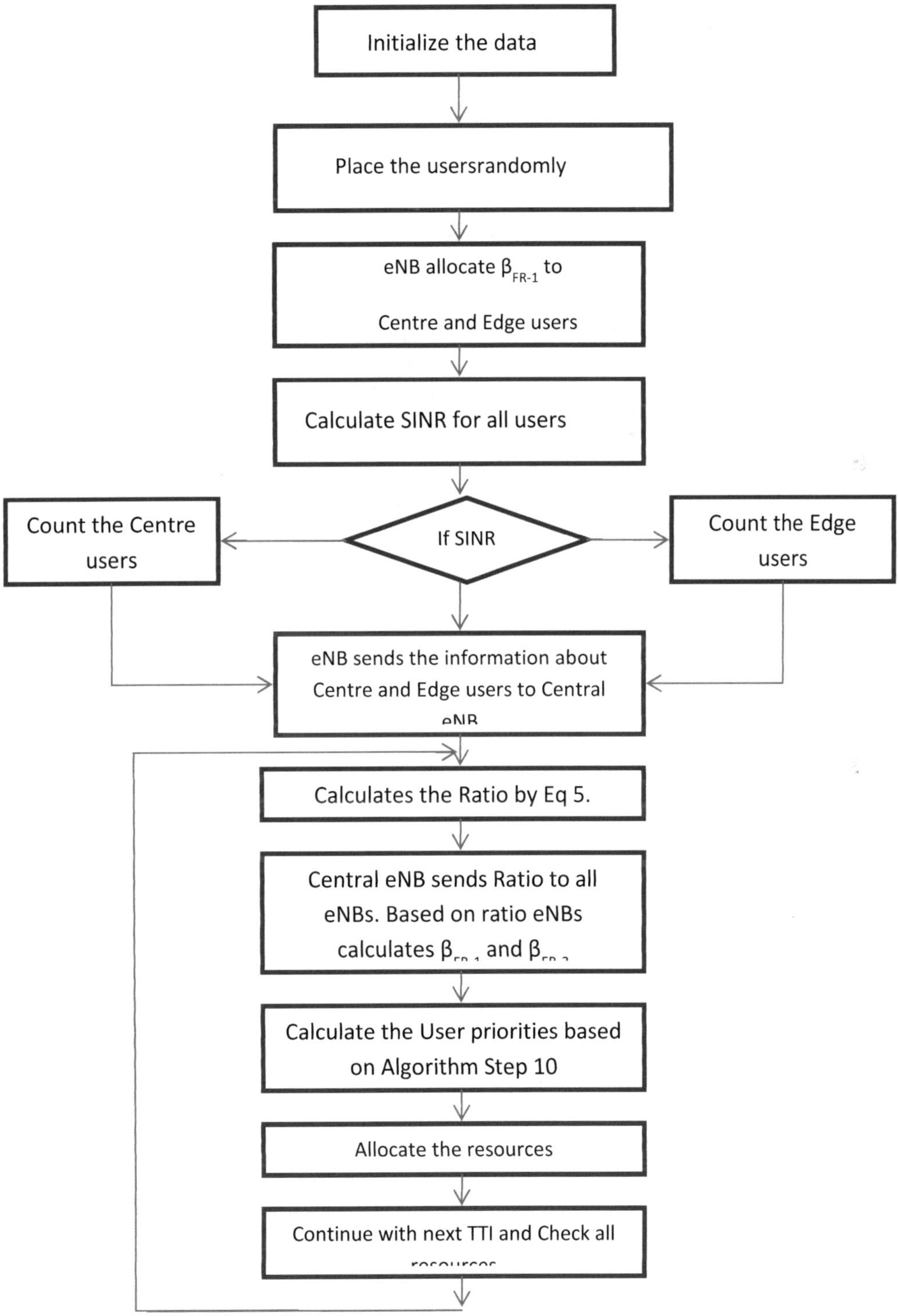

Fig.1. Flowchart of Algorithm

3. Simulation Results

The resource allocation efficiency is evaluated via MATLAB and is correlated with other reuse factors. Simulation parameters are chosen as Bandwidth-20 MHz, Carrier Frequency-2.6GHz, eNB power-46dBm, Channel Model-LTE Model, Number of Resource Block-100 RB per slot/7Symbols, Modulation Scheme-16QAM, Total Number of Subcarriers-1200.

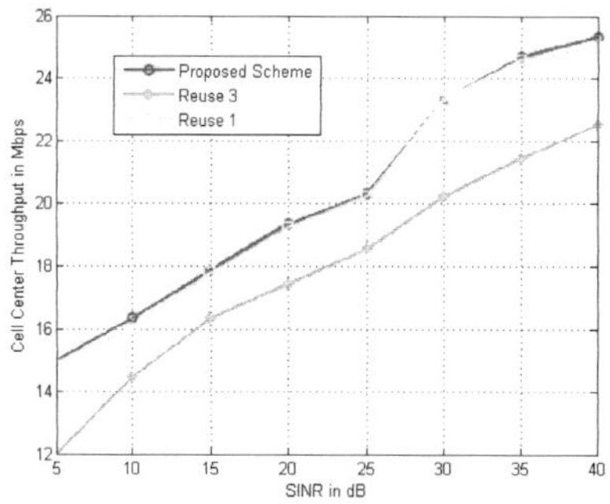

Fig.2. Cell Center Throughput Evaluation of Proposed Scheme with other Reuse Factor

From fig.2, it could be found that, as compared with other reuse methods, the proposed strategy has greater performance.For example, when SINR is 20dB the proposed scheme achieves 0.57% decreases when compared to reuse 1 and it achieves 9.7% increases when compared to reuse 3. Reuse 1 provides better throughput to cell center users.

From fig.3, as related to those other reuse factors, the proposed methodology provides greater throughput as it dynamically assigns the reuse factor to users on the basis of SINR. Reuse-3 provides better throughput to the edge users. For example, when SINR is 20dB the proposed scheme achieves 23% increases when compared to Reuse-3. Fig.4 and Fig.5 shows the total cell throughput comparison and fairness comparison of suggestedmethod with other reuse factor.

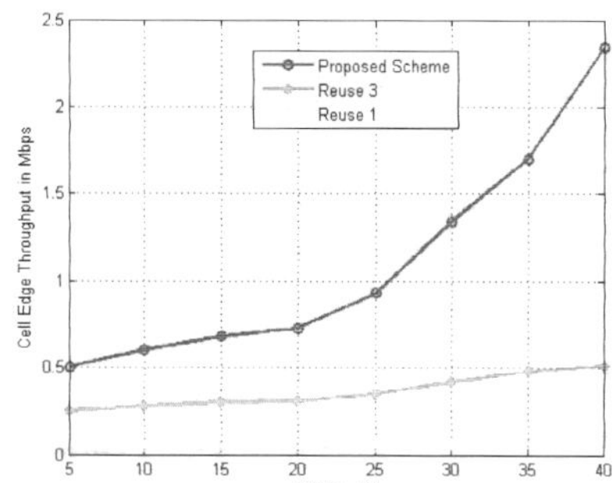

Fig.3. Cell Edge Throughput Evaluation of Proposed Scheme overother Reuse Factor

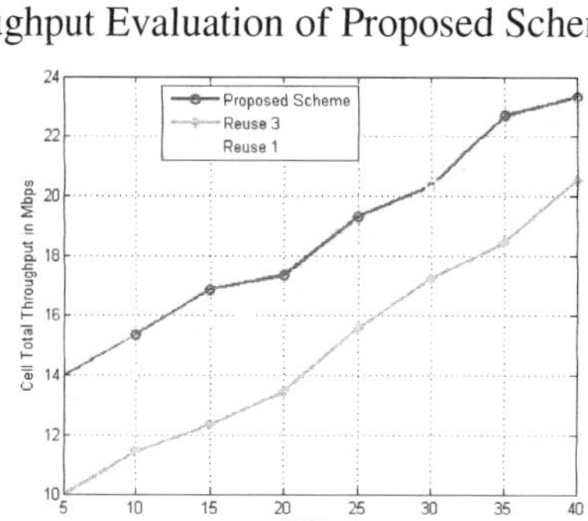

Fig.4 Total Cell Throughput Evaluation of Proposed Scheme with other Reuse Factor

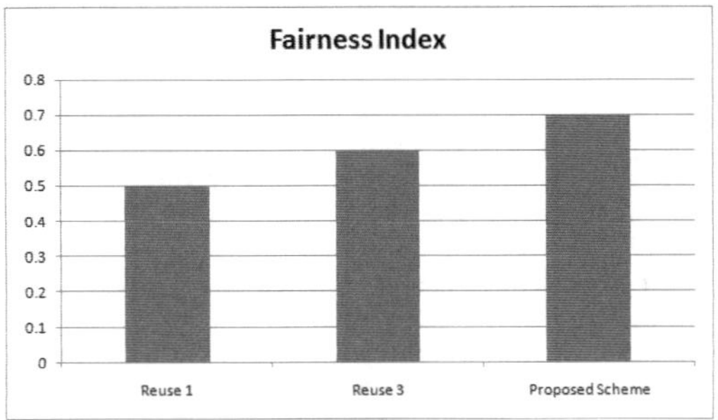

Fig.5 Fairness Comparison of Proposed Method with other Reuse Factor

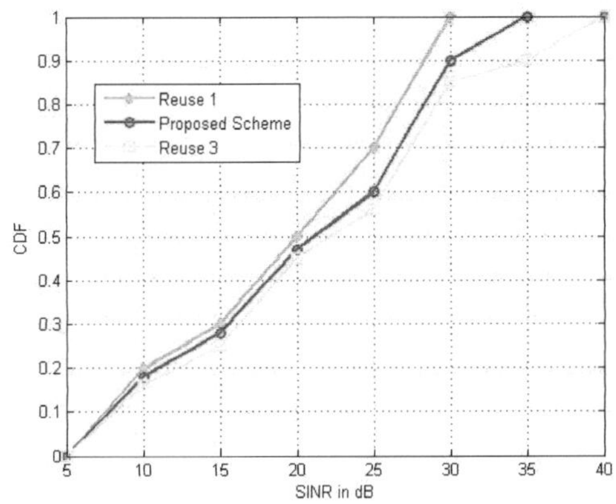

Fig.6 Cumulative Distribution Function of SINR for all users

Fig.6 depicts that the proposed scheme attainsimproved performance in comparison with other schemes because it allocates reuse factor in an organized manner.As per Fig.7, center users provides better performance due to reuse-1 while coming to both users such as center and edge users proposed scheme performs well.

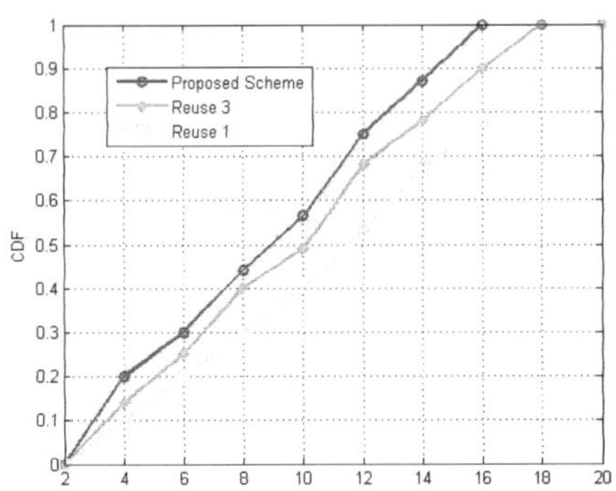

Fig.7 Cumulative Distribution Function of SINR for edge users

4. Conclusions

LTE-Advanced's tasks are to deliver services in a fair way to edge/center cell users, and to offer high SINR and low ICI in LTE-A. An algorithm is formulated in this work to deliver optimal efficiency for both edge/center users. In this algorithm, edge and center users are assigned with separate frequency reuse factors. Hence, in case of edge users ICI is found to be minimum. This scheme balances efficiency for edge/center cell users. The simulation results indicate that the proposed methodology offers maximum performance towards edge/center cell throughput. ICI is eliminated in the proposed system with the application of FR methodology. The proposed scheme is assessed with Reuse-

1/Reuse-3. Results showed thatthisdevelopedmethodologyoffersexcellent reasonable throughput for edge/center cell users.

References

[1] MSternad TOttosson AAhlen ASvensson Attaining both coverage and high spectral efficiency with adaptive ofdm downlinks in *Vehi Tech Conf* 20034: 2486-2490

[2] Chang R Tao Z Zhang J Kuo C A graph approach to Dynamic Fractional Frequency Reuse (FFR) in Multi-Cell OFDMA Networks *IEEE Int Conf on Comm* 1-6 June 2009

[3] Mohammad Assaad Optimal Fractional Frequency Reuse (FFR) in Muticellular OFDMA Systems *IEEE Veh Tech Conf* 1-5 Sep 2008

[4] Chen L Yuan D Generalizing and optimizing fractional frequency reuse in broadband cellular radio access networks *EURASIP J on W Comm &Netw* 2012 4(4): 1-15

[5] KMS Huq SMumtaz JRodriguez and RLAguiar A novel energy effiicient packet-scheduling for CoMP Comp Comm 50: 53-63 Feb 2014

[6] Ikuno JC Taranetz M Rupp M A fairness-based performance evaluation of fractional frequency reuse in LTE Int ITG Workshop on Smart Antennas Mar 2013

[7] Janarthanan R Doss S and Balamurali R 2020 Robotic-based nonlinear device fault detection with sensor fault and limited capacity for communication Journal Of Ambient Intelligence And Humanized Computing

[8] Bilios Dimitrios Selecting the Optimal Fractional Frequency Reuse Scheme in Long Term Evolution Networks *Wireless Personal Comm* 71(3): 2693-2712 2013

Efficient Communication Routing Through WiMax Network During Disaster

[1]*U Vanitha, [2]S Sivaramakrishnan, [3]S Karthika, [4]R G Babukarthik

[1,3]Associate Professor, Department of ECE, Sri Krishna College of Engineering and Technology, Coimbatore, Tamilnadu, India.
[2,4]Associate Professor, Dayananda Sagar University, Bangalore, India.
Corresponding mail id: kiruthiga.s@skct.edu.in

Abstract: During emergency situation like flood, Volcano, tsunami etc., the communication has to reach all the people in that area within a short span of time ensuring the necessary quality of service (QoS). The time at which the information is conveyed save a lot of life. In this paper we propose a communication through a Wi-max technology which broadcast information to all the nodes within the location. The node cooperates among them so that the information is conveyed to all the nodes within the coverage range of the Wi-max. Many cities are affected by some natural disaster where many lives are lots. The implementation of this technology cannot prevent the natural disturbance and certainly give warnings to people and save the valuable human lives. The WiMax network is implemented in Qualnet and various network parameters are measured when the node is mobile and non-mobile.
Keywords: Broadcast, Cooperative nodes, QoS, WiMax.

1. Introduction

Wireless communication plays a major role in modern communication. The Ad hoc network can be readily formed and that does not require any infrastructure. These advantages are considered for conveying the information or warning the people about the dangerous situation like floods in coastal area or about volcano occurrence or about a land slide in hilly region etc. For the emergency situation like these the best solution is to convey the information through the wireless medium as constructing a wired medium will not be possible and there are chance that the wired connection will get damaged during the natural disaster like these events

Since we have to cover a geographical area of few kilometers this best technology available for us is the Wi-Max technology. Wi-Max is an alternative technology to digital subscriber line and cable broadband access. Wi-Max technology is based on the IEEE 802.16 standard. Wi- Max stands for Wireless microwave access. It uses the frequency in terms of gigahertz.

The Wi-Max 802.16a standard can cover an area of 30 miles. It supports the data rate of 70 Mbps with frequency of operation at 10 Ghz and 66 Ghz spectrum. The Wi-Max 802.16e operates in the range of 2 Ghz to 6 Ghz. The biggest problem faced in the wireless communication is the communication over the non-line of sight region and multipath issues. The use of orthogonal frequency division multiplexing (OFDM) helps it to overcome these issues. Wi-Max also provides duplex communication with Time division multiplexing or frequency division multiplexing (FDM).

Wi-Max uses OFDM which helps is transmitting more amounts of data without any interference. OFDM technique uses multi carrier modulation with a frequency division multiplexing technique. At the receiver the desired carrier signal alone has the maximum value and the adjacent carrier signal magnitude is zero. So the chance of interference is nil due to the principle of orthogonality. Also the use of OFDM technique helps in using the spectrum very efficiently compared to FDM technique. The figure 1 represents the spectrum of OFDM with multiple carriers. The figure 2 indicates how effectively the spectrum is utilized in OFDM when compared to the FDM technique.

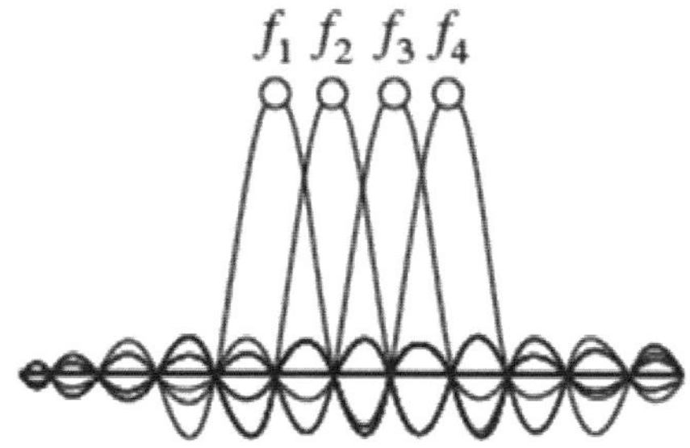

Fig.1. Spectrum of OFDM with multiple carriers

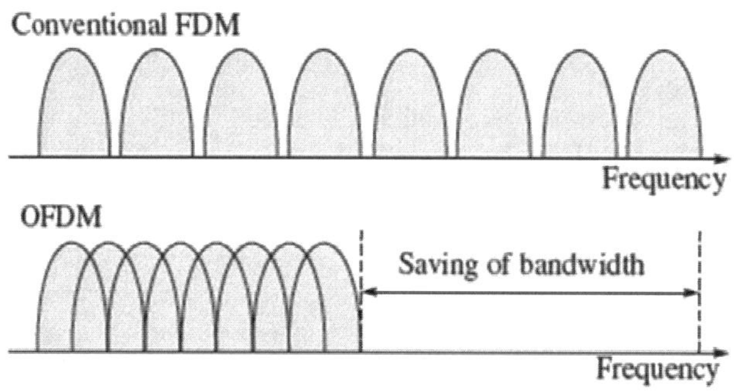

Fig.2. Spectrum usage comparison of FDM and OFDM

The network communicates from the station to the access point. There will be stationary nodes gaining the access from the access point. Under the emergency situation the information is conveyed to the fixed node in each access point. The fixed node then cooperates with the other nodes in its access point to broadcast the information to all the nodes. As the information about a dangerous situation is got by the authority he can immediately convey the information in the dangerous region. So with a few seconds the information will reach all the nodes over there.

The geographical location will be passed through the Wi-Max network there by people will be well aware of the location of danger and can save guard themselves from staying away from the location. This technology not only warns the people but also help the rescue team by giving the indication of the affected place and where there have to move so that they can rescue more people.

2. Related work

The performance improvement in IEEE 802.16e is done concentrating the physical layer by improving the Bit Error Rate (BER) [1]. Further improvement in the performance is achieved by using error control codes. Low density parity check codes and convolution codes are used to improve the performance. To overcome the effect of fading orthogonal frequency division multiple access technique is used, where there is multiple carrier signals. The WiMax also provides non line of sight coverage and this is beneficial as most of the technology provides only line of sight coverage [2].

The WiMax can also be connected to PSTN network and internet as shown in the WiMax architecture diagram in figure 3 [3]. The quality of service (QoS) has to be maintained when the data are transmitted between these networks. The WiMax has the advantage of connecting to the existing network also by providing better QoS. For the transmission of information using the direct sequence spread spectrum technique, there interference produced by a group of ultra wide band transmitter and for IEEE 802.16 these effects is carried out for various parameter like receiver bandwidth, number of user etc., and the

level of interference is found out to be below the threshold level and it was concluded the effect of the interference will not disturb the Wi-Max system [4][5].
A cognitive system is used to make the physical layer of Wi-Max to adapt to the change in the wireless channel. It is based on Support Vector Machine technique that is used for adaptation of Wi-Max physical to the wireless channel. The effect of BER for cognitive improves the performance of the system [6][7].To improve the quality of service Handover decision and execution layer (HDEL) is used to provides seamless connectivity when the node moves across the heterogeneous network[1]. It provides the quality of service when the node moves from Universal Mobile Telecommunication System to Wi-Max network or vice versa. Some important parameter like handover latency is also considered so as to improve the quality of service. Handover latency is the time taken by a node to get access to a new network or time taken to get access while moving from one Wi-Max to another Wi-Max. The handover is done for two reasons, the one when the node moves out of range from its connected network. The second reason is when there is multiple access point the node switchover to a new one so as to improve the some of its performance.
The routing of the data from the WiMax to the nodes should be energy efficient routing. The nodes under the coverage of the WiMax should receive the signal without any compromise in the QoS.[8][9][10][11] The receiving nodes act as a sink node for the WiMax transmitted data. For sending the confidential information key is used at the transmitter and the receiver. The payload increases as the data has to be encrypted at the transmitter and decrypted at the receiver.
The modulation technique and the coding technique are adaptive in nature. The reason behind this is to ensure that the network bandwidth is used effectively. The data is encoded in different layers. This ensures fairness in the overall network and thereby the loss in the data is minimized.

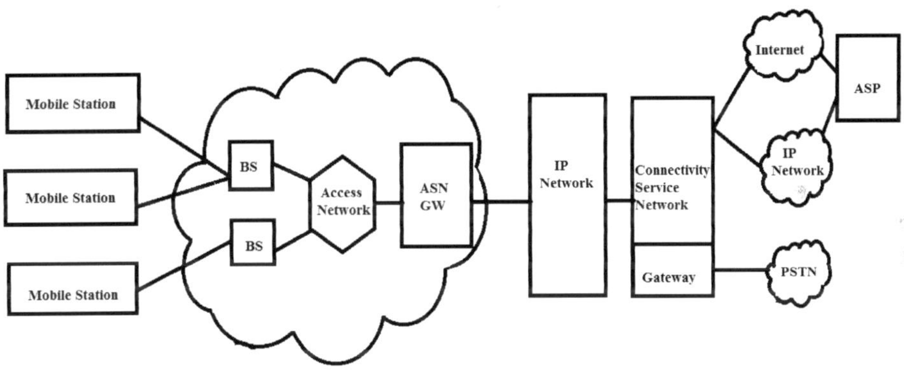

Fig.3. WiMax Architecture

3. QOS Requirement
Quality of Service plays a major role in a real time application. That too in an emergency situation the quality of service plays a major role. The protocol stack of 802.16 is shown in the figure 4a. The physical layer as shown in the figure is responsible for modulation of the signal, Whereas Medium access layer (MAC) takes care of security and uniform distribution of services to all the nodes. The following are the important parameter to be considered to offer a better QoS. Figure 4b shows the flow chart for handing over the connection.

- Handover Latency.
- Offering required Bandwidth.
- Congestion free transmission.

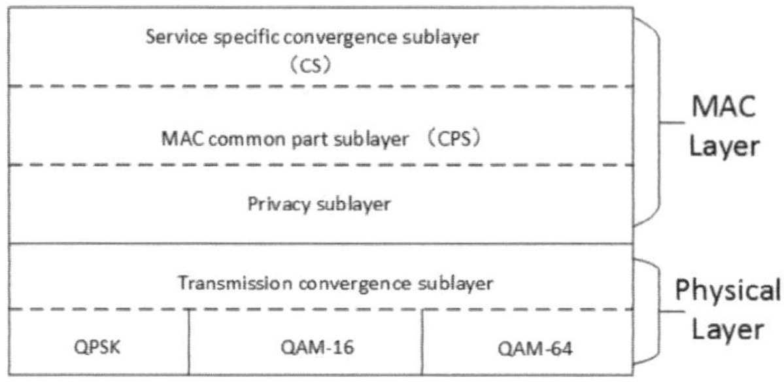

Fig.4a. Protocol Stack of 802.16

A. *Handover Latency*

When the node moves from one place to another the connectivity of the node has to be maintained. For this purpose handover mechanism is used. When the node moves away from an access point the receiving power of the node decreases which is inversely proportional to the square of the distance. When the received power moves below the threshold level then the node searches for another access point from which the received power is above the threshold. The process of handing over the control from one access point to another is termed as the handoff. Now the handover latency is the time taken to get access whiling moving from one Wi-Max to another Wi-Max

B. *Offering Required Bandwidth*

The bandwidth requirement is very important and it is application dependent. For the application like video transmission or video conference the bandwidth should be high and if the quality of video transmission is better then more amount of bandwidth should be allocated based on the video quality. For the transmission of data the bandwidths requirement of the channel is less. For the application like conveying the information during the emergency situation using the Wi-Max technique the bandwidth requirement is less as the data is alone transmitted.

C. *Congestion Free Transmission*

The data transmitted over the channel should reach the destination at the correct time. Any delay in the reach of the information will result in major disaster. So for on time delivery of the data required bandwidth should be provided.

4. Functionality in handover process

The functionality in handover process is divided into three stages. The Stages are as follows

- The signal Strength of the receiving node.
- Managing Handover.
- Taking the control over the node.

A. *The Signal Strength of the receiving node*

The Signal strength of the receiving node is monitored continuously and when the received signal strength falls below the threshold level, the receiving node searches for the next access point where the signal is well over the threshold level. As soon as the node identifies the next access point then the access point takes over the control of the node.

B. *Managing Handover*

As the node moves away from one Wi-max the other Wi-max takes over the control. The first Wi-max leaves the control only after ensuring that the second Wi-max has taken over the control. This is the best policy as the chances of connection being dropped is less.

C. *Taking the control over the node*

In some case there will be more number of Wi-max, so the node will choose a particular Wi-max during handover based on the fact that the performance does not degrade. The node choses a particular Wi-max for which the packet drop is less and where the quality of service is good.

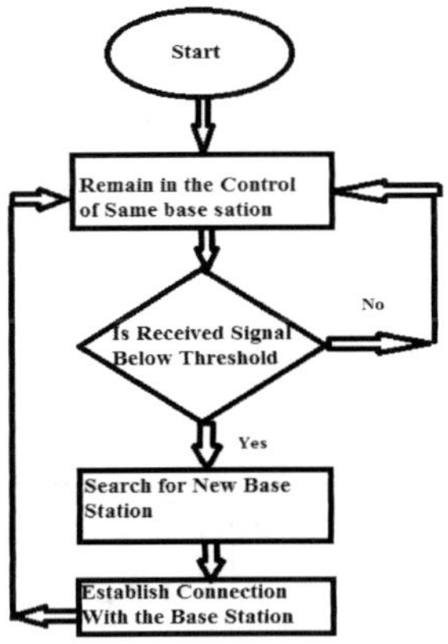

Fig.4b. Flow chart for Handing over the connection

5. Simulation layout

The simulation output is shown in the figure 5 below. The WiMax transmits the data to the surrounding nodes in the area considering 1500 X 1500 m^2. The Simulation will perform reasonably well when we consider large area as the WiMax is meant to cover a large area.

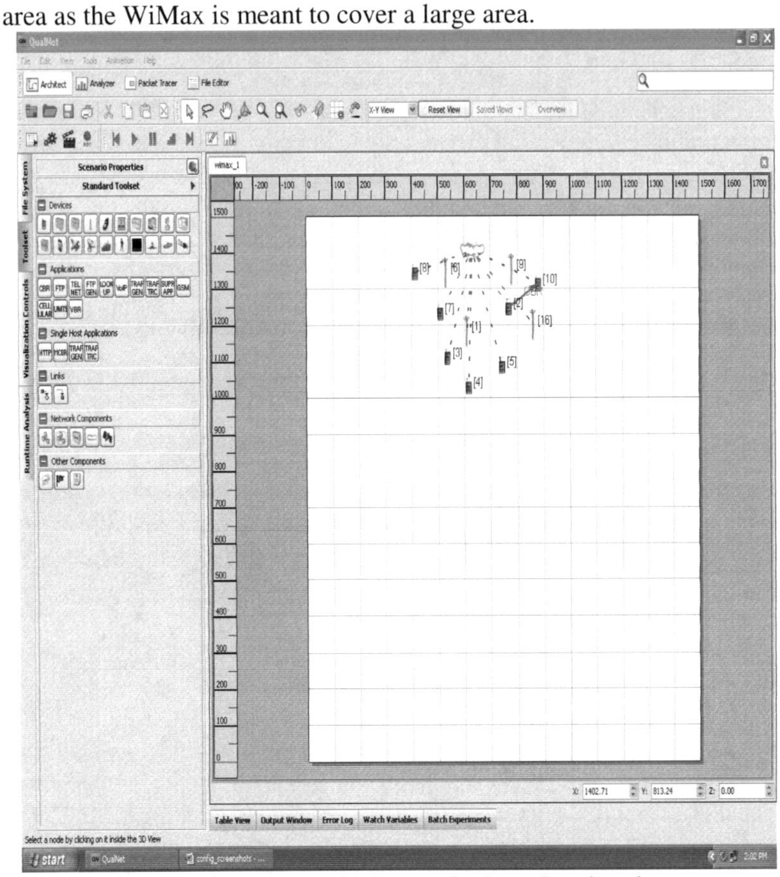

Fig.5. Simulation of Wi-max for broadcasting data

6. Protocols used in wimax

Ad Hoc on Demand Distance Vector is a on demand routing protocol. AODV can route the data for one to one or one to many communication. The term on demand is used as this protocol finds the route between the source and the sink only when the source node needs to transmit the data. The established route is maintained until the source finishes the transmission of the data. To initiate the communication the source node transmits a route request message (RREQ). When there is a valid route available the destination nodes send back a route reply message to the source node (RREP). If there is any link failure then RERR is used to inform the node about the failure in the link.

7. Simulation parameter

The table I shown below gives the area, no of nodes considered and other parameter considered for the simulation in Qualnet.

Table 1. Simulation Parameter

Parameter	Value
Area	1500 x1500m
No of nodes	11
Physical Layer	IEEE 802.16
MAC	IEEE 802.16
Routing Protocol	AODV
Mobility	None / Random waypoint
Simulation time	300 sec
Packet size	512tes

8. Simulated results

The Simulation result is done in Qualnet simulation tool for WiMax network. The simulation is carried out with and without mobility as practically there will be node in movement. The parameters like Total byte received, throughput, packet delivery ratio, average end to end delay and average jitter are compared for Wi-max network with and without mobility.

A. *Total bytes received*

The total bytes received are around 980000 bytes when there is no mobility and during mobility the received bytes is only 400000 as shown in the figure 6 below. So during mobility there is loss of data.

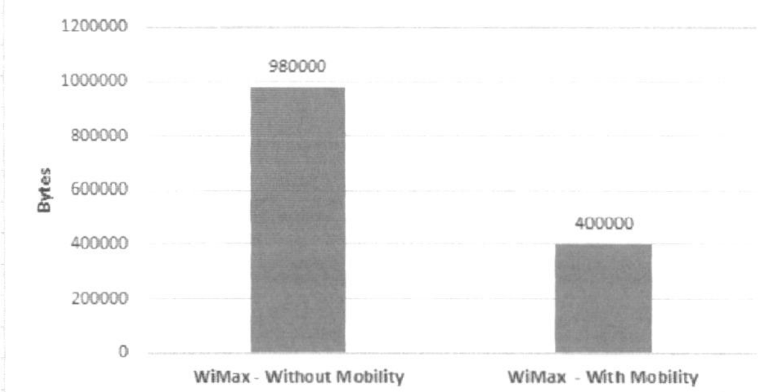

Fig.6. Comparasion of Total Bytes Received

B. *Throughput*

The number of successfully transmitted packet in a network is given by the parameter called throughput of the network. The throughput of the network decreases drastically to 1000 from 4200 whenever there is mobility in the network. Due to mobility the data loss is more and hence throughput decreases as shown in the figure 7 below

IOP Conf. Series: Materials Science and Engineering **994** (2020) 012035 doi:10.1088/1757-899X/994/1/012035

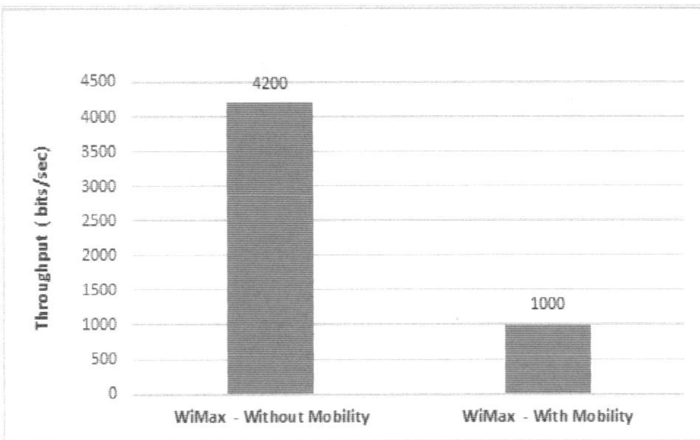

Fig.7. Comparasion of Throughput

C. *Packet Delivery Ratio (PDR)*

The packet delivery ratio is the ratio of successfully between the total packets successfully received to the total packet transmitted. Without mobility PDR is 95 percent and with mobility it reduces to 25percent (figure 8).

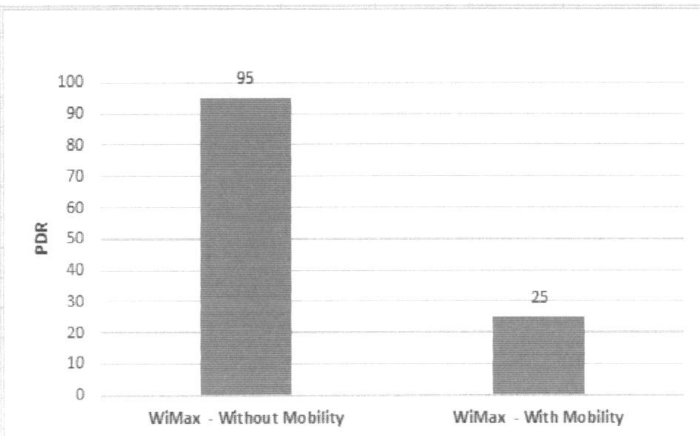

Fig.8. comparasion of Packet delivery Ratio

D. Average end to end delay

The average end to end delay is the time taken for the packet to reach the destination node. The graph 8 shown below indicates how the network performs with and without mobility. It is evident that the delay is more (0.14) when the nodes are mobile in nature compare to 0.04 when nodes are not mobile (figure 9).

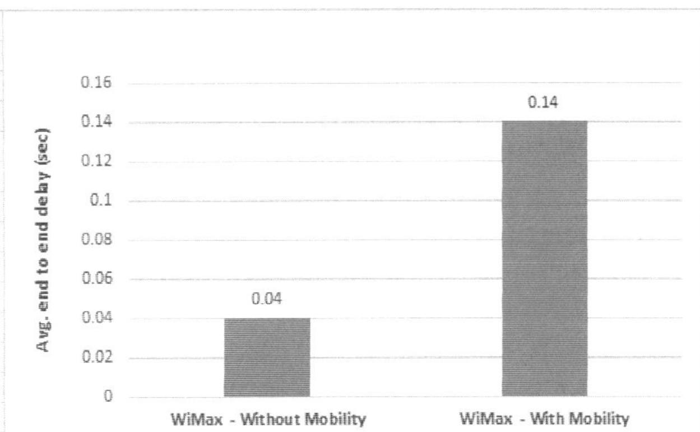

Fig.9. Comparision of Average End to End Delay

E. Average jitter

Jitter represents the variance in the time delay for the packets transmitted in the network. The figure 10 shown below indicates the average jitter for WiMax network with and without mobility. And with mobility the average jitter value is 0.02 than without mobility which comes around 0.05.

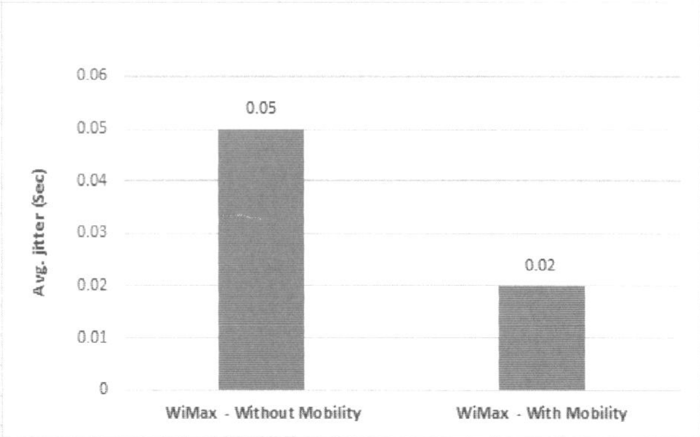

Fig.10. Comparision of Average Jitter

9. Conclusions

In this work, an effort has been made to convey information to the emergency area like flood, land slide etc. The various parameters like total byte received, throughput, packet delivery ratio, Average end to end delay, Average jitter is compared for the Wi-Max network with mobility and without mobility. It is also evident that the network provides better PDR without mobilty. This Work if implemented in the emergency areas will certainly help people living over there. Our work will certainly help them in warning about the situation and will make them to act quickly and stay away from the danger and thus saving the valuable human lives.

References

[1] Rengaraju Perumalraja Chung-Horng Lung Anand Srinivasan and R H M Hafez "Qos Improvements in mobile WiMAX networks" AHU J of Engineering & Applied Sciences 3 no 1 (2010): 107-118

[2] Jindal Shilpa Alka Jindal and Neena Gupta "Grouping WI-MAX 3G and WI-FI for wireless broadband" In 2005 1st IEEE and IFIP International Conference in Central Asia on Internet pp 5-pp IEEE 2005

[3] Wright David J "Maintaining QoS during handover among multiple wireless access technologies" In International Conference on the Management of Mobile Business (ICMB 2007) pp 10-10 IEEE 2007

[4] C Eklund R Marks and K Stanwood "IEEE Standard 80216: A Technical Overview of the WirelessMAN Air Interface for Broadband Wireless Access" IEEE Communications Magazine Jun 2002

[5] K Sarfaraz SA Ghorashi M Ghavami and AH Aghvami "Performance of Wi-Max receiver in presence of DS-UWB System" ELECTRONICS LETTERS 8th December 2005 Vol 41 No 25

[6] Jian Tao Fuqiang Liu Zhihui Zeng Zhangxi Lin "Throughput enhancement in WiMax mesh networks using concurrent transmission" Proceedings 2005 International Conference on Wireless Communications Networking and Mobile Computing 2005

[7] Msadaa Ikbal Chammakhi Pasquale Cataldi and Fethi Filali "A comparative study between 80211 p and mobile WiMAX-based V2I communication networks" 2010 Fourth International Conference on Next Generation Mobile Applications Services and Technologies IEEE 2010

[8] Sivaramakrishnan S and T Kesavamurthy "Emseep--Effective Modelling of Scalable Energy Efficient Protocol In Wireless Sensor Network" Advances in Natural and Applied Sciences 11 no 12 (2017): 18-26

[9] Sivaramakrishnan S and T Kesavamurthy "Identifying Cluster Head and Data Transmission Through Them for Efficient Communication in Wireless Sensor Network" Journal of Computational and Theoretical Nanoscience 14 no 8 (2017): 4014-4020

[10] Maheswaran G Balamurali R Sundarambal B Soundarya Priyadharshini S Vinodhini P A Smart Plug for Controlling and Automation Anywhere Using Iot International Journal of Advanced Science and Technology 29(3)(2020) pp8570 - 8577

[11] Sivaramakrishnan S and T Kesavamurthy " Optimum Routing of Data through Cluster Head in Wireless Sensor Network" International Journals of Control Theory and Applications 10 no 31 (2017): 27-35

An Image Processing based Fault Detection in Fabrics

V Jacintha[1], K H Shakthi Murugan[2], Karanam Arun Kumar[3], S Devi [3],G Saravanan [3] and D Shyam Ganesh[3] Research Scholar[1], Assistant Professor[2], Student[3]
Sathyabama Institute of Science & Technology[1], Jeppiaar Maamallan Engineering College[2],
Veltech High Tech Dr Rangarajan and Dr Sakunthala Engineering College[3]
jacinthaece@gmail.com[1]
Abstract. There are several defect that are occurring in fabrics they occur in the form of a hole, mark, improper stitch, oil stains, missed threads etc. The common thing about these defect are, they cannot be viewed by the naked eye. Hence we firmly say that they defects cannot be effectively identified by manual inspection. However, a much-automated method of inspection is essential. Hence, we are going for computer vision based defect detection. The primary requirement about computer vision is a full-fledged camera that can capture even minute defects. We use K-means clustering along with FCM segmentation in order to segment the flaws efficiently. The classifier used here is SVM classifier, with sensible classification rates of up to 98.9%.

Keywords. Fabrics, K-Means, FCM, SVM.

1. Introduction

The final produce in textile trade is critically suffering from the quality of yarn as well as weaving flaws. It is calculable that the material worth is diminish by 45%-65% because of the appearance of flaws. Therefore, flaw identification could be a very important phase for quality control in cloth yield that aids in reducing prices as well as enhancing product life. The types of defect include a hole, set mark (coarse), dropped stitch, oil stain, streak, and tight end [1]. Cloth review is carry out through human inspection or mechanically using computerized methods. In human inspection, several flaw area unit incomprehensible, and therefore the review is discrepant, having its output based on the coaching and therefore the technical ability of the inspecting individual. The review method being AN procedure with great level of persistence and repetition, it is not possible to try it on-line with inspection personnel. Sometimes this work is carry out victimization offline stations in a review phase that is separate from the assembly line. This method positively disrupts assembly speed. Computer-vision-based system perform an important task to scale back the review time and subdue the inspector observing flaws. These flaws are due to many reasons like fatigue, boredom, or perhaps carelessness. In industrial production plants, automated visual inspection systems are extensively, utilize to localize defects in products based on various digital images [2]. Computer-vision• based mostly review systems are progressively applied to interchange the human-based systems. The investment in this field is more than economical when reduction in labor cost and associated benefits are consider [3]. During this resolution, the human examiner is substitute using a camera that is assemble right away upon the weaving machinery. The photographic camera is employ for realistic identification of cloth flaws. In computer-vision-based systems, camera specification area unit vital since the camera has to be synchronic with the weaving machinery velocity. However, there are few classification issues having appreciable amount of vagueness. [4].

2. Existing System

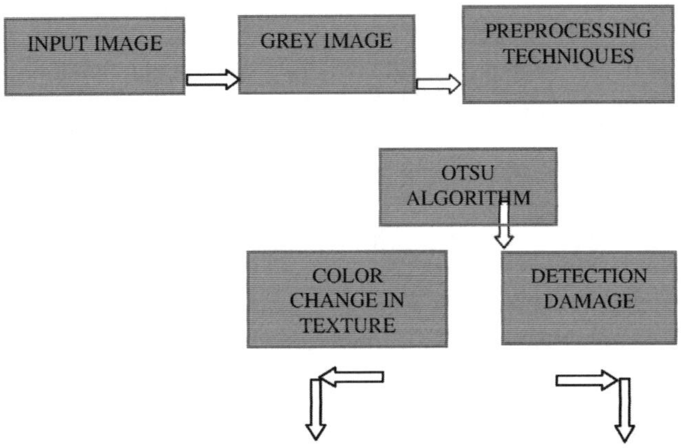

Figure 1. Block diagram of existing system

2.1 Preprocessing technique

Any scenes, watermarks as well as commotion, third, page division to isolate examples from content, fourth, character division to set apart characters from each other and finally morphological preparing to reinforce. There are a few stages required, first, few image improvement strategies to correct the complexity inside the image, second, limit to encourage block the foundation comprising the characters in situations where edge as well as other preprocessing procedures dissolved pieces of the characters or add pixels to them. The above methods present not many of these, which can be, utilize in character acknowledgment frameworks and in certain applications; few or kind of those strategies or others could even be utilize at various phases of the OCR framework.

2.2 Image Enhancement Technique

Picture upgrade improves the nature of pictures for human discernment by expelling commotion, diminishing obscuring, expanding contrast and giving more detail. This segment will give a portion of the strategies utilized in picture upgrade.

2.3 OTSU Algorithm

On-line surface distortion acknowledgment was attempted normally by separating surface pictures got by an automated camera [5]. In this paper another surface disfigurement area computation reliant on neighborhood homogeneity and numerical morphology is demonstrated .In an underlying advance ,the close by homogeneity of every pixel is figured to build up another homogeneity picture implied as (H-image).Then an old style histogram is enrolled for the H-picture to select a perfect thresholding motivating force to convey a looking at matched image, that would be used to expel the perfect size and the condition of the sorting out element(SE) for logical morphology. In a second step the image is presented to a movement of morphological exercises with this SE to make out the surface flaw. Re-enactment results depict exact distortion revelation having low counterfeit alerts. This strategy contains in separating the primary picture with a squared window and preparing the close by homogeneity for every pixel and building up the (H picture). In this work, we have utilized a window size of (11) ×(11) pixels. After, each H-picture is histogram leveled and a short time later edge to make twofold picture. By then a movement of morphological exercises, for instance, opening and closing using the Structuring Element (SE) is applied to the equal picture. Otsu's breaking point method remembers for rehashing through all potential edges and calculating the range of spread for the pixel levels each side of the edge, or example the pixels that either fall in bleeding edge or establishment .The aim is to determine the edge regards where the aggregate of frontal zone and back ground spreads is at its base. Otsu's procedure, named after its maker nobuyukiotsu, is one of various binarization computation. Otsu edge techniques were as regularly as conceivable used in various

IOP Conf. Series: Materials Science and Engineering **994** (2020) 012036 doi:10.1088/1757-899X/994/1/012036

fields. Two dimensional (2-D)otsu methodology, continues well in dividing, pictures of low, sign to commotion extent than one dimensional(1-D). The results demonstrated achievement in recognizable proof and request of most flaws [6] yet it gives great results exactly when the amount of pixels in each class are close to each other else it gives wrong results. The delayed consequences of the arranged system were differentiated and those of human vision [7].The endorsement tests exhibited that the structure performed well [8].

3. Proposed System

Mechanical sewing is an intricate creation process. Of all innovative modern weaving methods, enormous roundabout sewing machines permit the most elevated efficiency contrasted with level and little round sewing machines. The improvement of a completely robotized web review framework requires vigorous and effective texture deformity identification algorithms [9]. Texture quality is emphatically impact by machinery parametric quantities; fundamentally yarn pressure, take-off strain, sinker tallness and revolution speed. Effectively little maladjustments of the machine parts might prompt texture abandons. While clear deformities, for example, gaps or laddering can be identified utilizing current programmed inline process observation strategies, stripes are at present just discernible via prepared specialists. A stripe is a sewing deformity that happens if at any rate one of the yarn taking care of frameworks is misadjusted and employs an alternate strain to the yarn, brought about by maladjustment of pressure or sinker tallness. The outcome is a neighborhood circle tallness decline in the subsequent texture.

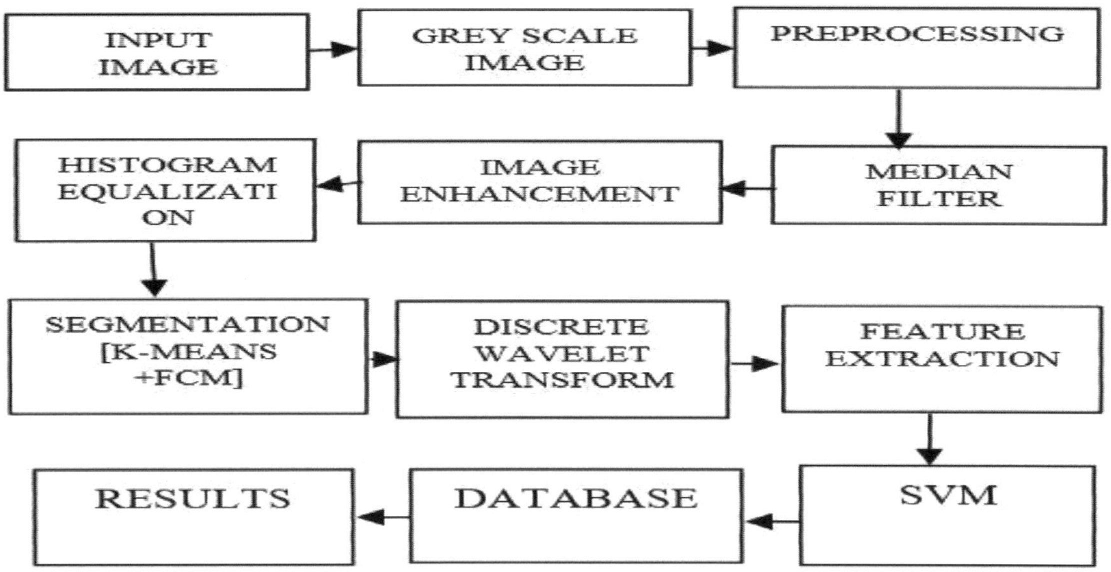

Figure 2. Block Diagram of proposed system

3.1 Gray Image

Dim scale image is one in which the estimate of each pixel in a solitary illustration speaking to just a measurement of light, i.e. it communicates mere force information. These kind of images, alternatively known as highly contrasting or monochrome, are made purely from shades of dark, fluctuating from dark at the most vulnerable force to white at the most grounded. Gray scale images are taken from the slightest bit of highly contrasting images, which w.r.t PC imaging are images with just two hues, dark and white (also termed bi-level or double images). Utilizing Binary image we can extract the commodious size and shape of the Structuring Element (SE) that is require for mathematical morphology [26].

Pre-handling is an apt term for activities with images at the lowest degree of reflection both information and yield are power images. Pre-preparing is a betterment of the image data that smothers unwanted contortions or uprates certain image highlights essential for additional handling.
Four classes of picture pre-preparing schemes as indicated by the size of the pixel neighborhood that is utilize for the figuring of novel pixel splendor: pixel brilliance changes, geometric changes, pre-handling techniques that utilization a nearby neighborhood of the prepared pixel, and picture reclamation that requires information about the whole picture. Different groupings of picture pre-preparing strategies exist.

3.2 Image Enhancement Techniques

Picture upgrade improves the nature of pictures for human discernment by expelling commotion, diminishing obscuring, expanding contrast and giving more detail. This segment will give a portion of the strategies utilized in picture upgrade.

3.3 Spatial Image Filtering Operations

In picture handling, channels are utilize to smother either the high frequencies in the image, for example smoothing the image, or the low frequencies, for example upgrading or detecting edges in the image. Picture rebuilding and improvement strategies are portray in both the spatial area and recurrence space, for example, Fourier changes.
Pictures caught frequently might be affect by clamor; nonetheless, the subsequent pictures may not give wanted pictures to investigation. Furthermore, in pictures with worthy quality, certain areas should be accentuate or featured. Spatial preparing is characterize into point handling and veils handling.

3.3.1 Median Filtering

The middle channel is a nonlinear computerized separating procedure, regularly utilize to remove clamor from an image or sign. Such commotion decrease is a run of the mill pre-processing venture to enhance the consequences of later preparing (for example, edge recognition on an image). Middle sifting is broadly utilize in computerized image handling in light of the fact that, under particular circumstances, it jelly edges while emptying commotion, likewise possessing applications in signal preparing. Middle separating is one kind of smoothing strategy, as is straight Gaussian sifting. For example: For little to direct degrees of (Gaussian) commotion, the middle channel is certifiably better than Gaussian haze at expelling clamor while protecting edges for assured, fixed window size.

3.4 Segmentation Method

Division process comprises of a few stages. The as a matter of first importance is input picture transformation to picked highlight space, which may depends of utilized bunching technique. For our situation is input picture changed over from RGB shading space to L*u*v* shading space and L*, u* and v* values are includes separately characteristics for fluffy c-implies grouping technique.
Our division strategy depends on the presumption that the picture comprises of patches of article surfaces which have uniform shading, i.e., the picture can be separated into numerous areas of uniform tone and immersion paying little heed to surface structure and wrapping up.
The division is perform uniquely in S space without utilizing any spatial data. Since isolated locales of comparable shading can be fragment together, division of progressively complex scenes requires parceling a picture in the spatial space. Along these lines, the utilization of spatial data is alluring to improve the division.

3.4.1 FCM Segmentation

Picture division assumes a significant job in clinical picture handling. Fluffy c-implies (FCM) is one of the famous grouping calculations for clinical picture division. FCM is exceptionally defenseless against clamor due to not thinking about the spatial data in picture division. The presentation of this strategy to get an ideal arrangement relies upon the underlying places of focuses of the bunches, the proportion of enrollment degree for every datum point, etc. FCM picture division approach, which

considers the dim element and overlooks different highlight is exceptionally delicate to clamor. This prompts some off-base orders. For instance, a few pixels ought to have a place with the homogeneous area are isolated. Therefore, to improve the counter - clamor and the impact of the division, we present the spatial element and the connection between the pixel and its neighbors to picture division.
Division of pictures is as yet a difficult issue since they region influenced by different factors, for example, (1) commotion caused in picture acquisition,(2) poor complexity and power in homogeneity genuinely connected to the radiofrequency MR signal,(3) incomplete volume impact being the blend of a few tissue flags in a similar pixel, actuated by the picture goals.

Fluffy c-implies bunching are filling in as follows:

1 .Convert picture into highlight space of grouping strategy (for the most part is utilized RGB shading space, however IHS, HLS, L*u*v* or L*a*b* shading spaces are utilized as well).

2. Run fluffy c-implies strategy on changed over picture.

3 Utilize few defuzzification rule or rules to order every pixel to section. Straightforward defuzzification rule depends on maximal participation evaluation of pixel to bunch. Expansion of highlight space of fluffy c-implies bunching strategy brings better division results.

3.5 K-Means Clustering

K implies bunching is a parceling technique. The capacity k-implies segments data into k fundamentally unconnected bunches, and gives back the file of the group to which it has doled out each perception. In contrast to various leveled bunching, k-implies grouping works on genuine perceptions and forms a solitary degree of groups. The differentiations states that k implies bunching is regularly more reasonable than various leveled grouping of more data.
K implies deals with each perception in your data as an item having an area in space. It finds a parcel where queries within each bunch are in close proximity to one another and as a long way from objects in varied groups as could be expect under the circumstances. You can browse five distinctive separation measures, depending on the kind of data you are bunching.
Each group in the parcel is characterize by its part protests and by its centroid, or focus. The centroid for each bunch is the point to which the entirety of good ways from all items in that group is limited. K-implies registers bunch centroid multifariously for each separation measure, to limit the total as for the measurement that you indicate.

3.6 SVM Feature Extraction

Bolster Vector Machine (SVM) is the most normally utilized grouping calculation for infection expectation. It is broadly utilize to anticipate diabetes, bosom malignancy, lung disease, coronary illness and so forth. It is a regulated learning method that is utilize for finding designs for grouping of information. SVMs was first present by Vapnik in 1960s for grouping of information. The two components utilized for the usage are the numerical programming and the portion capacities. The part work permits it to scan for an assortment of the speculation spaces. In SVM, characterization is perform by drawing hyperplanes. In two-class order, this hyperplane is equidistant from both the classes. The information occasions, which are utilize to characterize this hyperplane are term as help vector. An edge is characterize in SVM, which is the separation between hyperplane and the closest help vector. For good partition by this hyperplane, the separation of edge ought to be as extensive as conceivable in light of the fact that enormous separation gives less mistake. On the off chance that the edge is close, at that point it is increasingly delicate to commotion.

4. Algorithm

The 3 phases of this algorithm are :

- Preprocessing Phase:

Filter choice in noise reduction

a. Threshold in edge identification

- Feature extraction Phase:

    a. Probability of choosing neighbor pixels;

    b. Profile width

    c. The distance tracking point to cross point

    d. The count of repeated times

Threshold binarization

    e. Threshold in spatial reduction

- Matching Phase :

    a. Displacement within two templates.

5.Results

The test results fetched demonstrate precise flaw identification with low false alarms, thus depicting the effectivity and robust nature of the proposed detection strategy [10].

5.1 Simulation Output For Small Defects

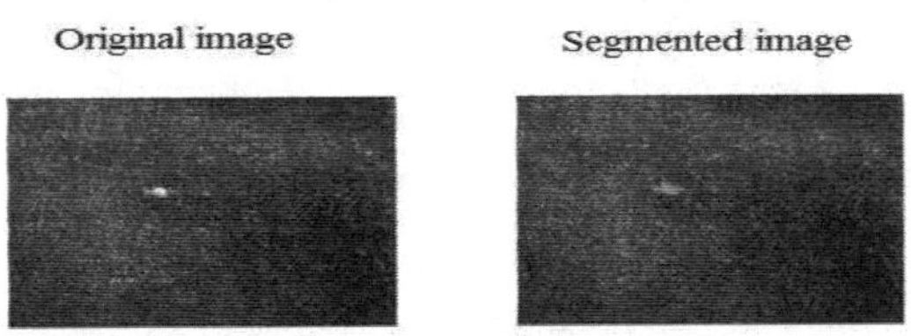

Figure 3. Segmented Image

IOP Conf. Series: Materials Science and Engineering **994** (2020) 012036 doi:10.1088/1757-899X/994/1/012036

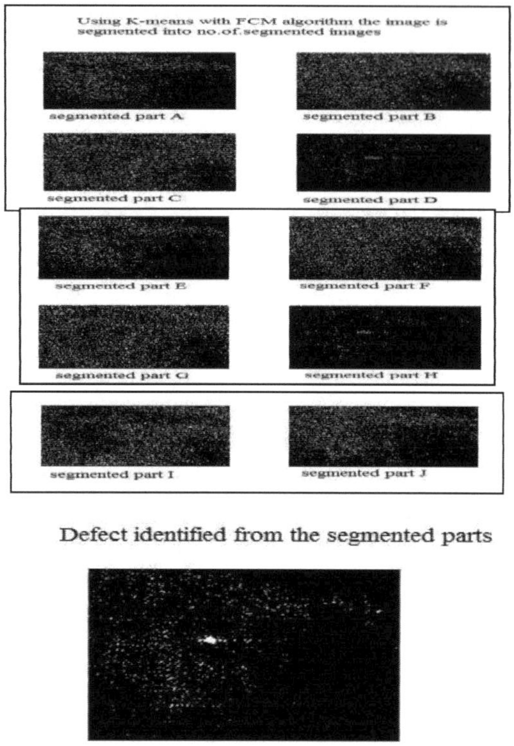

Figure 4. Defect identified from Segmented Part

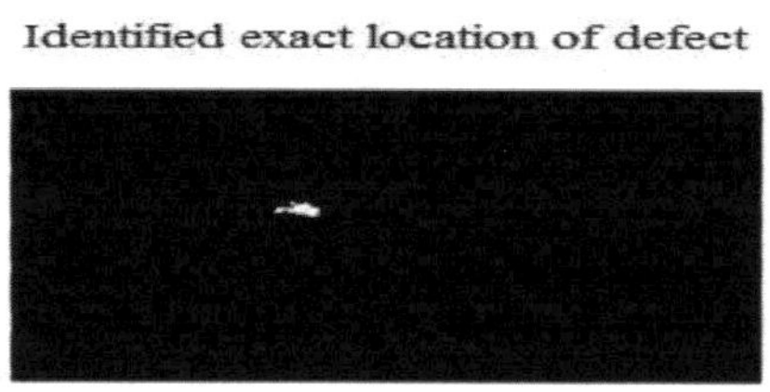

Figure 5. Identified Exact Location of Small Defect

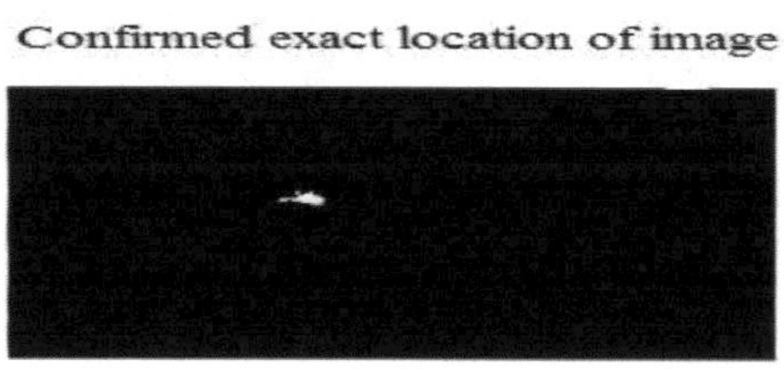

IOP Conf. Series: Materials Science and Engineering **994** (2020) 012036 doi:10.1088/1757-899X/994/1/012036

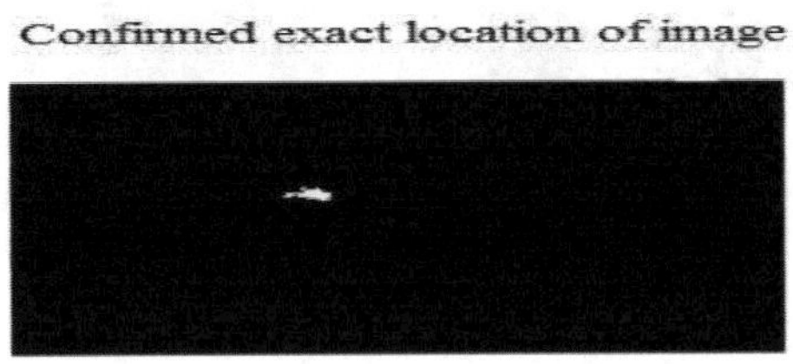

Figure 6. Confirmed Exact Location of Small Defect

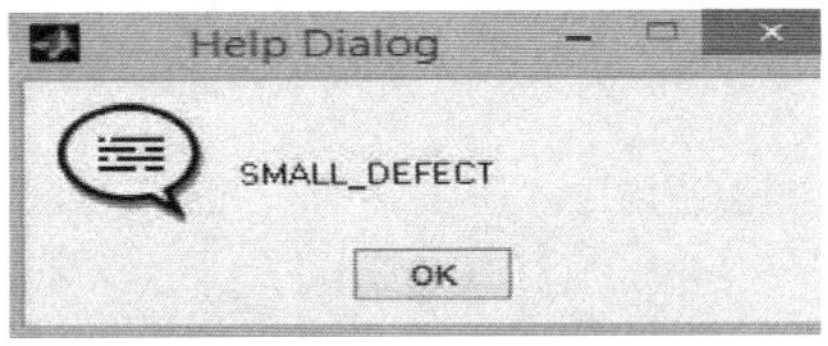

Figure 7. Output Dialog Box

Figure 8 Calculated Defected Area in a LCD Display

5.2 Simulation Output For Large Defects

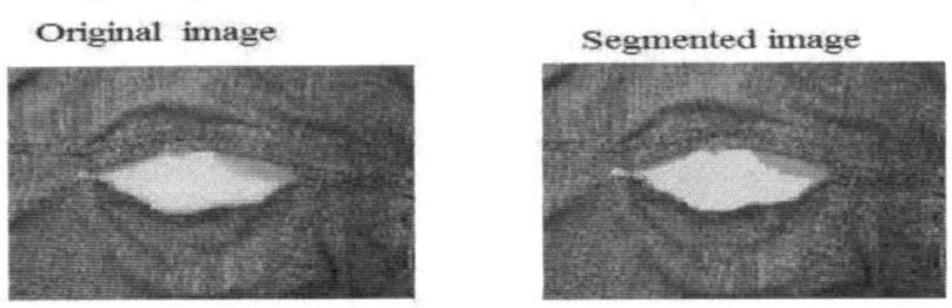

Figure 9 Segmented Image

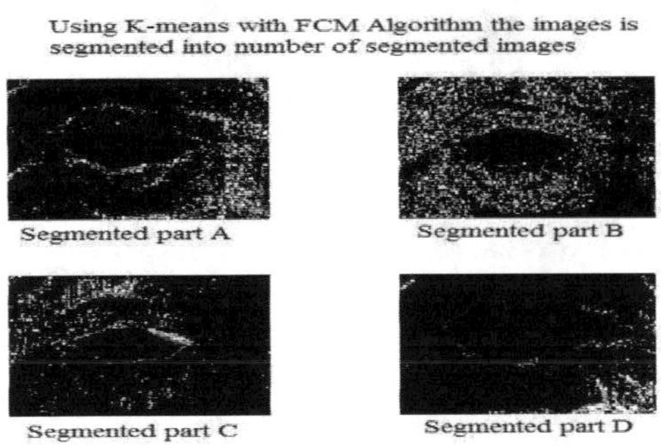

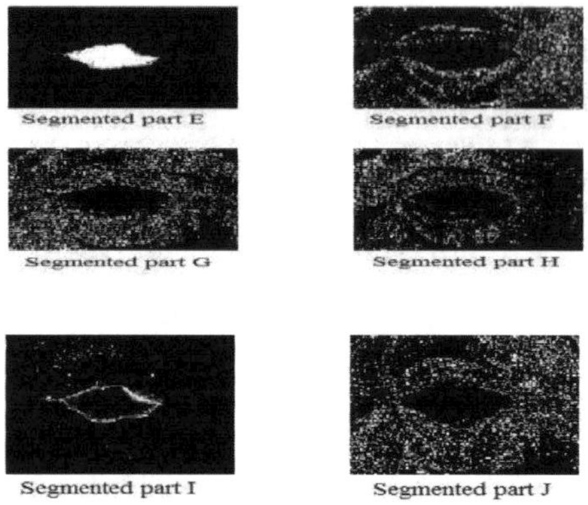

Figure 10. Number of Segmented Images

Figure 11. Defect identified from Segmented Part

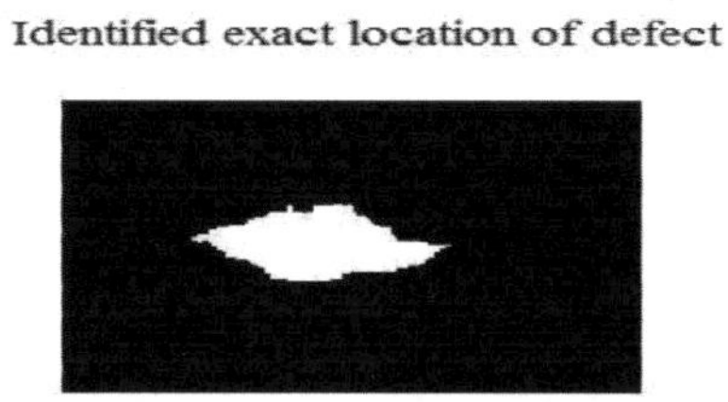

Figure 12. Identified Exact Location of Large Defect

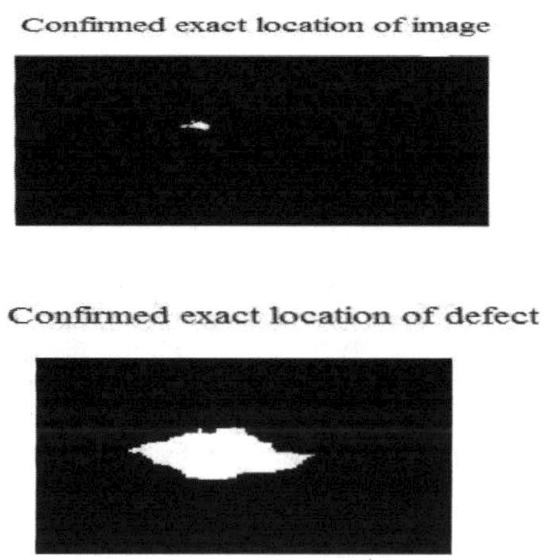

Figure 13. Confirmed Exact Location of Large Defect

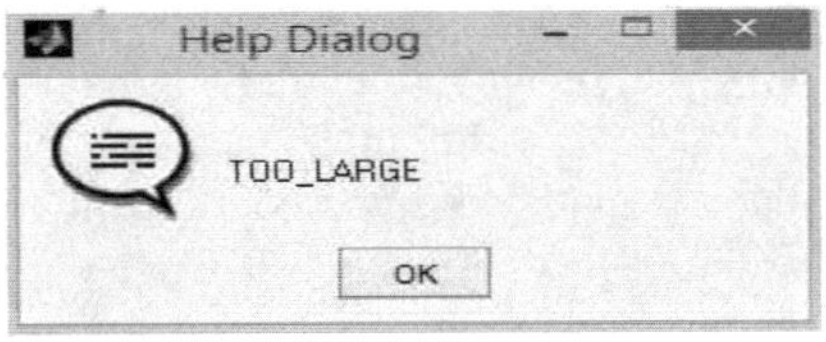

Figure 14. Output Dialog Box

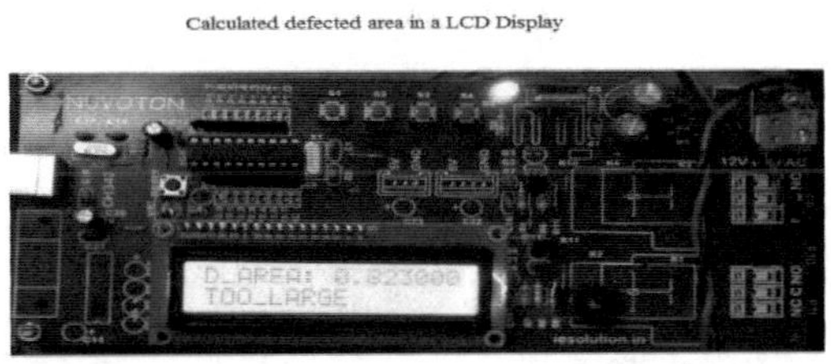

Figure 15. Calculated Defected Area in a LCD Display

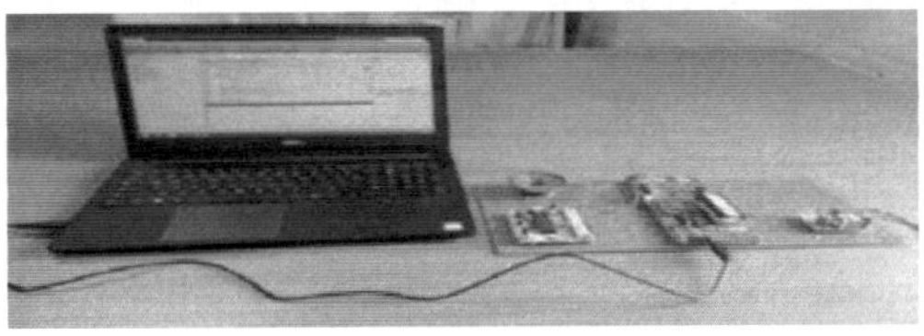

Figure 16. Hardware Setup

When the sensor value surpasses beyond normal,

- The mail is sent
- Relay turns ON.
- Chloroform is switch automatically.
- Permits call via GSM modem.
- Status is tweeted.

6 Conclusion

It is anything but difficult to recognize blames on texture pictures and procedure by utilizing this technique. The manual material quality control generally goes over the natural eye investigation. Human visual assessment is dull, tiring and exhausting errand, including perception, consideration and experience to identify effectively the flaw event. This framework is fit for recognizing textures abandons with more precision and productivity. We introduced a technique for computerized upgrade and identification of inconspicuous occasional deformities, for example, stripes in sewed texture. It was demonstrate that the recommended pipeline permits both a visual upgrade of imperfection appearance for manual assessment just as preparing of an AI based classifier for identifying absconds consequently. Finally, we train and verify the classifier. The technique shows great arrangement rates on the present examples.

7 Future Enhancement

In textile industry, we can detect the fault in real-time images but in future can be used new software like & the fault can be remove by using advanced control system. In this project, we use MATLAB software SCILAB, Virtual LAB & Computer Vision. In addition, in future, this work can be extend by implementing it through IOT that provides an online automate inspection system and the concerned person can get an alert through E-mail that is of great help in case of mobile signal or network problems.

References

[1] C-F Jeffrey Kuo C-Y Shih C-C Huang and Y-M 2015 Wen Image inspection of knitted fabric defects using wavelet packets Textile Research Journal

[2] Rebhi A Abid S and Fnaiech F 2015 Fabric defect detection using line variances of the local homogeneity images Computer System Science & Engineering vol 30 No 4 pp 1-11

[3] Rebhi A Benmhamed I Abid S and Fnaiech F 2015 Fabric defect detection using local homogeneity analysis and neural network Journal of photonics vol 2015 ID 376163

[4] Tabassian R Ghaderi R and Ebrahimpour 2011 Knitted fabric defect classification for uncertain labels based on Dempster–Shafer theory of evidence Expert Systems with Applications vol 38 pp 5259-5267

[5] Abou-Taleb HA and ATM Sallam 2008 On-line fabric defect detection and full control in a circular knitting machine in AUTEX Research Journal vol 8 No 1

[6] Shady E Gowayed Y Abouiiana M Youssef S and Pastore C 2006 Detection and classification of defects in knitted fabric structures Textile Research Journal vol 76 pp 295–300

[7] Saeidi M Latifi S Najar S and Saeidi AG 2005 Computer vision aided fabric inspection system for on-circular knitting machine Textile Research Journal vol 75 pp 492–497

[8] Sun Y and H-R 2011 Long Adaptive detection of weft-knitted fabric defects based on machine vision system Journal of the Textile Institute vol 102 pp 823–836

[9] Kumar A 2008 Computer-vision-based fabric defect detection: A survey IEEE Transactions on Industrial Electronics vol 55 pp 348–363

[10] Mak KL Peng P and Yiu KFC 2009 Fabric defect detection using morphological filters Image and Vision Computing vol 27 pp 1585–1592

[11] Shakthi Murugan KH and Jasmine Vijithra A 2020 A 40 Gbps Ro-FSO transmission system for 5 G applications Optoelectronics and Advanced Materials - Rapid Communications vol 14, pp 7-8 pp 306-309

[12] Shakthi Murugan KH et al, 2020 A novel rescuebot for borehole accidents IOP Conf Ser Mater Sci Eng 872 012044

[13] Shakthi Murugan K H and Sumathi M 2020 Performance Analysis of Millimeter waves over FSO System under Fog Weather Conditions. *International Journal of Advanced Science and Technology* vol *29*(8s) pp 3462-3477

[14] Shakthi Murugan and KH Abisheik P 2019 Advanced Smart Home Security Alert System International Journal of Scientific Research in Computer Science Applications and Management Studies vol 8 pp 1-5

[15] Murugan KH and Manickam, Sumathi 2018 Design and implementation of 5G optical Communication System for various Filtering operations using wireless optical transmission Results in Physics vol 12 10.1016/j.rinp.2018.10.064

[16] Shakthi Murugan KH and Sumathi Maqsooda Naaz K Judy Simon S 2018 Low Cost Micro Strip Patch Antenna International Journal of Pure and Applied Mathematics, vol 119 pp 573-579

[17] Shakthi H and Manickam Sumathi 2018 Millimetre Waves Over Free Space Optics System for 5G Application. Journal of Optical Communications 10.1515/joc-2018-0159

[18] Shakthi Murugan KH and Sumathi K 2018 Advanced Smart Home Security Alert System International Journal of Emerging Technology in Computer Science & Electronics (IJETCSE) ISSN: 0976-1353 vol 25 pp 96-101

[19] Shakthi Murugan KH and Jacintha Judy Simon V 2017 Safety System For Gold Mining Process Using Visible Light Communiction International Journal of Pure and Applied Mathematics vol 117 pp 371-376

[20] Manickam Sumathi and Shakthimurugan KH & Sumathi K (2016) Performance study of 4:1 multiplexer CMOS logic structures. International Journal of Control Theory and Applications, vol 9 pp 565-574

[21] Shakthimurugan KH 2015 Multi-carrier frequency offset estimation for carrier aggregation-orthogonal frequency division multiplexing communication 10. 31035-31041

[22] Shakthimurugan KH 2014 Hardware implementation of TCP/IP over visible light communication on FPGA (A concept of light fidelity). 10. 31095-31101

[23] S M K H S T J M Jerold P S A. Peeran A and Wearable AP 2019 ECG Electrodes for Detection of Heart Rate and Arrhythmia Classification 2019 IEEE International Conference on Electrical Computer and Communication Technologies (ICECCT) Coimbatore India pp 1-6

[24] Shakthi Murugan KH and Sumathi M BER and Eye 2019 Pattern Analysis of 5G Optical Communication System with Filters 2019 IEEE International Conference on Electrical Computer and Communication Technologies *(ICECCT)*, Coimbatore, India, pp 1-6 doi: 10.1109/ICECCT.2019.8869396

[25] Jacintha Judy Simon V Tamilarasu S Tamizhmani R Thanga Yogesh and K Nagarajan J 2019 A Review On Facial Emotion Recognition Techniques IEEE International Conference on Communication & Signal Processing (ICCSP)

[26] Suresh Kumar R and Manimegalai P 2020 2020 Near lossless image compression using parallel fractal texture identification Biomedical Signal Processing and Control vol 58 101862

[27] Dhanagopal R Krishanamurthi V and Rajaram A 2017 A low power robust cascade error detection in high level function verification - RCED International Journal of Control Theory and Applications vol 10, pp 15-24

[28] Jacintha VS and Karthikeyan 2019 A Survey On Various Fabric Defect Detection Methods IEEE International Conference on Communication & Signal Processing (ICCSP)

Design And Implementation Of Iterative Dct Using Concurrent Loading For Image Encoding

Dr. D Jessintha[1*], P Aishwaryalakshmi[2],UDivya[3], H Gayathri[4]

[1-4]Department of Electronics and Communication Engineering,Easwari Engineering College, Ramapuram, Chennai, India.

Corresponding Authour: jessintha@eec.srmrmp.edu.in

Abstract: Discrete Cosine Transform (DCT) is the most widely used transformforsignal compression in manyimage and video processing applications. Spatial signals are transformed to frequency domain in which the signal is broken down into different frequency componentsThrough DCT. DCT is generally used for image compression but also for noise reduction in images. But various digital architectures developed for computing DCT for images require large computations resulting in a complex and costly design. So a design of an advanced model of parallel DCT with iterative looping structureis proposed to process images. The benefit of such 8-point iterative DCT architecture is that it improves the resolution of the input image thereby utilizing low power principle and finally transforms the image data with reduced errors. The iterative DCT used in this system results in reduced processing time. The digital architecture used for the image processing follows distributed arithmetic algorithm. As a result, the proposed systemachieves an area reduction of 47% and an energy reduction of 65.55%.
Keywords:Compression, Discrete Cosine Transform (DCT), distributed arithmetic, noise reduction, resolution.

1. Introduction

Discrete Cosine Transform (DCT) is a processing technique of the signal into various frequency components. It is mostly employed in image compression modalities like JPEG compression. DCT modifiesthe picture element of an image into its corresponding frequency component values. JPEG process is the one which is a broadly used typefor lossy image compression that targetsDCT. The lossy image compression is utilized in digital images to increase the storage space [1]. DCT divides the image into various frequency sections. Another major feature of the DCT is the ability to quantify the DCT coefficients using weighted quantization values.During the quantization stage,the lower range of frequencies is discardedand only the dominantfrequencies are used to get the original image[2,4].

The need for image compression is to compress the size of the image and to keep most of the information incorporated within the original image. DCT based coding/decoding systems play a crucial role in real-time applications in the proposed system; the design of an advanced model of parallel DCT with iterative looping structure is used. The design of the architecture includes an algorithm which is of distributedarithmetic innature to achieve very high speed with low power. Also the architecture aims to achieve high accuracy with efficient hardware realization. Colored Objects of an image is identified by its hue, saturation and brightness [9].

The output obtained for DCT operation improve the pixel values to round off the quantized value and applied to the filter. The iteration is done to adjust the weights of the coefficients especially to remove the noise in the image. The iterative DCT used in this system results in reduced processing time in computingand also the processed image is of high resolution. In this architecture low power design uses parallel processing units that allow power savings from depletion in clock speed.

2. Existing System

In the existing system an approximated DCT is applied to the image for encoding it. The approximation is done in order to reduce the hardware cost. To have a good coding efficiency the approximation is done without changing the transform matrix. The main methods used are:

- LSB Truncation (LT)
- MSB Truncation(MT)
- Zero-Column Truncation(ZCT)

According to Chen's algorithm for 32-pointDCT [3], it can be splitted into three stages as shown in Fig.1.

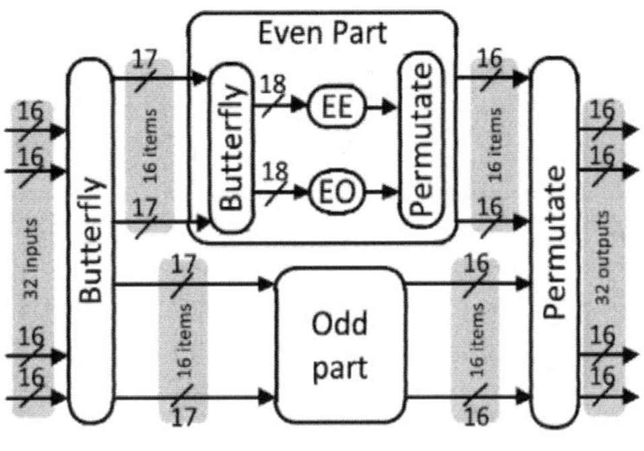

Fig. 1. Chen's Algorithm for 32-point DCT

A butterfly structure is used which divides the input terms into odd and even parts where each has 16 inputs and outputs. Even part is further divided into even-even (EE) and even-odd (EO) parts and the results are permutated. The EO and EE parts do 64 multiplications so the total even part does 128 multiplications and the odd part does 256 multiplications to produce 16 outcomes. It aims to minimize the hardware cost that occurs from the odd part and the EO part as it is similar to the odd part.

2.1LSB Truncation

In the odd part two main operations take place: Sum of product (SOP) and right-shift. Most of the LSB (Least Significant Bit) operations of SOP do not have much effect on the final result after the right shift operation. So LSB truncation is implemented. In the SOP, operation done in the Odd part each product can be written as in Equation (1).

$$Q_j = S_j Y_j \tag{1}$$

where Y_j are the inputs and S_j equals transform coefficients.

In Equation (2), the final output (summation) can be achieved where M is the number of products.

$$\sum_{j=0}^{M-1} S_j Y_j = \sum_{j=0}^{M-1} Q_j \tag{2}$$

After truncating LSBs, an accurate product term $Q\,j$ will be approximated as the term $\hat{Q}j$. Hence, the final result can be given by Equation(3)

$$\hat{y} = \sum_{j=0}^{M-1} \widehat{Q_J} \tag{3}$$

The truncation error and the overall errors are described in Equation (4) and (5) respectively as follows:

$$E_{trun} = \sum_{i=0}^{M-1} (Q_j - \widehat{Q_J}) \leq 2^{log_2^M - p1} \tag{4}$$

where, $p1$ is the precision for each product term and

$$E = E_{trun} + E_{round} \leq 2^{log_2^M - p1} + 2^{-p2-1} \tag{5}$$

*p2*is the precision for final output and $E_{round}$is the rounding error.

2.2 MSB Truncation

The wide set of high frequency elements for the conversion is generally small and so truncation of most significant bits (MSBs) is done. DCT results follow Gaussian Central Limit Theorem distribution. The variance of the pixels in the block is proportional to the variance of the distribution and mean equals zero. The DCT result *G(w, z)* is given in the probability density function (pdf) to change in the variance of pixels in a block as follows in the equation (6):

$$p(G_{w,z}) = \int_0^{\infty} p(G_{w,z}|\sigma^2)p(\sigma^2)d(\sigma^2) \tag{6}$$

Where ,$p(G(w,z)| \sigma^2)$follows Gaussian distribution as given in Equation (7).However the variance of pixels is noted as exponential distribution form[9] given in Equation (8).

$$p(G_{w,z}|\sigma^2) = \frac{1}{\sqrt{2\pi}} e^{-\frac{G_{w,z}^2}{2\sigma^2}} \tag{7}$$

$$p(\sigma^2) = \lambda e^{-\lambda} \tag{8}$$

On substituting (7) and Equation (8) in Equation (6), it gives the result that laplacian distribution followed by the variance of pixels as given by Equation (9).

$$p(G_{w,z}) = \frac{\sqrt{2\lambda}}{2} e^{-\sqrt{2\lambda|G_{w,z}|}} \tag{9}$$

Since the probability of small values are greater than the large values in Laplacian Distribution, the MSB truncation can be done. But the coefficient of index 0 is the lowest frequency component in the Odd part so that the probability of the value is to be high. To inhibit from the loss of performance, the elements at index 0 does not undergo MSB truncation.

2.3Zero Column Truncation

After encoding, quantization is done where some high-frequency coefficients become zero resulting in zero columns which can be removed to reduce the computation power consumption.Thus a less expensive approximate DCT architecture is developed with maximum of 6 LSBs truncation with MSB truncation done with respect to the index number (except 0) and the zero values are skipped for further processing.

3. Proposed System

In the proposed system, an advanced model of 32-point concurrent DCT is employed to remove noise from the image as well as compression is done. The concurrent architecture is developed using distributed arithmetic algorithm thereby introducing an iterative scheme to adjust the weight of the DCT coefficients to make DCT act as a noise filter.

3.1 Phases In The Proposed System

- ***Step I*:**The input image is vectorised into matrix. A specified type of image noise is added to it.
- ***Step II*:**Then the concurrent DCT architecture is established through VHDL (Verilog Hardware Definition Language) coding. The iterative scheme continues to apply DCT to every pixel in the input matrix and continues the process for the next inputs until it is stopped by the programmer.
- ***Step III*:**The cosine transform matrix is calculated by distributed arithmetic algorithm. This results to obtain the output with reduced processing time. This process is represented in Fig.2.

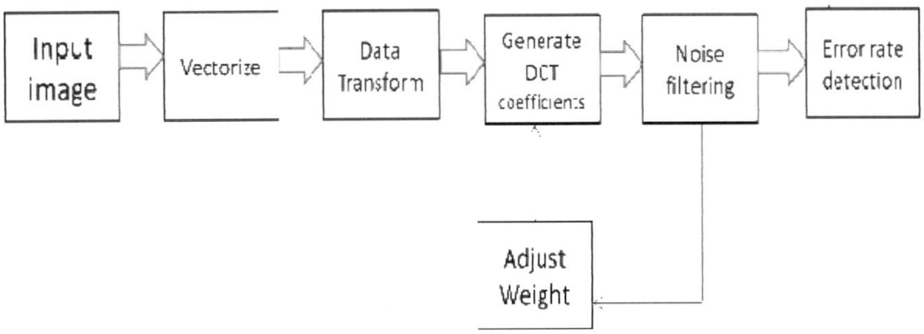

Fig. 2. Block Diagram representing the phases of Proposed System

This type of architecture does not involve approximation of Discrete Cosine Transform (DCT) and does not require the complex method of bit truncation. Moreover, the transformed output has less errors when compared to the approximated DCT output. In addition to that, the area and power consumption is also nominal. The input image is vectorised using MATLAB software and the Digital architecture is developed .

The three main techniques used in our system are:

- Iterative DCT
- Distributed Arithmetic (DA) algorithm
- Concurrent Loading Architecture

3.2Iterative DCT

DCT helps in breaking in the image into pixels by mainly focusing on the image quality. Commonly DCT is similar to DFT. DFT focuses on transforming signal, whereas DCT focuses on transforming the images to get a better resolution.Iteration is the repetition process of data which helps in generating series of outcomes. All the process is a single iteration and the output of one sequence is the input of another sequence. This process is repeated until the required result is obtained. It is a computational process. An iterative process should be convergent. Iteration in DCT image compression helps in getting more accurate pixel of the image.It is the method of computing images through iterative technique. It represents data in a cosine form. The transformation is orthogonal and energy is preserved. Also the algorithm processing rate is faster compared to other techniques. It uses real computations and it's hardware is easier to understand.

3.3Distributed Arithmetic Algorithm

Discrete Cosine Transform (DCT) used in image and video processing needs a huge amount of data transmission. Mostly in all works, 8x8 2-Dimensional DCT is performed for image compression. DCT is an operation whichrequires many number and stages of arithmetic computation. Many adders and multipliers are required to implement the hardware transformation. Multiplier blocks consume more power. Distributed Arithmetic (DA) is also used to perform a multiplicator free operation [5]. DA is used in the proposed architecture for low hardware and energy consumption. On computing DCT on an image a data array is created that can be compressed by data compaction algorithms. Thus the resultant data is stored in a compacted form. The stored image has its quality dependent on the quantization levels used in the compaction algorithm. On reconstructing the original image the data stored in the memory is retrieved and inverse DCT is performed on it.

Distributed Algorithm is a rearrangement of a multiply accumulate to mask the multiplications and implements this at bit level. This is a very effective approach to minimize the size of a multiplying parallel hardware that is suitable for FPGA designs. DA is extended to be used for complex multiplications of fourier transforms. DA acts as an effective technique to calculate inner products. Look up Tables (LUTs) and accumulators are used instead of multipliers. Since DA replaces the

multiplication blocks with adders and it has wide range of application in Very large Scale Integration (VLSI) implementations of many Signal Processing (DSP) applications.
DA simplifies the intensive arithmetic operations with multiply/accumulate (MAC) unit which is being the predominant operation in DCT.DA has many advantages such as computational efficiency, multiplier less architecture, less utilization of arithmetic computing resources. The significant featureof DAtechnique is that it gives an assumption of using both adders and multipliers for the DCT operation. Designs with DA can be implemented with standard available ICs.

3.4Concurrent Loading Architecture
A modern VLSI architecture for N multiplied by N Discrete Transform Cosine (DCT). This architecture uses the advantages of a bit serial architecture, which is bit parallel and distributed. The architecture that resulted contains only memories, adders and registers. No multipliers are used. It is also most suitable for VLSI implementation. With the proposed architecture, the design uses a concurrent architecture that has distributed arithmetic to achieve high performance with low power.

4. Results

The system is evaluated for the error between the input pixels and the transformed output. It is observed that the error remains negative which illustrates that the transformed output for each pixel has higher value than the input pixel.
The area consumption is obtained from the macro cells and pin usage is 50% lesser than the original DCT and the power utilized in this system is 23.97mW which is 64.95% lesser than the existing power consumption of 68.4mW. The existing system for approximated 32-point DCT the power consumption and the area consumption is compared to the original DCT system as given by Table 1. In comparison with the existing model the proposed system gives efficiency in the means of area at higher rate as shown in Fig. 3. The Table 2 shows the power reduction among the various designs of the system.

Table 1. Power and area consumption by various systems

Technique used	Area Consumption (no.of gates used)	Power Consumption(mW)
Original DCT [8]	167	68.4
Existing System (approximate DCT)[8]	103	26.95
Proposed System	86	24.97

Table 2. Power Reduction and Design of the system

Design	Power Reduction (%)
[6]	-23.91
[7]	-31.6
Existing[8]	-60.59
Proposed	-65.55

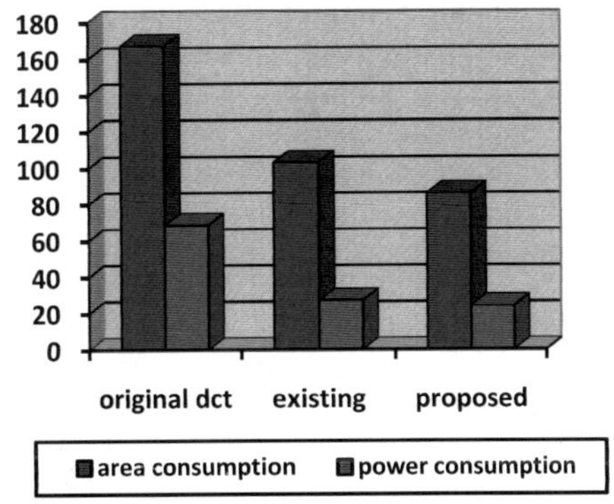

Fig. 3. Comparison of area utilization and power consumption by various techniques

5. Conclusion

Thusthe proposed model of DCT operations used MATLAB for preprocessing the image pixels and the architecture is implemented using HDL coding. At first the input image is vectorisedand then 8-point DCT is computed for the vectorised image resulting in image compression.Noise reduction has been done and so in addition to that it provides enhanced output image which can be inferred from the error determined between the input and output image. Moreover,the proposed architecture achieves a low power design with the other existing architectures andalso a reconfigurable structure.With further improvement,the architecture can be modified with various frequencies of operation and tothe next level. With the designed architecture, 46.9% areareduction and 65.55% power minimization are achieved with the same accuracy level.

References

[1] Anitha S" Image Compression Using Discrete Cosine Transform & Discrete Wavelet Transform"*International Journal of Scientific &Engineering Research Volume* 2 Issue 8 August-2011
[2] ChammanLalSahu Chandra Shekhar Singh Thakur Manoj Kumar Xalxo Mohit Thakur Miss Roshni Rathour "Survey on JPEG Image Compression Using DCT" Volume 4 Issue X October 2016 IC Value: 1398 ISSN: 2321-9653
[3] W-H Chen C H Smith and S Fralick "A fast computational algorithm for the discrete cosine transform" IEEE Trans Commun vol COM-25 no 9 pp 1004–1009 Sep 1977
[4] Balamurali R and Chandrasekar A 2019 Multiple parameter algorithm approach for adult image identification Cluster Computing 22(5) pp11909-11917
[5] VenkataVinethaKasturiYSyamalaInternational"VLSI Architecture For DCT Based On Distributed Arithmetic"*Journal of Engineering Research & Technology (IJERT)*Vol 2 Issue 5 May – 2013
[6] P K Meher S Y Park B K Mohanty K S Lim and C Yeo "Efficient integer DCT architectures for HEVC" *IEEE Trans Circuits Syst Video Technol* vol 24 no 1 pp 168–178 Jan 2014
[7] M Masera M Martina and G Masera "Adaptive approximated DCT architectures for HEVC" *IEEE Trans Circuits Syst Video Technol*vol 27 no 12 pp 2714–2725 Dec 2017
[8] Heming Sun MemberZhengxue Cheng Student Member Amir MasoudGharehbaghi Shinji Kimura and Masahiro Fujita"Approximate DCT Design for Video Encoding Based on Novel Truncation Scheme" IEEE Transactions on circuits and systems–iDec2019
[9] E Y Lam and J W Goodman "A mathematical analysis of the DCT coefficient distributions for images" IEEE Trans Image Process vol 9 no 10 pp 1661–1666 Oct 2000

IOP Conf. Series: Materials Science and Engineering **994** (2020) 012037 doi:10.1088/1757-899X/994/1/012037

[10] NR Deepa N Nagarajan D PalanivelRajan R Balamurali 2019 July Image Processing System for Detecting and Recovering from Cotton Leaf Disease using The Android Application International Journal of Innovative Technology and Exploring Engineering (IJITEE) 8(9) pp 3366 – 3381

Medical Image Retrieval Using Convolution Neural Networks

N Kalaivani[1], D Devi[2], Dr. S Sophia[3], N Manimaran[4], P Kowsalya[5]
[1,2]Assistant Professor, [3]Professor, Sri Krishna College of Engineering and Technology, Coimbatore.
[4]Assistant Professor, SACS MAVMM Engineering College, Madurai.
[5]Assistant Professor, Sri Venkateswara College of Engineering and Technology, Chittoor.
Corresponding author mail Id : kalaivani@skcet.ac.in

Abstract: In this paper, medical image retrieval is done in an effective manner by using convolution neural network (CNN). This proposed system locates reference tags and classes the DICOM images using image processing techniques and retrieval of images are done by using GUI panel. First, the deep learning technique is used to extract powerful features of the image for tag description. Conversely, this technique performs tag matching directly by passing suitable parameters which recognizes the classes as queries. The comparison features are able to capture the general form of the input image and its class based on image tags. Here, we have used a collection of 22 classes of database to demonstrate the efficiency of our system. The experimental results show the classification of images by deep learning algorithm which is used to gain the rate of retrieval accuracy by using MATLAB.
Keywords: Medical image retrieval, DICOM images, Deep Learning.

1. Introduction

The development of computerized medical imaging uses Picture Archiving and Communication Systems (PACS) such as CT, MRI and X-ray has led to the need for efficient classification and visual data search in hospitals worldwide. The use of standards such as digital imaging and Medical Communication (DICOM) enables an easy and reliable exchange of information in the medical environment. DICOM Structured Reporting (SR) is a standardized report encoding concept. This allows further data extraction and automatic interpretation. Due to the advantages that DICOM posses, the integration of the healthcare company (IHE) uses DICOM SR as a number of integration profiles. DICOM SR has become the standard format for the exchange of clinical CAD results. An effective method of deep learning along with Deep Convolution network and image processing techniques are used to handle the DICOM images.

CNN has many convolution layers and fully connected layers. For the activation function, the standard way to activate the output of a neuron is by hyperbolic tangent function or Logistic function. These nonlinear activation functions may lead to saturation and cause vanishing gradient problem. These saturating nonlinear functions show a much slower convergence than non- saturating nonlinear functions, such as max(0, x) for training optimization with gradient descent. If the input x is greater than zero, the function returns x itself, otherwise the function returns 0. The nonlinear function is referred to as Rectified Linear Unit (ReLUs). Deep CNNs with ReLUs is trained several times faster than their equivalents with hyperbolic tangent units, which is crucial for training large models in large data sets. In view of the output of convolution layers, we add it to a range of fully connected layers.

The output of the last fully connected layer is regarded as a softmax classifier input to further transform the feature vector to probabilities.

Thus, deep learning consists of three layers of input, hidden and output layers. In this system, input layer reads the file and detects the directory. The convolutionary layer (hidden layer) applies a mask or a filter to the input and the dot products of the values in the filter and the input is calculated. The convolution layer contains many filters and it produces different activation maps and proceeds to the respective fields such as rigid and non-rigid transforms of input images. Rigid transforms are pixel-oriented processes such as gray scaling, edge detection, image enhancement, image quantization and gray to RGB conversion. All layers can be called as hidden layer except the input and output layers. These hidden layers are used to store the different types of data for pooling layer purposes. Although hidden layers can be used with the high number of neurons, each pixel is relatively variable. pooling layer will down sample along the spatial dimensions. The depth dimension remains unchanged. The pooling layer accepts and produces the size volume of the image. The output layer connects all layers together to drive the output such as the plot and the numerical data driver.

2. Literature survey

In the recent past, the use of Neural Networks for data processing has been very expensive. The Deep Convolution Neural Network improves the geometric invariance of compact CNN descriptors [1]. A content- based image recovery system for gray images, RGB color space, Y C b Cr color space, images are grouped into various clusters and measured at the precision-recall crossover point [2,5]. In the field of retrieval, 3D matching is done by a 3D shape feature method of learning to extract high - level shape features which are insensitive to geometric shape deformations and uses a deep auto encoder to deform the invariable shape features as the hidden layer in this network [3]. A bidirectional model of learning representation is used to represent cross-modal text image retrieval. Since deep learning is used in Image and natural language processing , it is effective to use the deep neural networks to extract details from both text and image [4] which are trained with user generated contents data directly without any labeling effort. Although it deals with a common representation over multimodal inputs, training samples are used in the absence of certain modalities. A structured classification model is used for both visual and click functions in distance learning (DML). In particular, images and their results are first collected in order to form the training set.

These images collect multimodal features which includes click and visual features. A method of sketch- based image recovery through image- assisted inter domain learning is used. DNN model is used to study the differential characteristics by using the generalized boundary. DCN with hash codes is used to learn a medical image. Therefore, a hash layer is added to the network which is used to represent the image information as binary hash codes. A deep network images are compressed to a bit stream and then trained to extract image features as a binary vector which are combined with the network for the CBIR task. Unsupervised- learning- based method using HNN (Hopfield Neural Network) is modeled to increase the retrieval efficiency of human visual memory and reduce the semantic gap.

3. Proposed system

In the proposed system, deep convolution neural network (DCNN) consists of several convolution layers, pooling layers and fully connected network. Here, 8 layers are used for retrieving the DCM files. The output of convolution layer is feature maps produced by convolution of the filter mask with the input images. Pooling layer is used to decimate the samples in the activation maps and also reduces the count of neurons and also gives translational invariance. CNN contains a feature extractor and a classifier.

IOP Conf. Series: Materials Science and Engineering **994** (2020) 012038 doi:10.1088/1757-899X/994/1/012038

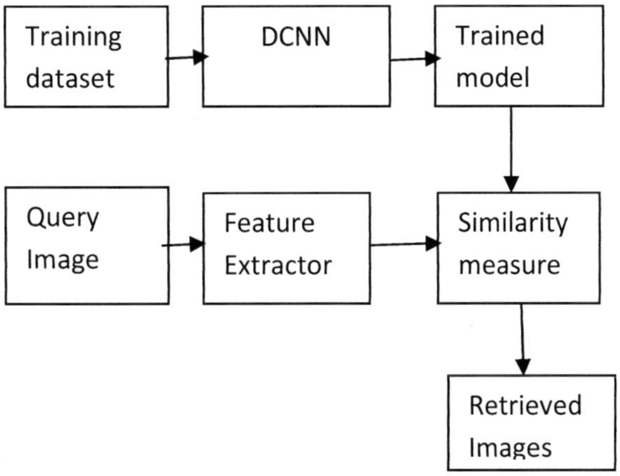

Fig.1. Block diagram of Proposed Algorithm

The deep learning structure also learn about the retrieval function with 8 layers of neural network and image processing techniques to increase the speed and efficiency of image analysis. The dataset consists of 22 classes of different parts of the body. Each part of the body consists of 300 DICOM images which are trained to collect from the DCM files.

4 Data set and simulation results

4.1 Data set

Database is trained using 22 classes of body parts of DCM tag files. The 22 classes consists of brain, breast, bladder, cervix, chest, colon, esophagus, head, head neck, heart, kidney, leg, liver, lung, ovary, pancreas, phantom, prostate, rectum, stomach, thyroid, and uterus. Each class contains 300 DICOM images as queries. The tag file is stored in the confusion matrix form. The preprocessing process is done for the model, which contains two sets of functions.

i. Training data set for model running.
ii. Testing data set for model evaluation.

4.2 Experimental results

An image is given to the system and the DICOM image matches the tag by passing the parameter for recognizing classes as queries. Comparison features can capture the general form of the query and its classes based on image's reference tags. Then, the similar images for the given feature are extracted and the image is retrieved in the MATLAB GUI panel.

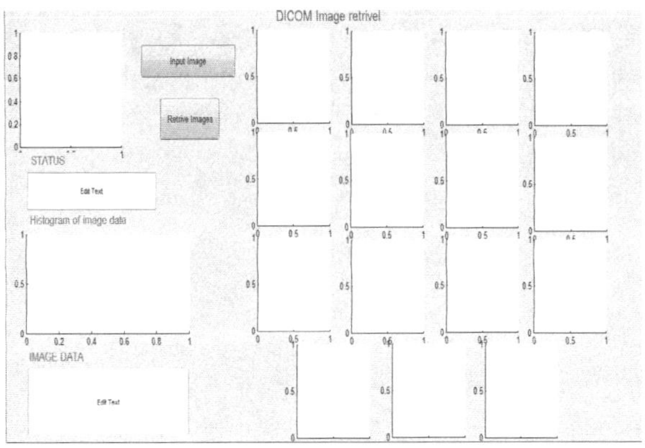

Fig.2. GUI panel

IOP Conf. Series: Materials Science and Engineering **994** (2020) 012038 doi:10.1088/1757-899X/994/1/012038

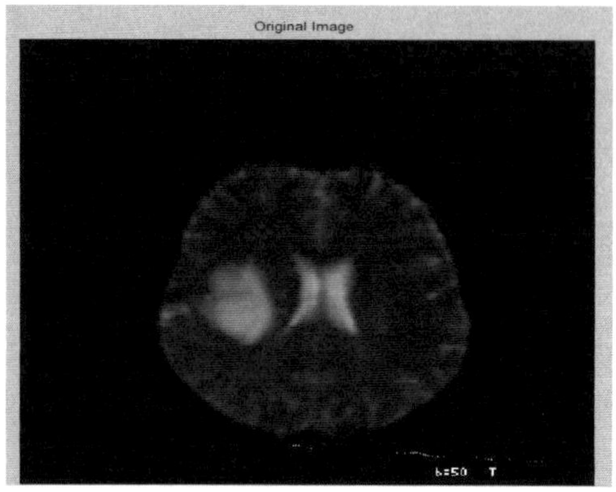

Fig. 3. Original image of the DICOM images

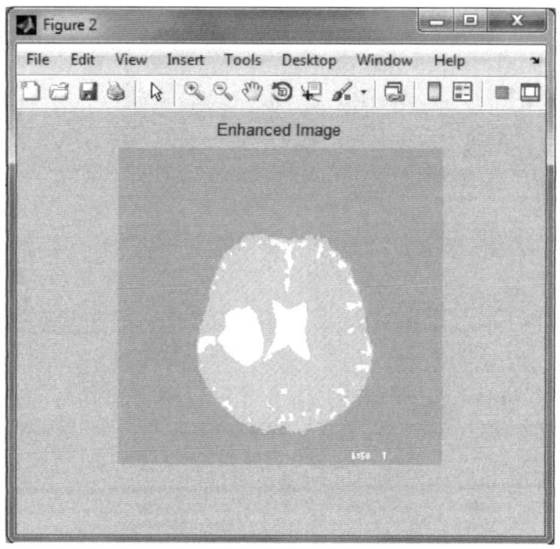

Fig.4. Enhanced image

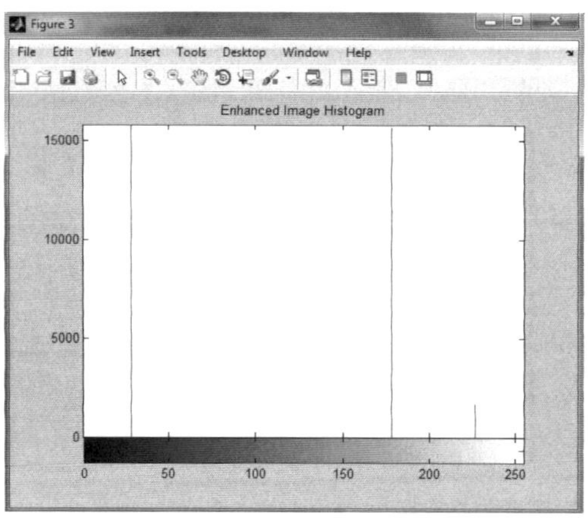

Fig.5. Enhanced image histogram

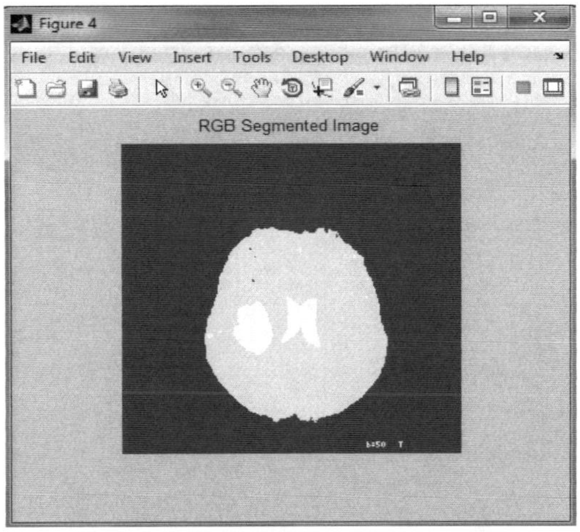

Fig.6. RGB segmented image

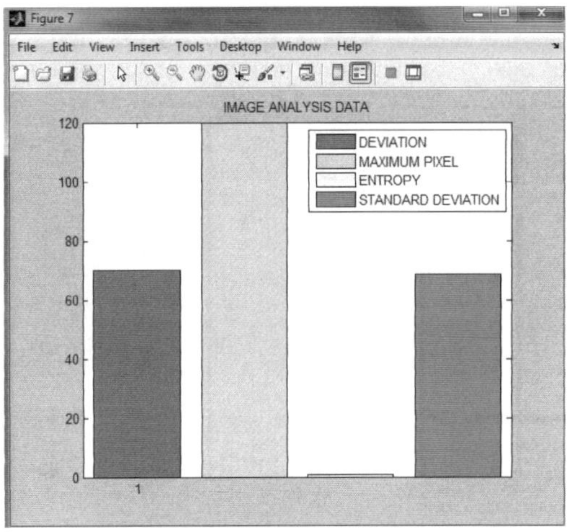

Fig.7. Image analysis data

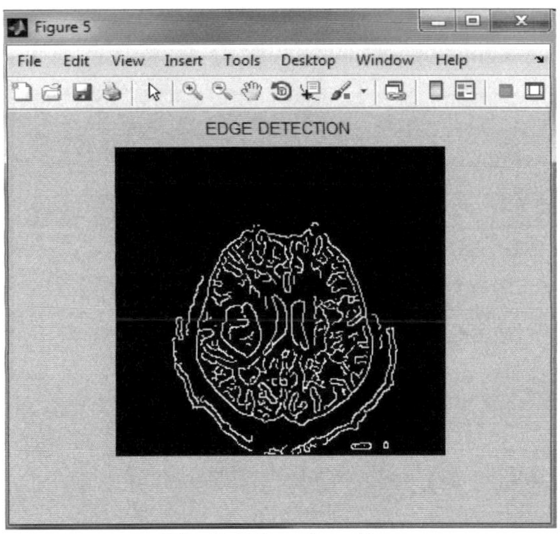

Fig.8. Edge detection

IOP Conf. Series: Materials Science and Engineering **994** (2020) 012038 doi:10.1088/1757-899X/994/1/012038

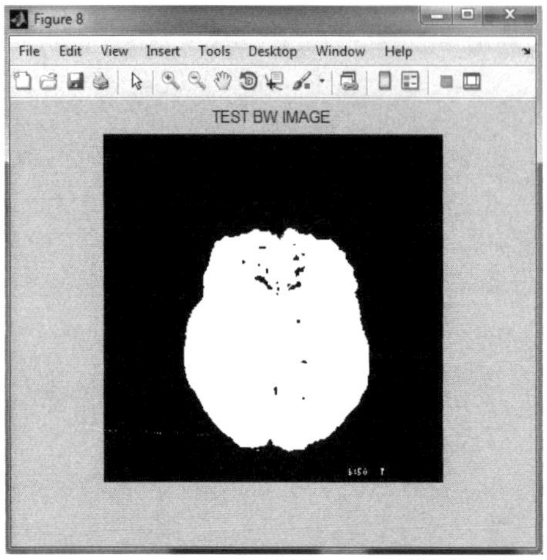

Fig.9. Black/White image

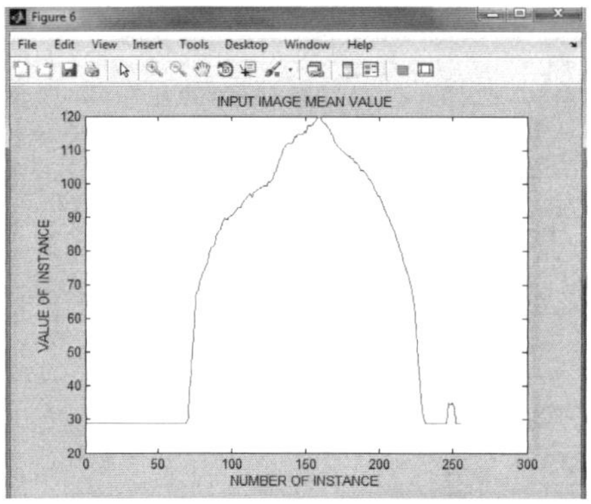

Fig.10. Input image mean value

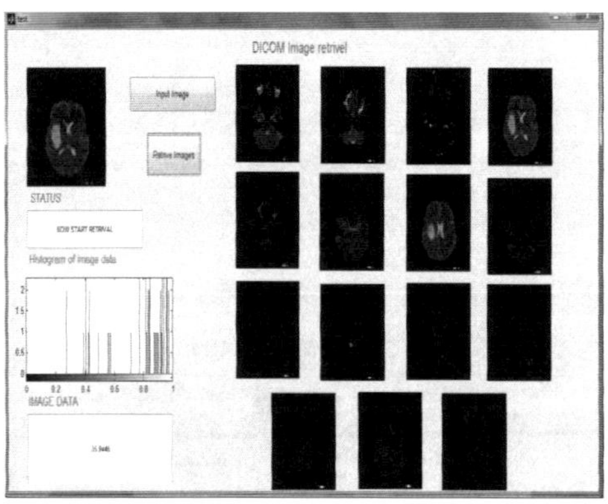

Fig.11. Retrieved images for brain

5. Conclusion

In this proposed model, the deep convolution neural network (CNN) is trained and tested using the DICOM images of 22 classes of body parts. The given Query image is predicted by the trained network and then relevant images are retrieved from the database. In the traditional neural networks, feature maps of the images are given as the input. If some features are missed in the input, then the image retrieval is not efficient.

But the proposed method improves the performance by extracting the features from the images by the convolution neural network itself. The depth of feature extraction is decided by the convolution neural network. So all the required features are extracted and given to the next layer which will improve the image retrieval rate. Increasing the data set and performing image retrieval for 3D applications are the future extension of this work.

References

[1] Aaron Chadha and Yiannis Andreopoulos" Voronoi based Compact Image Descriptors: Efficient Region-of-Interest Retrieval" *IEEE Transaction on multimedia* volume 2017

[2] Pukale D D S G Bhirud and V D Katkar "Content-based Image Retrieval using Deep Convolution Neural Network" *IEEE International Conference on Computer and Communications* 2017

[3] JinXie Guoxian Dai Fan Zhu Edward KWong and Yi Fang" Deep Shape: Deep-Learned Shape Descriptor for 3D Shape Retrieval" IEEE Transaction on Pattern Analysis and Machine Intelligence 2016

[4] Yonghao He Shiming Xiang Cuicui Kang Jian Wang and Chunhong Pan" Cross-Modal Retrieval via Deep and Bidirectional Representation Learning" *IEEE Transactions on Multimedia* NO 2017

[5] Balamurali R and Chandrasekar A 2019 Multiple parameter algorithm approach for adult image identification Cluster Computing 22(5) pp11909-11917

REAL TIME EMOTION ANALYSIS USINGBRAINWAVESIN COMPARISON TO FACIAL EXPRESSION

*S.DEVI, **Dr. M.RAJALAKSHMI, ***S.SARANYA, #B.JEEVANANDAN, #A.RAMYA
devis.aimit@gmail.com
*Assistant Professor, Department of Information Technology
**Associate Professor, Department of Information Technology
***Assistant Professor, Department of Computer Science and Engineering
#B.E Student, Department of Information Technology
Coimbatore Institute of Technology

***Abstract*:-** Emotions play a vital role in human living and are correlated with every single word which is being spoken, without them life will be saturated and irrelevant. Understanding of the real apparent facial expressions might be difficult because part of the emotion might not be revealed outside by everyone. In the proposed method, the emotions of the human are analyzed in two different ways so that the real hidden emotion of the subject can be detected. Every human has unique brain waves, which gives better accuracy than other techniques.It ensures that the facial expression exhibited outside is real or not.The emotion analyzed will be useful in treating the person with high mental stress and also infinding the emotion of differently abled person who requires medical treatment.

Keywords: Emotion Analysis, EEG, Emotivepoc+, Convolution neural network(CNN)

I. INTRODUCTION

Emotions are analysed using facial expressions and EEG signals, which are directly collected from the brain [1]. The emotions detected using these techniques are used to identify the real time emotion of the subject [2]. To record facial expressions, we used deep learning approaches. EEG signals are recorded using emotive and machine learning approaches.

A. FACIALEXPRESSIONS

DETECTION

Facial expressions of human will vary every minute and every second. Some of the emotions through facial expressions are happiness, sadness, anger, surprise, disgust, fear, confusion, excitement, desire and contempt. Facial expressions are movements of muscles beneath the skin of the face. These movements convey the emotional state of the individual. These expressions are used to convey different types of meaning in various contexts. Facial expressions are recorded using the movement of the facial parts. Some of the movements are given below:

Eyebrow raise distance is the distance between the junction point of the upper and the lower eyelid and the lower central tip of the eyebrow,upper eyelid to eyebrow distanceis the distance between the upper eyelid and eyebrow surface, inter eyebrow distance is the distance between the lower central tips of both the eyebrows, upper eyelid and lower eyelid distance is the distance between the upper eyelid and lower eyelid,top lip thickness is the measure of the thickness of the top lip, lower lip thickness is the measure of the thickness of the lower lip,mouth width is the distance between the tips of the lip corner and mouth opening is the distance between the lower surface of top lip and upper surface of lower lip [3].

B. EEG BRAIN WAVES

Brain waves are nothing but brain rhythms of different frequencies. There are five different brain frequencies from which we can find thoughts and mental state of a person. The are Gamma waves, Beta waves, Alpha waves, Theta waves and Delta waves. Gamma waves: Gamma waves have frequency of greater than 40Hz.It is mainly associated with Higher mental activity, problem solving and consciousness. Beta waves: Beta waves have frequency of between 13Hz and 39Hz.It is mainly associated with active processing, busy thinking and active concentration. Alpha waves: Alpha waves have frequency of between 7Hz and 13Hz.It is mainly associated with calm relaxed yet alert state. Theta waves: Theta waves have frequency of between 4Hz and 7Hz.It is mainly associated with deep meditation/relaxation, REM sleep. Delta waves: Delta waves have frequency of less than 4Hz.It is mainly associated with deep dreamless sleep, loss of body awareness. Through these frequencies, emotions are classified [7]. These signals are recorded using electroencephalogram (EEG) [9].

EEG is important in the diagnosis and treatment of mental and brain neuro-degenerative diseases. The role of the EEG is to identify the mental state of a human being. Electroencephalography (EEG) is an efficient tool that helps to acquire brain signals that represents various states from the scalp surface area. In this paper brain waves are collected using EEG.

C. BRAIN COMPUTER INTERFACE (BCI)

Brain computer interface (BCI) is a new way of human computer interaction. BCI facilitates the controlling of computers or related devices without physical interaction with them. BCI is the capturing and analysis of brain signals [6]. EEG isone of the most promising techniques because it is easy to capture EEG signal and analyze those signals. In this paper, we use arduino board to read the EEG signals via bluetooth and those signals are transmitted to system via wired connectivity such as USB cable. The system will display the results.

D. MACHINE LEARNING

Machine Learning is a system that can learn through self-improvement and without coded by programmer. A machine can singularly learn from the data to produce accurate results. Machine learning combines data with statistical tools to predict expected output. This output is then used by corporate. Machine learning is like data mining. The machine receives data as input and use algorithm to formulate answers. Here, the various levels are analyzed based on the accuracy of the image pixels[5] and then they are classified into convolution layers.

DEEP LEARNING

Deep learning is a machine learning technique that helps to teach the computers to do something to humans. It is the key to control voice in consumer devices like phones, TVs, and speakers.Deep learning was introduced long years before but nowadays it becomes very popular.The results of deep learning are not possible before.

In deep learning, a computer system learns to perform classification works directly from images, text, or sound which are provided in the database [8]. Deep learning models can achieve accuracy, sometimes it exceeds human-level performance. Models are trained using a large set of data and neural networks that contain various layers [10].

Deep learning is a subset of machine learning. It is based on learning and improving on its own by examining corresponding algorithms. Deeplearning works with artificial neural networks, which are formed to imitate how humans think and learn. This helps to learn the facial movements of various live images using the pixel parameters from the data information [4].

II. PROPOSED METHODOLOGY

The brain waves being recorded from the brain,they are converted from its analog form to digital and then sent to the Arduino through the Bluetooth module. The received digital signals are processed in the Arduino using the code that is provided within it. After the analyzing the signals, the result is displayed in the monitor on what emotion that the subject is currently experiencing whether the subject is angry or calm, sad or happy, amazed or frightened. Another analyzation on the emotion is made by capturing the facial expression of the subject using the lens then splitting the video into frames in which the frames are used to identify the type of emotion in which the subject is experiencing right now is analyzed.Finally the comparison is made on both the results to determine the accuracy of the analysis.

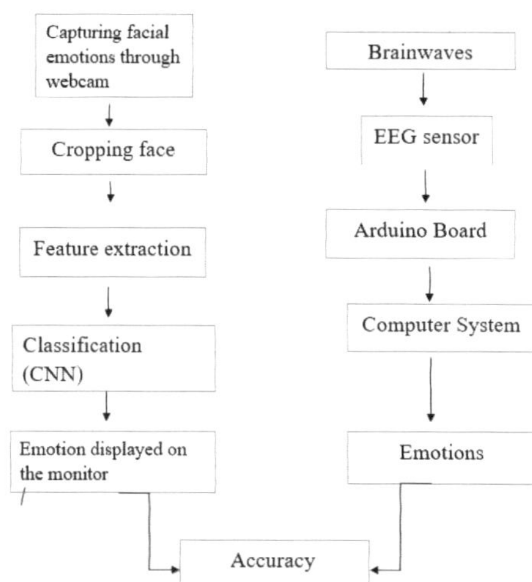

Fig 1. Architectural diagram for emotional analysis using brainwaves with facial expressions

The EMOTIV EPOC+ is a portable, high resolution, 14-channel, EEG system. Itis quick and easy to fit and take measurements in practical research applications. It is compatible with all EMOTIV software products.The Emotiv EPOC is used to distinguish emotional states.Thisability is also used to control music type, volume, light intensity, and for indicating distress. The headset also provides head turns commands such as "yes" or "no".Another important application of the Emotiv EPOC is essential for medical use.It is possible that the Emotiv EPOC can aid medical needs.

Fig 2. Emotiv EPOG

Fig 3. Arduino

Arduino is an open-source electronics platform.It is easytouse hardware and software.Arduino boards read inputs and turn it into an output. Arduino software is easy to use for beginners. It runs on Windows,Mac and Linux. In this paper the Arduino reads signals from emotivepoc+ via Bluetooth and sends the output to the system via wired connectivity like USB cable.

III. IMPLEMENTATION

A. BRAINWAVE ANALYSIS

The neuro signals from the brain are collected by placing the dry electrodes of the brainwave sensor on the scalp above the emotion sensitive part of the brain which is the frontal lobe. Those signals are then transmitted to the Arduino using air as the medium in which a embedded Bluetooth is used to do such transmission. After receiving the signal from the brain, the Arduino performs the programmed function which is already imported in it through the system. Then the signals are processed to detect the emotion based on the classified analogy on the emotion according to its frequency types such as alpha, beta, gamma and theta waves.

B. FACIAL ANALYSIS

In this part, the subject is focused in the lens by making the required arrangements. The system starts recording the expression given by the subject when the emotion stimulus videos are played. Then the videos are processed to extract the frames from the continuous recording in order to detect the emotion in that specific period. Here the emotion which is expressed by the subject is detected only based on capturing the facial movements.

Ubuntu as Operating system: In Ubuntu, the python programhas been runin the terminal as this operating system is more flexible and supports other free sourcesto be implemented easily. *Anaconda platform:*In association with ittensorflow, an open source platform for machine learning and*keras*, an open source neural networkwhich supports CNN (Convolution Neural Network) to classify the data. Together,they are used to make the suitable environment and opencv is used to import the libraries which make the processing of data easier. Finally, the result will be displayed in the monitor. The CNN algorithm is used for the following processes:

*Feature Extraction:*CNN performs a series of convolutions and pooling operations during which the features are detected.

*Classification:*The connected layers act as a classifier on the top of the extracted features. There is a probability for the object on the image being predicted by the classification algorithm.

IV. EXPERIMENTAL RESULT

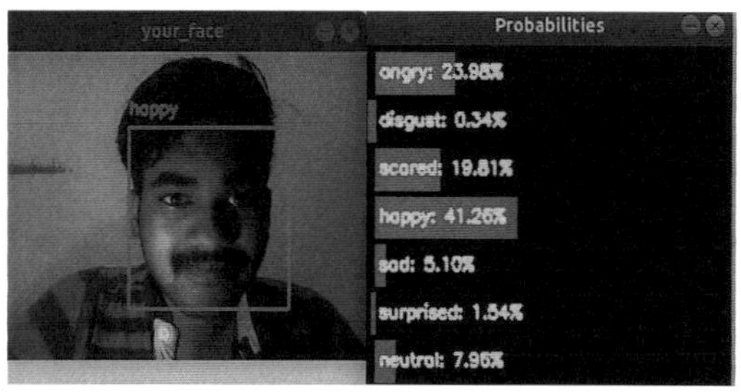

The emotion detected directly in the real time is described above. The given figure shows the emotion specific to happiness. Similarly the other emotions can also be detected. From the analysis, the captured emotions are considered to have more accuracy with EEG signals than the facial expressions. The accuracyachieved by EEG signals is 83% which is more accurate than 66% accuracy by facial expressions.

V. CONCLUSION

The analyzed emotion through the method of sensing the brain signals are accurate and are nearer to the expected expression which must be exhibited as a result of the given stimulating video inputs. Hence this paper concludes that the EEG based emotional analysis has higher accuracy than that of the facial expression analysis.

VI. REFERENCES

[1] Soleymani Mohammad SadjadAsghari-Esfeden Yun Fu and Maja Pantic Analysis of EEG signals and facial expressions for continuous emotion detection *IEEE Transactions on Affective Computing* 7 no 1 (2015): 17-28

[2] Matlovic Tomas Peter Gaspar Robert Moro Jakub Simko and Maria Bielikova Emotions detection using facial expressions recognition and EEG In *2016 11th international workshop on semantic and social media adaptation and personalization (SMAP)* pp. 18-23 IEEE 2016

[3] Chakraborty Aruna Amit Konar Uday Kumar Chakraborty and Amita Chatterjee. Emotion recognition from facial expressions and its control using fuzzy logic *IEEE Transactions on Systems Man and Cybernetics-Part A: Systems and Humans* 39 no 4 (2009): 726-743

[4] Aleksic Petar S and Aggelos K Katsaggelos Automatic facial expression recognition using facial animation parameters and multistream HMMs." *IEEE transactions on information forensics and security* 1, no. 1 (2006): 3-11

[5] Yeasin Mohammed Baptiste Bullot and Rajeev Sharma Recognition of facial expressions and measurement of levels of interest from video *IEEE Transactions on Multimedia* 8 no 3 (2006): 500-508

[6] Alarcao Soraia M and Manuel J Fonseca Emotions recognition using EEG signals: A survey *IEEE Transactions on Affective Computing* (2017).

[7] Petrantonakis Panagiotis C and Leontios J Hadjileontiadis Emotion recognition from EEG using higher order crossings *IEEE Transactions on information Technology in Biomedicine* 14 no 2 (2009): 186-197

[8] Koelstra Sander Christian Muhl Mohammad Soleymani Jong-Seok Lee Ashkan Yazdani Touradj Ebrahimi, Thierry Pun Anton Nijholt and IoannisPatras Deap: A database for emotion analysis; using physiological signals." *IEEE transactions on affective computing* 3 no 1 (2011): 18-31

[9] Lahane Prashant and Mythili Thirugnanam A novel approach for analyzing human emotions based on electroencephalography (EEG) In *2017 Innovations in Power and Advanced Computing Technologies (i-PACT)* pp 1-6 IEEE, 2017.

[10] Ackermann, Pascal Christian Kohlschein JóÁgilaBitsch Klaus Wehrle, and Sabina Jeschke EEG-based automatic emotion recognition: Feature extraction, selection and classification methods In *2016 IEEE 18th international conference on e-health networking applications and services (Healthcom)* pp 1-6 IEEE 2016.

[11] C R Rathish A Rajaran Sweeping Inclusive connectivity based routing in wireless sensor networks ARPN Journal of Engineering and Applied Science 13(5) March 2018

IOP Conf. Series: Materials Science and Engineering **994** (2020) 012040 doi:10.1088/1757-899X/994/1/012040

Application of hyperoperations for engineering practice

Konstantin A. Rubtsov, Igor S. Konstantinov, Sergey A. Lazarev, Konstantin A. Polshchykov, Vladimir E. Kiselev

Belgorod State University 85 Pobedy str., Belgorod, Russia, 308015,
rubcov@mail.ru

Abstract. This article presents some of the main points of the study for the practical application of various hyperoperations in engineering practice. The authors examined the use of hyperoperations in mathematical formalization of analytical solutions of branching algorithms, the use of new number formats for encoding information based on large numbers. The methods of applying hyperoperations considered by the authors allow them to be effectively used on modern microprocessor-based computers with a built-in mathematical coprocessor, which is an integral part of the processor core, which allows implementing algorithms based on operations of the order higher than "addition" and "subtraction".
Keywords: hyperoperations, homomorphism, number formats, superlogarithm, tetration, zeration, Ackermann function.

1. Introduction

In connection with the mass distribution of microprocessor computing in recent decades, increasing its productivity and lowering prices, mathematical modeling and digital processing of information are widespread in engineering practice. The use of mathematical modeling as a special case of sign modeling, in which the description of the modeling object is carried out in the mathematical language, and model research is carried out on the basis of mathematical methods, is currently one of the most productive and frequently used methods of scientific knowledge. The low cost of high-performance microcontrollers, signal processors and FPGAs (programmable logic device, PLD) allowed for mass production of various industrial and household equipment based on digital algorithms for processing information and signals. The emergence of new powerful hardware information processing allows the use of algorithms, previously difficult to implement. One of the ways to create new algorithms is homomorphism [1]. Such algorithms are widely used in mathematical logic and cybernetics [2]. This article considers another aspect of the application of mathematical formalism in engineering practice - the application of new functions based on the expansion of the classical mathematical basis: addition-subtraction, multiplication-division, power-exponential function, root function and logarithmic function.

2. Methods

The works [3, 4] present a mathematical description of homomorphism. The work [5] describes a homomorphic method for constructing individual mathematical objects: numbers, operations, functions, differentials, derivatives, integrals, series, numerical methods, and differential equations. Each object can have an infinite number of homomorphisms. Based on the homomorphism, it is possible to create new methods of studying functions on the basis of existing methods, new numerical methods, and expand the set of classes of solved differential equations [6]. The basis for the construction of hyperoperations is the recursive Ackermann function [7]. Ackermann function finds

IOP Conf. Series: Materials Science and Engineering **994** (2020) 012040 doi:10.1088/1757-899X/994/1/012040

practical application for testing the compiler's ability to optimize recursion. The first who used the Ackermann function for this purpose was Yngve Sundblad [8].

3. Problem statement

This article considers the main directions of research on the use of hyperoperations for engineering practice: the construction of number formats that can encode large numbers, the use of the hyperoperation zeration for writing in the form of a single formula of functions previously defined as a system with conditions.

4. Main part

4.1 Ackermann function and hyperoperator

In 1928, William Ackermann, a student of David Hilbert, published his function $\varphi(m,n,p)$, which was defined for $p=0,1,2$ and corresponded to the well-known operations of addition, multiplication, raising to a power: $\varphi(m,n,0)=m+n$, $\varphi(m,n,1)=m\cdot n$, $\varphi(m,n,2)=m^n$ [7]. In 1948, Rafael Robinson presents a function with two arguments constructed on the basis of the Ackerman function [9]. Currently, this variant of the Ackerman function is the most popular. According to Robinson, Ackerman's function is defined recursively for $m,n\in\mathbf{N}$:

$$A(m,n)=\begin{cases} n+1, m=0; \\ A(m-1,1), m>0, n=0; \\ A(m-1, A(m,n-1)), m>0, n>0. \end{cases} \quad (1)$$

In 1976, Donalt Knuth expanded the Ackermann function $\varphi(m,n,p)$ for $p>2$, who proposed to write down a sequence of operations higher in rank than degree in the form of arrows [10],[11]. This notation was called Knuth's up-arrow notation. Its meaning was that multiplication can be represented as a repetition several times of addition, and raising to a power is a multiplication performed several times. This notation made it possible to compactly record the repeated exponentiation, called super-degree or tetration. The idea of the existence of tetration, an operation with a rank higher than the degree was described as far back as 1844, and in 1901 a record of the tetration was proposed as a "degree to the left" [12, 20]. In fact, the Ackeramann function $\varphi(m,n,p)$ at $p>2$ allows describing an infinite number of operations with a rank higher than a degree - tetration, pentation, hexation, etc. The whole set of operations formed using the Ackermann function and denoted by a single operator is called a hyperoperator. Operations obtained with the help of a hyperoperator are called hyperoperations. In an extended version, the hyperoperator also describes the inverse operations. In 1989, K.A. Rubtsov proposed writing such a hyperoperator ${}_p^iR_m^n$ as a function of four arguments ($i=1,2,3$), which in the case of non-commutativity has two inverse functions ($i=2,3$) [13]. For degree, this is the root and logarithm, for tetration (superdegree) it is the superroot and superlogarithm [13]. In the same work, the notation of the function of superlogarithms as " slog " was proposed. Tetration and its inverse functions are currently still at the stage of research and development of algorithms for their exact calculation [14, 15].

4.2 Zeration and its application

The Ackerman function $\varphi(m,n,p)$ for $p=0$ corresponds to addition. In Robinson's notation according to formula (1), $m=1$ corresponds to addition, $m=2$ - to multiplication, $m=3$ - to degree, $m=4$ - to tetration, etc. It should be noted that formula (1) is defined for $m=0$. It is logical to assume that formula (1) for $m=0$ gives the result of computing an unknown operation with a rank less than addition. Such an operation was called "zeration" and was first described in 1989 as a "zero-action" paper [13]. Ackerman's function in Robinson's record gives only the result of this operation, and to determine it, it is necessary to consider addition as a result of repeating zeration:

$$\underbrace{m \circ m \circ \ldots \circ m}_{n} = m+n \tag{2}$$

Formula (2) for natural numbers produces the definition of zeration [13, 15]:

$$m \circ n = \begin{cases} m+1, \ m>n; \\ n+1, \ m<n; \\ m+2=n+2, \ m=n. \end{cases} \tag{3}$$

Formula (2) does not contradict the Ackermann function (1) for $n>2$. In table 1, zero rank should be written as $2\circ(n+3)-3$. Given that the level of operation of zeration is lower than addition, then its priority is lower than addition and subtraction. Then $2\circ(n+3)-3=2\circ n+3-3=2\circ n$. From (3) it follows that for $n>2$: $2\circ n=n+1$. It was proposed to supplement definition (3) so that it corresponds to the value of the Ackermann function $n+1$ for $n=0,1,2$. However, such additions contradict the definition of the operation of zeration according to the formula (2).

Table 1. Ackermann function values and operations used.

Level	Ackermann function	Value	Operation
0	$A(0, n)$	$n+1$	Zeration
1	$A(1, n)$	$2+(n+3)-3$	Addition
2	$A(2, n)$	$2\cdot(n+3)-3$	Multiplication
3	$A(3, n)$	$2^{n+3}-3$	Degree
4	$A(4, n)$	${}^{n+3}2-3$	Superdegree (tetration)
…	…	…	…
k	$A(k, n)$	${}^{k}_{1}R_2^{n+3}-3$	k-rank operation [1]

It is now accepted that elementary arithmetic operations are addition and subtraction. The remaining arithmetic operations are their repetition. Addition and subtraction refer to operations of the first level; multiplication and division is the second level; exponentiation, root extraction and logarithms - the third level; tetration, superroot, and superlogarithm - the fourth level. Formulas (2, 3) determine the arithmetic operation of the zero level. In fact, addition and subtraction cease to be elementary arithmetic operations.

Addition and subtraction are the most commonly used arithmetic operations. Even the simplest microprocessor contains integer addition instructions. After determining zeration, the questions arise: "Where can zeration be used?", "Why is it needed if there is mathematics and computer technology?".

In practice, most processes cannot be described by a single mathematical formula without the use of systems with conditions. For example, the relationship of the transition of a car from a state of rest to movement and subsequent changes in its speed to a stop. Using a finite number of arithmetic operations, it is impossible to write a single formula for the process of motion. To solve this problem, systems with the condition were introduced into mathematical formalism. When creating algorithms for solving practical problems in computer technology, it was necessary to use Boolean algebra operations (priority below addition and subtraction), since the arithmetic operations available are not enough to describe real processes. In mathematical formalism a lot of functions have also appeared that cannot be written by a finite number of arithmetic operations without using conditional systems.

For example, the module function $|x|$, the function of the sign of the number "$\operatorname{sgn} x$", the maximum and minimum of two numbers, and others. Zeration solves this problem in the framework of arithmetic operations. We shall consider examples of writing some functions [15].
Real value module:

$$|x| = (1 - (-x \circ 0)) \cdot ((x \circ 0) - 2) + ((x \circ 0) - 1) \cdot (2 - (-x \circ 0)) . \tag{4}$$

Number sign function:

$$\operatorname{sgn}(x) = (2 - (-x \circ 0)) \cdot (2 - (1 - (x \circ 0) \circ 0)) - ((x \circ 0) - 2) \cdot ((1 - (-x \circ 0) \circ 0) - 2) . \tag{5}$$

Within the framework of arithmetic operations, zeration describes the elements of Boolean algebra - the basis of modern computing systems. We shall consider the description of the basic set of operations: disjunction $(a \vee b = c)$, conjunction $(a \wedge b = c)$, negation $(\neg b = c)$ [15].

$$c = \neg b = (b \circ 0) - (b + 1) , \quad c = a \vee b = a + b - (a + b \circ 2) + 3 , \quad a \wedge b = \neg((\neg a) \vee (\neg b)) . \tag{6}$$

In some cases, zeration is not enough. For example, to describe the Dirac function $\delta -$. In this case, it is possible to use the inverse operation - "deltation" [15].

4.3 Notation of numbers based on homomorphism

The use of hyperoperations of the fourth level of tetration, superroot and superlogarithm is difficult due to problems with limiting the format of numbers like $D = d \cdot a^n$, where $d -$ is the mantissa, $a -$ is the base of the number system, and $n -$ is the exponent. For compact coding of large numbers in the case of applying level 4 hyperoperations, new number formats are needed [16, 17].
In [18],[19], the use of number formats based on the hyperoperations RRH1, RRH2, RRH3, and RRH4, called hyperformats, was proposed.

RRH1 format $D_{\text{RRH1}} = {}^{(n+\operatorname{slog}_a d)}a = d \overset{[a]}{*} ({}^{n}a)$. For calculations, the base $a = 10$ or $a = 2$ is used. The main feature of this format is its similarity with the well-known floating point format: $D = d \cdot a^n$. RRH1 hyperformat is constructed by increasing the rank of operations of the classical floating-point format: multiplication is replaced by the addition homomorphism based on the superlogarithm, and the degree by tetration. The algorithm for converting a standard real number $D = d \cdot a^n$ to RRH1 is very simple. To do this, it is necessary to perform a logarithm or antilogarithm with a base a until we get a number $D \in [1, a[$, then $d := D$. Initially $n := 0$, with each logarithm, numbers D should be incremented $n := n + 1$, and with antilogarithm $n := n - 1$. Thus, we can write numbers in RRH1 as follows: $-100 = 1{,}00000000\ 000000 * {}^{-2}10$, $-10 = 1{,}00000000\ 023026 * {}^{-2}10$, $-1 = 1{,}25892541\ 179417 * {}^{-2}10$, $-0{,}5 = 2{,}07122732\ 048046 * {}^{-2}10$, $0 = 1 * {}^{-1}10$, $0{,}5 = 3{,}16227766\ 016838 * {}^{-1}10$, $1 = 1 * {}^{0}10$, $5 = 5 * {}^{0}10$, $10 = 1 * {}^{1}10$, $100 = 2 * {}^{1}10$, $10^{10} = 1 * {}^{2}10$, $10^{100} = 2 * {}^{2}10$, $5 \cdot 10^{100} = 2{,}00302502\ 84114 * {}^{2}10$, $1 \cdot 10^{1000} = 3 * {}^{2}10$, $5 \cdot 10^{1000} = 3{,}00030345\ 277606 * {}^{2}10$.
RRH1 can encode negative numbers without the additional sign bit. However, this case poses problems with accuracy. For negative numbers, the RRH2 format is used, which has the sign of a number.

RRH2 format $D_{\text{RRH2}} = \{+,-,\cdot,:\}^{(n+\operatorname{slog}_a d)}a = \{+,-,\cdot,:\}d \overset{[a]}{*} ({}^{n}a)$. In fact, RRH2 is the RRH1 format with the addition of one or more characters in front of the number D_{RRH1}. A number in RRH2 is encoded according to the rules of RRH1 if $D \geq 0$. If $D < 0$, then RRH2 is presented as $D_{\text{RRH2}} = -d \overset{[a]}{*} ({}^{n}a)$, and the calculation d and n is performed similarly to RRH1, after calculating the module of the number $|D|$. The use of the sign "$-$" eliminates the problem of accuracy of the RRH1 format for negative numbers. However, if one or more signs of the number are used (advanced RRH2), the accuracy of RRH1 increases for given ranges of numbers. For RRH2, the range $0 < D < 1$ is written in the reciprocal of the number and the sign $":"$ is put. For example, a number $0{,}5$ can be written as $0{,}5 = 1/2 = 1:2 = :2$, by analogy with $0 - 2 = -2$. For positive numbers, by default, the sign $"+"$, for numbers $D \geq 1$, by default, the sign "multiplication": $5 = 0 + 5 = +5$ and $5 = 1 \cdot 5 = \cdot 5$. Such a mathematical formalism made it possible to increase the accuracy of RRH2 in the range of numbers $-1 < D < 1$.

RRH3 hyperformat of the number D is a complex number. This format consists of two numbers: the real part is a floating point number or in the RRH1 format, the imaginary part is a pseudo-character number, i.e. calculation history of the function $\ln|x|$. The algorithm for converting a real number to RRH3 format will be performed according to the scheme given for RRH1. With a negative argument in the function $\ln x$, we perform the calculation $\ln|x|$, while incrementing the imaginary part of the number and multiply by 2. If the argument is positive, then we calculate $\ln x$ and multiply the imaginary part of the number by 2 without increment. Thus, the imaginary part of the number accumulates information about the history of negative arguments in the function $\ln|x|$.

For example, we need to calculate ${}^{-60{,}34}10$ using an approximation based on the approximation $\mathrm{sln}(x) \approx \ln(x)$ for $x \in [1; e[$. In this case, we can write: $\mathrm{sln}(x) \approx \mathrm{sln}(e^x) - 1$ at $x \leq 0$, $\mathrm{sln}(x) \approx x - 1$ at $0 < x \leq 1$, $\mathrm{sln}(x) \approx \mathrm{sln}(\ln(x)) + 1$ at $1 < x$ [6, 14].

In Mathematica we get ${}^{-60{,}34}10 \approx \lg(\lg \ldots (\lg(1 - 0{,}34))) \approx -0{,}119193 + 0{,}750583 \cdot i$. In the inverse problem, we obtain $\mathrm{slg}(-0{,}119193 + 0{,}750583 \cdot i) \approx -60{,}334776 - 0{,}0161977 \cdot i$ instead of the exact value $-60{,}34$. Relative module error does not exceed $8{,}7 \cdot 10^{-3}$ %. The increase in bit depth does not lead to a significant increase in accuracy. For the RRH3 format, we get ${}^{-60{,}34}10 \approx (-0{,}348672 + 571674577919191807 \cdot i)_{\mathrm{RRH3}}$. RRH3 is resistant to imaginary rounding. For example, we can write $(-0{,}348672 + 571674577919191807 \cdot i)_{\mathrm{RRH3}} \approx (-0{,}34867 + 5{,}7167 \cdot 10^{17} \cdot i)_{\mathrm{RRH3}}$. The inverse solution to $\mathrm{slg}(-0{,}34867 + 5{,}7167 \cdot 10^{17} \cdot i)_{\mathrm{RRH3}} \approx -60{,}3398918065$, i.e. the relative error does not exceed $1{,}8 \cdot 10^{-4}$ % and the result of the inverse transformation without the imaginary part.

In the RRH3 format, complex numbers are used only for storage and compatibility with existing program libraries. The rules for working with RRH3 numbers for logarithms and antilogarithms differ from Euler's formulas. For example, $\lg(-1) = (\pi i + 2\pi n i)/\ln(10)$ is known, where $n \in \mathbf{Z}$, for RRH3: $\lg(-1) = (0 + 1 \cdot i)_{\mathrm{RRH3}} = i_{\mathrm{RRH3}}$. Standard calculation: $\lg(\lg(-1)) \approx 0{,}134934 + 0{,}682188\pi i$, and calculation in the RRH3 format: $\lg(\lg(-1)) \approx (-4{,}34294 + 2 \cdot i)_{\mathrm{RRH3}}$.

The RRH3 format reduces the error of multiple logarithms using the properties of tetration and superlogarithm.

RRH4 hyperformat of a number D is a positive number constructed according to the rules of RRH1 and the positional number system. In fact, in RRH4, the pseudo-sign (imaginary) part of RRH3 was converted to a fractional part of the number and an integer part was formed on the basis of the real part of RRH3. The algorithm for converting a regular number to RRH4 contains a bitwise calculation of the division homomorphism based on the superlogarithm with the accumulation of the result. The resulting number is positive real and compatible with all mathematical libraries [18],[19].

As an example, we shall write some numbers in RRH4. Translation algorithms are implemented in Mathematica: $10000! \approx 2{,}846 \cdot 10^{35629} \approx (455{,}21747)_{\mathrm{RRH4}} \approx (100011{,}0001010111001)_{2/\mathrm{RRH4}}$, where $2/\mathrm{RRH4}$ – is the designation RRH4 for the binary number system.

$-15 \approx (0{,}0100010001010100010101)_{2/RRH4}$. The above calculation example $^{-60,34}10 \approx (-0{,}3478362 + 5{,}71674577912 \cdot 10^{17} \cdot i)_{RRH3}$ in RRH4: $^{-60,34}10 \approx (4{,}57095 \cdot 10^{-61})_{RRH4}$. This number can be rounded without serious loss of accuracy:

$$(4{,}57095 \cdot 10^{-61})_{RRH4} \approx (4{,}57 \cdot 10^{-61})_{RRH4}, \quad \mathrm{slg}(4{,}57 \cdot 10^{-61})_{RRH4} \approx \lg(4{,}57 \cdot 10^{-61}) \approx -60{,}3400838 \approx -60.34.$$

The RRH4 format allows compact encoding of large numbers and the creation of unique numerical identifiers in sorting algorithms for large data arrays.

5. Conclusion

The hyperoperator and the hyperoperations obtained on its basis are a relatively new direction in mathematics. The use of the new hyperoperation - zeration - makes it possible to expand the set of basic arithmetic operations and the mathematical apparatus as a whole when modeling complex physical processes. Using the fundamental principles of mathematics (the concept of numbers and operations), zeration unifies the recording of a whole class of functions and algorithms. On its basis, one can get a single method for studying special functions, the formation of computing systems and algorithms directly related to basic mathematical elements. In the future, such unification will create the ability to effectively carry out research on mathematical models of complex objects and processes. Tetration is useful for building new number formats. The considered examples of new number formats provide a compact record of large numbers with the possibility of using standard formats for real and complex floating-point numbers.

6. Summary

The authors examined several special cases of the use of hyperoperations of the zero and fourth levels. The operation of the fourth level (tetration) is currently being actively studied, but has limited application due to the complexity of its calculation. The hyperformats of the considered numbers make it easier to apply tetration in practice. Zero-level hyperoperation (zeration) is implicitly used everywhere in practice in computing systems in the form of Boolean algebra, condition functions, conditional operators in programming languages, and mathematics to describe systems with conditions. The authors show the possibility of expanding arithmetic operations to solve the problems of engineering practice. In the long term, zeration can be included in the basic instruction set of microprocessor technology, since it has the simplest calculation algorithm. This will allow solving various classes of problems in engineering practice, using only arithmetic operations.

References

[1] Mathematical encyclopedia 1977 Soviet encyclopedia Vol 1: 1152 p
[2] Parmar PV 2014 Survey of various homomorphic encryption algorithms and schemes Intern J of Computer Applications Vol 91 no 8
[3] Maltsev AI 1970 Algebraic systems Nauka: 392 p
[4] Chang CC Keisler HJ 2013 Model Theory Courier Corporation: 672 p
[5] Rubtsov KA 1996 New mathematical objects BelGTASM NPP INFORMAVTOSIM: 251 p
[6] Rubtsov KA Konstantinov IS Lazarev SA Romerio GF 2018 On Application of Homomorphism in Engineering Practice and Scientific Research International Journal of Pharmaceutical Research Vol 10 Issue 1 : 283-288
[7] Ackermann W 1928 Zum Hilbertschen Aufbau der reellen Zahlen 99 : 118-133
[8] Sundblad Y 1971 The Ackermann function A theoretical computational and formula manipulative study BIT journal Vol 11 : 107-119
[9] Robinson RM 1948 Recursion and double recursion Bulletin of the American mathematical society Vol 54 № 10 : 987-993
[10] Knuth DE 1976 Mathematics and Computer Science: Coping with Finiteness Science journal Vol 194 : 1235-1242

[11] Maheswaran G Balamurali R Sundarambal B Soundarya Priyadharshini S Vinodhini P 2020 A Smart Plug for Controlling and Automation Anywhere Using Iot International Journal of Advanced Science and Technology 293 pp8570 - 8577

[12] Maurer H 1901 Uber die Funktion fur ganzzahliges Argument Abundanzen Mittheilungen der Mathematische Gesellschaft in Hamburg 4: 33-50

[13] Rubtsov KA 1989 Algorithmization of ingredients in a variety of algebraic operations Cybernetics No 3: 111-112

[14] Rubtsov KA Romerio GF 2006 Hyper-operations as a tool for science and engineering International Congress of Mathematicians ICM-2006 Abstracts Posters Short Communications Mathematical Software Other Activities: 22-23

[15] Rubtsov KA Romerio GF 2011 Hyperoperations for Science and Technology: New Algorithmic Tools for Computer Science Lambert Academic Publishing: 185 p

[16] Rubtsov K Giovanni FR 2010 Applications of a number notation hyperformat for science and engineering International Congress of Mathematicians ICM-2010

[17] Rubtsov KA Romerio GF 2010 The hyperformat of numbers in the homomorphism of numerical methods for solving differential equations Proceedings of the international scientific conference MMTT-23 Vol 1: 76-78

[18] Rubtsov KA Romerio GF 2012 Algorithmization of the hyperformat of numbers in scientific research Proceedings of the international scientific conference MMTT-25 Vol 10: 118-122

[19] Janarthanan R Doss S and Balamurali R 2020 Robotic-based nonlinear device fault detection with sensor fault and limited capacity for communication JOURNAL OF AMBIENT INTELLIGENCE AND HUMANIZED COMPUTING

[20] Eisenstein G 1844 Entwicklung von $a^{a^{a^{\cdots}}}$ J reine angew Math 28: 49-52

COGNITIVE ASPECT OF DESIGNING ETHNICITY

Valeriya Leonidovna Vorontsova1, Maxim Vladimirovich Salimgareev [2] & Dmitry Vladimirovich Salimgareev 3

[1]Associate Professor at the Department of General Mathematics Institute of Mathematics and Mechanics Lobachevsky,Orcid 0000-0001-5736-0646, ID Scopus 15066528300
e-mail:milen99@yandex.ru
[2]Associate Professor at the Department of GMUS of Kazan National Research Technological University, Orcid 0000-0002-7068-671X, ID Scopus 57210818965
e-mail: msalimga.dis@mail.ru
[3]Assistant at the Department of History, Philosophy and Cultural science of Kazan State University of Culture and Arts, Orcid 0000-0003-4349-9369,ID Scopus 57194637803

e-mail: demiyrgiys@mail.ru

[1]Kazan Federal University
[2]Kazan National Research Technological University
[3]Kazan State University of Culture and Arts
e-mail: milen99@yandex.ru
Tel.: +79172549100

Abstract. The problem of cognitive construction of consolidating mechanisms by analyzing the phenomenon of common sense as the cognitive basis of solidarity, leading to a stable integrity of ethnic structures, is considered. The work uses a system analysis, in which the content of intelligence and ethnic structures are understood as integral systems capable of deep integration. It was found that the presence of ethnicity at the borders of social and biological transforms it into a phenomenon with a high degree of transformation, and especially in the imaginary space of identity. A cognitive mechanism for constructing identification, which ultimately generates certain models of ethnic behavior, seems to be a systemic element that allows any ethnic images to function. Common sense is one of the methods of constructing ethnic, goal-oriented activity. It can potentially act not only as a source of everyday communication, but also initiate both internal and external cooperative activities of ethnic entities united by an understanding of the common social good, including the good of the planetary. The authors make an attempt to comprehend the cognitive conditions for the functioning of ethnicity through the prism of the concept of common sense - the universal anthropological basis of a conscious social good. Common sense defines historical memory, operational reactions in the moment of the present, and a joint vision of the future. This important ability is based on a common basis for any ethnic group - the biopsychosocial nature of the cognitive capabilities of most members of the group. It is in this process of collective self-awareness that social intuition manifests itself, that is, such an integral socio-psychological characteristic of the subject of social action, which includes the ability to effectively predict the development of social situations.
Keywords: common sense, consciousness, social organization, solidarity, ethnicity, culture, interaction, social situation

1. Introduction

The problem of ethnicity, as a phenomenon of social consciousness and social interaction, is one of the key ones in the existing ethnological discourse. For scientific comprehension, the problem of building consolidating mechanisms leading to a stable integrity of ethnic communities seems urgent. Obviously, important components in this process are such components as language, cultural preferences and genetic identity. The finding of ethnicity at the borders of social and biological transforms it into a phenomenon with a high degree of transformation, and above all in the realm of imaginary identity space. [1]. A cognitive mechanism for constructing identification, which ultimately generates certain models of ethnic behavior, seems to be a systemic element that allows any ethnic images to function.

At the moment, there are several scientific directions studying those or other aspects of the problems of ethnic groups and ethnicity. Most of them are within the framework of sociological research, which operate within the framework of established cognitive horizons [2]. For example, in English ethnology there is an established paradigm for which it is typical to perceive ethnic groups as forms of cultural differences social organization formed under the influence of the environmental factor. [3]. A systematic option of ethnicity is recognized as identity, which constructs its positive space of meanings through negation. According to the Chicago school, ethnic groups are nothing but forms of situational identities social institutionalization occurring in an act of free choice [4]. Representatives of the Manchester school are focused on researching the resource competition of those who are gaining "ethnicity." So, for example, Barth is sure that the primary thing is not ethnicity, but demarcation, in the framework of which there is a process of cultural content polarization of ethno-differentiating attributes [5]. The Vienna school generally does not recognize the ethnicity ontology, from the point of view of the general historical meaning [6]. Smith E. believes that the ethnos is such a "community of people, having a name that shares myths about ancestors, has a common history and culture, is associated with a specific territory and has a sense of solidarity" [7]. Horowitz considers the common origin, in the form of historical memory expressed in the "myth of a single root", as the most important, system-forming aspect of all ethnicity [8],[9].

In modern sociology, the first to study the cognitive factors that shape the dynamics of social development was Sorokin P. [22]. In the second half of the 20th century, sociologists such as E. Goffman [10], Loomis [11], Axelrod [12]. For a long time, various approaches to the phenomenon of the relationship of cognitiveness and sociality have been formed, which, to one degree or another, push researchers following them to understand social action beyond just social conditionality [13].

In the diverse intellectual space of the opinions and assessments that have arisen regarding the social nature of ethnicity, the question of cognitive methods of constructing it has not been resolved. For example, within the framework of the research areas studied, the question remains as to how intuitively rational recognition of the best solution (the common sense method) promotes ethnic solidarity.

2. Methods

We rely on the provisions of system analysis, in the framework of which we accept the content of intelligence and the social structures of both systems, therefore, we use the following concepts convention in the cognitive mechanisms study of the ethnic solidarity emergence.

A system is a set of interconnected elements, which is considered as a whole.

An element is indecomposable further in this system with this method of reviewing and analyzing the components of complex objects, phenomena, processes

Structure - a relatively stable fixation of connections between elements of the system.

The integrity of the system is its relative independence from the environment and other similar systems.

Emergence - irreducibility is the degree of irreducibility of system properties to the system elements properties.

Note that the above definitions are more of a nature of meaningful explanations. All of them are interconnected, one clarifies the meaning of the other, and together they give us the basis of our research.

Since both cognitive and social systems are always dynamic, we introduce the following basic definitions.
By the behavior (functioning) of a system we understand its action in time.
The change in the structure of the system over time can be considered as the evolution of the system.
The purpose of the system is its preferred state.
Purposeful behavior - the desire to achieve a goal.
Feedback is the effect of the results of the system functioning on the nature of this functioning.
If feedback enhances the results of functioning, it is called positive, if it weakens negative. Positive feedback can lead to unstable states, while negative feedback ensures the stability of the system using negative feedbacks. Ethnicity and intellect as systems strive to maintain their vital activity and vitality. In the analysis of ethnic communities and structures of intelligence, we proceed from the position that, as systems, they are constituted as integral systems. Their structure i.e. interaction of elements may vary. We can say that both the cognitive and ethnic systems carry out structural conjugation with other systems and the external environment. It is important to emphasize that the systems themselves choose whether to respond to changes in the external environment and how. Through their organization, these dynamic systems determine the area of all interactions into which they can enter without losing their own identity, therefore they are represented by self-referential systems.

3. Results and discussion

Aristotle was one of the first to speak about common sense, pointing to the mechanism of reason as the fundamental component of the social nature of man, which underlies any social associations. The thinker attributed sanity to the basic foundation of self-organization, since it builds the radical subjectivity of sensory given, embodied in the individual truth of each individual subject of social action, into an objectively universal and therefore vital for everyone. It is this universal foundation that allows us to comprehend the reality of a world that is common to all.
One can really agree with the ancient author, because universal, a priori acceptance of life as a fundamental value in the public ethnic consciousness can serve as a cognitive foundation for the search for ways to achieve rapidly arising social tasks.
Today, the space of cognitive sociology in solving this issue is heterogeneous. At one extreme, those who seek to consider the problem of consciousness balancing between science and social studies. At another extreme, those who seek to consider the problem of consciousness balancing between science and social studies.
For example, T. Lawson calls for a very clear and careful attitude to the concept of “knowledge”, with which we completely agree. In his opinion, the statement that “cognition is a set of processes by which we know the world” and that “cognition is an act of a social being” looks not quite correct [14]. And here we are faced with difficulties associated with the applied scientific methods [15]. K. Serulo brings to the agenda the adaptation of sociological knowledge to natural science conclusions regarding the functioning of the human brain, which cannot be recognized as a rational way to solve the problem [16]. E. Zerubavel too radically denies modern science the ability to analyze large patterns of social and cultural life through the synthesis of different areas of the human sciences, which in our opinion is not quite productive [17]. However, both cognitive approaches emphasize the fundamental gap between the natural and social sciences.
However, modern cognitive discourse in sociology inspires confidence in the possibility of solving a number of complex and unresolved issues, including the connection between intelligence and social interaction. K. Knor-Tsetina has a very balanced position, she believes that the cognitive approach puts at the forefront the knowledge of the subject, which allows him to act effectively in various situations [18]. J. Padioleau is convinced that the cognitive approach solves the problem of the social order, since the essence of any social action is cognitive. Since the social subject is the initial source of the cognitive, it is obvious that it encodes social routes, it determines the quality of certain cultural symbols [19]. B Barnes, believes that every social order is a cognitive order. The knowledge of social subjects is distributed in social systems that structure the entire normative setting of the system [20],[21]. E. Zerubavel suggested that the analysis of social conventions inevitably leads us to the problem of information perception by those who, as the subject of social action, ultimately constructs

the meanings and meanings of social memory. The considered approaches only bring us closer to the study of the cognitive aspects of the functioning of social structures.

We are in the position of a sociologization model. In our opinion, ethnic unity as a product of a social convention is impossible without a unified spiritual foundation, a common field of common sense, shared by most members of the group. In this situation, we find that the category of common sense serves as the universal anthropological basis of the perceived ethnic good. It is by its nature based on a single basis for the group — the rational nature of thinking, which acts as the foundation of principles, norms and rules that operate at the level of social reflexes and "instincts" that lead ethnic communities to acts of solidarity.

Common sense is a universal, heuristic tool given to a person outside his ethnic identity as a "default option", a certain initial intellectual ability to adapt to reality, which at a certain stage of social organization turns into a way of mutual coordination of interests in social groups. On the other hand, common sense is equivalent to spontaneously arising understanding, self-evident and extremely simple, on the basis the life forms and dynamics of society are obvious. [21]. To possess this knowledge, people do not need to have special professional qualifications, powers and political convictions; it exists outside of any identifications. This knowledge found in the depths of one's consciousness does not require scientific research, laboratory tests, or the opinions of any experts. It does not need any logical evidence.

Intuitive knowledge is the basis of rational thinking, but it is neither thinking nor its product. Performing an operational function, it is irrational. So, without "common sense" rational thinking cannot exist, however, the definition of common sense is not amenable to exhaustive rationalization. The term is open to all semantic modifications, but due to its essential uncertainty, it should be used with caution. The fact is that cultural norms cease to apply in a situation of violation of human dignity. In such a reversal, the mechanisms of common sense can act as elements of the destabilization of social structures. Although common sense limits the reflexivity inherent in human consciousness, the principle of doubt underlying philosophy and science, social life is turning into a source of natural universal human solidarity. The latter is based on the knowledge of reality, how it is seen by people regardless of their identity, including ethnic and religious.

4. Summary

The common sense method allows you to focus not on the elite circle of consumers and producers of skills and abilities, meanings and knowledge, but on society as such and can be used in the process of education (translation of knowledge and recommendations). Enlightenment, on the other hand, gives impetus and sets the paradigm of integration on the foundation of generally understood values. The "common cause" arising due to common sense concerns everyone, which means that conditions arise for the transparency of control, which can be carried out in the course of reproducible intragroup interaction.

Common sense is the source of all, including ethnic, communication, which, for all its rational basis, is spontaneous. It initiates the cooperative activity of people united by the idea of the common good as a fundamental value, rooted in the bowels of the social nature of the human race. And although humanity is organized into small and large ethnic, political and other groups, with diverse local and often overlapping interests, solidarity based on common sense is the foundation of any human organization.

A separate important segment of the manifestation of common sense is the relationship between ethnic groups. Ethnic groups are subjects of social action and are able to influence the adoption and adjustment of state and regional decisions, the course of spontaneous or negative impulses of destructive behavior introduced from outside. As a rule, members of ethnic groups have similar goals, motivation and are active participants in the social process. They have a sense of ownership and solidarity, understood as a shared and practiced strategy based on tolerance or xenophobia.

The connecting thread of common sense and solidarity in relation to ethnic relations is the construct of historical memory as one of the basic coordinates of sociocultural identity and mentality.. The memory of the intersection landmark events of the historical destinies of ethnic groups carries an emotionally-shaped estimated load and at the same time can be corrected when the attitude to establish cooperation-interaction relations of corresponding categories communities is consciously introduced.

5. Conclusions

If we consider an ethnic group as a specific discursive strategy of cultural direction finding in the "friend or foe" range, then we can assume that on the one hand it is a natural reaction to the complicated conditions of interethnic interaction, when contact with the external ethnic environment becomes constant. Ethnic unity, is strengthened, which occurs naturally, since in the space of ethnic discourse imaginary meanings such components of social convention are catalyzed, based on an intuitive understanding of the general field of meanings shared by most group members . Here we have before us a mental and epistemological attitude called common sense. It is she who serves as the universal anthropological foundation of the perceived ethnic good. From it arise a historical memory and a joint vision of a future for all. This important ability is based on a common basis for any ethnic group - the biopsychosocial nature of the cognitive capabilities of most members of the group. It can be assumed that this is how social intuition manifests itself, such an integral socio-psychological characteristic of the social action subject, which includes the ability to effectively predict the development of social situations. In other words, the common sense method is based on principles, norms and rules that operate at the level of social reflexes and "instincts", social intuition, which allow ethnic communities to maintain their integrity.

Acknowledgements
The work is performed according to the Russian Government Program of Competitive Growth of Kazan Federal University.

BIBLIOGRAPHY

[1] AA Kibrik Kognitivnye issledovaniya po diskursu // Vopr YAzykoznaniya №5 pp 126-145 1994
[2] SN Bogolyubova Povsednevnost': prostranstvo social'noj identichnosti Moscow pp 63-64 2011
[3] S Hall The local and the global: Globalization and ethnicity // Culture Globalization and the World System London pp131-138 1991
[4] R Park Collection of translations / collect and transl by V G Nikolaev; ed by D V Efremenko Moscow: Institute of Scientific Information on Social Sciences of the Russian Academy of Sciences p 38 2011
[5] F Barth The analisis of culture in complex societies // Ethnos Stockholm Vol 54 pp120-142 1989
[6] I Wood Conclusion: Strategies of Distinctions // Strategies of Distinctions The Construction of Ethnic Communities 300 - 800 Ed by W Pohl with H Reimitz Leiden–Boston–Köln p 298 1998
[7] A Smith The Ethnic Origins of Nations Oxford; NY: B Blackwell 1987 p 30 1986
[8] D Horowitz Ethnic Groups in Conflict Los Angeles p 52 1985
[9] R Rameshbabu R Karthikeyan R Balamurali 2016 FPGA Implementation of Synchronized Non Identical Hyperchaotic Systems International Journal Of Control Theory And Applications 95 pp 2633 – 2642
[10] E Goffman Forms of Talk Philadelphia: University of Pennsylvania Press 320 p 1981
[11] C Loomis Social Systems: Essays on Their Persistence and Change Princeton N J: D Van Nostrand 349 p 1960
[12] R Axelrod The Complexity of Cooperation Princeton University Press 248 p 1997
[13] I B Bovina Teoriya social'nyh predstavlenij: istoriya i sovremennoe razvitie // Sociologicheskij zhurnal № 3 p 5-20 2010
[14] T E Lawson «Cognition» In: Braun W McCutcheon R T Eds Guide to the study of religion London - New York: Cassell pp75-85 2000
[15] EN SHul'ga Problematika predponimaniya v germenevtike fenomenologii i sociologii Moscow: Institute of Philosophy the Russian Academy of Sciences 166 p 2004

[16] K A Cerulo «Establishing a Sociology of Culture and Cognition» In: Cerulo K A Ed // Culture in Mind: Toward a Sociology of Culture and Cognition New York: Routledge p 34 2002
[17] E Zerubavel Social mindscape An invitation to cognitive sociology Cambridge: Harvard University Press p 84 1997
[18] Knorr Cetina The practice turn in contemporary theory / ed by T Schatzki [et al] New York: Routledge p 56 2001
[19] Z H Pod'olo Social'nyj poryadok: principy sociologicheskogo analiza // Sovremennaya zapadnaya sociologiya: klassicheskie tradicii i poiski novoj paradigmy Moscow: Progress p 34 1990
[20] O V Semenec Psihologicheskie determinanty social'noj intuicii lichnosti v prognozirovanii situacij: dissertaciya kandidata psihologicheskih nauk: 190005 Saratov p 79 2010
[21] Janarthanan R Balamurali R Annapoorani A and Vimala V 2020 Prediction of rainfall using fuzzy logic Materials Today: Proceedings
[22] P Sorokin Contemporary Sociological Theories New York London: Harper and Brathers XXIII 785 p 1928

Vorontsova Valeriya Leonidovna — candidate of Physico-Mathematical Sciences, associate Professor at the Department of General mathematics of Institute of Mathematics and Mechanics Lobachevsky of Kazan Federal University. Author of more than 65 scientific papers, 2 monographs, 20 textbooks. Research interests — the dynamics of space cable systems, the stability of the movement, modeling and calculation of heat exchangers with heat exchange elements in the form of twisted pipes and spring-twisted channels, system analysis, information management and processing, mathematical modeling of complex systems, social philosophy.
Salimgareev Maxim Vladimirovich — candidate of History Sciences, *Associate Professor at the Department of* ***GMUS*** of Kazan National Research Technological University. Author of more than 45 scientific papers, 4 monographs, 3 textbooks. Research interests — social philosophy, anthropology, intellectual history.
Salimgareev Dmitry Vladimirovich — candidate of philosophical Sciences, *Assistant of the Department of Philosophy* of Kazan State University of Culture and Arts. Author of more than 14 scientific papers. Research interests — social philosophy, anthropology, cultural science, sociocultural situation,

GEOMETRIC CHARACTERISTICS OF THE PIPE IN PIPE TYPE DEVICE WITH A SCREW HEAT EXCHANGE ELEMENT

***Valeriya Leonidovna Vorontsova*[1], *Alfiya Gizzetdinovna Bagoutdinova*[2]**
[1]*Associate Professor at the Department of General Mathematics* Institute of Mathematics and Mechanics Lobachevsky ***Orcid 0000-0001-5736-0646, ID Scopus 15066528300***
e-mail:milen99@yandex.ru
[2]*Associate Professor at the Department of General Mathematics* Institute of Mathematics and Mechanics Lobachevsky ***Orcid 0000-0001-6591-2914, ID Scopus 57195479523***
e-mail: bagoutdinova@rambler.ru
[1]Kazan Federal University
e-mail: milen99@yandex.ru
Tel.: +79172549100

Abstract: To increase the thermal and hydrodynamic efficiency of tubular apparatuses, it is necessary to develop new designs of heat-exchange elements, which are distinguished by high thermal efficiency, manufacturability and low cost of production.
In the work, it is proposed to use screw pipes as heat-exchange elements of tubular heat exchangers, since they have a higher degree of surface development compared to smooth-tube elements.
The manufacturing technology of such pipes can be carried out in the cold state by the method of rotational forging (rotational compression), which is one of the varieties of the forging process and implemented on special rotary forging machines.
To determine the heat transfer coefficients in the screw heat exchange elements mounted a special experimental setup. The main part of the installation is a pipe-in-pipe type heat exchanger, the heat exchange element of which is made in the form of a screw pipe.
A experimental research technique has been developed, and the results of physical experiments on heat transfer, as is known, are generalized by criterion equations in which an equivalent diameter is used as a characteristic size.
In this regard, the authors proposed a method for calculating the basic geometric characteristics of a screw heat-exchange element, and also obtained formulas for calculating the equivalent diameter of the pipe and annular space in a tube-in-pipe apparatus with a screw heat-exchange element.
The results of this work can be used in the calculations and design of new promising heat exchange equipment.
Keywords: equivalent diameter, screw pipe, heat exchanger, heat exchanger hydraulic diameter, volume, surface.

1. INTRODUCTION

Currently, one of the most common types of heat exchange equipment is tubular heat exchangers.
To intensify heat transfer in such devices, artificial flow turbulization is often carried out using spiral pipes, twisted tapes, screws [1-7], as well as changing the pipe wall from smooth to profiled by deforming it [9-12]. Various methods of intensifying heat transfer in channels due to flow swirling are described in detail in [13]. Screw tubes [14], which the authors propose to use as promising heat-exchange elements of tubular apparatuses, also belong to modern designs of twisting devices.
In [15], a comparison was made of the development of the surface of helical heat-exchange elements with the surface of a smooth pipe. It is established that this parameter, with certain geometric ratios, takes values from 1.4 to 2. Since when the surface is considered effective, screw pipes can be successfully used as heat exchange elements of tubular heat exchangers [16].
The most promising technology for producing twisted pipes is the method of rotational forging, implemented on special rotary forging devices [17]. Such technologies make it possible to obtain a high-quality pipe with a given geometry of the flowing part. In fig. 1 shows a sample of a twisted pipe manufactured using a rotary forging machine SSK-14 from GFM.

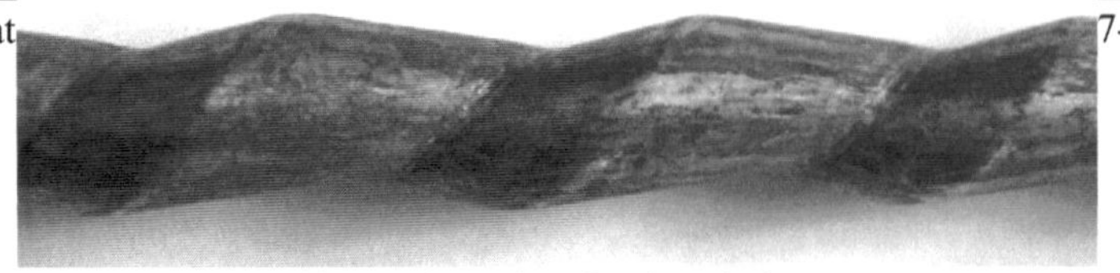

Fig. 1. Sample of twisted pipe

In [18], the results of computer simulation of hydrodynamics and heat transfer in a turbulent flow of a viscous fluid in screw pipes are presented. Modeling was performed using the SolidWorks Flow Simulation software package. The objects of study were screw pipes with various geometric sizes. As a result of the simulation, the fields of velocities, temperatures, and pressures in the computational domain are determined. To confirm the results of computer simulation and determine the heat transfer coefficients in a screw heat exchange elements, a special experimental setup was mounted (Fig. 2). The main unit of the installation is a pipe-in-pipe type heat exchanger, the heat exchange element of which is made in the form of a screw pipe.

Fig. 2. A special experiment setup

The concept of equivalent diameter was first introduced in the derivation of the Darcy-Weisbach equation to determine the hydraulic resistance of pipelines [19].

2. METHODS

The hydraulic radius is understood as the ratio of the area of the free section of the channel through which the fluid flows to the wetted perimeter

$$r = \frac{S}{P},$$

where is the cross-sectional area of the fluid flow, is the wetted perimeter.
For a round pipe with an inner diameter d and free cross-sectional area

IOP Conf. Series: Materials Science and Engineering 994 (2020) 012042 doi:10.1088/1757-899X/994/1/012042

$S = \frac{\pi d^2}{4}$, when it is completely filled with liquid $P = \pi d$, whence the hydraulic radius

$$r = \frac{S}{P} = \frac{\frac{\pi d^2}{4}}{\pi d} = \frac{d}{4}.$$

The diameter expressed in terms of the hydraulic radius is the equivalent diameter: $d = d_{eqv} = 4r$.
Therefore,

$$d_{eqv} = \frac{4S}{P}. \qquad (1)$$

Thus, the equivalent diameter is equal to the diameter of a hypothetical round pipe, for which the ratio of area to wetted perimeter is the same as for a given non-circular pipe.
Multiplying the numerator and denominator by the length of the pipe, formula (1) can be written as

$$d_{eqv} = \frac{4V}{F}, \qquad (2)$$

where V is the volume of the tube space, F is the area of the wetted surface.

3. RESULTS AND DISCUSSION

The equivalent diameter of the pipe space. The surface of a helical pipe (Fig. 3) formed by the movement of a triangle (Fig. 4) along a helical line located on a circular cylinder of radius r_0 is written as [20]:

$$\vec{r} = (r_0 + f(s))\cos t \cdot \vec{\mathrm{i}} + (r_0 + f(s))\sin t \cdot \vec{\mathrm{j}} + (bt + s)\cdot \vec{\mathrm{k}}, \quad 0 \le t \le 2\pi n, \ -l_d \le s \le l_k,$$

where n is the number of turns of the helix, $l = l_d + l_k$ is the length of one pipe module, $b = l/(2\pi)$ is the parameter of the helix, $f(s) = \alpha s + h$ is the generatrix,

$$\alpha = \begin{cases} \frac{h}{l_d}, & -l_d \le s \le 0 \\ -\frac{h}{l_k}, & 0 \le s \le l_k \end{cases}$$

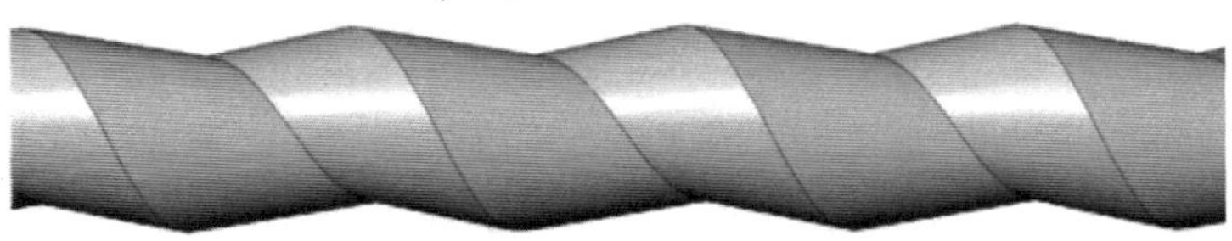

Fig. 3. Screw pipe

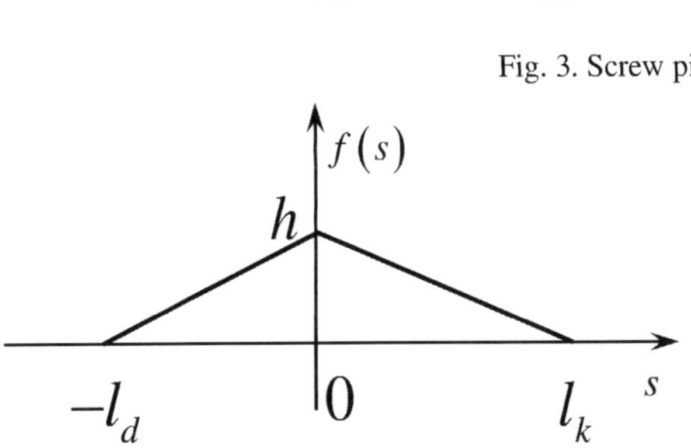

We calculate the geometric characteristics of one module of a screw pipe. The parametric equations of the screw pipe module are written as:

$$\begin{cases} x = r\cos t \\ y = r\sin t \\ z = bt + s \end{cases},$$

$$0 \le r \le r_0 + \alpha s + h, \quad 0 \le t \le 2\pi, \quad -l_d \le s \le l_k.$$

The body volume is calculated by the formula [21]:

$$V = \iiint\limits_V dxdydz.$$

We calculate the Jacobian of the transformation of Cartesian coordinates into curvilinear:

$$J(r,s,t) = \begin{vmatrix} \cos t & 0 & -r\sin t \\ \sin t & 0 & r\cos t \\ 0 & 1 & b \end{vmatrix} = -r.$$

Then,

$$V = \iiint\limits_V |J(r,s,t)| drdsdt = \int_0^{2\pi} dt \int_{-l_d}^{l_k} ds \int_0^{r_0+\alpha s+h} rdr = \int_0^{2\pi} dt \int_{-l_d}^{0} \frac{(r_0+\alpha s+h)^2}{2} ds +$$

$$+\int_0^{2\pi} dt \int_0^{l_k} \frac{(r_0+\alpha s+h)^2}{2} ds = \frac{\pi}{3}(l_d + l_k)(3r_0^2 + 3r_0h + h^2) = \frac{\pi l}{3}(3r_0^2 + 3r_0h + h^2). \quad (3)$$

Formula (3) can be written as

$$V = \frac{\pi l}{3}\left(r_0^2 + (r_0+h)^2 + r_0(r_0+h)\right).$$

This means that the volume of one module of a helical pipe is equal to the volumes of truncated cones forming a direct confuser-diffuser element.

To calculate the surface area given by parametric equations

$$\begin{cases} x = x(s,t) \\ y = y(s,t) \\ z = z(s,t) \end{cases}$$

formula [21] applies

$$S = \iint\limits_D \sqrt{EF - G^2}\, dsdt,$$

where

$$E = \left(\frac{\partial x}{\partial s}\right)^2 + \left(\frac{\partial y}{\partial s}\right)^2 + \left(\frac{\partial z}{\partial s}\right)^2, \quad F = \left(\frac{\partial x}{\partial t}\right)^2 + \left(\frac{\partial y}{\partial t}\right)^2 + \left(\frac{\partial z}{\partial t}\right)^2,$$

$$G = \frac{\partial x}{\partial s}\frac{\partial x}{\partial t} + \frac{\partial y}{\partial s}\frac{\partial y}{\partial t} + \frac{\partial z}{\partial s}\frac{\partial z}{\partial t}.$$

Since the surface of a screw pipe is described by the equations

$$\begin{cases} x=(r_0+\alpha s+h)\cos t \\ y=(r_0+\alpha s+h)\sin t \\ z=bt+s \end{cases}$$

then

$$E=\alpha^2+1,\ F=(r_0+\alpha s+h)^2+b^2,\ G=b.$$

Accordingly,

$$S=\iint_D \sqrt{(\alpha^2+1)\left((r_0+\alpha s+h)^2+b^2\right)-b^2}\,dsdt=$$

$$=\int_0^{2\pi}dt\int_{-l_d}^{l_k}\sqrt{(\alpha^2+1)\left((r_0+\alpha s+h)^2+b^2\right)-b^2}\,ds=$$

$$=\frac{2\pi\sqrt{\alpha^2+1}}{\alpha}\int_{-l_d}^{l_k}\sqrt{(r_0+\alpha s+h)^2+\frac{\alpha^2b^2}{\alpha^2+1}}\,d(r_0+\alpha s+h).$$

We use the formula:

$$\int\sqrt{x^2+a^2}\,dx=\frac{1}{2}\left(x\sqrt{x^2+a^2}+a^2\ln\left|x+\sqrt{x^2+a^2}\right|\right)+C$$

Then

$$S=\frac{\pi\sqrt{\alpha^2+1}}{\alpha}\left((r_0+\alpha s+h)\sqrt{(r_0+\alpha s+h)^2+\frac{\alpha^2b^2}{\alpha^2+1}}\right)\Bigg|_{-l_d}^{0}+$$

$$+\frac{\pi\alpha b^2}{\sqrt{\alpha^2+1}}\ln\left|r_0+\alpha s+h+\sqrt{(r_0+\alpha s+h)^2+\frac{\alpha^2b^2}{\alpha^2+1}}\right|\Bigg|_{-l_d}^{0}+$$

$$+\frac{\pi\sqrt{\alpha^2+1}}{\alpha}\left((r_0+\alpha s+h)\sqrt{(r_0+\alpha s+h)^2+\frac{\alpha^2b^2}{\alpha^2+1}}\right)\Bigg|_{0}^{l_k}+$$

$$+\frac{\pi\alpha b^2}{\sqrt{\alpha^2+1}}\ln\left|r_0+\alpha s+h+\sqrt{(r_0+\alpha s+h)^2+\frac{\alpha^2b^2}{\alpha^2+1}}\right|\Bigg|_{0}^{l_k}.$$

Given that $\alpha=\frac{h}{l_d}$, when $-l_d\le s\le 0$ and $\alpha=-\frac{h}{l_k}$ when $0\le s\le l_k$, we get

$$S=\frac{\pi\sqrt{h^2+l_d^2}}{h}\left((r_0+h)\sqrt{(r_0+h)^2+\frac{h^2b^2}{h^2+l_d^2}}-r_0\sqrt{r_0^2+\frac{h^2b^2}{h^2+l_d^2}}\right)+$$

$$+\frac{\pi\sqrt{h^2+l_k^2}}{h}\left((r_0+h)\sqrt{(r_0+h)^2+\frac{h^2b^2}{h^2+l_k^2}}-r_0\sqrt{r_0^2+\frac{h^2b^2}{h^2+l_k^2}}\right)+ \quad (4)$$

$$+\frac{\pi hb^2}{\sqrt{h^2+l_k^2}}\ln\left|\frac{r_0+h+\sqrt{(r_0+h)^2+\frac{h^2b^2}{h^2+l_k^2}}}{r_0+\sqrt{r_0^2+\frac{h^2b^2}{h^2+l_k^2}}}\right|+\frac{\pi hb^2}{\sqrt{h^2+l_d^2}}\ln\left|\frac{r_0+h+\sqrt{(r_0+h)^2+\frac{h^2b^2}{h^2+l_d^2}}}{r_0+\sqrt{r_0^2+\frac{h^2b^2}{h^2+l_d^2}}}\right|.$$

When b=0 formula (4) is converted to

$$S=\pi(2r_0+h)\left(\sqrt{h^2+l_d^2}+\sqrt{h^2+l_k^2}\right),$$

which coincides with the well-known formula of the surface area of the direct confuser-diffuser element.

We introduce the following notation: $R_0=r_0+h,\ L_d=\sqrt{h^2+l_d^2},\ L_k=\sqrt{h^2+l_k^2}$.

Then formulas (3) - (4) can be written as

$$V=\frac{\pi l}{3}\left(r_0^2+r_0R_0+R_0^2\right); \tag{5}$$

$$S=\frac{\pi R_0}{h}\left(\sqrt{R_0^2L_d^2+h^2b^2}+\sqrt{R_0^2L_k^2+h^2b^2}\right)-\frac{\pi r_0}{h}\left(\sqrt{r_0^2L_d^2+h^2b^2}+\sqrt{r_0^2L_k^2+h^2b^2}\right)+$$
$$+\frac{\pi hb^2}{L_k}\ln\left|\frac{R_0+\sqrt{R_0^2+\frac{h^2b^2}{L_k^2}}}{r_0+\sqrt{r_0^2+\frac{h^2b^2}{L_k^2}}}\right|+\frac{\pi hb^2}{L_d}\ln\left|\frac{R_0+\sqrt{R_0^2+\frac{h^2b^2}{L_d^2}}}{r_0+\sqrt{r_0^2+\frac{h^2b^2}{L_d^2}}}\right|. \tag{6}$$

Equivalent diameter of the tube space:

$$d_{eqv}=\frac{4V}{S},$$

where V are S calculated by formulas (5), (6).

The equivalent diameter of the annulus. The wetted surface area, taking into account the wall thickness, is determined by the formula

$$S_0=2\pi Rl+S_1,$$

where S_1 is calculated by the formula (6) with $r_0=r_1,\ R_0=R_1$, where $r_1=r_0+\delta,\ R_1=R_0+\delta$:

$$S=\frac{\pi(R_0+\delta)}{h}\left(\sqrt{(R_0+\delta)^2L_d^2+h^2b^2}+\sqrt{(R_0+\delta)^2L_k^2+h^2b^2}\right)-$$
$$-\frac{\pi(r_0+\delta)}{h}\left(\sqrt{(r_0+\delta)^2L_d^2+h^2b^2}+\sqrt{(r_0+\delta)^2L_k^2+h^2b^2}\right)+$$
$$+\frac{\pi hb^2}{L_k}\ln\left|\frac{R_0+\delta+\sqrt{(R_0+\delta)^2+\frac{h^2b^2}{L_k^2}}}{r_0+\delta+\sqrt{(r_0+\delta)^2+\frac{h^2b^2}{L_k^2}}}\right|+\frac{\pi hb^2}{L_d}\ln\left|\frac{R_0+\delta+\sqrt{(R_0+\delta)^2+\frac{h^2b^2}{L_d^2}}}{r_0+\delta+\sqrt{(r_0+\delta)^2+\frac{h^2b^2}{L_d^2}}}\right|. \tag{7}$$

The volume of the annulus taking into account the wall thickness δ:

$$V_0=\pi R^2l-\frac{\pi l}{3}\left(3(r_0+\delta)^2+3(r_0+\delta)h+h^2\right). \tag{8}$$

Then the equivalent diameter of the annulus

$$D_{eqv} = \frac{4V_0}{S_0},$$

where S_0 and V_0 are calculated by formulas (7), (8).

Calculation of equivalent diameters in a tube-in-tube apparatus with a screw heat exchange element. The geometric dimensions (mm) of the apparatus are shown in Fig. 5.

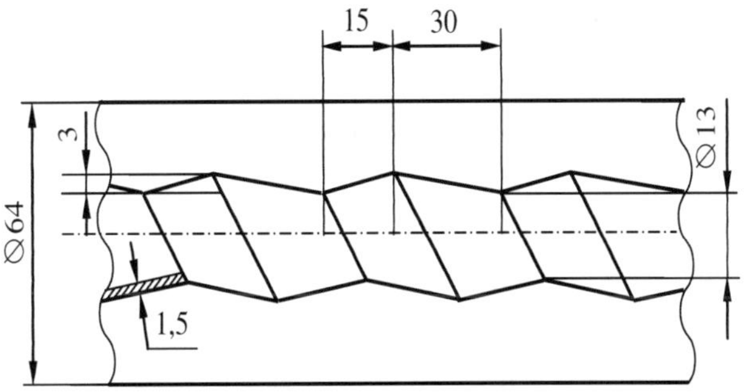

Fig. 5. Tubular heat exchanger

We have $R = 0{,}032$ m; $r_0 = 0{,}0065$ m; h=0,003 m; δ=0,0015 m; $l_d = 0{,}015$ m; $l_k = 0{,}03$ m.

We calculate $l = l_d + l_k = 0{,}045$ m; $R_0 = r_0 + h = 0{,}0095$ m;

$$L_d = \sqrt{h^2 + l_d^{\,2}} = 0{,}0153 \text{ m};$$

$$L_k = \sqrt{h^2 + l_k^{\,2}} = 0{,}0301 \text{ m};$$

$$b = \frac{0{,}045}{2\pi} \approx 0{,}007166 \text{ m};$$

$$V = \frac{\pi l}{3}\left(r_0^{\,2} + r_0 R_0 + R_0^{\,2}\right) = 9{,}1538 \cdot 10^{-6} \text{ м}^3;\ S = 0{,}0023 \text{ m}^2;$$

$$V_0 = \pi R^2 l - \frac{\pi l}{3}\left(3\left(r_0 + \delta\right)^2 + 3\left(r_0 + \delta\right)h + h^2\right) = 1{,}319 \cdot 10^{-4} \text{ m}^3;$$

$$S_1 = 0{,}0027 \text{ m}^2;$$

$$S_0 = 2\pi R l + S_1 = 0{,}0118 \text{ m}^2.$$

Then the equivalent diameter of the tube space:

$$d_{eqv} = \frac{4V}{S} = 0{,}016 \text{ m}.$$

Equivalent annulus diameter:

$$D_{eqv} = \frac{4V_0}{S_0} = 0{,}045 \text{ m}.$$

4. SUMMARY

Based on the integral calculus, the basic geometric characteristics of the heat-exchange element, made in the form of a twisted pipe, are determined. Formulas are proposed for calculating the area of the wetted surface and the volume of the flow part of the heat exchange element.

Formulas are obtained for calculating the equivalent diameter of the tube and annular space in a tube-in-tube apparatus with a screw heat-exchange element.

5. CONCLUSIONS

A experimental research technique has been developed, and the experimental results, as is known, are generalized by criteria equations in which an equivalent diameter is used as the determining geometric size.

In this regard, it became necessary to obtain formulas for calculating the equivalent diameter of the pipes and annular space in a device of this type. Currently, an equivalent diameter is used to calculate a wide class of heat exchangers, as well as to determine the hydraulic radius of elements and nozzles of mass transfer equipment.
It is established that the results can be applied not only to calculate the geometric parameters of twisted pipes, but also for direct "confuser-diffuser" pipes.

6. ACKNOWLEDGEMENTS

The work is performed according to the Russian Government Program of Competitive Growth of Kazan Federal University.

BIBLIOGRAPHY

[1] Piriyarungrod, S. Eiamsard, C. Thianpong, M. Pimsarn, K. Nanan, "Heat transfer enhancement by tapered twisted tape inserts", *Chem. Eng. Process. Process Intensif.*, Vol. 96, pp. 62-71, 2015.

[2] Y. Vazifeshenas, M.A. Delavar, "Investigation of thermo-hydraulic performance of circular tube fitted with center cleared twisted tape using CFD modeling", *International Journal of Engineering, Trans.*, Vol. 28, pp. 476–482, 2015.

[3] E.Z. Ibrahim, "Augmentation of laminar flow and heat transfer in flat tubes by means of helical screw-tape inserts Energy Convers", *Energy Convers. Manag.*, Vol. 52, pp. 250-257, 2011.

[4] S.D. Salman, A.A.H. Kadhum, M.S. Takriff, A.B. Mohamad, "CFD simulation of heat transfer augmentation in a circular tube fitted with alternative axis twisted tape in laminar flow under a constant heat flux", *Heat Transfer Asian Res.*, Vol. 43, pp. 384–396, 2014.

[5] A. Dewan, P. Mahanta, K.S. Raju, P.S Kumar, "Review of passive heat transfer augmentation techniques", *Journal of Power and Energy*, Vol. 218(7) , pp. 509-527, 2004.

[6] S. N. Sarada, A. V. S. R. Raju, K. K. Radha, L. S. Sunder, "Enhancement of heat transfer using varying width twisted tape inserts", *International Journal of Engineering, Science and Technology,* Vol. 2, pp.107-118, 2010.

[7] W. Noothong, S. Eiamsa-ard, P. Promvonge, "Effect of twisted-tape inserts on heat transfer in a tube", *Proceedings of the 2nd Joint International Conference on Sustainable Energy and Environment (SEE 2006)*, Bangkok, Thailand, 21–23 November, pp. 1–5, 2006.

[8] Bhadouriya R., Agrawal A., Prabhu S. V., "Laminar flow heat transfer studies in a twisted square duct for constant wall heat flux boundary condition" //*Sadhana*. Vol. 40, №. 2, pp. 467-485, 2015.

[9] Li Zhang, Sheng Yang, Hong Xu, "Experimental study on condensation heat transfer characteristics of steam on horizontal twisted elliptical tubes", *Applied Energy*, Vol. 97, pp. 881-887, 2002.

[10] Junlin Cheng, Zuoqin Qian, Qiang Wang, "Analysis of heat transfer and flow resistance of twisted oval tube in low Reynolds number flow", *School of Energy and Power Engineering, Wuhan University of Technology, China, International Journal of heat and mass transfer*, Vol.109, pp. 761-777, 2017.

[11] Anurag Agrawal, Sandeep Saha, Nilesh Kumar Sharma, "Comparative CFD Analysis of Circular and Elliptical Twisted Tube and Shell Heat Exchanger", *IJSRD - International Journal for Scientific Research & Development*, Vol. 5, Issue 04, pp. 1119-1124, 2017.

[12] Anurag Agrawal , Sandeep Saha , Nilesh Kumar Sharma, "A Review on Tube and Shell Heat Exchanger with Elliptical Twisted Geometry", *International Journal for Research in Applied Science & Engineering Technology (IJRASET)*, Vol. 5, Issue VI, pp. 2393-2398, 2017.

[13] C. Maradiva, J. Vad har, R. Agarwal, "The heat transfer enhancement techniques and their thermal performance factor", *Journal of Basic and Applied Sciences*, Vol.7 (1) , pp. 1-21, 2018.

[14] Pat. 119452 Russian Federation. "The heat exchange element" / Zolotonosov A.Ya., Zolotonosov Ya.D., Bagoutdinova A.G., Osika I.I. № 2012109355/06; appl. 12.03.12; publ. 20.08.2012, Bull. № 23. (in Russian)

[15] Bagoutdinova A.G., Zolotonosov Ya.D., Posokhin V.N., "Mathematical modeling of helical heat exchange elements", *Stroitelstvo*, № 8, pp. 41–46, 2014.

[16] Pismennyi E. N., Baranyuk M. M, Voznyuk M. M., "Equalized heat transfer surfaces and the technique of numerical studies of their thermo hydraulic characteristics", *Promischlennaya teplotechnika*, Vol. 34, № 1, pp. 45–54, 2012.

[17] Pat. 2542865 Russian Federation. "Method of production of twisted pipes of the confuser-diffuser type" / Zolotonosov Ya.D., Shaipov N.M., Minigareev D.Z., Matushko A.A., Bagoutdinova A.G., Yahnev M.N. № 2013116659/02; appl. 04.11.13 publ. 20.10.14, Bull. № 6. (in Russian)

[18] V.L. Vorontsova, A.G. Bagoutdinova, A.F. Galemzianov, "Numerical modeling of hydrodynamic processes in screw tubes", *Journal of Advanced Research in Dynamical and Control Systems*, Vol.11, Is.5. pp.38-43, 2019.

[19] Vorontsova V.L, Bagoutdinova A.G, Galemzianov A.F., "Calculation of the geometrical parameters of the heat exchanger type «Pipe in Pipe« with a spiral-coiled channel", *Journal of Computational and Theoretical Nanoscience*, Vol.16, Is.11. pp. 4513-4518, 2019.

[20] Vorontsova V.L, Bagoutdinova A.G, Gilemzianov A.F., "Mathematical models of the ocurved spring tubes surfaces", *Journal of Computational and Theoretical Nanoscience*, Vol.16, Is.11, pp.4554-4559, 2019.

[21] Brown G.O. "The history of the Darcy–Weisbach equation for pipe flow resistance", *Proceedings of the environmental and water resources history*, pp. 34–43, 1994.

[22] Bagoutdinova A.G., Vachagina E.K., Zolotonosov Ya.D. "Mathematical modeling of pipes with a helical heat exchange surface", *Izvestiya KGASU*, № 4 (42), pp. 237–244, 2017.

[23] Fichtenholz G.M. "A course of differential and integral calculus".- Moscow, Fizmatlit, 864 p., 2003.

Vorontsova Valeriya — candidate of Physico-Mathematical Sciences, associate Professor at the Department of General mathematics of Institute of Mathematics and Mechanics Lobachevsky of Kazan Federal University. Author of more than 65 scientific papers, 2 monographs, 20 textbooks. Research interests — the dynamics of space cable systems, the stability of the movement, modeling and calculation of heat exchangers with heat exchange elements in the form of twisted pipes and spring-twisted channels, system analysis, information management and processing, mathematical modeling of complex systems.

Bagoutdinova Alfiya— candidate of Technical Sciences, associate Professor at the Department of General mathematics of Institute of Mathematics and Mechanics Lobachevsky of Kazan Federal University. Author of more than 90 scientific papers, including 15 inventions. 2 monographs, 5 textbooks. Research interests — modeling and calculation of heat exchangers with heat exchange elements in the form of twisted pipes and spring-twisted channels.

INTELLIGENT ORIGINS OF EUROPEAN GLOBALISM: EARLY STOICISM AND CICERON

***Valeriya Leonidovna Vorontsova[1], Maxim Vladimirovich Salimgareev [2]*& Dmitry Vladimirovich Salimgareev [3]**

[1]Associate Professor at the Department of General Mathematics Institute of Mathematics and Mechanics Lobachevsky, **Orcid 0000-0001-5736-0646, ID Scopus 15066528300**
e-mail:milen99@yandex.ru
[2] Associate Professor at the Department of **GMUS** of Kazan National Research Technological University, Orcid 0000-0002-7068-671X, ID Scopus 57210818965
e-mail: msalimga.dis@mail.ru
[3]Assistant at the Department of History, Philosophy and Cultural science of Kazan State University of Culture and Arts, Orcid 0000-0003-4349-9369,*ID Scopus* 57194637803
e-mail: demiyrgiys@mail.ru
[1]Kazan Federal University
[2]Kazan National Research Technological University
[3]Kazan State University of Culture and Arts *e-mail: milen99@yandex.ru*
Tel.: +79172549100

Abstract:The origin of socio-philosophical ideas about universal unity in the tradition of Stoicism and the writings of Cicero in the context of the cultural and historical conditions of the Roman intellectual reception of the ideas of cosmopolitanism of the early Stoics is considered. Performed the analysis of the continuity and development of the concept of "primary inclination" in the picture of the emergence of society and its structures developed by thinkers. The work uses the general scientific inductive method of ascending from the particular to the general, which allows a heuristic approach to the analysis of the intellectual practice of constructing a philosophical picture of the world of ancient authors. Authors used historical analysis of ideas, aimed at revealing the specific cultural context of their origin and development in space and time. The authors reveal the specifics of the stoic representation of the manifestations of the original, nature-like solidarity in the historical and philosophical discourse of ancient authors.The key provisions of the doctrine of the early Stoics and Cicero cosmopolis are analyzed. When considering the cultural and historical specificity of the indicated elements of the intellectual tradition, attention is drawn to the dependence of the Roman speaker views as a social philosopher on the political teachings of ancient authors. The article considers in a new way of he cultural and historical formation of ideas about universal unity in ancient anthropological discourse. The work is based on the analysis of the reception of the stoic idea of "primary inclination" in the writings of Cicero.
Keywords: early stoicism, Cicero, moral consciousness, cosmopolis, social nature, man.

1.Introduction

The discovery of the metaphysical foundations of the human race unity is an important intention of the whole of ancient philosophy. One of the significant attempts to solve the problem of aspiration for universal unity was made by representatives of Stoicism. Broadcasting their ideas over the six centuries of school tradition, the Stoics had a strong influence not only on their contemporaries, but also on subsequent thinkers. Thus, the developed idea of cosmopolis had deep philosophical significance for those European thinkers who tried to approach the idea of the universalist integrity of mankind. One of the first was Cicero. Such an impact on European moral consciousness from the side of stoicism was due to the desire to expand social space, and at the same time to individual self-discipline in the name of the common good, which at critical turning points in the development of Europe became a kind of intellectual movement vector.

At the moment, there is a certain amount of research devoted to both stoic cosmopolitanism and Cicero. According to stoicism, the most significant, in the context of the declared theme, can be recognized as works [1-5]. Authors such as [6-8] examine in depth the most key aspects of stoic anthropology and ethics, some [9-10] dwell in more detail on cosmopolitanism, indicating that stoic cosmopolitanism was an important element in the philosophical system of the early Stoics, since it built human life into outer space, which corresponded to the new changed cultural and geopolitical changes of the ancient world. However, all the considered researchers practically do not make serious attempts to discover the reception of stoic ideas on Roman soil.

A somewhat different situation is observed in relation to the problem field associated with the socio-philosophical views of Cicero. One of the first serious works on this subject was publication [11], which, exploring the socio-philosophical roots of humanism and anthropocentrism, considers Cicero to be one of the first prominent representatives of these ideological philosophical directions. The tradition of studying the heritage of the Roman thinker from the perspective of philosophy was continued [12] by performing a detailed study of almost all of his major philosophical treatises. A comprehensive attempt to consider the socio-political system of Cicero in the context of his general social views was the work [13]. This work brings the information of disparate reasonings of the Roman thinker on socio-political themes into a clear integrity. In fact, for the first time in the scientific literature, the author draws attention to aspects of Cicero's work The trend of socio-political analysis of the ideas of Cicero develops [14] in a work devoted to the genesis of the ideas of republicanism in the works of Cicero. Of the recent studies devoted to our problem, the work [15] is of interest, since it is devoted to the ideological sources of the views of Cicero. Thus, not one of the represented authors made the subject of his study the problem of the development and continuity of the universal unity idea in the works of the early Stoics and Cicero, including in the context of the cultural and historical reality of the ancient world. Meanwhile, this aspect of the problem seems to be very interesting.

2. Methods

The work uses a set of procedures related to the understanding of the text, which we call the interpretation. The term "interpretation" we use in the broad sense of the word - as a way of understanding the text. In this sense, the interpretation includes a number of procedures, the main ones being interpretation, description and explanation.

Explanation and interpretation are the key methodological basis of our study.

As a result of the interpretation, we establish the meaning of certain terms and statements, which allows us to finally describe the object systematically.

The explanation allows us to reveal the essence of the objects under study, allows us to answer the question why the thinkers we have taken came to just this, and not to a different solution of the theoretical problem.

In the work, we proceed from the methodological installation that the studied texts of ancient authors, carrying an exposition of ideas characteristic of the current trend of philosophical thought, act as a "trace" along which we try to "look for" a phenomenon of interest to us in the process of reconstruction.

3. RESULTS AND DISCUSSION

The origin of the Stoic school and its doctrine coincided with the moment of large-scale civilizational transformations. The crisis of the polis system that arose during the Hellenistic era testified that the former narrow-band foundation of social consolidation did not correspond to sociocultural and political challenges. A synthesis of the social structures of Greece and the Middle East, that is, the Greek polis and the Middle East monarchy, has become a historical phenomenon that has emerged as a platform for the emergence of new solidarity foundations. In the new state formations, the West merged with the barbarian East into a single whole. The development of society and the strengthening of rationalism led the ancient monarchies to the search for a new ideological foundation.

Torn out of the familiar world, a man in a border situation acutely felt the significance of his own personality. At this moment, there was an intensive process of adapting it to a new cultural situation, a special intellectual desire appeared to purposefully “grasp” the world with the mind as a whole, manifested in a certain set of separate parts A completely different, broad basis was needed, corresponding to the diversity and universality of emerging political and cultural phenomena.

If Alexander the Great and Caesar tried to literally eliminate national borders, the Stoics did it intellectually. The latter preached the supremacy of nature and reason, the single essence of everything that exists, and the insignificance of all artificial and historical divisions and boundaries. The idea of the universal characteristic of the Hellenistic era, supported by the concept of “general speech” in a language, is established in philosophy and receives an anthropological sound in the concept of Stoics. The Stoics' reasoning about the “universal” as the fundamental principle of the world stimulated the birth of the idea of universal history, which will be formalized already in the Roman era in the person of the stoically thinking historian Polybius and the Stoic Posidonius.

Stoicism, in the face, first of all, of its early representatives, began to develop the idea of the coexistence of man and the world that gave birth to him. The Stoics saw the world in the form of a system, and it was in this sense of the word that they used the term "cosmos", already known to the ancients. “The Universe is like a city,” said Chrysippus, “containing gods and people” [16]. That is why there is an image of a common for all inhabited space - the system of the world organized on the same principles. Zenon and his students recognized that the world is intelligent and animated. Their call to live in accordance with nature meant to be in agreement with the rational arrangement of the cosmos. Reason (λόγος) was a universal divine law that formed a single cosmos and constituted the essence of man [17]. If the nature of all intelligent beings is identical, then, guided by it, considering it as a paradigm for judgments and actions, we enter the macro community of all intelligent beings. Politicizing the cosmological picture, the Stoics argued that the world is “one large state that uses a single device and a single law. And this law is the mind prescribing what should be done, and turning away from what should not be." [16]. Understanding of one’s nature leads drains to solidarity with all their homogeneous, rational beings. Moreover, the principle of universal organization for all peoples at all times is a common leader and overlord, it is the basis for universal communication of all with everyone, it determines the mutual participation of any rational creature, estranged from the animal world and joining the world of gods. This awareness arises among the Stoics only when they are clearly understood and adopted a single law, the "voice" of which does not sound from outside the person, but from the inside [16]. Cicero, as you know, belonged to the Roman intellectual elite. He received a worthy education during his stay in Greece with the outstanding philosophers and rhetoricians of his time. Having a good command of the Greek language, it was not difficult for him to delve into the foundations of Greek philosophy, among which was Stoic. According to some researchers, it was the thought of these Hellenistic philosophers that had a special effect on him [15]. Obviously, Cicero’s close acquaintance with the ideas of one of Zeno’s disciples - Kitai Chrysippos [16]. An important fact in the formation of Cicero, not only as an interpreter of Greek wisdom, but rather an original philosopher, was played by such adherents of stoic philosophy as Lucius Elius Stilon and Publius Nigidiy Figul [18]

The representative of the Middle Stoi Posidonius had some influence on Cicero. He met him during his studies in Rhodes [19]. The first two books of Cicero’s last treatise, “De officiis”, according to his own testimony, are a creative revision of the famous work of Panethius [20]. It is worth paying attention to the fact that thanks to Cicero, some fragments of early Stoic thinkers have reached us. At the same time, with all this intellectual stoic plume, Cicero does not directly identify with adherents of stoicism. However, well acquainted with the teachings of the Stoics in his writings, he tried to intellectually substantiate and develop a number of their postulates, for example, the idea of primary inclination.

In the philosophical treatise "On the Limits of Good and Evil" (45 BC), which is crucial for our study, and devoted to the interpretation of Stoic philosophy, Cicero relies on the imperative of duty and responsibility of man to society (naturalis quaedam hominum quasi congregatio). On the issue of the determinants of civil solidarity, he considered the key factors contributing to the unification of human unions to be "consensus on issues of law" and "community of interests" expressed in "inborn need to live together" (naturalis quaedam hominum quasi congregatio). The last thesis in the retelling of Lactantius is conveyed by the words: "they united because human nature avoids loneliness and seeks communication and alliance" [21].

Following the early Stoics, Cicero attached great importance to the mind, which he considered a key force that promotes the unification of people. All representatives of the human race differ from animals in that, he wrote, that "they are endowed with the mind, the nature that they gave them, sharp and powerful thought". It is he who "makes a person strive for people, uniting by nature with speech and life experience, starting with love for loved household members, go on and enter a community, first civil, and then universal, and, as Plato writes to Architects, remember that he is born not only for himself, but also for the fatherland, and for relatives, and only a very small part of it goes to him himself " [22]. Thus, Cicero interprets the inherent quality of a person as a primary inclination and as a social feeling that draws him to a community with his neighbors.

In our opinion, giving the whole thing a property of close cohesion, the Stoics saw in it an important foundation on which the content of all differences is based, denoting external coherence, which determines the attraction of initially heterogeneous entities according to the principle of functional similarity of the individual parts as elements of the whole. This, in turn, does not deny on the way of self-preservation of the risk existence of breaking the connection between self-identity and the appearance of inconsistency. On this path, the Stoics began the search for special foundations of an inescapably abiding change in space. This allowed them to discover a special component of relations in which the basis of a person's identity was groped, his deep, total correlation with nature. Movement to unity is possible due to attraction by the call of one's own inclination, but this inclination is seen by the Stoics in connection with the individual with the general, it implements self-development outside. It follows from this that the presence of general coherence and conformity is an important condition, but by no means limiting, not sufficient. And if each of us is a particle of this world, then this is an actual condition for self-realization, which, as a natural gift, anyone can use or not use. Therefore, the natural tendency to self-preservation on the basis of universal affinity (instinctive impulse), first at the level of immediate environment (relatives, tribesmen, citizens of the same policy), and then - with the development of the mind - at the level of all mankind does not guarantee universal unity, because the world does not consist from the wise men. This is the internal contradiction of the human community, revealed by the Stoics, which, on the one hand, cannot but stem from the Logos, the rational cosmic principle and has a potential and stable connection with it, but at the same time, it can deviate from the original source in reality.

Roman thought, taking the intellectual baton from the Hellenistic philosophers, used, as you know, mainly ethics, which was quite natural. Rome in the era of the 1st century BC. gradually loses the polis tradition, before us is no longer a classic policy, but, at the same time, not a totality of subjects of the classical empire. These changes affect not only socio-political practice, but also determine the specificity of the Romans ideas about republican values. These changes affect not only socio-political practice, but also determine the specificity of the Romans ideas about republican values. Let us pay attention to the fact that Roman jurisprudence, having accepted the stoic idea of a natural, universal cosmic mind, placed it in its foundation, as a system-forming principle of nature and society from which the property of a common virtue and equal, natural law flows. Therefore, Cicero derives the revealed primary tendency not only from the mind, he considers it in the context of love, especially parental. It is obvious to him that it is common for animals to care about their offspring, and since man is superior to every animal in his talents and endowed with the ability to think intelligently, this nature-like intention, transferred to the entire human community, becomes a striking feature of human aspirations. Including in political life: "therefore, we are naturally capable of unification, assembly, and life in the state" (itaque natura sumus apti ad coetus, concilia, civitates) [22]. The highest virtue and duty of every person is to move along a path that indicates the primary inclination expressed by people in the service of the common good and, if necessary, sacrifice themselves for the sake of the homeland (state), for the homeland is more valuable for a citizen than his own life. According to the philosopher, a person lives his sociality so sharply and deeply that for him a lot of anguish is in complete isolation. Therefore, "... we were born in order to live in a natural community, union and communication with other people" [22]. From this attitude, the collectivist thesis, popular in its stoic tradition, follows about the dominance of social interests over individual interests: "a wise and decent person who obeys the laws and recognizes his civic duty takes care of the common good more than the benefit of any single person his own [22]. In our opinion, the disclosed aspects of Cicero's view on the "mechanics" of sociality are vaguely reminiscent of the paternalistic approach, because in the sphere of political life he puts love, conditioned by natural causes, at the base of the mechanism of social integration that holds back the decay. It is obvious that unifying aspiration can form not only a narrow-clan or intrastate, but also a super-clan relationship, on the basis of which the idea of universal unity of the human race arises. However, this emphasis in the thinker manifests itself in an implicit form.

As a speaker, Cicero could not help but discover that it is human speech that is one of the most important tools for gaining human unity. "When this mind discovered that people randomly make slurred and inarticulate sounds, he divided these sounds, broke them into pieces and, so to speak, stamped words on objects and people previously separated from each other, tied them with pleasant ties of speech" [23]. And although the mind is recognized as an important component of the appearance of the language, it is precisely speech that is indicated by him as the fundamental reason for social unity.

As a professional lawyer, the thinker understood that various economic and social motives were not the last centripetal factor of solidarity. For example, people tend to seek commonwealth among themselves at the time of protecting their property. At the same time, Cicero clearly differentiates private and public property. The emergence of the state is connected with the provision of mutual assistance, for "... people were born for the sake of people, so that they could be useful to each other ... serve common interests, exchanging services, giving and receiving, and their knowledge, work and ability to firmly bind human society" [23]. This public intention makes people look like bees who "come together not for the sake of sculpting a honeycomb, but sculpting a honeycomb because they are naturally inclined to get together" [23].

It should be noted that revealing the rational foundations of human solidarity in the context of the stoic idea of primary inclination developed by him, which lies at the foundation of the emergence of society and its structures, Cicero discovers its most important psychological and socio-economic dominants: firstly, the ability of representatives of the human species to rational activity, speech interaction; secondly, the need to adapt to natural conditions; thirdly, the desire to integrate with their own kind for a reasonably organized life; fourthly, the need for legal protection of private property.

The idea of cosmopolis is based on the concept of "primary inclination", which was adopted by Cicero. It asserted the existence of an intention for the universal in the organization of the universe, which integrated all levels of existence and directed toward a goal common to all. According to A.A. Stolyarova, the doctrine of the "primary inclination" as a whole is a "bunch" of extremely important meanings, from which then the doctrine of the goal, the good, society and the state unfolds [1].

A.S. Stepanova believes that an important spatial development of the concept of "primary inclination" in the Hellenistic era was the increasing spatial dominance (this is especially noticeable in Posidonius), which required the expansion of untranslatable borders. The idea of a common dwelling, where all living creatures live, the gods, brought to life an understanding of the special proximity of all things Thus, an important conceptual turnaround took place. Genetic unity was secondary, while the concept of a common living environment, where everything is in the process of coexistence, becomes key, kinship gives way to affinity, which is commensurate with the process of ascension from nature to culture [24].

Long AA argues that much of the modern interest in Stoicism is directly related to the Stoics' attempt to "reconcile a highly developed moral theory with objective facts, facts that take into account the innate human motives, environmental influences and laws governing all natural phenomena" [25].

From the point of view of M. Forschner, the Stoics could glean the idea of primary inclination from medical practice, where it takes the form of opposing the right (fair, natural) to all artificially-forced, deviating from the natural course of things. In this case, nature acts as a self-sufficient principle that underlies things and does not tolerate external interference. The task of medicine is to bring things into a state of agreement with nature. Cicero sees in the "primary inclination" the foundation on which civil duties are built [6].

4. Summary

So, it is obvious that the early representatives of Stoicism, primarily Zenon and Chrissip, became interpreters of the new worldview, where there was no place for social and political borders. And although the available reality indicated the facts of the existence of different peoples, countries and states, all this was insignificant in comparison with the fact that all people as rational creatures are very different from any animal. They have the same physical and spiritual structure, common inclinations and aspirations for all, universal destiny for an exalted, virtuous way of life. The cosmopolitan interpretation of social reality proposed by the Stoics was of great importance for the modern attitude of the Hellenes, and then the Romans. Perceiving the present political organization of the era undergoing through the prism of the "great policy", the Stoics not only discursively made an attempt to remove the problem of the coexistence of individual states, but carried out an unprecedented existential breakthrough beyond the boundaries of a specific socio-political situation. In this expansion act, a significant role was played by religious experiences, for the first time giving a person consciousness of not only his god-like greatness, but also the cosmic responsibility arising with it.

5. Conclusions

Roman thought in the person of Cicero in the conditions of painful social transformations of the republican system, although it was aimed at the socio-philosophical substantiation of the idea of the unity of the human race, but tended to a certain narrowing of the cosmopolitan concept. The most important goal of Cicero was the justification of the formal legal overcoming of the emerging socio-political and economic fragmentation of Roman society, plunged into the abyss of civil wars. The thinker, imitating his teachers, grounded the universal aspiration by appealing to the totality and ubiquity of the system-forming principle of the organization of space. In his interpretation of the idea of "primary inclination", he tried to deduce universal aspiration from nature conformity, revealing its mechanisms on the psychological and social series of reality, justifying the legal expediency of social unity.
Both the early Stoics and Cicero, despite the naivety of their views, meanwhile, came to important conclusions for their eras about the systemic foundations of social integration that could be applied by nascent imperial ideologies that were looking for ways to justify a new type of consolidation of society.

6. Acknowledgements

The work is performed according to the Russian Government Program of Competitive Growth of Kazan Federal University.

Bibliography

[1] A A Stolyarov Standing and Stoicism Moscow: Kami Group 441 p 1995
[2] S K Strange J Zupko eds Stoicism: Traditions and Transformations Cambridge 345 p 2004
[3] A S Stepanova Stand philosophy as a phenomenon of Hellenistic-Roman culture SPb : Publishing house "Petropolis" 400 p 2012
[4] J Sellars Stoic Cosmopolitanism and Zeno's Republic // History of Political Thought Spring Vol XXVIII № 1 p 30-35 2007
[5] M Polenz Standing The history of the spiritual movement SPb: Publishing project "Quadrivium" 1040 p 2015
[6] M Forschner Die stoische Ethik: Über den Zusammenhang von Natur- Sprach- und Moralphilosophie im altstoischen System Stuttgart 200 p 1981
[7] B Inwood Ethics and Human Action in Early Stoicism Oxford Clarendon press 356 p 1985

[8] Ph Mitsis The Stoic Origin of Natural Rights // Topics in Stoic Philosophy / ed K Ierodiakonou – Oxford: Clarendon p 153-177 1999
[9] D Obbink 'The Stoic Sage in the Cosmic City' in Topics in Stoic Philosophy ed K Ierodiakonou Oxford pp 178–95 1999
[10] T Scaltsas A S Mason The Philosophy of Zeno Larnaca 430 p 2002
[11] Hunt HA The Humanism of Cicero Melbourne 221 p 1954
[12] P Mackendrick K Singh The Philosophical Books of Cicero L 429 p 1989
[13] N Wood Cicero's Social and Political Thought Berkeley; Los Angeles 288 p 1988
[14] R Radford Cicero: A Study in the Origins of Republican Philosophy NY 142 p 2002
[15] K Mayet Chrysipps Logik in Ciceros philosophischen Schriften Tübingen: Gunter Narr Verlag 340 p 2010
[16] Fragments of the early Stoics Volume III Part 1 Chrissip from Sol Ethical fragments Translation AA Stolyarova Moscow: Greek-Latin cabinet 310 p 2007
[17] Fragments of the early Stoics Volume 1 Zeno and his students Translation AA Stolyarova Moscow : Greek-Latin cabinet 250 p 1998
[18] Bowen AC Nigidius Figulus // The Encyclopedia of Ancient History / Ed by Bagnall RS Brodersen K Oxford John Wiley & Sons Inc pp 4783–4784 2013
[19] H Taylor Cicero: a Sketch of His life and Works Chicago 615 p 1918
[20] A A Dyck Commentary on Cicero De Officiis Ann Arbor 760 p 1996
[21] Lactantius Divine Establishments / Per with lat entry Art and approx VM Tyulenev SPb: Publishing house of Oleg Abyshko 512 p 2007
[22] Mark Tullius Cicero On the Limits of Good and Evil Paradoxes of the Stoics Moscow : Russian state humanity Univ 474 p 2000
[23] Mark Thulius Cicero On the State About the laws About old age About friendship About duties Speech Letters Moscow: Thought 782 p 1999
[24] A S Stepanova "Stand Anthropology: The Communicative Aspect" Izvestia RGPU im AI Herzen pp 134–144 2005
[25] A A Long ed Problems in Stoicism London 277 p 1971

Vorontsova Valeriya — candidate of Physico-Mathematical Sciences, associate Professor at the Department of General mathematics of Institute of Mathematics and Mechanics Lobachevsky of Kazan Federal University. Author of more than 65 scientific papers, 2 monographs, 20 textbooks. Research interests — the dynamics of space cable systems, the stability of the movement, modeling and calculation of heat exchangers with heat exchange elements in the form of twisted pipes and spring-twisted channels, system analysis, information management and processing, mathematical modeling of complex systems, social philosophy .
Salimgareev Maxim — candidate of History Sciences, *Associate Professor at the Department of* ***GMUS*** of Kazan National Research Technological University. Author of more than 45 scientific papers, 4 monographs, 3 textbooks. Research interests — social philosophy, anthropology, intellectual history.
Salimgareev Dmitry — candidate of philosophical Sciences, *Assistant of the Department of Philosophy* of Kazan State University of Culture and Arts. Author of more than 14 scientific papers. Research interests — social philosophy, anthropology, cultural science, sociocultural situation,

Network simulation of the sewage treatment system in machine-building enterprises

Valeriya Leonidovna Vorontsova [1], *Svetlana Nikolaevna Savdur* [2]
[2]Associate Professor at the Department of General Mathematics Institute of Mathematics and Mechanics Lobachevsky, **Orcid 0000-0001-5736-0646, ID Scopus 15066528300**
e-mail:milen99@yandex.ru
[2] Associate Professor at the Department of
of Merchandising and Technology of Public Catering Kazan Cooperative Institute (branch) "Russian University of Cooperation" **Orcid 0000-0002-3261-2154, ID Scopus 56088680200** e-mail: savdur.svetlana@yandex.ru

[1]Kazan Federal University
[2]Kazan Cooperative Institute (branch) "Russian University of Cooperation"
e-mail: milen99@yandex.*ru*
Tel.: +79172549100

Abstract: The article examines the technological complex of wastewater treatment (WWS) of machine-building production. The wastewater treatment complex (WWS) of machine-building production has a complex multi-level structure; therefore, modern methods of system analysis, computer modeling, and the theory of Petri nets are used to study, model, and analyze such systems. The apparatus of the theory of Petri nets (SP) is selected from the main methods for modeling discrete-continuous chemical-technological systems (CTS) for modeling the process of SAL of engineering production. Petri nets are used, focused on modeling and analysis of discrete-continuous CTS, by including priority transitions, delay time of labels in positions and transitions. A model of a typical apparatus that implements the OSV technological process was synthesized in the form of a modified Petri net (SME). The software package for the OSV technological process control system was created using the TRACE MODE SCADA technology.
Keywords: modified Petri nets, wastewater treatment of machine-building production, simulated systems, chemical-technological system, computer simulation, emergency situations, technological process

1. Introduction

Engineering enterprises are a powerful source of environmental pollution. One of the main factors affecting the environment, flora and fauna, as well as humans, is the wastewater of these enterprises, containing mechanical impurities, petroleum products, motor oils, sand, coolant, surfactants, salts of heavy metals, various types of fuel. In this regard, wastewater with high toxicity must be treated before it enters open water. The treatment facilities of machine-building enterprises have a complex multi-level structure, they can be considered as complex cybernetic systems [1]. Therefore, wastewater with high toxicity cannot be discharged into open water without proper treatment. The effectiveness of the functioning of such systems can be achieved using new methods of information processing, using a system analysis of complex objects based on mathematical models of the technological process [2,3]. The industrial OSV installation is a chemical-technological system that includes a set of

IOP Conf. Series: Materials Science and Engineering **994** (2020) 012044 doi:10.1088/1757-899X/994/1/012044

interconnected material, heat and information flows of the apparatus. The wastewater treatment complex (WWS) can be divided into interconnected subsystems, with various management tasks at each level of the hierarchy. The overall objective is to treat wastewater to standard levels or to the level of recycled water supply. An information approach based on mathematical modeling of an object [4,5] is the main direction of the study of complex WWS systems. In order to create control systems, consider the behavior of an object in emergency situations, evaluate its structure and control laws, and also take into account the stochastic nature of disturbing influences [6,23], one can model and conduct computer experiments with a model - a substitute for the object. When modeling real objects, two approaches are distinguished. In the first of these, the object is represented as a dynamic system with a continuous variable. This approach is more often used in modeling chemical-technological systems with the continuous organization of the technological process, provided it is stationary and the physicochemical parameters are unchanged [7, 8]. In the second approach, the object is represented in the form of a dynamic system with discrete events (DSDS). These include production systems, assembly lines, computer networks, as well as discrete-continuous biochemical and technological systems. To solve the problems of organization of control of such discrete dynamic systems, special mathematical methods are required. Usually, methods of finite automata, logical-linguistic and simulation modeling, apparatus of the theory of graphs and networks, and SP are used [9]. We have chosen the apparatus of the theory of SP [9] as the main apparatus of mathematical modeling. Petri nets allow modeling discrete parallel asynchronous processes, obtaining a graphical representation of the network, describing systems at various levels of abstraction, representing the system hierarchy, and analyzing models using application software packages [10]. There are several definitions of a joint venture. We will use the terminology introduced in [11].

2. Methods

To analyze various modifications based on the theory of SP, a definition is formulated of a common Petri net - SP, on the structure of the behavioral properties of which there are no restrictions. This class of networks is central to many modifications of the joint venture.

Researchers studying and developing methods for analyzing Petri nets (methods based on the reachability tree, algorithmic methods based on matrix equations, decomposition methods for SPs) have come to the conclusion that, being a powerful tool for studying SPs, these methods turn out to be suitable for Petri nets with imposed restrictions .

Therefore, the following subclasses of SPs were distinguished, reflecting its behavioral (defined by labeling) and structural (defined by GSP topology) properties: ordinary SPs, non-reflexive SPs, simple SPs, live SPs, safe SPs, conflict-free SPs, stable SPs, automatic SPs, marked graphs, Free choice joint venture (CER), asymmetric choice networks (CAB) or correct joint ventures.

Using subclasses of SP in modeling, analysis and synthesis of real systems allows you to create objects with previously known properties (correct algorithms, microcircuits, parallel programs).

Modifications aimed at enhancing the modeling capabilities of Petri nets by introducing structural features (new types of transitions and positions, special transition functions, special arcs) are called extensions of the joint venture. There are a significant number of them, which is constantly growing, increasing the possibilities of modeling Petri nets of various classes of complex systems.

3. Results and discussion

The possibilities for modeling Petri nets of real systems are limited. This explains the appearance of a tendency to expand the model. Researchers using Petri nets developed their modifications of the joint venture for a narrow range of their tasks.

Stable SPs [12] - SPs are such that, for any two allowed transitions, starting one of them does not inhibit the operation of the other. In a stable network, any transition, having become allowed, retains this state until it works.

The concept of stability is useful in the study of parallel computing circuits and asynchronous circuits.

Conflict-free SP [13] - there is SP for which either for each of its positions $p_i \in P$ there is at most one outgoing arc $|O(p_i)| \leq 1$, or for all $t_j \in O(p_i)$ performed $t_j \in I(p_i)$ (any position that is input for more than one transition is simultaneously output for each such transition).

Networks with an asymmetric choice (the correct joint ventures) are those networks in which each transition needs to have no more than one input position, which is shared with another transition and therefore serves to limit the possibility of conflicts. In solving problems of determining the correctness (correctness) of the algorithm and software of logical control systems, the correct Petri nets are widely used in ACS. Typically, the object of study is modeled by the joint venture, and the properties of the object are investigated in the analysis of the properties of the joint venture. The analysis of the joint venture can be omitted if we knowingly build the correct Petri nets. The technique for forming regular Petri nets is based on substituting a special kind of well-formed blocks into the original regular Petri net, the correctness of which was investigated on the basis of structural properties. This direction was developed in [14].

Networks with asymmetric choice (correct SPs) are those networks in which each transition needs to have no more than one input position, which is shared with another transition and therefore serves to limit the possibility of conflicts [14].

One of the extensions of the joint venture is the joint venture containing inhibitor arcs [15]. A more general case of the expansion of a joint venture introduced by Patil [16] is a joint venture with domains of restriction.

Another extension introduced in - joint venture with the transition "exclusive OR". The exclusive-OR transition can be triggered when only one of its inputs has labels, and all the others do not have them, and when it starts, it deletes the label only from the input with labels. A similar extension is a joint venture with switches. A switch is a transition with a special input called a switching and exactly two switching outputs. The allowed switching transition is triggered according to the following rule: the label is placed in the output marked with the e symbol if the switching input is empty, or in the output marked with the f symbol if the switching input is not empty.

Priority SPs are networks whose elements are partially ordered by some relation (less than or equal to), and priorities np (t) are associated with each transition t of the SP. Then we supplement the transition triggering rule by the following condition: the transition t can work, if for any other transition t of this network that can work according to the standard condition: np (t) $\leq$ np (t), that is, if several transitions are ready to work, then that whose priority is not less than the priorities of the remaining transitions ready for triggering. Control SPs can be used for algorithmic description of a wide range of tasks and modeling control algorithms in the construction of parallel programs and their parallel algorithms. Thus, we have described the main classes of Petri nets.

The range of Petri nets modifications is constantly supplemented and expanded. Each researcher introduces his extensions for his range of tasks.

The apparatus of the theory of Petri nets is the most convenient tool for modeling the structure and functioning of CTS. To simulate CTS, we decided to use temporary deterministic Petri nets that allow inhibitor arcs and priority transitions. It should be noted that these modifications were used by us for the convenience of modeling and software implementation of network models. The apparatus of classical Petri nets allows simulating the functioning of the CTS, but the SP - model will be cumbersome, beloved and large.

4. Summary

Figure 1 shows the control system of the sewage treatment plant of machine-building production (Fig. 1), in which a mathematical model based on the joint venture is used.

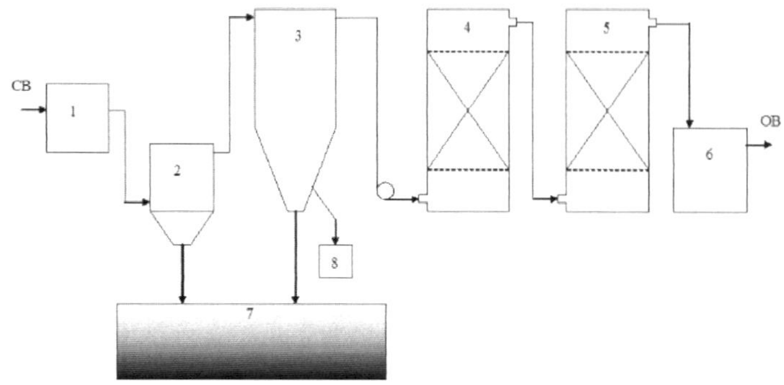

Fig. 1. The technological scheme of the installation of wastewater treatment engineering

Figure 1 shows: 1 - capacity; 2 - sand trap; 3 - hydrocyclone; 4 - coarse filter; 5 - fine filter; 6 - capacity of purified water; 7 - sludge collector; 8 - oil pan. SV - waste water; OV - purified water.

To describe the system, N – schemes are used, based on the mathematical apparatus of Petri nets, one of the advantages of which is the possibility of representing the network model both in analytical form, with the ability to automate the analysis process, and in graphical form, which allows you to visualize the model.

It is necessary to take into account the main limitation of the formalism of N-schemes, which consists in the fact that they do not take into account the temporal characteristics of the simulated systems, since the response time of the transition is assumed to be zero. Therefore, we proposed modified Petri nets. SME - Petri net of the form: $C=<P,T,I,O,M,L,\tau 1,\tau 2>$,

where T={tj} a finite nonempty set of characters called transitions is estimated based on the number of conditional servings of products when continuously supplied to the apparatus of the technological scheme;

P={pi}- a finite nonempty set of symbols called positions (in our case, this is a set of devices of the technological scheme);

I: PxT→{0, 1} – input function, which for each transition ti defines a set of its positions pi ∈ I (tj);

O: PxT→ {0, 1} - an output function that displays the transition to multiple output positions pi ∈ I (tj);

M: P → {1, 2, 3...} - the function of marking (marking) the network, which associates with each position a non-negative integer equal to the number of labels in this position, which changes during the operation of the network;

$\tau 1$: T→N and $\tau 2$: P→N functions that determine the delay time when the transition is triggered and the delay time in the position.

Thus, for each transition, you can define a set of input positions I (tj) and output positions O (tj) as:

$I(tj) = \{pi \in P / I(pi, tj) = 1\}; \; O(tj) = \{pi \in P / O(pi, tj) = 1\}$ (1)

The transition triggering instantly changes the markup M (p)=(M (p1), M (p2), M (p3)...M (pn)) to the markup M /(p) according to the following rule:

$M/(p) = M(p) - I(tj) + O(tj)$ (2)

Writing equation (2) means that the transition tj removes one label from each of its input positions and adds one label to each of the output.

The dynamics of the implementation of SMEs is determined by the movement of labels simulating the balance of discrete flows of intermediates under the conditions of specified restrictions on the volume of the OSV apparatus.

The described modification of the joint venture allows an analysis of the functioning of the apparatus of the system in emergency situations, control switching at the network level, as well as technological schemes for discrete-continuous production to ensure a stable, stable state of the cleaning system.

A mathematical model of the technological scheme and its software implementation for controlling the WWS process have been developed. The implementation of the mathematical model of the WWS system developed in the form of SMEs made it possible to study the systemic relationships and laws of the operation of the installation as a whole. Models of the main apparatuses that implement the WWS

technological process were also constructed [17, 18, 19]. The model of the entire installation (Fig. 2) was synthesized from SP - models of typical devices.
The software package of the OSV technological module system simulating the operation of cleaning in virtual time was developed using the SP model. The software package of the OSV technological process control system [20, 21,22] was modeled using the TRACE MODE SCADA technology. Using the process control system, you can perform supervisory control of the main elements of the control system, stop the WWS system and analyze its state to predict the development of emergency situations [4,5, 17].

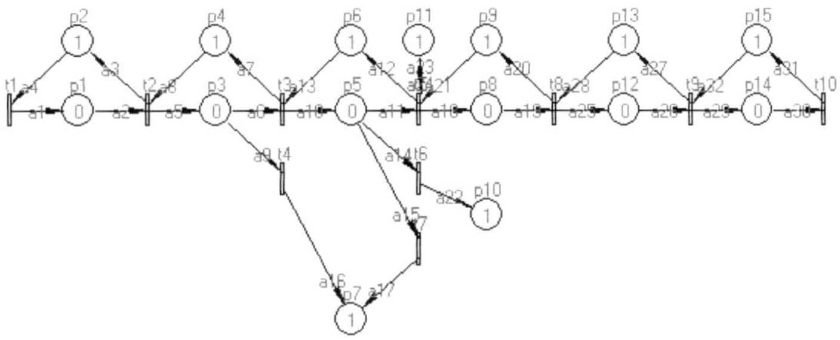

Fig. 2. The model of the technological module in the form of SMEs

Analytical description of the general Petri net:
15 Positions, 10 Transitions

O(t1)={ p1*1 }	I(t1)={ p2*1 }
O(t2)={ p2*1 p3*1 }	I(t2)={ p1*1 p4*1 }
O(t3)={ p4*1 p5*1 }	I(t3)={ p3*1 p6*1 }
O(t4)={ p7*1 }	I(t4)={ p3*1 }
O(t5)={ p6*1 p8*1 p11*1 }	I(t5)={ p5*1 p9*1 p11*1 }
O(t6)={ p10*1 }	I(t6)={ p5*1 }
O(t7)={ p7*1 }	I(t7)={ p5*1 }
O(t8)={ p9*1 p12*1 }	I(t8)={ p8*1 p13*1 }
O(t9)={ p13*1 p14*1 }	I(t9)={ p12*1 p15*1 }
O(t10)={ p15*1 }	I(t10)={ p14*1 }

5. Conclusions

The main limitation of the formalism of N-schemes, which consists in the absence of taking into account the temporal characteristics of the simulated systems, was revealed in the analysis of chemical-technological systems. Therefore, it is necessary to use the SP modification, focused on modeling and analysis of discrete-continuous CTS, by including priority transitions, as well as the delay time of labels in positions and transitions. The constructed mathematical model of the functioning of the treatment systems for technological effluents of machine-building production, implemented in the form of a modified Petri net, allows us to study the systemic relationships and laws of the operation of the installation as a whole. Using the created software package of the OSV system, it is possible to analyze the state of the OSV system as a whole and predict the development of emergency situations.

6. Acknowledgements

The work is performed according to the Russian Government Program of Competitive Growth of Kazan Federal University.

BIBLIOGRAPHY

[1] Fesina E Savdur S Modeling of Sewage Bioremediation as a Modified Petri Net //World Applied Sciences Journal 31(6) 2014 pp 1191-1197

[2] Hunt J C R TimoshkinaY Baudains P J Bishop S R System Dynamics Applied to Operations and Policy Decisions // European Review 20(3) 2012 pp 324–342

[3] Motameni H Movaghar A Shirazi B Aminzadeh M Samadi H Analysis Software with an Object-Oriented Petri Net Model //World Applied Sciences Journal 3 (4) 2008 pp 565-576

[4] Huilinir C Aspe E Roeckel M Modeling of the denitrification/anaerobic digestion process of salmon fishery wastewater in a biofilm tubular reactor // Journal of Environmental Management 92 2011 pp 1591-1608

[5] Janarthanan R Doss S and Balamurali R 2020 Robotic-based nonlinear device fault detection with sensor fault and limited capacity for communication JOURNAL OF AMBIENT INTELLIGENCE AND HUMANIZED COMPUTING

[6] Ruiz M Sin G Berjaga X Colprim J Puig S Colomer J Multivariate Principal Component Analysis and Case-Based Reasoning for monitoring fault detection and diagnosis in a WWTP // Water Science Technology 64 (8) 2011 pp 1661–1667

[7] Peter P Determination of Biological Degradability of Organic Substrates // Water Research 10 1976 pp 231-235

[8] Buswell AM Mueller MF Mechanisms of Methane Fermentation // Industrial and Engineering Chemistry 44 1952 pp 550-552

[9] Guest Editors MengChu Zhou Li ZhiWu Special issue on «Petri nets for system control and automation» //Asian Journal of Control 12 (3) 2010 pp237-239

[10] Barzegar B Motameni H Modeling and Simulation Firewall Using Colored Petri Net //World Applied Sciences Journal 15 (6) 2011 pp826-830

[11] Albert W Yao L Zhiming Ji Liao H T A Petri nets-based process planning system for wastewater treatment //Asian Journal of Control 12 (3) 2010 pp 281 – 291

[12] Nasby G Phillips M SCADA standardization: Modernization of a municipal waterworks with SCADA standardization: Past present and planning for the future // InTech 58 (5-6) 2011 pp1-5

[13] SN Savdur and EI Kadochnikova Network Modelling of Biological Wastewater Treatment System of Paint Manufacture //International Business Management 10 2016 pp 5629-5632

[14] Morozov NV Savdur SN Ganiev IM Process module of oil-containing wastewater treatment of local sewage system and its controlling mathematical model // International Journal of Pharmacy and Technology Vol8 №4 2016 pp 24212 – 24221

[15] Azimov Yu I Savdur SN Network Modeling of Biological Treatment Complex of Polymer Production Waste Water / Yu I Azimov SN Savdur // Asian Social Science Vol 11 No 11 2015 pp 50 – 55

[16] Yu I Azimov I I Ismagilov S N Savdur Network Modelling of Functioning System of the Process Module of Oil-Contaminated Wastewater Treatment // Asian Social Science Vol 11 No 11 2015 pp 313 – 318

[17] Azimov Yu I Savdur S N and Fesina E L Ensuring environmental safety based on the modeling of biological process of oily SEWAGE // Mediterranean Journal of Social Sciences Vol 5 (24) 2014 pp 372-377

[18] Kadochnikova E Savdur S Sungatullina L Khisamova E Modeling marginal factors of credit institutions: eliminating a collinearity problem // 4th International Multidisciplinary Scientific Conference on Social Sciences and Arts / SGEM Vienna Book 1 Vol 3 2017 pp471-478

[19] Savdur S N Fessina Ye L Stepanova Yu Kostromin A V; Kodolova I A Dynamics of cancer in Republic of Tatarstan and approaches to development of mathematical methods for analysis of clinical trials//PROCEEDINGS OF THE INTERNATIONAL CONFERENCE ON TRENDS OF TECHNOLOGIES AND INNOVATIONS IN ECONOMIC AND SOCIAL STUDIES 2017 Vol38 Is 2017 pp594-599

[20] Savdur SN Stepanova YV Kodolova IA Fesina EL Modeling of wastewater treatment system of car parks from petroleum products //Journal of Physics: Conference Series 1015 (3) article № 032121 2018 pp1-5

[21] Vorontsova VL Savdur SN Gilemzianov AF Network modeling of systemwastewater treatment at the enterprises of the metal processing industry //Journal of Advanced Research in Dynamical and Control Systems - 2019 - Vol11 Is5 - pp1-6
[22] Maheswaran G Balamurali R Sundarambal B Soundarya Priyadharshini S Vinodhini P 2020 A Smart Plug for Controlling and Automation Anywhere Using Iot International Journal of Advanced Science and Technology 29(3) pp8570 – 8577
[23] Haroonabadi A Teshnehlab M Movaghar A A Novel Method for Behavior Modeling in Uncertain Information Systems // World Applied Sciences Journal 3 (5) 2008 pp797-805

Vorontsova Valeriya— candidate of Physico-Mathematical Sciences, associate Professor at the Department of General mathematics of Institute of Mathematics and Mechanics Lobachevsky of Kazan Federal University. Author of more than 65 scientific papers, 2 monographs, 20 textbooks. Research interests — the dynamics of space cable systems, the stability of the movement, modeling and calculation of heat exchangers with heat exchange elements in the form of twisted pipes and spring-twisted channels, system analysis, information management and processing, mathematical modeling of complex systems.
Savdur Svetlana — candidate of Technical Sciences, associate Professor at the Department of of Merchandising and Technology of Public Catering Kazan Cooperative Institute (branch) "Russian University of Cooperation". Author of more than 80 scientific papers, 3 monographs, 5 textbooks. Research interests —system analysis, information management and processing, mathematical modeling of complex systems.

IOP Conf. Series: Materials Science and Engineering **994** (2020) 012045 doi:10.1088/1757-899X/994/1/012045

Design of 12T SRAM cell for low power dissipation

R Suresh Kumar[1], K Ranjani[2] and A Rasheedha[3]

[1]Associate Professor,Department of ECE, Chennai College of Technology, Chennai, TamilNadu, India
[2]Assistant professor, Department of EIE, SNS College of Technology, Coimbatore, TamilNadu, India
[3]Assistant professor, Department of BME, SNS College of Technology, Coimbatore, TamilNadu, India

E-mail: sureshkumarr@citchennai.net

Abstract. The present semiconductor industry is developing quite quickly. Mobile and portable devices are becoming small for each day and moreover there is a rising demand for extended battery power. With these demands it is significant for researchers to emphasis on the leakage capacity in standby mode. For the exact communication, SRAM was intended with DSP, processor andapplications with lowpower devices. The design engineer emphasesmostly on the making of low energy consuming, higher bandwidth and large memory capacity. Memory is an essential part of majority of these systems as well asit is also reduced as the system size reduces. High processing architecture as well as low power is consequently a foremost concern. The robustness of static random access memory cells is other important factor. This paper proposed the architecture design of 12T SRAM for the reduction minimization of power leakages or power consumption during the sleep transistor. This supports in the minimization of the leakage in the CMOS transistor. Based on the parameters like write delay, read delay, write voltage, read voltage, and power consumption at various temperature are compared to the existing SRAM architecture of 6T, 7T and 8T, we acquire reduced power consumption in the proposed 12T SRAM architecture..

1. Introduction

The need for low power / energy consumption of the integrated ICs, is increasing rigidlythrough the success of wearable and portable electronic componentsas well as the omnipresent implementation of sensor networks in wireless communication. The supply voltage of the integrated circuits has been reduced in the near/sub-threshold values if the frequency is not a majorissue in order to decrease the energy/ power consumption of these circuits. The embedded chips include essential elements which make a major contribution to overall power / energy use [16] in the static random access memory devices. Nonetheless, owing to the severe damage in the noise margin with static read and the capacity with write enable, in addition to the rigorous impacts of the leakage current with read bitline[17], a regular six transistor (6 T).

SRAM track usuallysupplied by the foundry has not beenranged to near/ sub-thresholdvalues. Designing an ultralow power Static Random Access Memory circuit is vital for embedded chips used in wearable or portable electronics components. The topology of the Static Random Access Memory cell is revamped to allow many numbers of transistors to maximize the noise gap with static read, also to boost write performance, and decrease the leakage present in read bitline [18]. On the other hand, the compilers of the memory must be implemented in a difficult and conventional way, given the presence of analog signal and pre-loaded circuitry with sensor amplifiers in the SRAM array as the analog signal. An alternative in performing the ultralow power operation for Static Random Access Memory circuits is the architecture of the standard cell based memory (SCM) elements [19]. To construct the Static Random Access Memory modules by using regular logic cells, all the inputs as well as outputs are made electrical signals.The automated model cells can be implemented fully through the peripheral devices. The current Electronic Design Automated software can therefore be utilized to conduct logical synthesis along with automated positioning, routing to produce the net list and configuration with the necessary arrays of SRAM [20]. For the applications of embedded system, the arrays of SRAM in rapid generation are highly desirable in which a huge number of different sized memory elements are required. In comparison, write and read operations are executed similarly through the use of modern standard cells as the transmission of signal in conventional combinational circuits thatassurances huge data reliability and good capacity of write operation [21]. The biggest problem with the SCM cells is the overhead in each cell with huge area. Compared to conventional 6T SRAM cell, each SCM cell has the layout region that could be greater for two to three times than 6T SRAM. Nevertheless, the area output of the total SRAM system has beenmuch higher compared with the SRAM regular cell-based circuit relative to the earlierimplementation of SRAM circuits with ultralow voltage. It happens while there is a memory array of relatively little size,so that peripheral circuit greatly supplied to the entire array area [7]. A novel robust 12T SRAM based on standard cell for ultralow voltage and ultralow power applications is proposed in this article. In Complementary Metal Oxide Semiconductor to minimize the leakage, a lower supply voltage is powered to the circuitwhich,inturn,the speed of the circuit is lowered. By using the lower threshold voltage for the transistors, the delaycanbereduced, however this further reduces the flow rate again (especially the flow cycle). Different conditions have presented and the implementation of a memory cell with a low standing leakagein addition to practical reliability necessitates a better optimization. The important data reliability loss in the cells occurs when a higher voltages and smaller measurements are induced. In order to achieve high range performance and low write energy consumption, three transistors are utilized on every column of Static Random Access Memory cells. During the three-stage reading cycle, reading delays and power consumption are minimized. In addition, parallel cross-coupling feedback with the proposed SRAM 12 T cell, ensures high data stability and writing ability.

This is how the article is arranged. Section II reviews the existing SRAM standard cell-based circuits. Sections III in which the problem statement was depicted and Section IV present and evaluate proposed novel 12 T SRAM based on standard cellcircuit. Section V depicts the summary of the paper.

2. Literaturesurvey

In [1], a modern 12 T memory cell that is energy-efficient is suggested to help architecture hardened radiation (RHD) in close-to-threshold voltage resistance multi-node (SEMNU) single-event upsets. Through monitoring cross-coupled PMOS devices of inverters through foolish access transistors, the radiation complexity of the proposed storage cell will be increased. In STMicroelectronics 6, we had validated the proposed memory cell. In [2] they propose an overall reduction in the amount of VMINand resources per service with a 12 ST-based SRAM-bit cell to promote the process variance sensitive write- capability. In [3] a halve select 12 T SRAM free cell features an input cutting tool to enhance the ability to write and separate reading paths in order to improve reading stability. The write and read capabilities are 2.84 and 1.95 respectively better than the normal 6 T cell at 0.4 V.

In [4] presents the SRAM, a compiler that targets small to medium array sizes and provides a solution with smaller area compared to traditional 6T-based SRAMs, with the one-port random access memory of ultralow voltage. The design uses a 12 T write and the read upset free bit-cell. Array architecture uses a read-modifying writing scheme to enable masking and multiplexing columns for bit-write (BW). The Built-inside Self-Test (BIST) and Synchronous Writing Through (SWT) methods have testability, while the Sleep-and Shutdown Methods Power Management (PM) alternative is included.In[5]a9-Transistor(TG9T)SRAMbitcellpower-efficienttransmissiongatehasbeenproposed.In[6] Optimized MTJ (Approximate Calculation Application (AC) Procedure, focused on resistive ternary information addressable memory

IOP Conf. Series: Materials Science and Engineering **994** (2020) 012045 doi:10.1088/1757-899X/994/1/012045

(ReTCAM). In [7] a new 12 T SRAM cell is introduced that is used to improve reliability with an Schmitt control circuit and to reduce the leakage current in standby mode with a transistor. In [8] discusses the phenomenon of crucial linear transfer of energy (LET) and the charge obtained by scaling the technology of each circuit. In [9] Propose modified 12 T SRAM cell (WWL12 T) FinFET and selfrefreshing. To increase the FinFET's reliability and performance, the double-k gate isolator and symmetric distance is used. In [10] the 12 T static random access memory (SRAM) cell dependent on the one-ended Schmitt signal (ST). Cross-point data-conscious writing framework is used in the planned cell and therefore promotes interlocking architecture to minimize soft errors. The proposed cell is not interrupted by reading because the bit line is completely isolated from the data storage node. In [11] presents a 12 T 2RW SRAM low voltage with parallel exposure and blocked interruption to improve the performance without degradation. The suggested SRAM cell reduces interference by splitting reading from internal nodes and minimizes the probability of worst case stability with a 6 percent region cost. Furthermore, hierarchical bitlines, a virtual ground technology and the minimum voltage and energy consumption will be further reduced. In [12] the design based on the 12 transistor (12 T) cell is proposed for the low power and variability-aware static random access memory (SRAM). In [13] an advanced 12 T Static Random Access Memory (SRAM) cell with the following advantages:decreasedleakagecurrentandincreasedefficiencyutilizing180NMTechnologywasproposed.In [14] Presents a 9 T static random access memory (SRAM) cell with a single-ended Schmitt signal. The architecture suggested offers improved reading stability when ST-based inverters are used. In [15] the introduction of the Schmitt Trigger (ST) enhanced switching functionality, reduced the power outages and increased the SNM. Upon applying the memistor to the SRAM trigger Schmitt, the cell was not fragile and the SNM improved more.

3. Problem statement

The SRAM technology suffered a lot due to leakage issues. There are several other existing technologies that are there, but they are all not useful. Hence there is a need for an effective method to overcome the leakage issues in the SRAM technology. We would propose a new SCM with parallel cross-coupling logic that will reduce expenses, time, power, and memorycomplexity.

4. Proposed methodology

This proposed method implemented to decrease write and read energy consumption in order to maintain high data reliability and write skills on the current 12 T SRAM low power circuit. With the use of advanced 65-nm CMOS materials, all transistors in SRAM cells use smallestchannel length (60 nm) and channel width (80 nm). The transistor has shared the every column of SRAM cells. For physical design, the proposed static random access cell has been implemented and described as a standard cell and are treated in ananalogous manner to the sleep transistors utilizedwhile routing and planning.

The proposed method can handle an upto 4 bit that consists of a block. Such four bits are interpreted by mask transistors, gate transistors, four pairs of intertwined cross coupled inverters, read buffer and writer entry transistors. Based on the stacked nMOS configuration theory, the read buffer is used.The leakage voltage has been avoided by this configuration. It contains a line to pick the signal you need, called as BPS [Block Pick Signal]. The typical 12T cell process is carried on as follows. BPS is set to 0V to enforce write operation. And therefore the word line is permitted. Likewise, BPS is set at 1V for a read operation, as well as the word line is stopped. Therefore the line of terms serves as an ON-OFFturn. Figure 1 shows the schematic representation of the proposed methodology for the design of 12T static random access memory.

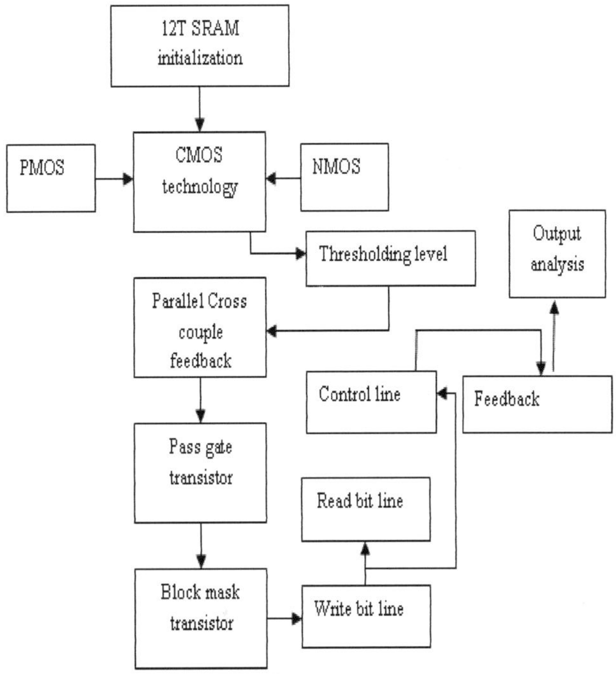

Figure 1. Schematic representation of the proposed methodology

For LBL to discharge, a supply voltage VDD is carried out. This will shut the read buffers off, in effect. In the meantime, the full VDD supply voltage has preloaded to RBLs and the word lines are retained at 0V.During the read process, the block mask transistors are enforced to persist in the OFF place. To hold the transistor in ON stage, the word line is restricted for the transfer accesses. As a consequence, through the pass gate transistor the data stored in a single cell is distributedto the RBL from LBL. When the data is stored, it discharges the RBLs. These RBLs are located on the blocks of the given column and row. After the read process ends, word lines are connected at zero voltage to switch OFF the transistors pass-gate. The best way to minimize energy consumption through the use of 12T SRAM is to do so. This novel configuration is built. It minimizes energy consumption as a result of standby, read wait, writing delays, etc. Experimentally demonstrated by implementing the parallel cross coupling logic in 12 T SRAM which is the best way to save energy use and operating delays.
The size of memory cell is also minimized with the lower processing nodes. The current memory cell leakage factor will be increased thereby. The intersection of leakage, gate leakage and multithreshold leakage current are various additives which triggers the 12 T static random access memory cell leakage. However in lower-technologies, multi threshold leakage current is predominant. When the leakage current is off, the voltage of transistor gate is less than the voltage at its highest and containsmostly of a diffusion stream; the transistor gate voltage is the lowest. It is represented by the followingequation (1),

$$L_{multi}= \mu_0 D_0(W/L)(KT/Q)\text{^}2(N-1)\ exp[q(V_{gs} - V_t)\ (1/nPT)[1 - exp(-qV_{cs}/PT)] \qquad (1)$$

The circuit consists of M6 transistors. RBL and BL are also utilized, which act as output as well as input lines respectively from the memory cell. The cell leakage element must be calculated for estimation. The cell ratio is the reader-sized ratio to the connection transistor through the processing, in addition the PR-to-access ratio is the load-transistor ratio across the whole write activity. Equation (2) and (3) gives the read and write operation respectively.

During Read operation,

$$CR = (W/l)n \quad (2)$$

During write operation

$$PuR = (W/l)p/(W/l)n \quad (3)$$

Where PuR is the pull ratio and CR is the cell ratio.
In the proposed architecture,we had analyzedthe 12T static random access memory cell, for every write operation the energy consumption of complex bit-lined has also estimated. To motivate this, we examine as adesign for two columns of static random access memory cells (i.e., four bit lines). Each of static random access memory cell has been connected to the ground (GND) until the write process. After the written service, two bit lines from GND to VDD has been paid. The estimation of 12T static random access memory cell for a complex energy consumption is carried out using the equation (4).

$$P_{12T} = P_{Bitline\_1} = (1/2CV_{DD})^2 = 4CV^{1/2}{}_{DD} \quad (4)$$

Regarding to our 12T-SRAM cells mentioned, after the written process the four bit voltages are GND, 1⁄2 VDD, 1⁄2 VDD, and VDD. As the expectation to charge is only from 1⁄2 VDD to VDD line, the robust 12T SRAM cell has prompted to the complex power consumption.

$$P_{12T\text{-}SRAM} = P_{Bitline} = ¼\, C*(V_{DD}/2)\text{^}2 = ¼\, CV^{1/2}DD \quad (5)$$

It can be seen from the above equation (5) that the planned system would reduce the electricity consumption. The projected power consumption is less than typical 6 T or 8 T owing to leakage of power consumption. Figure 2 shows the design of proposed 12T SRAM in cadence tool.

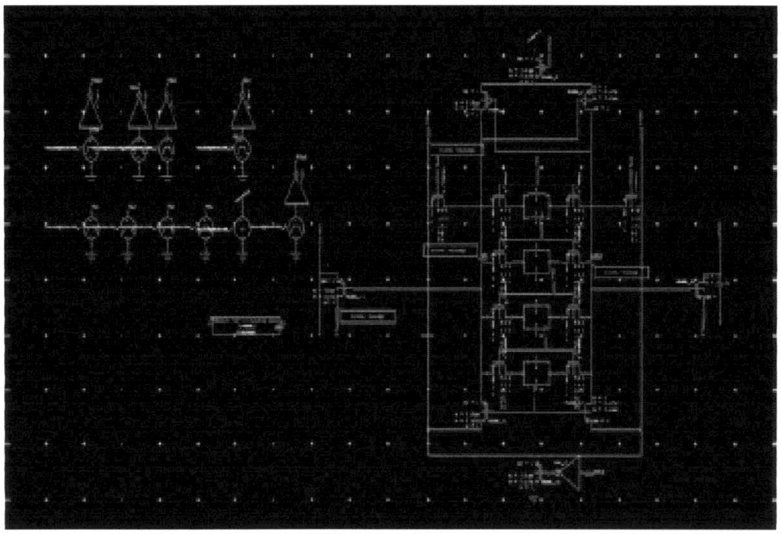

Figure2. Proposed 12T SRAM architecture

4.1. Leakage power calculation

Multi-threshold leakage and gate leakage are the main sources of SRAM leakage capacity. This is where a technique for the elimination of multi-threshold leakage and the leakage of robust 12T-SRAM cells is implemented and suggested. However, even if the body distortion is presented, the leakage current present is less than 8 percent of total power. Therefore, eliminating leakage is not our original design intent. The potential energy usage of 12 T leakage is less than the conventional system. The small energy consumption difference was found owing to the various pre-loaded bitline voltages as given in equation (6). In a Robust 12- T SRAM data manager, bit lines BL1 and BL1 N are preloaded onto "1" and word line write on "0" while the gate-leakage is induced from the source to the gate by the CMOS transistorswitch.

$$LC_{12T\,SRAM} = C_{wncntl} + C_{rdcntl} + C_{dec} + C_{wn\,col} + C_{rd\,col} + C_{mem\,core} \quad (6)$$

5. Result and discussion

The new method results in an effective 12 T SRAM design for power consumption, and the research that is being developed further shows that 12T cells can be extremely valuable for modeling SRAM cells relative to standby, write and read operation. The tool used to design this system is Cadence EDA.

Different temperature conditions in reserve mode are recognized in the capacity with power dissipation of the suggested cell along with recommended robust 12T SRAM Cell. The estimated dissipation of the current rely upon the various temperature. Figure 3 showsthe study of the temperature in different static random access memory cell types.

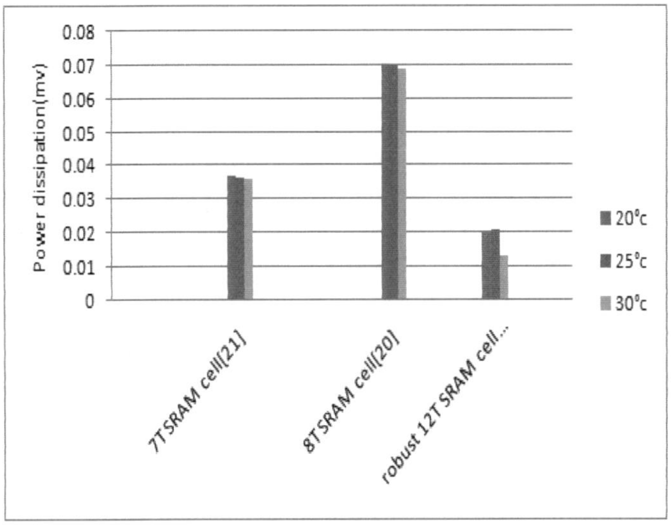

Figure 3. Power Dissipation at Different Temperature

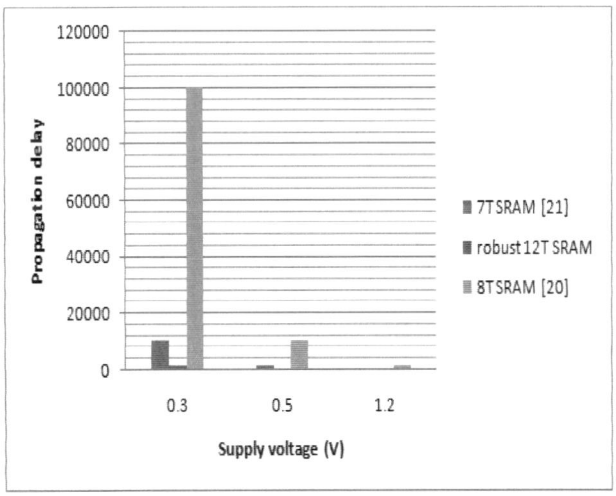

Figure 4. Read and write delays of the 12T SRAM arrays

From the figure 4, note that when compared to the previous methodologies, the 12T SRAM circuit has still shorter propagation delay owing to the considerablylengthier read access delay of the existing 12T SRAM circuit.

Figure 5 shows the memory array in idle mode for the leakage power consumption. The minimum channel length is utilized for every transistors in order to reduce the area layout of the memory cell, for the proposed robust 12T SRAM cell. When compared to the existing array of memory, the leakage power consumption of the proposed 12T SRAM array is consequentlylittle higher. While the supply voltage VDD brings down to 0.2 V, there is an increase in 12T SRAM array with the leakage overhead rises from 0.8% to 8.8%, compared to the existing array of memory.

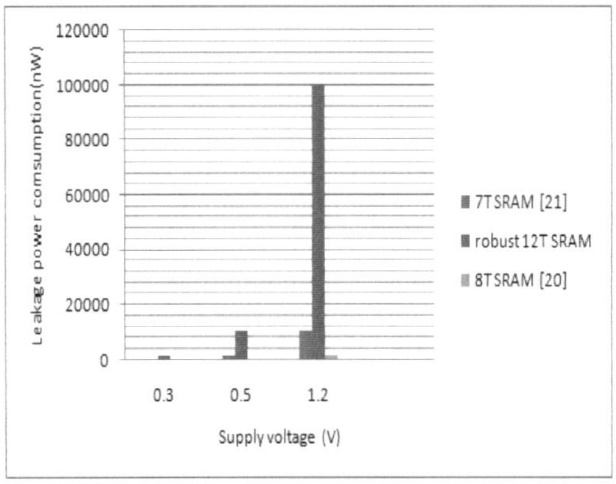

Figure 5.Leakage power consumption

6. Conclusion

The Cadence EDA used for the design of SRAM cells has been developed and modeled here. The results from the simulation showed that the 12 T SRAM cell circuits proposed has decreased leakage power consumption. With increasing supply voltages and increased temperatures with specific supply voltage, leakage power increases. Total power used is also substantially reduced for all planned loops. Nevertheless, the trustworthiness of the proposed circuits is jeopardized because leakage current is minimized for the individual states in the proposed circuit. Thus the low power 12T SRAM has been used.

References

[1] Kumar C I and Anand B 2019 A Highly Reliable and Energy Efficient Radiation Hardened 12T SRAM Cell Design IEEE Transactions on Device and Materials Reliability

[2] Sharma P Gupta S Gupta K and Pandey N 2020 A low power subthreshold Schmitt Trigger based 12T SRAM bit cell with process-variation-tolerant write-ability *Microelectronics Journal* vol 104703

[3] Sharma V Bisht P Dalal A Gopal M Vishvakarma S K and Chouhan S S 2019 Half-select free bit-line sharing 12T SRAM with double-adjacent bits soft error correction and a reconfigurable FPGA for low-power application *AEU-International Journal of Electronics and Communications* vol 104 pp 10-22

[4] Sinangil M E Lin Y T Liao H J and Chang J 2019 A 290 mV 7 nm Ultra Low Voltage One Port SRAM Compiler Design Using a 12T Write Contention and Read Upset Free Bit-Cell *IEEE Journal of Solid-State Circuits* vol 54 pp1152-1160

[5] Pal S Bose S Ki W H and Islam A 2020 A highly stable reliable SRAM cell design for low power applications *Microelectronics Reliability* vol 105 p113503

[6] Suresh Kumar R and Manimegalai P 2019 implementation of Neural Network with ALE for EEG Signal Processing *Current Sig Trans Therapy* vol 14 pp1-7

[7] Upadhyay G Rajput M A S and Saxena M N 2017 An Analysis of Novel 12T SRAM Cell with Improved Read Stability *Int J Innov Res Eng Appl Sci* vol 3

[8] Yusop N S Nordin A N Khairi M A and Hasbullah N F 2018 The impact of scaling on single event upset in 6T and 12T SRAMs from 130 to 22 nm CMOS technology *Radiation Effects and Defects in Solids* vol 173 pp 1090-1104

[9] Yadav N Shah A P and Vishvakarma S K 2017 Stable reliable and bit-interleaving 12T SRAM for space applications: a device circuit co-design *IEEE Transactions on Semiconductor Manufacturing* vol 30 pp276-284

[10] Suresh Kumar R and Manimegalai P 2020 Near lossless image compression using parallel fractal texture identification *Biomedical SigProc and Contl* vol 58 p101862

[11] Wang B Zhou J and Kim T T H 2017 A 0 4 V 12T 2RW dual-port SRAM with suppressed common-row-access disturbance *Microelectronics journal* vol 69 pp78-85

[12] Yadav A and Nakhate S 2016 Low standby leakage 12T SRAM cell characterisation *International Journal of Electronics* vol 103 pp 1446-1459

[13] Thakare P V and Tembhurne S 2016 A power analysis of SRAM cell using 12T topology for faster data transmission *Int J Sci Technol Eng* vol 2 pp 441-446

[14] Suresh Kumar R, Dhanagopal R, Menaka R and Ramesh R 2020 IoT based Drowsiness detection, Monitoring and Controlling System *International Journal of Advanced Science and Technology*, Vol. 29, No. 03, pp. 8543 – 8549.

[15] Nikitha L Bhargavi N S and Kariyappa B S 2019 Design and Development of Non-volatile Multi-threshold Schmitt Trigger SRAM Cell *In Emerging Research in Electronics Computer Science and Technology* Springer Singapore pp. 877-884

[16] Alioto M 2012 Ultra-low power VLSI circuit design demystified and explained: A tutorial *IEEE Transcations on Circuits and System* vol 59 pp 3-29

[17] Sinangil M Verma N and Chandrakasan A 2008 A reconfigurable 65 nm SRAM achieving voltage scalability from 0.25-1.2 V and performance scalability from 20 kHz-200 MHz*Proc. of the IEEE European Solidstate Circuits Conference* pp.282-285

[18] Dhanagopal R and Muthukumar B 2019 A Model for Low Power High Speed and Energy Efficient Early Landslide Detection System Using IoT *Wireless PersCommun* https://doi.org/10.1007/s11277-019-06933-7

[19] Meinerzhagen P Sherazi S M Y Burg A and Rodrigues 2011 J N Benchmarking of standard-cell based memories in the sub-VT domain in 65-nm CMOS technology *IEEE Journal of Emerging and Selected Topics in Circuits and Systems* vol 1 pp 173-182

[20] Muthukumar B Dhanagopal R and Ramesh R 2019 KYP modeling architecture for cardiovascular diseases and treatments in healthcare institutions *J Ambient Intell Human Comput.* https://doi.org/10.1007/s12652-019-01653-z

[21] Dhanagopal R KrishanamurthiV and Rajaram A 2017 Mitigating the Soft Errors and Controlling the Voltage Level of VLSI Devices – MSCV *Journal of Computational and Theoretical Nanoscience (CTN)* Vol 14 pp. 3209-3219 https://doi.org/10.1166/jctn.2017.6618.

AUTHOR INDEX